NEURAL CELL BIOLOGY

NORMAL CELL BIOLOGY

NEURAL CELL BIOLOGY

Editors

Cheng Wang

Division of Neurotoxicology
National Center for Toxicological Research (NCTR)
Food and Drug Administration (FDA)
Jefferson, AR
USA

William Slikker, Jr.

National Center for Toxicological Research (NCTR)
Food and Drug Administration (FDA)
Jefferson, AR
USA

CRC Press
Taylor & Francis Group
Boca Raton London New York

CRC Press is an imprint of the
Taylor & Francis Group, an **informa** business

A SCIENCE PUBLISHERS BOOK

Cover illustration has been provided by the editors of the book and is reproduced with their kind permission.

CRC Press
Taylor & Francis Group
6000 Broken Sound Parkway NW, Suite 300
Boca Raton, FL 33487-2742

First issued in paperback 2021

© 2017 by Taylor & Francis Group, LLC
CRC Press is an imprint of Taylor & Francis Group, an Informa business

No claim to original U.S. Government works

ISBN-13: 978-0-367-78230-6 (pbk)
ISBN-13: 978-1-4987-2600-9 (hbk)

Library of Congress Cataloging-in-Publication Data

Names: Wang, Cheng, 1954- editor. | Slikker, William, Jr., editor.
Title: Neural cell biology / editors, Cheng Wang, William Slikker, Jr.
Description: Boca Raton, FL : CRC Press, Taylor & Francis Group, [2016] | "A Science Publishers book." | Includes bibliographical references and index.
Identifiers: LCCN 2016028128| ISBN 9781498726009 (hardback : alk. paper) | ISBN 9781498726016 (e-book)
Subjects: | MESH: Neurons--physiology | Neurons--drug effects | Neuroglia--physiology | Neuroglia--drug effects | Nervous System Physiological Processes | Models, Biological
Classification: LCC QP363.2 | NLM WL 102.5 | DDC 612.8/1046--dc23
LC record available at https://lccn.loc.gov/2016028128

Visit the Taylor & Francis Web site at
http://www.taylorandfrancis.com

and the CRC Press Web site at
http://www.crcpress.com

Preface

Because of the complexity and temporal features of the manifestation of function and structure in the nervous system, neural cells in the central nervous system (CNS) and peripheral nervous system (PNS), especially in developing brains, are typically thought to be more susceptible to toxic insults than cells in mature systems. The study of neural cell biology has great potential for helping to advance the understanding of brain-related biological processes, including neuronal plasticity, neurodegeneration/ regeneration, toxicity and the efficacy of therapeutic or other interventions.

This book delineates how neural cells, systems biology, pharmacogenomic and pathophysiological approaches, as applied to neural cell biology, provide a structure around which to arrange information in a biological platform. Specific attention has been directed to some of the more recent methodologies, including genomics, proteomics and metabolomics, applied in the evolving field of neurobiology. This book also presents cross-cutting research tools and models and discusses applications involving neural stem cell system and 3D cultures that closely mimic native cellular environments. In addition, the topics of developmental neurotoxicity and risk assessment, along with the challenges and achievements in the field of neural cell biology are discussed. This book is a valuable addition to the current literature and an important reference for students, neurobiologists, pathologists, pharmacologists, and toxicologists in academia, industry and government settings.

This book provides the reader with technical background and knowledge of current test systems and models recently introduced and utilized in the study of neuronal development and neural cell biology.

Achieving the goals set forth for this book was most challenging and we wish to thank the contributing authors for conveying the excitement of recent findings and for identifying remaining knowledge gaps. We would appreciate any comments you wish to offer about *Neural Cell Biology*.

Cheng Wang, M.D., Ph.D.
William Slikker, Jr., Ph.D.

Contents

1

Neurons–Nerve Cells

Fang Liu,[1,a] *Tucker A. Patterson,*[2,d] *Jingshu Zhang,*[3] *Merle G. Paule,*[1,b] *William Slikker, Jr.*[2,e] and *Cheng Wang*[1,c]

Introduction

The nervous system is composed of two major types of cells—nerve cells, or neurons, which process information and rapidly convey information; and—glial cells which serve as supporting cells of the nervous system. The human nervous system has approximately 100 billion neurons, as many as 10 times the number of glial cells. Neurons are the key components of the nervous system, which is composed of the central nervous system (CNS)—and the peripheral nervous system (PNS). Neurons create neural networks by connecting to each other to facilitate communication.

Each neuron is an independent unit in terms of its anatomy and function. Morphologically, a neuron generally consists of 3 parts: the cell body (soma), dendrites, and an axon. The cell body provides nutrition to the entire cell and is responsive to stimuli. The dendrites are processes that receive information from various sources, such as the sensory epithelial cells, or other neurons. The axon is a single process that generates and spreads electrochemical changes used to communicate with other cells (nerves, muscles, and glands). Because the distal portion of the axon is branched, it is called the terminal arborization. Each branch of this arborization communicates with adjacent cells via synapses on button-like end bulbs (boutons).

[1] Division of Neurotoxicology, HFT-132, National Center for Toxicological Research (NCTR), Food and Drug Administration (FDA), 3900 NCTR Rd, Jefferson, AR 72079.

[a] E-mail: fang.liu@fda.hhs.gov; [b] E-mail: merle.paule@fda.hhs.gov; [c] E-mail: Cheng.Wang@fda.hhs.gov

[2] National Center for Toxicological Research (NCTR), Food and Drug Administration (FDA), 3900 NCTR Rd, Jefferson, AR 72079.

[d] E-mail: tucker.patterson@fda.hhs.gov; [e] E-mail: william.slikker@fda.hhs.gov

[3] Department of Toxicology, School of public Health, Nanjing Medical University, 101 Longmian Avenue, Nanjing 21166, P.R. China.
E-mail: jingshuzh@njmu.edu.cn

Neurons and their processes are extremely variable in size and shape. The dynamic depolarization of the cell with concomitant transmission of a nerve impulse (action potential) depends on highly specialized structures called synapses, which are classically defined as the contacts of one axon with the dendrites or perikaryon and, rarely, the axon of other neurons. Thereby, neurons usually receive information through their dendrites and cell bodies and transmit it via their axons to other neurons via their synapses. The vast majority of vertebrate synapses are variations on a common theme. Typically, an enlargement (the presynaptic element of the synapse) of a distal axonal branch abuts part of another neuron (the postsynaptic element), separated from it by a synaptic cleft 10 to 20 nm wide. The presynaptic ending contains membrane-bound packets (synaptic vesicles) containing neurotransmitter molecules (Figure 1.2); some vesicles release their contents into the synaptic cleft in response to cellular electrical activity (action potentials). The neurotransmitter diffuses across the synaptic cleft, binds to receptor molecules in the postsynaptic membrane, and causes electrical and/or chemical signaling in the postsynaptic neurons (Figure 1.3). These can be either excitatory or inhibitory in nature. Most synapses have an axonal ending as the presynaptic element and part of a dendrite as the postsynaptic element, but in fact any part of a neuron can be presynaptic or postsynaptic to any part of another neuron (or sometimes even to itself).

Neurons are in the business of conveying information. They do so by a combination of electrical and chemical signaling mechanisms: electrical signals are used to convey information rapidly from one part of a neuron to another (Figure 1.4) (Yu *et al.*, 2002),

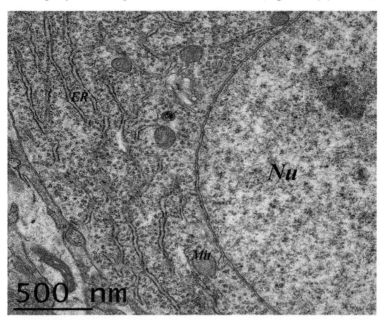

Figure 1.1. An electron micrograph of a representative pyramidal neuron from the frontal cortex of a six day old rhesus monkey shows intact cytoplasmic, mitochondrial and nuclear membranes. The nucleus *[Nu]* and prominent nucleolus are apparent, as are other organelles typical of healthy neuronal cell bodies—Golgi cisternae, rough endoplasmic reticulum *[ER]* with interspersed clusters of free ribosomes, and mitochondria *[mit]*.

whereas chemical messengers are typically used to carry information between neurons. Hence, there are anatomically specialized zones for collecting, integrating, conducting, and transmitting information.

Most neurons fall into 3 categories based upon the characteristics of their processes: multipolar neurons, which have one axon and multiple dendrites; bipolar neurons, which have one dendrite and one axon; and pseudo-unipolar neurons which have only one process that divides into two axonal branches. Neurons can also be classified according to their connections and functions. Sensory neurons are involved

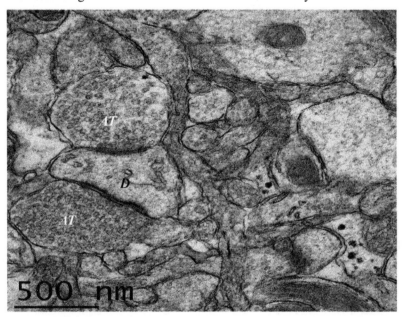

Figure 1.2. An electron micrograph of the neuropil in layer II or III of the frontal cortex of a 6 day old rhesus monkey. In this field, the presynaptic elements are axonal terminals *[At], filled with* round synaptic vesicles and abutting postsynaptic elements, such as dendrite spines *[D]* of an adjacent neuron. The two elements are separated by a synaptic cleft and the postsynaptic membrane is thickened, an indication of the presence of specialized molecules in and near the membrane at this site. Both axons and dendrites can contain mitochondria.

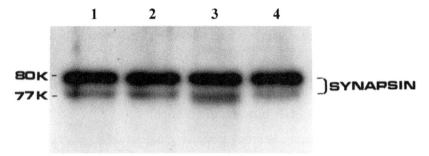

Figure 1.3. Western blot analysis of synapsin expression in neuronal cell cultures of frontal cortical cells after (1) vehicle or PCP (phencyclidine; NMDA-type glutamate receptor antagonist)-exposure (2), and from cultures of cells from a striatum-nucleus accumbens complex after vehicle (3) or PCP exposure (4). Bands of 80 and 77 kDa were identified as synapsin.

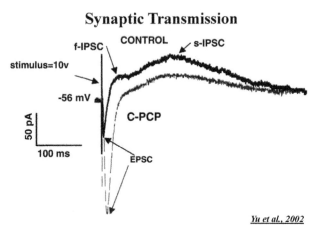

Synaptic Transmission

Yu et al., 2002

Figure 1.4. Synaptic transmission measured as glutamate ionotropic receptor-mediated excitatory post synaptic current (EPSC) and GABA receptor-mediated inhibitory post synaptic currents (IPSCs). Fast (f) or slow (s) currents are altered in neurons from rats treated chronically with PCP compared with those from control rats. Top (darker tracing) vs. bottom (lighter tracing): triphasic synaptic currents after chronic treatment with PCP (C-PCP) results in an enhanced (downward) EPSC (after the stimulus artifact), with diminished f-IPSCs and s-IPSCs (upward traces). Both recordings are at –56 mV (No drugs were present in the superfusion medium during the recording of these responses).

in the reception of sensory stimuli by being either directly responsive to stimuli such as touch or temperature changes, or receiving direct connections from non-neuronal receptor cells. Motor neurons control effector organs and synapse directly on muscles, glands, or other neurons or ganglia in the peripheral nervous system (PNS). Interneurons establish relationships among neurons, forming complex functional chains or circuits. In general, most of the neurons in the CNS can be classified by their morphology, position, synaptic inputs and/or targets, axonal projections, molecular requirements and electrophysiology.

Neurons utilize special mechanisms to manage their electrical and chemical signaling functions. Neurons can be localized and identified using a variety of techniques including those that label specific neurotransmitters, their cellular synthesizing enzymes or their surface receptors. By using these approaches, the majority of neurons in the central nervous system (CNS) can be characterized *in vivo* or *in vitro* by the neurotransmitters they use such as: glutamatergic, GABAergic (γ-aminobutyric acid), dopaminergic, serotoninergic, etc. Additionally, cellular function, behavior, electrical activity, ion channel activity and receptor type can be used to characterize neurons. Approaches used in the characterization of neurons that provide high content, high impact and high value will also be discussed in this chapter.

Glutamatergic Neurons

Glutamatergic neurons are neurons that use glutamic acid, glutamate, as a neurotransmitter. Glutamate is a non-essential amino acid which also serves as arguably the most important excitatory neurotransmitter in the central nervous system (CNS) (Orrego and Villanueva, 1993). The majority of excitatory neurons in the CNS are glutamatergic and over half of all brain synapses release glutamate. Glutamate receptors are located mainly on the neuronal membrane. There is plenty of glutamate (glutamic acid) in the human body and it is particularly abundant in the nervous system, especially in the brain. Glutamate is also a precursor of GABA, the brain's main inhibitory neurotransmitter. Glutamate receptors can be ionotropic, forming an ion channel that is activated upon neurotransmitter binding, or metabotropic, which when activated trigger a G-protein-coupled signaling cascade. Ionotropic receptors are classified according to the affinity of specific agonists: *N*-methyl-D-Aspartate (NMDA), α-amino acid-3-hydroxy-5-methyl-4-isoxazol (AMPA) and kainic acid (KA). Prolonged neuronal excitation by glutamate with its associated ion influx can be toxic to neurons. Such excitotoxicity is the main cause of neuronal death in stroke and other forms of CNS trauma. Glutamate receptors are also involved in various diseases of the CNS. Glutamate receptor-mediated excitotoxicity is thought to contribute to the expression of a variety of neurodegenerative diseases such as Alzheimer's and Parkinson's disease.

NMDA-type glutamate receptors are excitatory in neurons that play key roles in many physiological and pathological processes. NMDA receptors are densely localized on neurons in most major brain areas and are physically connected to proteins involved in cell-signaling cascades (Arundine *et al.*, 2003; Arundine and Tymianski, 2003; Liu *et al.*, 2013). These receptors are well-known mediators of neuronal cell death in numerous neuropathological conditions (Bubenikova-Valesova *et al.*, 2008; Kari *et al.*, 1978; Liu *et al.*, 2013; Medeiros *et al.*, 2011; Wakschlag *et al.*, 2010). NMDA receptors are heteromeric complexes composed of obligatory NR1 subunits as well as subunits from the NR2 subfamily (NR2A, NR2B, NR2C, NR2D) or the NR3 subfamily (NR3A or NR3B) (Furukawa *et al.*, 2005; Laube *et al.*, 1998; Liu *et al.*, 2013; Premkumar and Auerbach, 1997; Ulbrich and Isacoff, 2007). Various combinations of subunits generate a large number of NMDA receptor subtypes with differing pharmacological and biological properties. NMDA receptors are a class of ion channel-forming receptors that are highly permeable to calcium. Cytosolic calcium is an important mediator of neuronal signal transduction, participating in diverse biochemical reactions that elicit changes in synaptic function, metabolic rate, and gene transcription (Liu *et al.*, 2013). Up-regulation of NMDA receptors can result in an excessive entry of endogenous calcium, triggering a series of cytoplasmic and nuclear processes such as loss of mitochondrial membrane potential, which ultimately results in neuronal cell death (Figure 1.5) (Liu *et al.*, 2013).

Since NMDA receptor-regulated ion channels are known to be highly permeable to calcium, it is important to determine whether intracellular calcium concentrations are altered by NMDA receptor subunit dysregulation that results from toxicant exposure (such as prolonged general anesthesia caused by NMDA receptor blockade). The application of NMDA to typical neurons (Figures 1.6A–C) in the presence of glycine (100 μM) and absence of Mg^{2+} produces an immediate increase in intracellular free Ca^{2+} ($[Ca^{2+}]_i$) in both control and ketamine (an agent commonly used as a pediatric general anesthetic)-exposed neurons (Liu *et al.*, 2013). In these studies, ethylene glycol tetra acetic acid (EGTA), a chelating agent with a high affinity for calcium, was used to sequester extracellular calcium (Liu *et al.*, 2013). To dissect the underlying mechanisms, NMDA was applied in the absence of extracellular Ca^{2+} (i.e., after chelation with EGTA; Figures 1.6A–B). Under these conditions no increase in NMDA-

NMDA Receptor NR1(Subunit)-labeled Neurons

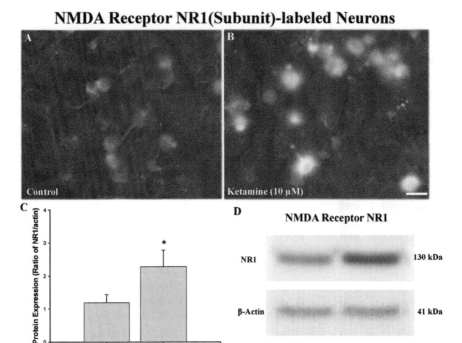

Figure 1.5. Cells from a control culture (A) and a ketamine (non-competitive NMDA receptor antagonist)-exposed culture (B) were double-immunostained with a mouse monoclonal antibody to the NMDA receptor NR1 subunit protein (green) and a rabbit polyclonal antibody to GFAP (red; astrocytes). NR1 immunoreactivity (green) was localized specifically on neurons and the fluorescent density was significantly up-regulated in ketamine-exposed cultures. Scale bar = 50 μm. The NMDA receptor NR1 protein levels were also evaluated using Western blot analysis. A major protein band at about 130 kDa was observed in both control and ketamine-exposed cultures (D). Ketamine administration produced a marked up-regulation of the NR1 protein compared with controls. No significant difference was detected in GFAP expression levels (astrocytes) between control and ketamine treated cultures. Densitometry measurements from three independent experiments were used to calculate a ratio of NMDA receptor NR1 protein to β-actin (C). The data are shown as means ± S.D. *$p < 0.05$ was considered significant compared to control (Liu *et al.*, 2013).

evoked $[Ca^{2+}]_i$ was observed. As a control, 25 μM glutamate, a concentration known to preferentially stimulate NMDA receptors, was used on the same cells without EGTA and produced an increase in $[Ca^{2+}]_i$ similar to that observed previously with NMDA (Figures 1.6A–B) (Liu *et al.*, 2013). In addition, the ketamine-exposed neurons showed a significantly greater increase in $[Ca^{2+}]_i$ than that seen in neurons in control cultures. The change in $[Ca^{2+}]_i$ resulting from NMDA stimulation was approximately 40% higher in the neurons from ketamine-exposed cultures (Liu *et al.*, 2013). Taken together, these observations are consistent with the view that prolonged/continuous exposure of developing neurons to NMDA antagonists such as ketamine causes a compensatory up-regulation of NMDA receptors (Liu *et al.*, 2013; Slikker *et al.*, 2007; Wang *et al.*, 2005; Wang *et al.*, 2006), and activation of these upregulated NMDA receptors generates $[Ca^{2+}]_i$ elevations that arise primarily from Ca^{2+} influx, not from release of intracellular stores (Liu *et al.*, 2013).

Changes in *[Ca²⁺]i* in Fura-2-Loaded Neurons

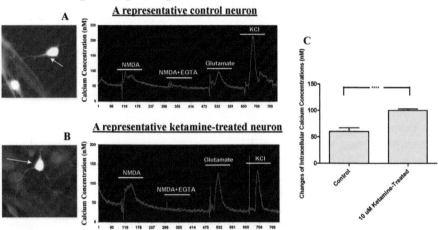

Figure 1.6. Dynamic changes in intracellular calcium concentrations $[Ca^{2+}]_i$ in a control neuron (A) and a ketamine-exposed neuron (B). Application of NMDA (50 µM) or glutamate (25 µM) caused an immediate elevation in intracellular free Ca^{2+} in both control and ketamine-exposed neurons. No increase in NMDA-evoked $[Ca^{2+}]_i$ was observed when the extracellular Ca^{2+} was chelated and, thus, unavailable for intracellular transport (50 µM NMDA + 200 µM EGTA in the perfusion buffer). A significant increase in intracellular free calcium $[Ca^{2+}]_i$ was detected in ketamine-exposed neurons compared to control neurons (C) after NMDA (50 µM) stimulation. Each condition was assessed at least in triplicate and experiments were repeated independently three times. Data are presented as means ± S.D. (Liu *et al.*, 2013).

In general, glutamatergic neuronal transmission, glutamate receptor expression levels and signal transduction (e.g., Ca^{2+} influx) play critical roles in normal neuronal functions in the CNS as well as in toxicant-induced neurotoxicity. Such information will be essential in order to increase the likelihood of the clinical success of attempts to develop effective rescue and prevention strategies (Wang, 2013).

GABAergic Neurons

γ-Aminobutyric acid (GABA) is the chief inhibitory neurotransmitter of the mammalian central nervous system. It plays a principal role in reducing neuronal excitability, and the inhibitory action of GABAergic neurons regulates synaptic integration. Impaired inhibitory function is thought to be one of the major underlying causes behind neuropathies characterized by neuronal hyper-excitability, such as epilepsy.

Notably, GABA is an inhibitory transmitter in the mature brain, but it may work as an excitatory transmitter in the developing brain (Ben-Ari *et al.*, 2007; Li and Xu, 2008). In immature neurons, the intracellular concentration of chloride is higher than that in extracellular environment, which is opposite to that in mature neurons (Wang et al., 2015b). Activated GABA-A receptors drive Chloride ions out of the cell, resulting in a depolarizing current (Wang et al., 2015b). During development, GABA is believed to contribute to the maturation of ion pumps (Ganguly *et al.*, 2001).

Previous work has demonstrated that GABAergic cell bodies can be found in the forebrain and all cortical regions by embryonic day 16 in rodents (Henschel *et al.*, 2008; Sanchez *et al.*, 2011). *In vivo* neural stem cells express GABA as well as other types of receptors. However, many of them including, N-methyl–D-aspartic acid receptors, are not yet functional at this stage of development (Boscolo *et al.*, 2012; Ge *et al.*, 2007; LoTurco *et al.*, 1995; Maher and LoTurco, 2009). In contrast to their inhibitory effect in adulthood (Ge *et al.*, 2007; Maher and LoTurco, 2009), stimulation of immature GABA receptors is excitatory early in development. GABA is also tonically released by neural stem cells and acts as a trophic factor, regulating their key developmental processes including proliferation and differentiation (Ben-Ari *et al.*, 2007; Ge *et al.*, 2007).

During development, for example, exposure of cultured neural cells to propofol (a commonly used general anesthetic agent which is thought to act primarily via stimulation of GABA$_A$ receptors), may enhance the excitatory action of endogenous GABA and lead to toxicity. Recent immunocytochemical data have shown that neural stem cells (identified using nestin staining) do not express GABA$_A$ receptors, as evidenced by a lack of receptor immuno-staining. However, strong immunoreactive staining for the GABA$_A$ receptor was observed on neurons that had been differentiated from the same neural stem cells that previously lacked GABA receptors (Figure 1.7) (Liu *et al.*, 2014). Therefore, the enhanced neuronal damage seen *in vivo* after propofol exposure is likely due to its actions on young neurons that have become more vulnerable to GABA stimulation because of newly expressed GABA receptors: more receptors allow more Ca^{2+} influx, followed by increased ROS generation and cell death. These findings are consistent with recent *in vitro* studies (Kahraman *et al.*, 2008; Pearn *et al.*, 2012) which have also demonstrated significant elevations in intracellular calcium levels in association with propofol-induced cell death.

Neurons Differentiated from Neural Stem Cells
GABA Receptor Expression

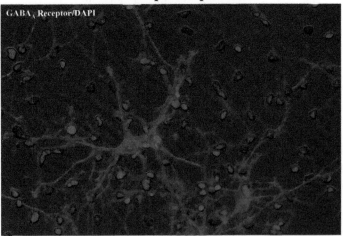

Liu et al., 2014

Figure 1.7. Representative photograph of immuno-staining for GABA$_A$ receptors on neurons that are differentiated from neural stem cells. Strong immuno-reactivity was detected on these immature neurons. Nuclei are counter-stained with DAPI (nuclear dye).

Dopaminergic Neurons

Dopamine has been extensively studied due to its involvement in mental and neurological disorders. The major sources of dopamine in the mammalian CNS are the dopaminergic neurons located at midbrain. Dopaminergic neurons in brain can be identified and located using the Falck-Hillarp histofluorescence method (Falck *et al.*, 1982), which makes dopamine cells visible through fluorescence after formaldehyde treatment. Dopaminergic neurons can be found in different regions of brain (diencephalon, mesencephalon, and the olfactory bulb) and have various functions (Bjorklund and Dunnett, 2007). The most remarkable cluster of dopaminergic neurons is in the ventral part of mesencephalon, which accounts for almost 90% of the dopaminergic neurons in brain. The dopaminergic neurons have been further nominally categorized into several sub-groups, and nigrostriatal system mainly mediates voluntary motor movement of the body.

The dopaminergic neurons located in the ventral tegmental area (VTA) extend their fibers mainly into the nucleus accumbens and olfactory tubercles, in addition to the septum, amygdala and hippocampus. This bundle of projections is known as the mesolimbic dopaminergic system. Projections from dopaminergic neurons of the medial VTA to the prefrontal, cingulate and perirhinal cortex represent the mesocortical dopaminergic system. There is a lot of overlap of projections from the VTA cells to their targets, which results in them being described as the mesocorticolimbic system (Wise, 2004) in reference to the two systems together. Thus, various groups of dopaminergic neurons in the CNS are located in different brain locations, have different projections, and play different crucial roles in support of specific functions. These neurons have obvious distinctions but all have the ability to synthesize the neurotransmitter dopamine (Chinta and Andersen, 2005).

Serotoninergic Neurons

Serotonin or 5-hydroxytryptamine (5-HT) is another monoamine neurotransmitter. The feelings of well-being and happiness are generally attributed to serotonin (Young, 2007). Many antidepressants are thought to exert their therapeutic action by adjusting the secretion and uptake of serotonin. The enterochromaffin cells in the GI tract are the biggest source of serotonin, accounting for 90% of serotonin in the human body. Here, serotonin functions as a regulator of intestinal movement (Berger *et al.*, 2009). Another source of serotonin is, of course, the central nervous system where it is generated by serotonergic neurons and contributes to regulating mood, appetite and sleep. Serotonin also plays roles in cognition and wound healing (Mukherjee *et al.*, 2013).

Neurons in Primary, Neural Stem Cell-Derived and Organotypic Cultures

The CNS is one of the earliest systems to differentiate during embryogenesis and this process can be modeled to a great extent using cell cultures *in vitro* (Wang, 2015b). Advances in our understanding of cell biology and neuroscience have enabled us to mimic the differentiation process *ex vivo*, opening up new avenues of research for understanding stress-induced developmental neurotoxicity and for the development of protective strategies against toxicant-induced neurotoxicity (Wang, 2015a). The utilization of primary neuronal cell cultures (Figure 1.8) and recently developed neural stem cell cultures, especially neural stem cells of embryonic origin (both from animals and more recently, humans), has provided valuable tools for evaluating developmental neurotoxic effects of xenobiotics *in vitro* (Bai *et al.*, 2013; Liu *et al.*, 2014; Wang, 2015b). This capability is due to the following attributes of neural stem cells: (1) neural stem cells can be obtained directly from humans, therefore the data obtained does not need interspecies translation; (2) with the capability of differentiation, a more extensive evaluation on xenobiotics is possible: from evaluating effects on neural stem cells, to differentiated neurons, astrocytes and oligodendrocytes; (3) utilization of neural stem cell cultures reduces animal use and shortens time needed for study completion; (4) neural stem cell cultures can facilitate the assessment of regeneration capacity after exposure to toxicants; (5) neural stem cell cultures have the potential to influence best clinical practices, such as pediatric general anesthesia (Wang *et al.*, 2013); and (6) neural stem cells can mimic particular developmental stages of CNS in animals and humans (Wang, 2015b). Thus, application of these advanced *in vitro* models combined with biological approaches can provide better capabilities for understanding mechanisms underlying the etiology of developmental neurotoxicity, leading to improved strategies for ameliorating adverse effects (Wang, 2015b).

Although neural stem cells *in vitro* can mimic CNS developmental process *in vivo* and provide an unlimited supply of cells for a variety of uses such as studying

Specific Neuronal Marker PSA-NCAM

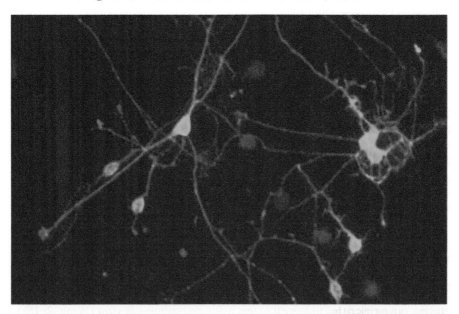

Figure 1.8. Immunofluorescence micrograph showing PSA-NCAM (polysialic acid neural cell adhesion molecule) staining on neurons from control cultures. PSA-NCAM is a specific marker for neurons. This slide illustrates intense immuno-staining of PSA-NCAM on the surface of neuronal cell bodies and processes.

the mechanisms underlying neurological diseases and evaluating developmental neurotoxicity (Bai *et al.*, 2013), it remains difficult to make direct comparisons between the developing brain with its three-dimensional aspects and neural stem cells in culture. Organotypic cell culture systems, which maintain the three-dimensional structure of CNS, do provide more relevant platforms to extrapolate preclinical findings to the human condition (Wang, 2015a). Organotypic cell cultures can be obtained from developing rodents and many other animals. Organotypic cell culture models can maintain critical anatomical structure and synaptic connectivity, in addition to providing the convenience of *in vitro* preparations (Figure 1.9). Using such systems, cell growth, migration and connectivity can be studied under environments that closely mimic *in vivo* conditions; and naturalistic brain structure can be maintained (Wang, 2015a). Organotypic cell cultures are also well suited for the rapid screening of the neurotoxic effects of drugs and other chemical in systems designed to model the most sensitive periods of brain development. Recently developed functional high-throughput three-dimensional model platforms and micro-fluidic cell culture chips

3D *In Vitro* Model - Organotypic Culture
(Frontal Cortical Slice from PND-7 Rat Pups)

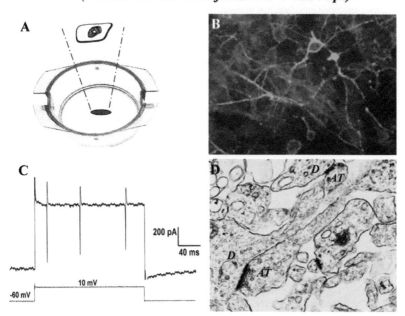

Modified from Wang & Slikker, 2008

Figure 1.9. Representation of neuropil from an organotypic slice culture (400 μm) obtained from the frontal cortex of a PND 7 rat pup brain. (A) General view of 3D *in vitro* organotypic culture system. (B) The neural network, viability of neurons (PSA-NCAM positive), and synaptic connections are well preserved in organotypic slices after 7 days of culture in control conditions. (C) Representative sodium current spikes in an organotypic rat brain slice. The sodium current spikes were evoked by applying a depolarizing voltage (200 ms, shown in the lower trace) when the neuron potential was held at –60 mV. (D) Electron photomicrograph showing axonal terminals *[At]* (presynaptic elements containing round synaptic vesicles) abutting a dendritic spine (postsynaptic element of another neuron).

will help improve studies in revealing the effects of chemical exposures using both embryonic and adult nervous systems.

Disclaimer

This document has been reviewed in accordance with United States Food and Drug Administration (FDA) policy and approved for publication. Approval does not signify that the contents necessarily reflect the position or opinions of the FDA. The findings and conclusions in this report are those of the authors and do not necessarily represent the views of the FDA.

References

Arundine, M., T. Sanelli, B. Ping He, and M.J. Strong. 2003. NMDA induces NOS 1 translocation to the cell membrane in NGF-differentiated PC 12 cells. Brain Research 976: 149–158.

Arundine, M., and M. Tymianski. 2003. Molecular mechanisms of calcium-dependent neurodegeneration in excitotoxicity. Cell Calcium 34: 325–337.

Bai, X., Y. Yan, S. Canfield, M.Y. Muravyeva, C. Kikuchi, I. Zaja, J.A. Corbett, and Z.J. Bosnjak. 2013. Ketamine enhances human neural stem cell proliferation and induces neuronal apoptosis via reactive oxygen species-mediated mitochondrial pathway. Anesthesia and Analgesia 116: 869–880.

Ben-Ari, Y., J.L. Gaiarsa, R. Tyzio, and R. Khazipov. 2007. GABA: a pioneer transmitter that excites immature neurons and generates primitive oscillations. Physiol Rev 87: 1215–1284.

Berger, M., J.A. Gray, and B.L. Roth. 2009. The expanded biology of serotonin. Annu Rev Med 60: 355–366.

Bjorklund, A., and S.B. Dunnett. 2007. Dopamine neuron systems in the brain: an update. Trends Neurosci 30: 194–202.

Boscolo, A., J.A. Starr, V. Sanchez, N. Lunardi, M.R. DiGruccio, C. Ori, A. Erisir, P. Trimmer, J. Bennett, and V. Jevtovic-Todorovic. 2012. The abolishment of anesthesia-induced cognitive impairment by timely protection of mitochondria in the developing rat brain: the importance of free oxygen radicals and mitochondrial integrity. Neurobiol Dis 45: 1031–1041.

Bubenikova-Valesova, V., J. Horacek, M. Vrajova, and C. Hoschl. 2008. Models of schizophrenia in humans and animals based on inhibition of NMDA receptors. Neurosci Biobehav Rev 32: 1014–1023.

Chinta, S.J., and J.K. Andersen. 2005. Dopaminergic neurons. Int J Biochem Cell Biol 37: 942–946.

Falck, B., N.A. Hillarp, G. Thieme, and A. Torp. 1982. Fluorescence of catechol amines and related compounds condensed with formaldehyde. Brain Res Bull 9: xi–xv.

Furukawa, H., S.K. Singh, R. Mancusso, and E. Gouaux. 2005. Subunit arrangement and function in NMDA receptors. Nature 438: 185–192.

Ganguly, K., A.F. Schinder, S.T. Wong, and M. Poo. 2001. GABA itself promotes the developmental switch of neuronal GABAergic responses from excitation to inhibition. Cell 105: 521–532.

Ge, S., D.A. Pradhan, G.L. Ming, and H. Song. 2007. GABA sets the tempo for activity-dependent adult neurogenesis. Trends Neurosci 30: 1–8.

Henschel, O., K.E. Gipson, and A. Bordey. 2008. GABA$_A$ receptors, anesthetics and anticonvulsants in brain development. CNS Neurol Disord Drug Targets 7: 211–224.

Kahraman, S., S.L. Zup, M.M. McCarthy, and G. Fiskum. 2008. GABAergic mechanism of propofol toxicity in immature neurons. J Neurosurg Anesthesiol 20: 233–240.

Kari, H.P., P.P. Davidson, H.H. Kohl, and M.M. Kochhar. 1978. Effects of ketamine on brain monoamine levels in rats. Res Commun Chem Pathol Pharmacol 20: 475–488.

Laube, B., J. Kuhse, and H. Betz. 1998. Evidence for a tetrameric structure of recombinant NMDA receptors. J Neurosci 18: 2954–2961.

Li, K., and E. Xu. 2008. The role and the mechanism of gamma-aminobutyric acid during central nervous system development. Neurosci Bull 24: 195–200.

Liu, F., T.A. Patterson, N. Sadovova, X. Zhang, S. Liu, X. Zou, J.P. Hanig, M.G. Paule, W. Slikker, Jr., and C. Wang. 2013. Ketamine-induced neuronal damage and altered N-methyl-D-aspartate receptor function in rat primary forebrain culture. Toxicol Sci 131: 548–557.

Liu, F., S.W. Rainosek, N. Sadovova, C.M. Fogle, T.A. Patterson, J.P. Hanig, M.G. Paule, W. Slikker, Jr., and C. Wang. 2014. Protective effect of acetyl-l-carnitine on propofol-induced toxicity in embryonic neural stem cells. Neurotoxicology 42C: 49–57.

LoTurco, J.J., D.F. Owens, M.J. Heath, M.B. Davis, and A.R. Kriegstein. 1995. GABA and glutamate depolarize cortical progenitor cells and inhibit DNA synthesis. Neuron 15: 1287–1298.

Maher, B.J., and J.J. LoTurco. 2009. Stop and go GABA. Nature Neuroscience 12: 817–818.

Medeiros, L.F., J.R. Rozisky, A. de Souza, M.P. Hidalgo, C.A. Netto, W. Caumo, A.M. Battastini, and I.L. Torres. 2011. Lifetime behavioural changes after exposure to anaesthetics in infant rats. Behav Brain Res 218: 51–56.

Mukherjee, V., N.P. Singh, and R.A. Yadav. 2013. Theoretical DFT study on spectroscopic signature and molecular dynamics of neurotransmitter and effect of hydrogen removal. Spectrochim Acta A Mol Biomol Spectrosc 107: 46–54.

Orrego, F., and S. Villanueva. 1993. The chemical nature of the main central excitatory transmitter: a critical appraisal based upon release studies and synaptic vesicle localization. Neuroscience 56: 539–555.

Pearn, M.L., Y. Hu, I.R. Niesman, H.H. Patel, J.C. Drummond, D.M. Roth, K. Akassoglou, P.M. Patel, and B.P. Head. 2012. Propofol neurotoxicity is mediated by p75 neurotrophin receptor activation. Anesthesiology 116: 352–361.

Premkumar, L.S., and A. Auerbach. 1997. Stoichiometry of recombinant N-methyl-D-aspartate receptor channels inferred from single-channel current patterns. J Gen Physiol 110: 485–502.

Sanchez, V., S.D. Feinstein, N. Lunardi, P.M. Joksovic, A. Boscolo, S.M. Todorovic, and V. Jevtovic-Todorovic. 2011. General anesthesia causes long-term impairment of mitochondrial morphogenesis and synaptic transmission in developing rat brain. Anesthesiology 115: 992–1002.

Slikker, W., Jr., X. Zou, C.E. Hotchkiss, R.L. Divine, N. Sadovova, N.C. Twaddle, D.R. Doerge, A.C. Scallet, T.A. Patterson, J.P. Hanig, M.G. Paule, and C. Wang. 2007. Ketamine-induced neuronal cell death in the perinatal rhesus monkey. Toxicol Sci 98: 145–158.

Ulbrich, M.H., and E.Y. Isacoff. 2007. Subunit counting in membrane-bound proteins. Nat Methods 4: 319–321.

Wakschlag, L.S., E.O. Kistner, D.S. Pine, G. Biesecker, K.E. Pickett, A.D. Skol, V. Dukic, R.J. Blair, B.L. Leventhal, N.J. Cox, J.L. Burns, K.E. Kasza, R.J. Wright, and E.H. Cook, Jr. 2010. Interaction of prenatal exposure to cigarettes and MAOA genotype in pathways to youth antisocial behavior. Mol Psychiatry 15: 928–937.

Wang, C. 2013. Critical regulation of calcium signaling and NMDA-type glutamate receptor in developmental neural toxicity. J Drug Metab Toxicol 4: 151.

Wang, C. 2015a. Application of *in vitro* models in developmental neurotoxicity and pharmaceutics research. J Mol Pharm Org Process Res 3: e122.

Wang, C., F. Liu, T.A. Patterson, M.G. Paule, and W. Slikker, Jr. 2013. Preclinical assessment of ketamine. CNS Neuros Ther 19(6): 448–53.

Wang, C., F. Liu, T.A. Patterson, M.G. Paule, and W. Slikker, Jr. 2015b. Anesthetic drug-induced neurotoxicity and compromised neural stem cell proliferation. JDAR 4: Article ID 235905, 8 pages.

Wang, C., N. Sadovova, X. Fu, L. Schmued, A. Scallet, J. Hanig, and W. Slikker. 2005. The role of the N-methyl-D-aspartate receptor in ketamine-induced apoptosis in rat forebrain culture. Neuroscience 132: 967–977.

Wang, C., N. Sadovova, C. Hotchkiss, X. Fu, A.C. Scallet, T.A. Patterson, J. Hanig, M.G. Paule, and W. Slikker, Jr. 2006. Blockade of N-methyl-D-aspartate receptors by ketamine produces loss of postnatal day 3 monkey frontal cortical neurons in culture. Toxicol Sci 91: 192–201.

Wise, R.A. 2004. Dopamine, learning and motivation. Nat Rev Neurosci 5: 483–494.

Young, S.N. 2007. How to increase serotonin in the human brain without drugs. J Psychiatry Neurosci 32: 394–399.

Yu, B., C. Wang, J. Liu, K.M. Johnson, and J.P. Gallagher. 2002. Adaptation to chronic PCP results in hyperfunctional NMDA and hypofunctional GABA(A) synaptic receptors. Neuroscience 113: 1–10.

2

Astrocytes

Cheng Wang,[1,a] *Qi Yin,*[1,b] *Shuliang Liu,*[1,c] *Xuan Zhang,*[1,d]
Fang Liu,[1,e] *Jingshu Zhang,*[2] *Tucker A. Patterson,*[3,g]
Merle G. Paule[1,f] and *William Slikker, Jr.*[3,h]

Introduction

The nervous system contains only two principal types of cells—nerve cells, or neurons, which are the information-processing and signalling elements, and glial cells, which play a variety of supporting roles. Several cell types found in the central nervous system (CNS) in association with neurons are classified as neuroglia, or glial cells. Cell-cell interactions—neuron-glia and glia-glia—are involved in initiating and orchestrating critical developmental steps (Stipursky *et al.*, 2012; White and Kramer-Albers, 2014). There are several types of neuroglia, each with their own morphological and functional characteristics. Astrocytes are a kind of glial cells that are derived from neural stem cells (Kriegstein and Alvarez-Buylla, 2009; Levitt and Rakic, 1980; Merkle *et al.*, 2004; Powell and Geller, 1999). Astrocytes (Astro from the Greek *astron* = star and *cyte* from Greek "kyttaron" = cell) physically interact with neurons and blood vessels via a variety of cellular processes that generally give them a starlike morphology.

[1] Division of Neurotoxicology, HFT-132, National Center for Toxicological Research (NCTR), Food and Drug Administration (FDA), 3900 NCTR Rd, Jefferson, AR 72079.
[a] E-mail: Cheng.Wang@fda.hhs.gov; [b] E-mail: qi.yin@fda.hhs.gov; [c] E-mail: shuliang.liu@fda.hhs.gov
[d] E-mail: xuan.zhang@fda.hhs.gov; [e] E-mail: fang.liu@fda.hhs.gov; [f] E-mail: merle.paule@fda.hhs.gov
[2] Department of Toxicology, School of public Health, Nanjing Medical University, 101 Longmian Avenue, Nanjing 21166, P.R. China.
E-mail: jingshuzh@njmu.edu.cn
[3] National Center for Toxicological Research (NCTR), Food and Drug Administration (FDA), 3900 NCTR Rd, Jefferson, AR 72079.
[g] E-mail: tucker.patterson@fda.hhs.gov; [h] E-mail: William.Slikker@fda.hhs.gov

#Astrocytes, *Oligodendrocytes and ^Neurons Derived from Human Embryonic Neural Stem Cells

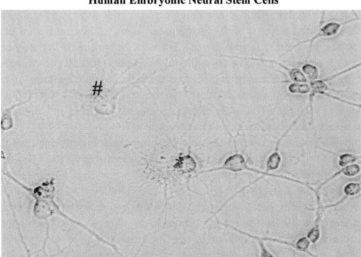

Figure 2.1. Astrocytes (#), oligodendrocytes (*), and neurons (^) differentiated from human embryonic neural stem cells (PhoenixSongs Biologicals, Inc. Branford, CT) obtained from the hippocampus of a fetal brain at 19-weeks of gestation. These cells were cultured in a defined neural differentiation medium for two weeks. This medium was supplemented with glutamine, laminin, brain derived neurotrophic factor (BDNF), and glial cell line-derived neurotrophic factor (GDNF) (PhoenixSongs Biologicals, CT).

During development, however, astrocytes are derived from heterogeneous progenitor cells with diverse functions (Kimelberg, 2004).

The proportion of astrocytes in the CNS increases as the interactions between glial cells and neurons become more sophisticated. In the rodent cortex, for example, the average astrocyte/neuron ratio is around 0.4:1, whereas in the human cortex it is about 1.4:1 (Nedergaard *et al.*, 2003). It is thought that, in addition to supporting neurons, astrocytes also modulate and control local neural networks.

Astrocytes have a well-developed cytoskeleton that is dominated by intermediate filaments but also includes microtubules and actin filaments, consistent with a role as structural support elements. Astrocytes are the largest of the neuroglia and elaborate numerous long processes and have spherical, centrally located nuclei. Many of their processes terminate in expanded pedicles (enlarged end-feet) which attach to the walls of blood capillaries (the outer surface of CNS capillaries). These pedicles, termed "vascular feet", completely surround and ensheath all vessels of the nourishing vascular network (Abbott *et al.*, 2006). Processes of the astrocytes also abut neurons, dendrites, synaptic endings, and nodes of Ranvier, and at the periphery of the brain and spinal cord form a layer under the pia mater. This layer, which also contains processes of other neuroglia, separates the connective tissue of the pia mater from CNS nerve cells. This carpeting of otherwise exposed surfaces with astrocytic processes suggests that these cells play a role in CNS metabolism and the regulation of extracellular ionic concentrations. Astrocytes are diverse in their origin, morphology, distribution, and function (Sovrea and Bosca, 2013). For example, the protoplasmic astrocytes mainly differentiate from the ventricular radial glia during development, whereas the fibrous astrocytes originate primarily from neonatal glial progenitors in the subventricular zone (Wang and Bordey, 2008). Classically, protoplasmic astrocytes are grey matter astrocytes, characterized by a bushy appearance owing to their numerous branched, short processes. In contrast, fibrous astrocytes are located primarily in the white matter and usually appear as elongated cells with long processes (Nag, 2011). More recently, the diversity of astrocytes has been classified according to their anatomy and include cerebellar Bergmann glia, retinal Müller cells, pituicytes of the neurohypophysis, etc. (Sofroniew and Vinters, 2010). In electron micrographs, astrocytes are typified by their relatively organelle-free, microfibril-rich cytoplasm (Peters *et al.*, 1991).

Astrocytes have long been regarded as the primary supportive cells of the CNS (Kimelberg, 2007): they maintain a viable milieu for neurons via their housekeeping functions. By active transport, astrocytes take up extrasynaptic neurotransmitters, including glutamate and GABA (Kimelberg, 2007). They also provide potassium ion buffering of the extracellular space and, thus, maintain neuronal excitability (Gerschenfeld *et al.*, 1959). The growth factors released by astrocytes are important for neuronal survival and maturation during development and the presence of astrocytes and the trophic factors they secrete promote synapstogenesise (Christopherson *et al.*, 2005; Mauch *et al.*, 2001; Pfrieger and Barres, 1997).

It has long been postulated that astrocytes provide nourishment to neurons since astrocytes project endfeet onto blood vessel walls. Glucose from blood vessels is, thus, thought to be provided to neurons directly to support their metabolism (Nehlig and Coles, 2007). In addition, low concentrations of glycogen, along with enzymes that synthesize and metabolize glycogen are present in astrocytes (Pfeiffer-Guglielmi *et al.*, 2005). Astrocytic glycogen provides energy for neurons under hypoglycemic conditions or when tissue energy demand is increased. Lactate is produced during aerobic glycolysis in astrocytes during the glutamate-glutamine cycle that is critical for neurotransmitter balance and maintenance (Pellerin and Magistretti, 1994). Lactate utilization is preferred over glucose in neurons during glutamatergic stimulation (Porras *et al.*, 2004). Importantly, astrocytes are a major part of the limited armamentarium available to the CNS in response to injury: they multiply, increase their production of intermediate filaments, and form dense, gliotic scars.

The Astrocyte Specific Biomarker—Glial Fibrillary Acidic Protein (GFAP)

Initially, GFAP was purified tissue from the brains of persons with Multiple Sclerosis that had been fixed in absolute ethanol (Bignami *et al.*, 1972; Eng *et al.*, 1971; Uyeda *et al.*, 1972). The amino acid composition of purified GFAP was first determined in the Neurochemistry Laboratory of Dr. Eric Shooter at the Stanford Medical Center. The biochemical properties of GFAP that have impeded its characterization are its insolubility in aqueous solvents, tendency to aggregate or polymerize, susceptibility to neutral proteases, highly specific and antigenic epitopes, and wide distribution of GFAP-containing astrocytes. It is now well established that GFAP is the principal 8–9 nm intermediate filament found in mature astrocytes of the CNS.

The *in vivo* functions of astrocytic intermediate filaments have been examined using gene knockout animal models (Galou *et al.*, 1997). Using gene targeting techniques, the GFAP gene in embryonic stem cells can be disrupted. In homozygous mice completely devoid of GFAP, the development of the CNS appeared normal in anatomy, and exhibited typical neuropathological changes following inoculation with scrapie prions (Gomi *et al.*, 1995). However, the data from further studies demonstrated the importance of GFAP in mediating astrocyte-neuron interactions and modulating synaptic activities via astrocytic processes (McCall *et al.*, 1996). In a mouse model of spinal cord injury induced by percussive impact on head, the GFAP-null mice were significantly more vulnerable to cervical cord contusion and subpial haemorrhage in comparison with wild type controls, suggesting the important role of GFAP in structural stability (Nawashiro *et al.*, 1998). In addition, in the mice lacking both GFAP and vimentin, the formation of intermediate filaments was severely impaired, suggesting some functional overlap between the intermediate filament proteins. Prior studies had suggested that protein filament from neurons shared similar chemical and immunologic properties with GFAP and glial filaments (Dahl and Bignami, 1976; Davison, 1975; Davison and Hong, 1977; Day, 1977; Goldman *et al.*, 1978; Lee *et al.*, 1977; Yen *et al.*, 1976). However, other reports (De Vries *et al.*, 1976; Eng *et al.*, 1976) indicated that GFAP from astrocytes was not related to neurofilaments, providing interpretations that were initially at odds with earlier reports but that were subsequently confirmed by others (Bignami and Dahl, 1977; Chiu *et al.*, 1980; Liem *et al.*, 1978; Schachner *et al.*, 1978; Schlaepfer *et al.*, 1979).

Only a very small pool of aqueous soluble GFAP can be detected throughout early development of the rat brain (Malloch *et al.*, 1987), in cultured astrocytes (Chiu and Goldman, 1984), and in the rat spinal cord (Aquino *et al.*, 1988). Early reports of

Astrocytes Derived from Human Embryonic Neural Stem Cells

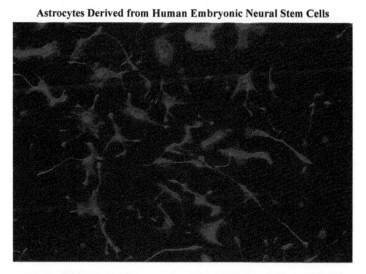

Figure 2.2. Cytoskeletal elements and immunochemical characterization of astrocyte. These astrocytic cells were differentiated from human embryonic neural stem cells collected from gestation week 19 fetuses, grown in defined neural differentiation medium for two weeks, and stained with an antibody directed against a specific protein present in astrocyte intermediate filaments (glial fibrillary acidic protein, or GFAP). The GFAP protein is present in large amounts in the intermediate filaments (red) of these cells.

detectable soluble GFAP found in human post-mortem and animal tissue were from studies in which the time between death and tissue analysis was not controlled. The low MW soluble forms of GFAP observed in these cases probably resulted from a calcium-activated neural proteinase, which has high substrate specificity for vimentin and desmin (Bigbee *et al.*, 1983; DeArmond *et al.*, 1983; Nelson and Traub, 1983; Schlaepfer and Zimmerman, 1981). Age-related increases in GFAP levels have been observed in astrocytes from rodents, nonhuman primates, and humans (Eng and Lee, 1998; Haley *et al.*, 2010; Nichols *et al.*, 1993). Astrocytic responses following CNS injuries can be assessed using quantitative immunocytochemical assays for GFAP (Dlugos and Pentney, 2001; Fix *et al.*, 1995a). Considering the age-related increases in GFAP and variations in sample collection procedures and tissue processing, treatment to inhibit protease activities of the aldehyde-fixed samples and the inclusion of age-matched controls had to be included in the staining protocol (Eng and Ghirnikar, 1994).

Astrocytes in Reactive Gliosis

It is now well known that GFAP is the major constituent of the intermediate filament proteins of the cytoskeleton in mature astrocytes (Pekny and Pekna, 2004). In many pathological CNS conditions including trauma, ischemia, and neurodegeneration, astrocytes respond by shifting from a quiescent to a reactive state characterized by an increased expression of GFAP (Pekny and Pekna, 2004). This astrocytic response that occurs with nearly all forms of CNS insult is defined as reactive astrogliosis and is a common pathological feature of CNS damages (Pekny and Pekna, 2014). Reactive astrocytes are characterized by enhanced expression of genes encoding intermediate filaments including GFAP, vimentin, and nestin, along with enlargement of cellular processes (Middeldorp and Hol, 2011).

Astrogliosis is an accepted universal marker of neurotoxicity but it is not unique to neurotoxicity. Astrogliosis results from an increase in the number of astrocytes and typically occurs in response to the destruction of nearby neurons, such as that which occurs in cases of hypoxia. Astrogliosis can also occur in response to infection and drug-induced neurotoxicity (Johnson *et al.*, 1998).

The pathological events that occur during astrocyte activation and the consequential influences of reactive astrogliosis for the neighboring neural tissue have been under extensive investigation. Reactive astrogliosis has been associated with neurodegenerative alteration induced by exposures to MK-801 or phencyclidine (a.k.a. PCP) in the cortico-limbic regions (Fix *et al.*, 1995b; Johnson *et al.*, 1998). PCP, a non-competitive NMDA receptor antagonist which was initially developed as a general anesthetic, is a commonly abused drug which causes schizophrenia-like symptoms in addition to euphoria (Allen and Young, 1978; Ban *et al.*, 1961; Cohen *et al.*, 1962; Lahti *et al.*, 1995). The association of PCP-induced neuronal cell death with surrounding reactive astrogliosis has been demonstrated using TUNEL-positive staining (indicating cell death, likely due to apoptosis) and GFAP immunoreactivity (Johnson *et al.*, 1998). Chronic PCP exposure leads to severe neuronal degeneration in the piriform cortex and olfactory tubercle (an area around the lateral olfactory tract), accompanied by a remarkable expansion of GFAP-stained cells indicating a proliferation of reactive astrocytes.

The influence of reactive astrogliosis on the outcome of brain lesions can occur at multiple levels. For example, during ischemic brain damage, astrocytes supply lactate from glycolysis in order to meet the metabolic demands of nearby neurons, a function also performed during periods of enhanced neural activity or glucose deprivation (Sorg and Magistretti, 1992; Wender *et al.*, 2000). During ischemic conditions, reactive astrocytes deliver metabolic substrate(s) to neurons to fuel their energy requirements. Astrocytes affected themselves by ischemia fail to support the energy requirements of neurons and, thus, contribute to neuronal degeneration (Rossi *et al.*, 2007; Takano *et al.*, 2009). Reactive astrogliosis contributes to the formation of distinct demarcations around CNS lesions and regulate the infiltration of inflammatory cells into the affected area (Li *et al.*, 2008; Okada *et al.*, 2006). In response to cerebral injury such as ischemic stroke, a fraction of the mature astrocytes surrounding the lesion proliferate, compensating for the loss of astrocytes at the lesion site (Barreto *et al.*, 2011). Stroke may also induce neurogenesis from periventricular neural progenitor cells that express GFAP (Ohab *et al.*, 2006). These GFAP-expressing neural progenitor cells proliferate as neuroblasts and migrate toward the lesions. The proliferation and migration of the neural progenitor cells are guided by environmental cues from neurovascular units and molecular mediators released by activated astroglia and infiltrating microglia (Kojima *et al.*, 2010; Marlier *et al.*, 2015). Functional recovery after an infarction also depends upon post-stroke neurovascular remodeling (Ardelt *et al.*, 2012). Therapeutic strategies of either employing exogenous neural progenitor cells or motivating endogenous neural progenitor cells have been investigated with the aim of replacing neuronal loss resulting from infarction (Marlier *et al.*, 2015).

Astrocytes are equipped with molecular sensors (i.e., pattern recognition receptors, PRRs) on their cellular surface and in their cytosolic compartment (Furr and Marriott, 2012). Astrocytes, along with microglia, respond to bacterial or viral CNS infections by generating inflammatory responses (Furr and Marriott, 2012; Kielian, 2004; Liu and Kielian, 2011). During acute infection, intercellular communications via glia-glia intercellular gap junctions are remarkably decreased among astrocytes immediately surrounding the abscessed areas induced by *Staphylococcus aureus* (Karpuk *et al.*, 2011). Such observations demonstrate the importance of astrocyte coupling and electrical communication during CNS inflammation. Astrocyte syncytial networks communicate via gap junctions and act as protective barriers to the adjacent normal parenchyma by sequestering dysregulated inflamed tissue and microbial pathogens (Sofroniew, 2005; Sofroniew, 2009).

There is a normal process of reactive astrogliosis and glial scar formation that provides a variety of benefits including the protection of neural cells and their function, restricting the spread of inflammation and infection, and promoting tissue repair. It is important to distinguish this normal process from those carried out by dysfunctional reactive astrocytes that contribute to CNS disorders.

Astrocytes in Blood-Brain Barrier (BBB)

The exchange of molecules between the blood and brain parenchyma is tightly controlled at the blood-brain interface. The endothelial cells lining the cerebral capillary vessels are the primary components of the all-important blood-brain barrier (BBB).

These capillary vessels are encircled by a basal lamina and astrocytic perivascular endfeet (Abbott *et al.*, 2006; Ballabh *et al.*, 2004). The physical barrier property of the BBB is bestowed by tight junctions formed between neighboring endothelial cells. Nevertheless, the endothelium provides several mechanisms by which substances can cross the BBB, including transcellular diffusion of small lipid-soluble molecules, transporter-mediated movement of specific compounds, and receptor-mediated transcytosis of some proteins (Abbott and Romero, 1996). Astrocytes provide the cellular bridge between neurons and associated microvessels. The close proximity of neural cells, including astrocytes, neurons, and microglia, to cerebral capillaries suggests possible roles for neural cells in the induction and maintenance of BBB properties (Abbott *et al.*, 2006). Co-cultures of brain capillary endothelial cells and astrocytes have shown that astrocytes promote tight junction formation (Dehouck *et al.*, 1990) and expression of specialized transporters and enzymes that regulate the permeability of the cerebrovascular endothelium (Abbott, 2002; Abbott *et al.*, 2006; Ballabh *et al.*, 2004; Beck *et al.*, 1984; Savidge *et al.*, 2007). The factors released by astrocytes which produce and maintain the barrier properties of the capillary endothelium include transforming growth factor-β (TGFβ), basic fibroblast growth factor (bFGF), glial-derived neurotrophic factor (GDNF) and angiopoietin 1 (Garcia-Segura and McCarthy, 2004; Igarashi *et al.*, 1999; Lee *et al.*, 2003). Other cells in the vicinity of the BBB, including pericytes, neurons, and perivascular macrophages, contribute to the production and health of the barrier (Ramsauer *et al.*, 2002; Schiera *et al.*, 2003; Zenker *et al.*, 2003). It had been demonstrated in co-culture studies that neural progenitor cells can affect the production of barrier properties in brain microvascular endothelial cells (Weidenfeller *et al.*, 2007).

The integrity of the BBB is affected in many CNS pathologies, including stroke (Tomas-Camardiel *et al.*, 2005; Tran *et al.*, 1999), trauma (Schwaninger *et al.*, 1999), infectious diseases (Gaillard *et al.*, 2003), multiple sclerosis (Minagar and Alexander, 2003), brain tumors (Davies, 2002), Alzheimer's disease (Berzin *et al.*, 2000), Parkinson's disease (Kortekaas *et al.*, 2005), and epilepsy (Abbott *et al.*, 2006; Marroni *et al.*, 2003). In conditions involving infection or inflammation, the normal interaction between endothelial cells and astrocytes is disturbed, resulting in an opening of BBB. For example, bradykinin, an inflammatory mediator released in cases of stroke or trauma, acts on endothelial cells and astrocytes via bradykinin receptors to cause the production of interleukin-6 (IL-6) in astrocytes, which leads to an increase in permeability of the BBB.

Normally, the BBB regulates ion transportation and fluid movement between the cerebral vascular space and the cerebral interstitial space, maintaining a steady environment for normal neurological functioning. Evidence of an early involvement of BBB dysfunction in a range of neurological diseases has been provided (Kamphuis *et al.*, 2015; Lee and Bendayan, 2004) and therapeutic strategies that improve BBB barrier functions can reduce neuronal damage secondary to, or aggravated by, loss of BBB integrity (Bauer *et al.*, 2005; Cucullo *et al.*, 2004). The interaction of astrocytic endfeet and cerebrovascular endothelial cells is a known requirement for maintenance of BBB function (Broux *et al.*, 2015; Lien *et al.*, 2012; Park *et al.*, 2015; Watkins *et al.*, 2014). Therefore, a gliocentric perspective would be valuable for studies of factors contributing to the maintenance or restoration of BBB health.

Summary

Astrocytes account for the majority of glial cells in the CNS. Physiologically, they are responsible for a variety of important functions, from constituting the architectural framework of neural tissue and maintaining the appropriate environmental conditions for healthy neurons, to modulating neurovascular activities. In response to a wide range of pathological conditions, astrocytes are activated by various molecular signals and transformed from quiescent to reactive astrocytes (Sofroniew and Vinters, 2010). Reactive astrocytes are involved in a number of important cellular events, including the secretion of neurotrophic factors, the buffering of neurotoxic agents, the restoration and maintenance of the BBB, and combating infections, all of which are greatly beneficial to the health of the CNS (Sofroniew and Vinters, 2010). The dysregulation of astrocytic activation contributes to the pathologies seen in a wide spectrum of CNS disorders from trauma and ischemia to neurodegenerative diseases (Bosch and Kielian, 2015; Rama Rao and Kielian, 2015).

Comprehensive knowledge of the roles that astrocytes play in the CNS under either normal or diseased states will clearly benefit the development of therapeutic strategies for a variety of neurological disorders. Modulation of reactive astrogliosis will likely be critical for the management of diverse CNS diseases with an aim toward achieving restoration of normal neurological function and attenuation of neural damage.

Disclaimer

This document has been reviewed in accordance with United States Food and Drug Administration (FDA) policy and approved for publication. Approval does not signify that the contents necessarily reflect the position or opinions of the FDA. The findings and conclusions in this report are those of the authors and do not necessarily represent the views of the FDA.

References

Abbott, N.J. 2002. Astrocyte-endothelial interactions and blood-brain barrier permeability. J Anat 200: 629–638.

Abbott, N.J., and I.A. Romero. 1996. Transporting therapeutics across the blood-brain barrier. Mol Med Today 2: 106–113.

Abbott, N.J., L. Ronnback, and E. Hansson. 2006. Astrocyte-endothelial interactions at the blood-brain barrier. Nat Rev Neurosci 7: 41–53.

Allen, R.M., and S.J. Young. 1978. Phencyclidine-induced psychosis. Am J Psychiatry 135: 1081–1084.

Aquino, D.A., F.C. Chiu, C.F. Brosnan, and W.T. Norton. 1988. Glial fibrillary acidic protein increases in the spinal cord of Lewis rats with acute experimental autoimmune encephalomyelitis. J Neurochem 51: 1085–1096.

Ardelt, A.A., R.S. Carpenter, M.R. Lobo, H. Zeng, R.B. Solanki, A. Zhang, P. Kulesza, and M.M. Pike. 2012. Estradiol modulates post-ischemic cerebral vascular remodeling and improves long-term functional outcome in a rat model of stroke. Brain Res 1461: 76–86.

Ballabh, P., A. Braun, and M. Nedergaard. 2004. The blood-brain barrier: an overview: structure, regulation, and clinical implications. Neurobiol Dis 16: 1–13.

Ban, T.A., J.J. Lohrenz, and H.E. Lehmann. 1961. Observations on the action of Sernyl-a new psychotropic drug. Can Psychiatr Assoc J 6: 150–157.

Barreto, G.E., X. Sun, L. Xu, and R.G. Giffard. 2011. Astrocyte proliferation following stroke in the mouse depends on distance from the infarct. PLoS One. 6: e27881.

Bauer, B., A.M. Hartz, G. Fricker, and D.S. Miller. 2005. Modulation of p-glycoprotein transport function at the blood-brain barrier. Exp Biol Med (Maywood) 230: 118–127.

Beck, D.W., H.V. Vinters, M.N. Hart, and P.A. Cancilla. 1984. Glial cells influence polarity of the blood-brain barrier. J Neuropathol Exp Neurol 43: 219–224.

Berzin, T.M., B.D. Zipser, M.S. Rafii, V. Kuo-Leblanc, G.D. Yancopoulos, D.J. Glass, J.R. Fallon, and E.G. Stopa. 2000. Agrin and microvascular damage in Alzheimer's disease. Neurobiol Aging 21: 349–355.

Bigbee, J.W., D.D. Bigner, C. Pegram, and L.F. Eng. 1983. Study of glial fibrillary acidic protein in a human glioma cell line grown in culture and as a solid tumor. J Neurochem 40: 460–467.

Bignami, A., and D. Dahl. 1977. Specificity of the glial fibrillary acidic protein for astroglia. J Histochem Cytochem 25: 466–469.

Bignami, A., L.F. Eng, D. Dahl, and C.T. Uyeda. 1972. Localization of the glial fibrillary acidic protein in astrocytes by immunofluorescence. Brain Res 43: 429–435.

Bosch, M.E., and T. Kielian. 2015. Neuroinflammatory paradigms in lysosomal storage diseases. Front Neurosci 9: 417.

Broux, B., E. Gowing, and A. Prat. 2015. Glial regulation of the blood-brain barrier in health and disease. Semin Immunopathol 37: 577–590.

Chiu, F.C., and J.E. Goldman. 1984. Synthesis and turnover of cytoskeletal proteins in cultured astrocytes. J Neurochem 42: 166–174.

Chiu, F.C., B. Korey, and W.T. Norton. 1980. Intermediate filaments from bovine, rat, and human CNS: mapping analysis of the major proteins. J Neurochem 34: 1149–1159.

Christopherson, K.S., E.M. Ullian, C.C. Stokes, C.E. Mullowney, J.W. Hell, A. Agah, J. Lawler, D.F. Mosher, P. Bornstein, and B.A. Barres. 2005. Thrombospondins are astrocyte-secreted proteins that promote CNS synaptogenesis. Cell 120: 421–433.

Cohen, B.D., G. Rosenbaum, E.D. Luby, and J.S. Gottlieb. 1962. Comparison of phencyclidine hydrochloride (Sernyl) with other drugs. Simulation of schizophrenic performance with phencyclidine hydrochloride (Sernyl), lysergic acid diethylamide (LSD-25), and amobarbital (Amytal) sodium; II. Symbolic and sequential thinking. Arch Gen Psychiatry 6: 395–401.

Cucullo, L., K. Hallene, G. Dini, R. Dal Toso, and D. Janigro. 2004. Glycerophosphoinositol and dexamethasone improve transendothelial electrical resistance in an *in vitro* study of the blood-brain barrier. Brain Res 997: 147–151.

Dahl, D., and A. Bignami. 1976. Isolation from peripheral nerve of a protein similar to the glial fibrillary acidic protein. FEBS Lett 66: 281–284.

Davies, D.C. 2002. Blood-brain barrier breakdown in septic encephalopathy and brain tumours. J Anat 200: 639–646.

Davison, P.F. 1975. Neuronal fibrillar proteins and axoplasmic transport. Brain Res 100: 73–80.

Davison, P.F., and B.S. Hong. 1977. Filaments in nervous tissue and muscle cells. Int Meet Int Soc Neurochem 6th, Abstracts. p. 106.

Day, W.A. 1977. Solubilization of neurofilaments from central nervous system myelinated nerve. J Ultrastruct Res 60: 362–372.

De Vries, G.H., L.F. Eng, D.L. Lewis, and M.G. Hadfield. 1976. The protein composition of bovine myelin-free axons. Biochim Biophys Acta 439: 133–145.

DeArmond, S.J., M. Fajardo, S.A. Naughton, and L.F. Eng. 1983. Degradation of glial fibrillary acidic protein by a calcium dependent proteinase: an electroblot study. Brain Res 262: 275–282.

Dehouck, M.P., S. Meresse, P. Delorme, J.C. Fruchart, and R. Cecchelli. 1990. An easier, reproducible, and mass-production method to study the blood-brain barrier *in vitro*. J Neurochem 54: 1798–1801.

Dlugos, C.A., and R.J. Pentney. 2001. Quantitative immunocytochemistry of glia in the cerebellar cortex of old ethanol-fed rats. Alcohol 23: 63–69.

Eng, L.F., G.H. DeVries, D.L. Lewis, and J.W. Bigbee. 1976. Specific antibody to the major 47,000 MW protein fraction of bovine myelin-free axons. Fed Proc Fed Am Soc Exp Biol 35: 1766.

Eng, L.F., and R.S. Ghirnikar. 1994. GFAP and astrogliosis. Brain Pathol 4: 229–237.

Eng, L.F., and Y.L. Lee. 1998. Glial response to injury, disease, and aging. pp. 71–89. *In*: H.M. Schipper (ed.). Astrocytes in Brain Aging and Neurodegeneration. R. G. Landes Co., Georgetown, TX.

Eng, L.F., J.J. Vanderhaeghen, A. Bignami, and B. Gerstl. 1971. An acidic protein isolated from fibrous astrocytes. Brain Res 28: 351–354.

Fix, A.S., K.A. Wightman, and J.P. O'Callaghan. 1995a. Reactive gliosis induced by MK-801 in the rat posterior cingulate/retrosplenial cortex: GFAP evaluation by sandwich ELISA and immunocytochemistry. Neurotoxicology 16: 229–237.

Fix, A.S., D.F. Wozniak, L.L. Truex, M. McEwen, J.P. Miller, and J.W. Olney. 1995b. Quantitative analysis of factors influencing neuronal necrosis induced by MK-801 in the rat posterior cingulate/retrosplenial cortex. Brain Res 696: 194–204.

Furr, S.R., and I. Marriott. 2012. Viral CNS infections: role of glial pattern recognition receptors in neuroinflammation. Front Microbiol 3: 201.

Gaillard, P.J., A.B. de Boer, and D.D. Breimer. 2003. Pharmacological investigations on lipopolysaccharide-induced permeability changes in the blood-brain barrier *in vitro*. Microvasc Res 65: 24–31.

Galou, M., J. Gao, J. Humbert, M. Mericskay, Z. Li, D. Paulin, and P. Vicart. 1997. The importance of intermediate filaments in the adaptation of tissues to mechanical stress: evidence from gene knockout studies. Biol Cell 89: 85–97.

Garcia-Segura, L.M., and M.M. McCarthy. 2004. Minireview: Role of glia in neuroendocrine function. Endocrinology 145: 1082–1086.

Gerschenfeld, H.M., F. Wald, J.A. Zadunaisky, and E.D. De Robertis. 1959. Function of astroglia in the water-ion metabolism of the central nervous system: an electron microscope study. Neurology 9: 412–425.

Goldman, J.E., H.H. Schaumburg, and W.T. Norton. 1978. Isolation and characterization of glial filaments from human brain. J Cell Biol 78: 426–440.

Gomi, H., T. Yokoyama, K. Fujimoto, T. Ikeda, A. Katoh, T. Itoh, and S. Itohara. 1995. Mice devoid of the glial fibrillary acidic protein develop normally and are susceptible to scrapie prions. Neuron 14: 29–41.

Haley, G.E., S.G. Kohama, H.F. Urbanski, and J. Raber. 2010. Age-related decreases in SYN levels associated with increases in MAP-2, apoE, and GFAP levels in the rhesus macaque prefrontal cortex and hippocampus. Age (Dordr) 32: 283–296.

Igarashi, Y., H. Utsumi, H. Chiba, Y. Yamada-Sasamori, H. Tobioka, Y. Kamimura, K. Furuuchi, Y. Kokai, T. Nakagawa, M. Mori, and N. Sawada. 1999. Glial cell line-derived neurotrophic factor induces barrier function of endothelial cells forming the blood-brain barrier. Biochem Biophys Res Commun 261: 108–112.

Johnson, K.M., M. Phillips, C. Wang, and G.A. Kevetter. 1998. Chronic phencyclidine induces behavioral sensitization and apoptotic cell death in the olfactory and piriform cortex. J Neurosci Res 52: 709–722.

Kamphuis, W.W., C. Derada Troletti, A. Reijerkerk, I.A. Romero, and H.E. de Vries. 2015. The blood-brain barrier in multiple sclerosis: microRNAs as key regulators. CNS Neurol Disord Drug Targets 14: 157–167.

Karpuk, N., M. Burkovetskaya, T. Fritz, A. Angle, and T. Kielian. 2011. Neuroinflammation leads to region-dependent alterations in astrocyte gap junction communication and hemichannel activity. J Neurosci 31: 414–425.

Kielian, T. 2004. Immunopathogenesis of brain abscess. J Neuroinflammation 1: 16.

Kimelberg, H.K. 2004. The problem of astrocyte identity. Neurochem Int 45: 191–202.

Kimelberg, H.K. 2007. Supportive or information-processing functions of the mature protoplasmic astrocyte in the mammalian CNS? A critical appraisal. Neuron Glia Biol 3: 181–189.

Kojima, T., Y. Hirota, M. Ema, S. Takahashi, I. Miyoshi, H. Okano, and K. Sawamoto. 2010. Subventricular zone-derived neural progenitor cells migrate along a blood vessel scaffold toward the post-stroke striatum. Stem Cells 28: 545–554.

Kortekaas, R., K.L. Leenders, J.C. van Oostrom, W. Vaalburg, J. Bart, A.T. Willemsen, and N.H. Hendrikse. 2005. Blood-brain barrier dysfunction in parkinsonian midbrain *in vivo*. Ann Neurol 57: 176–179.

Kriegstein, A., and A. Alvarez-Buylla. 2009. The glial nature of embryonic and adult neural stem cells. Annu Rev Neurosci 32: 149–184.

Lahti, A.C., B. Koffel, D. LaPorte, and C.A. Tamminga. 1995. Subanesthetic doses of ketamine stimulate psychosis in schizophrenia. Neuropsychopharmacology 13: 9–19.

Lee, G., and R. Bendayan. 2004. Functional expression and localization of P-glycoprotein in the central nervous system: relevance to the pathogenesis and treatment of neurological disorders. Pharm Res 21: 1313–1330.

Lee, S.W., W.J. Kim, Y.K. Choi, H.S. Song, M.J. Son, I.H. Gelman, Y.J. Kim, and K.W. Kim. 2003. SSeCKS regulates angiogenesis and tight junction formation in blood-brain barrier. Nat Med 9: 900–906.

Lee, V., S.H. Yen, and M.L. Shelanski. 1977. Biochemical correlates of astrocytic proliferation in the mutant Staggerer mouse. Brain Res 128: 389–392.

Levitt, P., and P. Rakic. 1980. Immunoperoxidase localization of glial fibrillary acidic protein in radial glial cells and astrocytes of the developing rhesus monkey brain. J Comp Neurol 193: 815–840.

Li, L., A. Lundkvist, D. Andersson, U. Wilhelmsson, N. Nagai, A.C. Pardo, C. Nodin, A. Stahlberg, K. Aprico, K. Larsson, T. Yabe, L. Moons, A. Fotheringham, I. Davies, P. Carmeliet, J.P. Schwartz, M. Pekna, M. Kubista, F. Blomstrand, N. Maragakis, M. Nilsson, and M. Pekny. 2008. Protective role of reactive astrocytes in brain ischemia. J Cereb Blood Flow Metab 28: 468–481.

Liem, R.K., S.H. Yen, G.D. Salomon, and M.L. Shelanski. 1978. Intermediate filaments in nervous tissues. J Cell Biol 79: 637–645.

Lien, C.F., S.K. Mohanta, M. Frontczak-Baniewicz, J.D. Swinny, B. Zablocka, and D.C. Gorecki. 2012. Absence of glial alpha-dystrobrevin causes abnormalities of the blood-brain barrier and progressive brain edema. J Biol Chem 287: 41374–41385.

Liu, S., and T. Kielian. 2011. MyD88 is pivotal for immune recognition of Citrobacter koseri and astrocyte activation during CNS infection. J Neuroinflammation 8: 35.

Malloch, G.D., J.B. Clark, and F.R. Burnet. 1987. Glial fibrillary acidic protein in the cytoskeletal and soluble protein fractions of the developing rat brain. J Neurochem 48: 299–306.

Marlier, Q., S. Verteneuil, R. Vandenbosch, and B. Malgrange. 2015. Mechanisms and functional significance of stroke-induced neurogenesis. Front Neurosci 9: 458.

Marroni, M., N. Marchi, L. Cucullo, N.J. Abbott, K. Signorelli, and D. Janigro. 2003. Vascular and parenchymal mechanisms in multiple drug resistance: a lesson from human epilepsy. Curr Drug Targets 4: 297–304.

Mauch, D.H., K. Nagler, S. Schumacher, C. Goritz, E.C. Muller, A. Otto, and F.W. Pfrieger. 2001. CNS synaptogenesis promoted by glia-derived cholesterol. Science 294: 1354–1357.

McCall, M.A., R.G. Gregg, R.R. Behringer, M. Brenner, C.L. Delaney, E.J. Galbreath, C.L. Zhang, R.A. Pearce, S.Y. Chiu, and A. Messing. 1996. Targeted deletion in astrocyte intermediate filament (Gfap) alters neuronal physiology. Proc Natl Acad Sci USA 93: 6361–6366.

Merkle, F.T., A.D. Tramontin, J.M. Garcia-Verdugo, and A. Alvarez-Buylla. 2004. Radial glia give rise to adult neural stem cells in the subventricular zone. Proc Natl Acad Sci USA 101: 17528–17532.

Middeldorp, J., and E.M. Hol. 2011. GFAP in health and disease. Prog Neurobiol 93: 421–443.

Minagar, A., and J.S. Alexander. 2003. Blood-brain barrier disruption in multiple sclerosis. Mult Scler 9: 540–549.

Nag, S. 2011. Morphology and properties of astrocytes. Methods Mol Biol 686: 69–100.

Nawashiro, H., A. Messing, N. Azzam, and M. Brenner. 1998. Mice lacking GFAP are hypersensitive to traumatic cerebrospinal injury. Neuroreport 9: 1691–1696.

Nedergaard, M., B. Ransom, and S.A. Goldman. 2003. New roles for astrocytes: redefining the functional architecture of the brain. Trends Neurosci 26: 523–530.

Nehlig, A., and J.A. Coles. 2007. Cellular pathways of energy metabolism in the brain: is glucose used by neurons or astrocytes? Glia 55: 1238–1250.

Nelson, W.J., and P. Traub. 1983. Proteolysis of vimentin and desmin by the Ca2+-activated proteinase specific for these intermediate filament proteins. Mol Cell Biol 3: 1146–1156.

Nichols, N.R., J.R. Day, N.J. Laping, S.A. Johnson, and C.E. Finch. 1993. GFAP mRNA increases with age in rat and human brain. Neurobiol Aging 14: 421–429.

Ohab, J.J., S. Fleming, A. Blesch, and S.T. Carmichael. 2006. A neurovascular niche for neurogenesis after stroke. J Neurosci 26: 13007–13016.

Okada, S., M. Nakamura, H. Katoh, T. Miyao, T. Shimazaki, K. Ishii, J. Yamane, A. Yoshimura, Y. Iwamoto, Y. Toyama, and H. Okano. 2006. Conditional ablation of Stat3 or Socs3 discloses a dual role for reactive astrocytes after spinal cord injury. Nat Med 12: 829–834.

Park, H.J., J.Y. Shin, H.N. Kim, S.H. Oh, S.K. Song, and P.H. Lee. 2015. Mesenchymal stem cells stabilize the blood-brain barrier through regulation of astrocytes. Stem Cell Res Ther 6: 187.

Pekny, M., and M. Pekna. 2004. Astrocyte intermediate filaments in CNS pathologies and regeneration. J Pathol 204: 428–437.

Pekny, M., and M. Pekna. 2014. Astrocyte reactivity and reactive astrogliosis: costs and benefits. Physiol Rev 94: 1077–1098.

Pellerin, L., and P.J. Magistretti. 1994. Glutamate uptake into astrocytes stimulates aerobic glycolysis: a mechanism coupling neuronal activity to glucose utilization. Proc Natl Acad Sci USA 91: 10625–10629.

Peters, A., S.L. Palay, and H. deF. Webster. 1991. The Fine Structure of the Nervous System, Third edn. Oxford University Press, New York.

Pfeiffer-Guglielmi, B., M. Francke, A. Reichenbach, B. Fleckenstein, G. Jung, and B. Hamprecht. 2005. Glycogen phosphorylase isozyme pattern in mammalian retinal Muller (glial) cells and in astrocytes of retina and optic nerve. Glia 49: 84–95.

Pfrieger, F.W., and B.A. Barres. 1997. Synaptic efficacy enhanced by glial cells *in vitro*. Science 277: 1684–1687.

Porras, O.H., A. Loaiza, and L.F. Barros. 2004. Glutamate mediates acute glucose transport inhibition in hippocampal neurons. J Neurosci 24: 9669–9673.

Powell, E.M., and H.M. Geller. 1999. Dissection of astrocyte-mediated cues in neuronal guidance and process extension. Glia 26: 73–83.

Rama Rao, K.V., and T. Kielian. 2015. Neuron-astrocyte interactions in neurodegenerative diseases: role of neuroinflammation. Clin Exp Neuroimmunol 6: 245–263.

Ramsauer, M., D. Krause, and R. Dermietzel. 2002. Angiogenesis of the blood-brain barrier *in vitro* and the function of cerebral pericytes. FASEB J 16: 1274–1276.

Rossi, D.J., J.D. Brady, and C. Mohr. 2007. Astrocyte metabolism and signaling during brain ischemia. Nat Neurosci 10: 1377–1386.

Savidge, T.C., P. Newman, C. Pothoulakis, A. Ruhl, M. Neunlist, A. Bourreille, R. Hurst, and M.V. Sofroniew. 2007. Enteric glia regulate intestinal barrier function and inflammation via release of S-nitrosoglutathione. Gastroenterology 132: 1344–1358.

Schachner, M., C. Smith, and G. Schoonmaker. 1978. Immunological distinction between neurofilament and glial fibrillary acidic proteins by mouse antisera and their immunohistological characterization. Dev Neurosci 1: 1–14.

Schiera, G., E. Bono, M.P. Raffa, A. Gallo, G.L. Pitarresi, I. Di Liegro, and G. Savettieri. 2003. Synergistic effects of neurons and astrocytes on the differentiation of brain capillary endothelial cells in culture. J Cell Mol Med 7: 165–170.

Schlaepfer, W.W., L.A. Freeman, and L.F. Eng. 1979. Studies of human and bovine spinal nerve roots and the outgrowth of CNS tissues into the nerve root entry zone. Brain Res 177: 219–229.

Schlaepfer, W.W., and U.P. Zimmerman. 1981. Calcium-mediated breakdown of glial filaments and neurofilaments in rat optic nerve and spinal cord. Neurochem Res 6: 243–255.

Schwaninger, M., S. Sallmann, N. Petersen, A. Schneider, S. Prinz, T.A. Libermann, and M. Spranger. 1999. Bradykinin induces interleukin-6 expression in astrocytes through activation of nuclear factor-kappaB. J Neurochem 73: 1461–1466.

Sofroniew, M.V. 2005. Reactive astrocytes in neural repair and protection. Neuroscientist 11: 400–407.

Sofroniew, M.V. 2009. Molecular dissection of reactive astrogliosis and glial scar formation. Trends Neurosci. 32: 638–647.

Sofroniew, M.V., and H.V. Vinters. 2010. Astrocytes: biology and pathology. Acta Neuropathol 119: 7–35.

Sorg, O., and P.J. Magistretti. 1992. Vasoactive intestinal peptide and noradrenaline exert long-term control on glycogen levels in astrocytes: blockade by protein synthesis inhibition. J Neurosci 12: 4923–4931.

Sovrea, A.S., and A.B. Bosca. 2013. Astrocytes reassessment—an evolving concept part one: embryology, biology, morphology and reactivity. J Mol Psychiatry 1: 18.

Stipursky, J., T.C. Spohr, V.O. Sousa, and F.C. Gomes. 2012. Neuron-astroglial interactions in cell-fate commitment and maturation in the central nervous system. Neurochem Res 37: 2402–2418.

Takano, T., N. Oberheim, M.L. Cotrina, and M. Nedergaard. 2009. Astrocytes and ischemic injury. Stroke 40: S8–12.

Tomas-Camardiel, M., J.L. Venero, A.J. Herrera, R.M. De Pablos, J.A. Pintor-Toro, A. Machado, and J. Cano. 2005. Blood-brain barrier disruption highly induces aquaporin-4 mRNA and protein in perivascular and parenchymal astrocytes: protective effect by estradiol treatment in ovariectomized animals. J Neurosci Res 80: 235–246.

Tran, N.D., J. Correale, S.S. Schreiber, and M. Fisher. 1999. Transforming growth factor-beta mediates astrocyte-specific regulation of brain endothelial anticoagulant factors. Stroke 30: 1671–1678.

Uyeda, C.T., L.F. Eng, and A. Bignami. 1972. Immunological study of the glial fibrillary acidic protein. Brain Res 37: 81–89.

Wang, D.D., and A. Bordey. 2008. The astrocyte odyssey. Prog Neurobiol 86: 342–367.

Watkins, S., S. Robel, I.F. Kimbrough, S.M. Robert, G. Ellis-Davies, and H. Sontheimer. 2014. Disruption of astrocyte-vascular coupling and the blood-brain barrier by invading glioma cells. Nat Commun 5: 4196.

Weidenfeller, C., C.N. Svendsen, and E.V. Shusta. 2007. Differentiating embryonic neural progenitor cells induce blood-brain barrier properties. J Neurochem 101: 555–565.

Wender, R., A.M. Brown, R. Fern, R.A. Swanson, K. Farrell, and B.R. Ransom. 2000. Astrocytic glycogen influences axon function and survival during glucose deprivation in central white matter. J Neurosci 20: 6804–6810.

White, R., and E.M. Kramer-Albers. 2014. Axon-glia interaction and membrane traffic in myelin formation. Front Cell Neurosci 7: 284.

Yen, S.H., D. Dahl, M. Schachner, and M.L. Shelanski. 1976. Biochemistry of the filaments of brain. Proc Natl Acad Sci USA 73: 529–533.

Zenker, D., D. Begley, H. Bratzke, H. Rubsamen-Waigmann, and H. von Briesen. 2003. Human blood-derived macrophages enhance barrier function of cultured primary bovine and human brain capillary endothelial cells. J Physiol 551: 1023–1032.

3

Oligodendrocytes [Myelin-related Glial Cells in the Central Nervous System (CNS)]

Shuliang Liu,[1,a,]* *Merle G. Paule,*[1,b] *Fang Liu,*[1,c] *Qi Yin,*[1,d] *Tucker A. Patterson,*[2] *William Slikker, Jr.*[3] and *Cheng Wang*[1,e]

Introduction

In the central nervous system (CNS), oligodendrocytes are the glial cells which myelinate axons (Bunge *et al.*, 1962). In addition, oligodendrocytes provide trophic factors for neuronal survival and axonal growth (Du and Dreyfus, 2002). Oligodendrocytes are the end products of oligodendrocyte lineage cells that arise through complex developmental processes under meticulous regulation.

---

[1] Division of Neurotoxicology, HFT-132, National Center for Toxicological Research (NCTR), Food and Drug Administration (FDA), 3900 NCTR Rd, Jefferson, AR 72079.
[a] E-mail: shuliang.liu@fda.hhs.gov
[b] E-mail: merle.paule@fda.hhs.gov
[c] E-mail: fang.liu@fda.hhs.gov
[d] E-mail: qi.yin@fda.hhs.gov
[e] E-mail: Cheng.Wang@fda.hhs.gov
[2] National Center for Toxicological Research (NCTR), Food and Drug Administration (FDA), 3900 NCTR Rd, Jefferson, AR 72079.
   E-mail: tucker.patterson@fda.hhs.gov
[3] Office of the Director, National Center for Toxicological Research (NCTR), Food and Drug Administration (FDA), 3900 NCTR Rd, Jefferson, AR 72079.
   E-mail: William.Slikker@fda.hhs.gov
* Corresponding author

The oligodendrocyte lineage cells proliferate, differentiate, migrate, and produce mature oligodendrocytes in a precisely controlled manner (Miller, 2002; Orentas and Miller, 1998; Raff *et al.*, 1998). At different developmental stages the oligodendrocyte lineage cells are characterized by their capacities to proliferate and migrate and by their cellular morphology and biochemical markers (Miller and Ono, 1998; Rogister *et al.*, 1999). The development of oligodendrocyte lineage cells is regulated by both extracellular signals and intracellular factors (Nguyen *et al.*, 2001). Disruption to the normal development of oligodendrocyte lineage cells is of significant consequence to the myelination or remyelination in the CNS (Casaccia-Bonnefil *et al.*, 1997; Deng and Poretz, 2003). The exposure of oligodendrocytes to environmental neurotoxicants including drugs may result in serious consequences from disruption of migration to cell death (Creeley *et al.*, 2014; Creeley *et al.*, 2013; Deng and Poretz, 2001).

Origin of Oligodendrocytes

Similar to precursor cells for neurons and astrocytes, the precursor cells for oligodendrocytes initially derive from the neuroepithelial cells of the neural tube. During development the increase in oligodendrocyte precursor cells (OPCs) in the CNS is closely related to that of motor neurons and astrocytes. The oligodendrocyte precursors arise from distinct regions of the ventricular zones of the neural tube (Miller, 2002). In the spinal cord, the majority of oligodendrocytes originate from the OPCs residing in ventral ventricular zone, characterized by their expression of the Oligodendrocyte Transcription Factor 2 (Olig2), which is encoded by the *Olig2* gene (Lu *et al.*, 2002; Lu *et al.*, 2000). A pool of OPCs arising later from dorsal region contributes 10–15% of the final oligodendrocyte population additionally (Fogarty *et al.*, 2005; Vallstedt *et al.*, 2005). The emergence of the oligodendrocyte precursors from the ventral ventricular zone depends on local signals which direct cells to assume the fate of the oligodendrocytes. For example, the notochord provides local signals for its neighboring cells, leading to the appearance of oligodendrocyte precursors in the spinal cord (Orentas and Miller, 1996). Sonic Hedgehog proteins, which are encoded by the *shh* gene, are known local signal molecules that promote the appearance of oligodendrocyte precursors through the induction of cell-type specific transcription factors. Blockade of all sonic hedgehog dependent signaling is reported to eliminate oligodendrocyte development (Tekki-Kessaris *et al.*, 2001). The bone morphogenetic proteins, members of the transforming growth factor beta (TGF-β) superfamily, induce astroglial cell development with concurrent suppression of oligodendrocyte lineage cell development (Mabie *et al.*, 1997; Sussman *et al.*, 2000).

Development of Oligodendrocyte Precursor Cells

The number of committed oligodendrocyte precursor cells (OPCs) increases greatly to accommodate the number of axons that need to be myelinated. The specification and proliferation of the oligodendrocyte precursors is dependent on a variety of signal molecules. Platelet-derived growth factor A (PDGF-A) is the major mitogen for immature OPCs which are characterized by the expression of PDGF receptor alpha

(PDGF-αR) (Richardson *et al.*, 1988). In addition, the homeodomain transcriptional factors, the regulatory DNA binding proteins encoded by the homeobox genes, are critical participants in directing the progenitor cells' specification to the oligodendrocyte lineage (Kim and Nirenberg, 1989). The arising of OPCs in forebrain takes place first in the ventricular and subventricular zones in the early stages and then mostly, in the developing white matter in later stages (Kessaris *et al.*, 2006; Yue *et al.*, 2006). In the mouse, the earliest OPCs are generated from PDGF-αR-positive precursors expressing Nirenberg-Kim homeobox 2.1 (Nkx2.1) (Kim and Nirenberg, 1989) in the median ganglionic eminence (MGE) and anterior entopeduncular region (AEP) at embryonic day 12 (E12). The second wave of PDGF-αR-positive cells derives from Glutathione Synthetase Homeobox 2 (Gsx2)-positive cells in the lateral ganglionic eminence (LGE), which proliferate after E16.5, eventually replacing the Nkx2.1-positive cells. The third wave of PDGF-αR-positive cells originates from the cortical Empty spiracles homeobox 1 (EMx1)-positive cells and produces oligodendrocytes in the neocortex and corpus callosum after birth (Kessaris *et al.*, 2006). The proliferation of immature OPCs is synergistically regulated by other factors including basic fibroblastic growth factor (bFGF) and the chemokine C-X-C Motif Ligand 1 (CXCL1) (McKinnon *et al.*, 1991; Robinson *et al.*, 1998). In the early stages of proliferation in the ventricular zone the immature oligodendrocyte precursors are bipolar and highly migratory and express a surface antigen specific to monoclonal antibodies A2B5 (A2B5-positive), which is considered as a marker for immature glial cell precursors (Dietrich *et al.*, 2002). As the cells mature the expression of antigens recognized by the monoclonal antibody O4 also increases (A2B5-positive/O4-positive). The differentiation of oligodendrocyte precursor cells into postmitotic premyelinating oligodendrocytes is accompanied by the expression of galactocerebroside (GalC), which is a major glycolipid in myelin and serves as a marker in the differentiation of myelin-producing cells. Mature oligodendrocytes express the major myelin proteins, including myelin basic protein (MBP) and proteolipid protein (PLP) and eventually assemble myelin (Miller, 2002).

In vivo, the proliferation, survival, and differentiation of oligodendrocyte lineage cells are dependent on growth factors, hormones, cytokines and neurotransmitters secreted by adjacent cells. *In vitro*, a combination of growth factors is needed to supplement the culture medium to direct and support the development of neural progenitor cells (Canoll *et al.*, 1999; McTigue *et al.*, 1998). Oligodendrocyte lineage cells cultured in growth factor-enriched medium retain their developmental characteristics observed *in vivo*. Cultured oligodendrocyte lineage cells have been used as well-defined systems in toxicological studies to determine the effects of neurotoxicants on cellular development (Davies and Ross, 1991; Deng and Poretz, 2001). Embryonic neural stem cells are pluripotent and can differentiate into neurons, astrocytes or oligodendrocytes in differentiation media supplemented with growth factors (Figure 3.1 and 3.2).

OPCs express a surface marker known as neural/glial antigen 2 (NG2). The NG2 proteoglycan is a type-I transmembrane protein which belongs to the protein family of chondroitin sulfate proteoglycans (CSPGs) (Sakry and Trotter, 2016). Sequential cleavage of NG2 by α- and γ-secretases occurs constitutively, and can be increased by neuronal activity acting on OPCs (Sakry *et al.*, 2014). The ectodomain and the intracellular domain released by cleavage exert multiple effects on migration, cytoskeleton interaction and target gene regulation of the OPCs (Sakry *et al.*, 2014). In both the developing and adult CNS up to 5–10% of the total glial population consists of NG2-positive cells. However, NG2 is not a specific marker for OPCs, since NG2-positive cells give rise not only to oligodendrocytes but also protoplasmic astrocytes as well as neurons under some circumstances during development (Trotter *et al.*, 2010). NG2 is expressed in immature Schwann cells in the peripheral nervous system as well. NG2-positive cells express different types of receptors for the neurotransmitters γ-aminobutyric acid (GABA) (class A receptor, $GABA_AR$) and glutamate (including α-amino-3-hydroxy-5-methyl-4-isoxazolepropionic acid (AMPA) and kainate subtype receptors). In the hippocampus, cortex, and other brain regions, NG2-positive cells were found to have intimate contact with neurons and form synaptic-like formations with the axons to be myelinated. Direct stimulation upon OPCs by the release of the neurotransmitters glutamate or GABA has been observed. Morphologically, mature oligodendrocytes are characterized by a multiprocessed appearance.

*Oligodendrocytes, #Astrocytes and ^Neurons Differentiated from Human Embryonic Neural Stem Cells

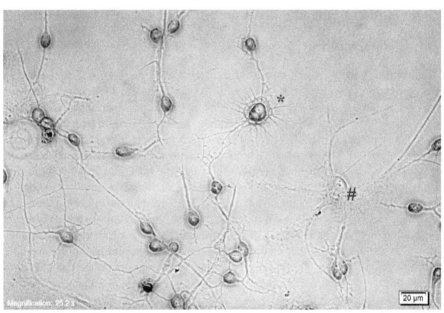

Figure 3.1. Oligodendrocyte (*), astrocyte (#), neurons (^) differentiated from human embryonic neural stem cells (PhoenixSongs Biologicals, Inc. Branford, CT). The human embryonic neural stem cells were from hippocampus of gestation 19-week fetus brain, cultured in differentiation medium for 2 weeks. Neural differentiation medium were supplemented with glutamine, laminin, brain derived neurotrophic factor (BDNF), and glial cell line-derived neurotrophic factor (GDNF) (PhoenixSongs Biologicals, CT).

**Oligodendrocytes Differentiated
from Embryonic Neural Stem Cells**

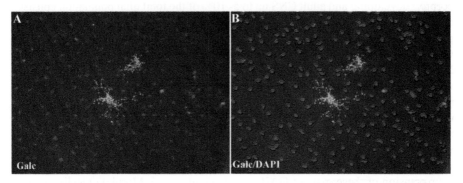

Figure 3.2. Immunochemical characterization of immature oligodendrocyte differentiated from human embryonic neural stem cells (PhoenixSongs Biologicals, Inc. Branford, CT). The human embryonic neural stem cells were from hippocampus of gestation 19-week fetus brain, cultured in differentiation medium for 2 weeks. The cells were stained for oligodendrocytic specific marker, galactocerebroside (GalC, 2A), biochemical precursor of galactolipids and sulfatide, as well as nuclei marker DAPI (2B).

Regulation of Oligodendrocyte Precursor Cells Differentiation

Extrinsic mechanisms

A multitude of mechanisms which either promote or inhibit the differentiation of oligodendrocyte precursors orchestrate the development of myelin. Extrinsic signaling mediated by axon surface ligands plays an important role in directing oligodendrocyte differentiation and myelination. Inhibitory or permissive signals from axon(s) contribute to the myelination of specific axons, rather than the indiscriminate myelination of all nearby axons. Inhibitory ligands expressed on axonal surfaces include Jagged-1, polysialylated neuronal cell adhesion molecule (PSA-NCAM), and leucine rich repeat and Immunoglobin-like domain-containing protein 1 (LINGO-1). During CNS development, Jagged-1 expressed on axons inhibits OPC differentiation via Notch receptors on OPCs (Wang *et al.*, 1998). In contrast to developmental myelination, the Notch and Jagged genes expressed following injury in adult animal did not have similar inhibitory effect on remyelination, suggesting that upregulation of the developmentally expressed genes per se do not imply similar function (Stidworthy *et al.*, 2004). The permissive signaling of axonal neuregulins in the CNS is not as essential as it is for myelination by Schwann cells in the periphery (Bozzali and Wrabetz, 2004).

Axonal electrical activity has been shown to modulate developing oligodendrocytes with respect to their proliferation, survival, terminal differentiation, and myelinogenesis (Barres and Raff, 1993; Demerens *et al.*, 1996). Neuronal electrical activity may regulate the expression of the axonal surface ligands mentioned above. In addition, activated axons release adenosine which acts on purinergic receptors on oligodendrocyte precursors and promotes their differentiation and myelination (Stevens *et al.*, 2002). The axonal release of ATP promotes myelination via stimulating astrocytes which release the promyelinating cytokine leukemia inhibitory factor (LIF) (Ishibashi *et al.*, 2006). OPCs are equipped with a large repertoire of neurotransmitter receptors, including the AMPA/kainate-type glutamate receptor. Exposure of OPCs to ligands, including kainate, AMPA, or L-glutamate, can elicit inward currents which may be reversibly blocked by the antagonist 6-cyano-7-nitroquinoxaline-2,3-dione (CNQX) (Borges *et al.*, 1994). *In vivo*, OPCs in the CA1 region of the hippocampus affect membrane potentials in neurons in the CA3 of the hippocampus via release of L-glutamate (Bergles *et al.*, 2000). The generation of postsynaptic potentials by OPCs in response to glutamate or GABA activation via synaptic-like structures with neurons or interneurons has been found in both white matter and grey matter (Lin and Bergles, 2002). The expression of voltage-gated sodium channels and ionotropic glutamate receptors and synapses with glutamatergic neurons are common in OPCs in different brain regions. Nevertheless, during the transition to premyelinating oligodendrocytes, the expression of these channels and receptors are down regulated (De Biase *et al.*, 2010). In the adult CNS glutamatergic synaptic formations onto OPCs are transient and down-regulated in mature oligodendrocytes during the process of remyelination following demyelination (Etxeberria *et al.*, 2010). The physiological effects of neuronal inputs onto OPCs via synaptic-like structures remain to be explored *in vivo*. *In vitro*, however, treatment of OPCs with selective glutamate receptor agonists has been shown to inhibit their proliferation and their progression from the oligodendrocyte-type2 astrocyte (O-2A)-positive stage to the O4-positive stage when cultured with mitogens (Gallo *et al.*, 1996); this suggests that the role of glutamatergic signaling to OPCs may be to limit, rather than promote, myelination possibly maintaining a pool of non-dividing NG2-positive cells in the adult CNS.

Intrinsic mechanisms

It has been observed that OPCs grown in culture in the absence of neurons are regulated by a number of mechanisms intrinsic to the OPCs which limit the number of cell divisions (Temple and Raff, 1986). Wnt (a blend name of *Int* (*Integration 1* gene) and *Wg* (*Wingless* gene)) signaling via a canonical pathway was transiently activated in OPCs during the initiation of terminal differentiation (Nusse, 2005). Subsequently, the expression of transcription factor 4 (Tcf4), a mediator in the Wnt pathway, and β-catenin activity was down-regulated in mature oligodendrocytes (Fancy *et al.*, 2009). In addition, extracellular ligands that modulate CNS myelination have been observed. For example, the orphan (which means natural ligands remain to be identified) G protein–coupled receptor (GPCR) Gpr17 is transiently expressed during oligodendrocyte differentiation. Overexpression of Gpr17 causes severe dysmyelination and suspension of oligodendrocyte differentiation at an early stage (Chen *et al.*, 2009).

At the level of transcription, the commitment of multipotent progenitors to the oligodendrocyte lineage cells is reliant on the transcription factor Olig2. Ventrally derived oligodendrocytes (and lower motor neurons) are derived from Olig2-expressing subventricular zone progenitors and the oligodendrocyte lineage is absent in Olig2-null mice (Rowitch *et al.*, 2002; Zhou *et al.*, 2001). Subsequently, the downstream induction of a number of transcription factors, most notably Oligodendrocyte Transcription Factor 1 (Olig1), Achaete-Scute Family BHLH Transcription Factor 1 (Ascl1), Nkx2.2, sex determining region Y box 10 (Sox10), Yin and Yang 1 (YY1), and Tcf4, is required for the generation of mature, postmitotic oligodendrocytes (Wegner, 2008). All these factors are present in OPCs as well as in postmitotic oligodendrocytes, with the exception of Tcf4, which is transiently expressed during differentiation. In addition, a number of transcription factors, most notably inhibitor of DNA binding 2(Id2), Id4, Hes Family BHLH Transcription Factor 5 (Hes5), and Sox6, have been identified that are active in maintaining OPCs in their undifferentiated state and repressing myelin gene expression. The dramatic advances in gene expression analysis—including acquiring the transcriptomes database (Cahoy *et al.*, 2008) and transcriptional profiling by polysomal RNA isolation (Heiman *et al.*, 2008)—have facilitated the identification of many oligodendrocyte-specific or regulated genes having likely roles in regulating the myelination process. This capability has enabled the identification of myelin gene regulatory factor (MRF), which is expressed when OPCs differentiate into postmitotic oligodendrocytes in the CNS. Inactivation of MRF by RNA interference within the oligodendrocyte lineage causes the oligodendrocyte lineage cells' failure to express myelin genes or to produce myelin (Emery *et al.*, 2009). These results demonstrate that the transition from an OPC into a myelinating oligodendrocyte requires the induction of promyelination factors, such as MRF, as well as the down-regulation of inhibitory factors.

Because gene expression is also controlled by accessibility of genomic DNA to regulatory transcription machinery proteins, chromatin remodeling, the dynamic modification of chromatin architecture, thus affects oligodendrocyte differentiation. Chromatin remodeling is dependent on the enzymatic activities of histone deacetylases (HDACs). Specific inhibition of HDAC activity by valproic acid in the early postnatal period in rats results in a delay in oligodendrocyte differentiation and a reversible hypomyelination (Shen *et al.*, 2005). More specifically, the oligodendrocyte lineage cells from mice with mutations in HDAC1 and HDAC2 suffer a loss of both OPCs and oligodendrocytes, suggesting that HDAC activity is required at multiple stages of the transition to oligodendrocytes (Ye *et al.*, 2009).

Post-transcriptional regulation of gene expression, including RNA interference by microRNAs (miRNAs), plays a pivotal role in regulating CNS myelination by arresting OPCs proliferation and promoting oligodendrocyte differentiating. The endoribonuclease Dicer, or Dicer, cleaves double-stranded RNA into short double-stranded RNA fragments, called microRNA. RNA interference occurs as a single miRNA or a miRNA complex represses translation of the complementary messenger RNA (mRNA) or promotes its degradation (Wu and Belasco, 2008). Mature miRNAs are required for the normal differentiation and function of neuron (Davis *et al.*, 2008). Mice which bear disrupted *Dicer1* allele in oligodendrocyte lineage cells demonstrate massive disruption of myelination (Dugas *et al.*, 2010; Zhao *et al.*, 2010). The interruption of the *Dicer* gene expression results in abnormal miRNAs processing, leading to a severe disruption of the postmitotic stage of the oligodendrocyte lineage cells. Several miRNAs, most notably miR-219 and miR-338, are induced concurrent with oligodendrocyte differentiation. Those key miRNAs which are induced early in differentiation act to inhibit the expression of genes that promote OPC maintenance, thus further inhibiting proliferation and promoting differentiation.

Migration of Oligodendrocyte Precursors

OPCs originate from the restricted ventricular zone of the neural tube during early development, whereas in adulthood, oligodendrocytes are widely distributed throughout the CNS. The spatial separation between the location of the origin of oligodendrocyte precursors and their final destination suggests that normal myelination is associated with the long distance migration of oligodendrocyte precursors. Oligodendrocyte precursors are, in fact, highly mobile and capable of long distance migration, whereas mature oligodendrocytes have lost this capacity. Studies have revealed that the oligodendrocytes which myelinate the optic nerve originate in the floor of the third ventricle and subsequently migrate along the length of the nerve (Ono *et al.*, 1997; Small *et al.*, 1987). In the spinal cord the generation of most of the oligodendrocytes in dorsal spinal cord is secondary to the arrival of ventrally-derived precursors (Ono *et al.*, 1995). In the forebrain oligodendrocyte precursors migrate from the basal forebrain into the overlying dorsal forebrain during the embryonic period (He *et al.*, 2001). The oligodendrocyte precursors in the subventricular zone (SVZ) migrate radially and tangentially after migration into white matter, cortex, and striatum (Kakita and Goldman, 1999). The cellular substrates and the molecular mechanisms mediating oligodendrocyte precursor migration have been described to some degree. In the optic nerve, the migrating oligodendrocyte precursors move to a close association with retinal ganglion cell axons and it has been suggested that this migration is axophilic (Ono *et al.*, 1997). In the optic axonal tracts, the migration and distribution of OPCs are controlled by the interactions between axons and OPCs mediated by eph receptors (for its first member was isolated from erythropoietin-producing human hepatocellular carcinoma line) and their ligands—ephrins (Prestoz *et al.*, 2004). For the ventral to dorsal migration in the spinal cord, pre-existing axon tracts formed earlier in development are utilized by migrating OPCs as pathways. Cell surface components, including adhesion molecules and extracellular matrix receptors also have been proposed to play a role in regulating migration (Kiernan *et al.*, 1996; Muller *et al.*, 1996; Wang *et al.*, 1994). Specific removal of polysialic acid (PSA) associated with neural cell adhesion molecules (NCAM) from the surface of O-2A progenitor cells completely blocks the dispersion of O-2A lineage population from the model explant (Wang *et al.*, 1994). In more recent studies, different factors and cellular substrates have been demonstrated to modulate OPC migration and differentiation (Biname *et al.*, 2013; Juliet *et al.*, 2009; Ortega *et al.*, 2012; Paez *et al.*, 2009a; Paez *et al.*, 2010). Voltage-operated Ca^{2+} channels play an important role in the development of OPCs. Increases in voltage-dependent Ca^{2+} influx may enhance OPC motility, which is believed to be essential for OPC migration. Protein kinase C (PKC) and tyrosine kinase receptors (TKrs) enhance Ca^{2+} influx induced by depolarization in OPCs, whereas protein kinase A (PKA) has an inhibitory effect. These kinases modulate voltage-operated Ca^{2+} uptake in OPCs and participate in the modulation of process extension and migration (Paez *et al.*, 2009b; Paez *et al.*, 2010). In experiments using optic nerve from the developing mouse embryo, neuregulin

(Nrg) 1 was shown to be a potent chemoattractant for the OPCs and this effect was mediated via the membrane associated receptor tyrosine kinase, ERBB4 (erb-b2 receptor tyrosine kinase 4). In contrast, OPCs colonizing the optic nerve postnatally do not respond to Nrg1-chemoattraction (Ortega *et al.*, 2012). It was also found that mouse embryos lacking ERBB4 display deficits in early OPC migration away from different oligodendrogliogenic regions *in vivo* (Ortega *et al.*, 2012). The sequential cleavage of the extracellular NG2 protein by α- and γ-secretases generated biologically active fragments which act as soluble extracellular matrix components with neural modulatory properties (Sakry and Trotter, 2016). NG2 stimulates small GTPases, Ras homolog family member A (RhoA) or Ras-related C3 botulinum toxin substrate (Rac) at the cell periphery, which favors formation of the bipolar shape of migrating OPCs and, thus, regulate cell migration (Biname *et al.*, 2013). Soluble blood components, particularly thrombin, have an adverse effect on maturing subventricular zone cells and OPCs derived from newborn rat brain (Juliet *et al.*, 2009).

The migration of OPCs in the developing CNS is also directed by secreted cues (de Castro and Bribian, 2005). Both chemoattractant and chemorepulsive signals guide the unidirectional migration of OPCs. For example, in the optic nerve the majority of the cell migration occurs from the optic chiasm to the retina, while the directional cues are released from the optic chiasm (Sugimoto *et al.*, 2001). The growth factors—including PDGF, bFGF, epidermal growth factor (EGF), and hepatocytic growth factor (HGF)—which promote their proliferation and survival, have chemotactic properties and direct the migration of OPCs in a gradient-related manner (Fruttiger *et al.*, 1999).

A number of chemotactic molecules, including the secreted netrins and semaphorins, influence the migration of OPCs (de Castro, 2003). The expression of the receptors for netrins and semaphorins mediate their effects upon cell migration. Cell surface receptors DCC ("deleted in colorectal cancer") and Unc5H (name derived from the "*Uncoordinated mutants*" induced in *C. elegans* (Brenner, 1973)), mediate netrins' chemoattractant or chemorepusive effects on the OPCs, respectively, in a time-dependent manner (Spassky *et al.*, 2002). Neuropilin-1 and -2 serve as co-receptors for semaphoring family members which play diverse roles in guiding cell migration. In the spinal cord, Netrin-1 exerts a chemorepulsive effect during the initial dispersion of OPCs from the ventral ventricular zone (Tsai *et al.*, 2003). In the early stages of optic nerve development Netrin-1 acts as an attractant, for OPCs expressing only DCC, the main receptor for these molecules involved in attraction of OPCs for growing axons or migrating neurons. After birth, when OPCs begin to express Unc5H, Netrin-1 acts as a repellent. Such a switch in receptor expression could serve as a stop signal for OPCs in the optic nerve, keeping them from migrating further across the nerve papilla. In addition, the chemokine CXCL1 secreted in the CNS participated in the regulation of oligodendrocyte precursors' migration. Signaling via its receptor CXCR2, CXCL1 inhibited the migration of oligodendrocyte precursors in the white matter of embryonic spinal cord (Tsai *et al.*, 2002). CXCR2, rather than other chemokine receptors, was detected in most of the A2B5+ or O4+ cells in white matter of developing spinal cord. In postnatal white matter of the spinal cord from CXCR2$^{-/-}$ mice, the density of differentiated oligodendrocyte was reduced and pattern of distribution altered, indicating the disturbance of oligodendrocyte migration. Vascular endothelial growth factor (VEGF)-A is known as a versatile growth factor in the CNS. VEGF-A treatment promotes the migration of OPCs cultured from rat neonatal cortex. The leading edge of the migration was associated with actin cytoskeleton reorganization. The stimulatory effect of VEGF-A upon OPC migration was demonstrated to be reactive oxygen species (ROS)- and focal adhesion kinase (FAK)-dependent, and did not affect OPCs' proliferation or differentiation (Hayakawa *et al.*, 2011).

Oligodendrocyte Maturation

The size of the population of oligodendrocyte precursors is controlled by several mechanisms. The proliferation of oligodendrocyte precursors is regulated, in part, by the availability of mitogen (Calver *et al.*, 1997). PDGF is a strong survival factor for oligodendrocyte lineage cells (Barres *et al.*, 1992) and oligodendrocytes survive through the competition for local survival factors including PDGF. Cell density-dependent regulation of cell proliferation has been observed in cultures of oligodendrocyte lineage cells. Increasing the cell density of oligodendrocyte precursors is associated with changes in the expression of cell cycle factors, including the upregulation of cell cycle inhibitor p27^{kip-1} and the reduction of cyclin A expression. Alterations in the expression of these cell cycle proteins could induce the progenitor cells to cease cell cycling and, thus, reduce proliferation (Nakatsuji and Miller, 2001). The proliferation and differentiation of the oligodendrocyte precursors are regulated in a coordinated fashion. Clonally-related cells derived from a single precursor cease cell proliferation and begin differentiating toward oligodendrocytes almost at the same time (Barres and Raff, 1994). The number of oligodendrocytes in any particular region of the CNS is the product of precursor differentiation and survival and becomes matched to the number of axons needing myelination.

The development of OPCs into a mature myelinating phenotype of oligodendrocyte is regulated by local axonal signals. Soluble signals derived from adjacent axons that drive this development include FGFs, and thyroid hormones. Thyroid hormone and its derivatives facilitate the differentiation of OPCs into myelinating phenotype oligodendrocytes (Baas *et al.*, 1997; Barres *et al.*, 1994; Calza *et al.*, 2002). Thyroid hormones act on their nuclear hormone receptors and promote the expression of oligodendrocyte-specific genes, producing myelin basic protein (MBP), myelin-associated glycoprotein (MAG), proteolipid protein (PLP) and cyclic nucleotide 3'-phosphodiesterase (CNP) (Baxi *et al.*, 2014; Calza *et al.*, 2002). In addition, the redox state of a particular glial precursor cell can affect the cell's propensity to proliferate or differentiate (Smith *et al.*, 2000). Axonal cell surface molecules, including L1 cell adhesion molecule (L1), MAG, NCAM and N-cadherin, participate in the regulation of myelin sheath formation. Neuronal signals also regulate the process of oligodendroglial plasma membrane rearrangement into the myelin sheath (Simons and Trajkovic, 2006). In myelinating culture prepared from embryonic cerebral hemispheres, the myelination of axons was inhibited when the firing of neuron action potential had been selectively blocked by tetrodotoxin. In contrast, the number of myelinated fibers increased following the incubation with α-scorpion toxin, which enhanced the neuronal electrical activity (Demerens *et al.*, 1996).

OPCs in CNS at Adulthood

OPCs are present in the grey and white matter, making up around 5% of the glial population in adult CNS (Levine *et al.*, 2001a). In mammalian CNS, myelination proceeds well beyond the postnatal stages into the adulthood, indicating the role of adult OPCs in maintaining the oligodendrocyte population and myelination (Young *et al.*, 2013). OPCs are well positioned to detect the alterations of CNS homeostasis, given their uniform distribution and constant surveillance via processes (Hughes *et al.*, 2013). In health, the OPCs abide in the adult CNS are remarkably different from their counterparts in the prenatal stages in cell differentiation and division. The O-2A-positive progenitor cells recovered from optic nerve of adult rats do not express any intermediate filament proteins, while their counterparts in perinatal animal do. The adult O-2A-positive progenitor cells were distinguished from the perinatal O-2A-positive progenitor cells by their morphology, antigenic phenotype, significantly longer cell cycle time and slower migration rate (Wolswijk and Noble, 1989). *In vitro*, exposure of adult cells to PDGF and bFGF together will bestow upon them characteristics of perinatal cells (Wolswijk and Noble, 1992). Following the contusive spinal cord injury to adult mouse, NG2+ OPCs were activated to proliferate and differentiate to oligodendrocytes (Lytle *et al.*, 2009). The ability of adult progenitors to remyelinate in a number of disease conditions, such as multiple sclerosis, appears to be relatively limited, although they may be capable of myelin repair under certain conditions (Chang *et al.*, 2000; Levine *et al.*, 2001b). Further understanding of the biology of oligodendrocyte precursors in the adult CNS should allow manipulation of them to effectively repair injured or diseased adult CNS tissue.

Oligodendrocyte Pathology

Oligodendrocytes form and maintain the myelin wrapping the adjacent axons. The lipid-rich myelin membrane is formed by extension of cell processes at a distance from the perikaryon. An oligodendrocyte usually elaborates and supports the myelin membrane enough to myelinate up to 50 internodes, which can be up to 100 times of its weight (Ludwin, 1997). These features of oligodendrocytes make them vulnerable to a wide spectrum of detrimental factors. During the peak of myelination, oligodendrocytes must maintain an extremely high metabolic rate to properly myelinate the axons. This need leads to the consumption of large amounts of oxygen and ATP and, thus, to the formation of a greater amount of hydrogen peroxide and ROS than in other neural cells. If anti-oxidant capabilities are overwhelmed, these reactive oxidants can—as has been shown *in vitro*—cause DNA degradation and apoptosis of oligodendrocytes (Bradl and Lassmann, 2010; Ladiwala *et al.*, 1999). In addition, the requirement of iron as a co-factor for the enzymes involved in myelin formation is greater than for other cells in the CNS. Larger intracellular iron storage could contribute to increased free radical formation and lipid peroxidation under pathological conditions (Braughler *et al.*, 1986). In concert, the content of glutathione in anti-oxidative enzymes in oligodendrocytes is low. Oligodendrocyte damage and loss attributed to oxidative stress have been noted in various pathological conditions including multiple sclerosis, Alzheimer's disease (AD), spinal cord injury, and CNS hypoxia and ischemia suggesting oxidative injury is a common cause of oligodendrocyte death or dysfunction (Bradl and Lassmann, 2010; McTigue and Tripathi, 2008).

Oligodendrocytes also exhibit an inventory of attributes rendering them vulnerable to excitotoxic cell death (Bradl and Lassmann, 2010; Domercq *et al.*, 2007). Oligodendrocytes carry AMPA (Tanaka *et al.*, 2000), kainate (Alberdi *et al.*, 2006), and NMDA (Salter and Fern, 2005) receptors which make them susceptible to glutamate excitotoxicity. In many CNS pathologies, including traumatic injury, stroke, AD, Parkinson's disease, amyotrophic lateral sclerosis, increased extracellular glutamate levels have been detected (McTigue and Tripathi, 2008). During the last two decades, there has been accumulating evidence from animal studies indicating that prolonged exposures (> 3 hours) to general anesthetic agents (including ketamine, isoflurane, propofol, etc.) at clinical concentrations during the brain growth spurt period can induce widespread neuronal apoptosis (Jevtovic-Todorovic *et al.*, 2003; Slikker *et al.*, 2007; Zou *et al.*, 2008). It has been observed in fetal and neonatal monkeys that exposure to isoflurane or propofol causes apoptosis of oligodendrocytes which were beginning to myelinate axons (Brambrink *et al.*, 2012; Creeley *et al.*, 2014; Creeley *et al.*, 2013). Another excitotoxic pathway associated with oligodendrocyte death in spinal cord injury is mediated by the APT receptor P2X7 expressed by oligodendrocytes and myelin (Matute *et al.*, 2007).

Several genetic mutations have been identified that affect the oligodendrocyte and result in CNS pathologies. The mutations of *GLAC* gene which encoding the enzyme galactosylceramidase underlie Krabbe disease, an inherited leukodystrophy. In the lysosome, galactosylceramidase hydrolyze galactosylceramide and psychosine. Without the enzymatic cleavage of the myelin metabolites, the toxic psychosine accumulate within the lysosome, resulting in demyelination and oligodendrocyte loss (Giri *et al.*, 2006; Haq *et al.*, 2003). In Pelizaeus–Merzbacher disease, mutations in *PLP1* gene encoding myelin PLP cause accumulation of misfolded PLP protein in the endoplasmic reticulum and eventually lead to dysmyelination and oligodendrocyte death (Gow *et al.*, 1994). The mutations in glial fibrillary acidic protein (*GFAP*) gene cause Alexander disease, another rare form of leukodystrophy. The abnormal glial fibrillary acidic protein adds up in astroglial cells, leading to the formation of Rosenthal fibers. The abnormal astrocyte to oligodendrocyte communication and the lack of astrocytic uptake of glutamate are thought to contribute to the disorder (Mignot *et al.*, 2004). X-linked adrenoleukodystrophy primarily occurs in male. It is caused by mutations in the ATP binding cassette subfamily D member 1 (*ABCD1*) gene encoding the adrenoleukodystrophy protein (ALDP), which is involved in transporting the very long-chain fatty acids (VLCFAs) into peroxisomes. Defective ALDP leads to the accumulation of VLCFAs, which make oligodendrocytes vulnerable to cell death (Moser *et al.*, 2004).

References

Alberdi, E., M.V. Sanchez-Gomez, I. Torre, M.A. Domercq, A. Perez-Samartin, F. Perez-Cerda, and C. Matute. 2006. Activation of kainate receptors sensitizes oligodendrocytes to complement attack. J Neurosci 26: 3220–3228.

Baas, D., D. Bourbeau, L.L. Sarlieve, M.E. Ittel, J.H. Dussault, and J. Puymirat. 1997. Oligodendrocyte maturation and progenitor cell proliferation are independently regulated by thyroid hormone. Glia 19: 324–332.

Barres, B.A., I.K. Hart, H.S. Coles, J.F. Burne, J.T. Voyvodic, W.D. Richardson, and M.C. Raff. 1992. Cell death and control of cell survival in the oligodendrocyte lineage. Cell 70: 31–46.

Barres, B.A., and M.C. Raff. 1993. Proliferation of oligodendrocyte precursor cells depends on electrical activity in axons. Nature 361: 258–260.

Barres, B.A., and M.C. Raff. 1994. Control of oligodendrocyte number in the developing rat optic nerve. Neuron 12: 935–942.

Barres, B.A., M.A. Lazar, and M.C. Raff. 1994. A novel role for thyroid hormone, glucocorticoids and retinoic acid in timing oligodendrocyte development. Development 120: 1097–1108.

Baxi, E.G., J.T. Schott, A.N. Fairchild, L.A. Kirby, R. Karani, P. Uapinyoying, C. Pardo-Villamizar, J.R. Rothstein, D.E. Bergles, and P.A. Calabresi. 2014. A selective thyroid hormone beta receptor agonist enhances human and rodent oligodendrocyte differentiation. Glia 62: 1513–1529.

Bergles, D.E., J.D. Roberts, P. Somogyi, and C.E. Jahr. 2000. Glutamatergic synapses on oligodendrocyte precursor cells in the hippocampus. Nature 405: 187–191.

Biname, F., D. Sakry, L. Dimou, V. Jolivel, and J. Trotter. 2013. NG2 regulates directional migration of oligodendrocyte precursor cells via Rho GTPases and polarity complex proteins. J Neurosci 33: 10858–10874.

Borges, K., C. Ohlemeyer, J. Trotter, and H. Kettenmann. 1994. AMPA/kainate receptor activation in murine oligodendrocyte precursor cells leads to activation of a cation conductance, calcium influx and blockade of delayed rectifying K+ channels. Neuroscience 63: 135–149.

Bozzali, M., and L. Wrabetz. 2004. Axonal signals and oligodendrocyte differentiation. Neurochem Res 29: 979–988.

Bradl, M., and H. Lassmann. 2010. Oligodendrocytes: biology and pathology. Acta Neuropathol 119: 37–53.

Brambrink, A.M., S.A. Back, A. Riddle, X. Gong, M.D. Moravec, G.A. Dissen, C.E. Creeley, K.T. Dikranian, and J.W. Olney. 2012. Isoflurane-induced apoptosis of oligodendrocytes in the neonatal primate brain. Ann Neurol 72: 525–535.

Braughler, J.M., L.A. Duncan, and R.L. Chase. 1986. The involvement of iron in lipid peroxidation. Importance of ferric to ferrous ratios in initiation. J Biol Chem 261: 10282–10289.

Brenner, S. 1973. The genetics of behaviour. Br Med Bull 29: 269–271.

Bunge, M.B., R.P. Bunge, and G.D. Pappas. 1962. Electron microscopic demonstration of connections between glia and myelin sheaths in the developing mammalian central nervous system. J Cell Biol 12: 448–453.

Cahoy, J.D., B. Emery, A. Kaushal, L.C. Foo, J.L. Zamanian, K.S. Christopherson, Y. Xing, J.L. Lubischer, P.A. Krieg, S.A. Krupenko, W.J. Thompson, and B.A. Barres. 2008. A transcriptome database for astrocytes, neurons, and oligodendrocytes: a new resource for understanding brain development and function. J Neurosci 28: 264–278.

Calver, A.R., A.C. Hall, M. Fruttiger, W.P. Yu, and W.D. Richardson. 1997. Control of cell number in the oligodendrocyte lineage. J Neurochem 69: S231–S231.

Calza, L., M. Fernandez, A. Giuliani, L. Aloe, and L. Giardino. 2002. Thyroid hormone activates oligodendrocyte precursors and increases a myelin-forming protein and NGF content in the spinal cord during experimental allergic encephalomyelitis. Proc Natl Acad Sci U S A 99: 3258–3263.

Canoll, P.D., R. Kraemer, K.K. Teng, M.A. Marchionni, and J.L. Salzer. 1999. GGF/neuregulin induces a phenotypic reversion of oligodendrocytes. Mol Cell Neurosci 13: 79–94.

Casaccia-Bonnefil, P., R. Tikoo, H. Kiyokawa, V. Friedrich, Jr., M.V. Chao, and A. Koff. 1997. Oligodendrocyte precursor differentiation is perturbed in the absence of the cyclin-dependent kinase inhibitor p27Kip1. Genes Dev 11: 2335–2346.

Chang, A., A. Nishiyama, J. Peterson, J. Prineas, and B.D. Trapp. 2000. NG2-positive oligodendrocyte progenitor cells in adult human brain and multiple sclerosis lesions. J Neurosci 20: 6404–6412.

Chen, Y., H. Wu, S. Wang, H. Koito, J. Li, F. Ye, J. Hoang, S.S. Escobar, A. Gow, H.A. Arnett, B.D. Trapp, N.J. Karandikar, J. Hsieh, and Q.R. Lu. 2009. The oligodendrocyte-specific G protein-coupled receptor GPR17 is a cell-intrinsic timer of myelination. Nat Neurosci 12: 1398–1406.

Creeley, C.E., K.T. Dikranian, S.A. Johnson, N.B. Farber, and J.W. Olney. 2013. Alcohol-induced apoptosis of oligodendrocytes in the fetal macaque brain. Acta Neuropathol Commun 1: 23.

Creeley, C.E., K.T. Dikranian, G.A. Dissen, S.A. Back, J.W. Olney, and A.M. Brambrink. 2014. Isoflurane-induced apoptosis of neurons and oligodendrocytes in the fetal rhesus macaque brain. Anesthesiology 120: 626–638.

Davies, D.L., and T.M. Ross. 1991. Long-term ethanol-exposure markedly changes the cellular composition of cerebral glial cultures. Brain Res Dev Brain Res 62: 151–158.

Davis, T.H., T.L. Cuellar, S.M. Koch, A.J. Barker, B.D. Harfe, M.T. McManus, and E.M. Ullian. 2008. Conditional loss of Dicer disrupts cellular and tissue morphogenesis in the cortex and hippocampus. J Neurosci 28: 4322–4330.

De Biase, L.M., A. Nishiyama, and D.E. Bergles. 2010. Excitability and synaptic communication within the oligodendrocyte lineage. J Neurosci 30: 3600–3611.

de Castro, F. 2003. Chemotropic molecules: guides for axonal pathfinding and cell migration during CNS development. News Physiol Sci 18: 130–136.

de Castro, F., and A. Bribian. 2005. The molecular orchestra of the migration of oligodendrocyte precursors during development. Brain Res Brain Res Rev 49: 227–241.

Demerens, C., B. Stankoff, M. Logak, P. Anglade, B. Allinquant, F. Couraud, B. Zalc, and C. Lubetzki. 1996. Induction of myelination in the central nervous system by electrical activity. Proc Natl Acad Sci U S A 93: 9887–9892.

Deng, W., and R.D. Poretz. 2001. Lead alters the developmental profile of the galactolipid metabolic enzymes in cultured oligodendrocyte lineage cells. Neurotoxicology 22: 429–437.

Deng, W., and R.D. Poretz. 2003. Oligodendroglia in developmental neurotoxicity. Neurotoxicology 24: 161–178.

Dietrich, J., M. Noble, and M. Mayer-Proschel. 2002. Characterization of A2B5+ glial precursor cells from cryopreserved human fetal brain progenitor cells. Glia 40: 65–77.

Domercq, M., M.V. Sanchez-Gomez, C. Sherwin, E. Etxebarria, R. Fern, and C. Matute. 2007. System xc- and glutamate transporter inhibition mediates microglial toxicity to oligodendrocytes. J Immunol 178: 6549–6556.

Du, Y., and C.F. Dreyfus. 2002. Oligodendrocytes as providers of growth factors. J Neurosci Res 68: 647–654.

Dugas, J.C., T.L. Cuellar, A. Scholze, B. Ason, A. Ibrahim, B. Emery, J.L. Zamanian, L.C. Foo, M.T. McManus, and B.A. Barres. 2010. Dicer1 and miR-219 are required for normal oligodendrocyte differentiation and myelination. Neuron 65: 597–611.

Emery, B., D. Agalliu, J.D. Cahoy, T.A. Watkins, J.C. Dugas, S.B. Mulinyawe, A. Ibrahim, K.L. Ligon, D.H. Rowitch, and B.A. Barres. 2009. Myelin gene regulatory factor is a critical transcriptional regulator required for CNS myelination. Cell 138: 172–185.

Etxeberria, A., J.M. Mangin, A. Aguirre, and V. Gallo. 2010. Adult-born SVZ progenitors receive transient synapses during remyelination in corpus callosum. Nat Neurosci 13: 287–289.

Fancy, S.P., S.E. Baranzini, C. Zhao, D.I. Yuk, K.A. Irvine, S. Kaing, N. Sanai, R.J. Franklin, and D.H. Rowitch. 2009. Dysregulation of the Wnt pathway inhibits timely myelination and remyelination in the mammalian CNS. Genes Dev 23: 1571–1585.

Fogarty, M., W.D. Richardson, and N. Kessaris. 2005. A subset of oligodendrocytes generated from radial glia in the dorsal spinal cord. Development 132: 1951–1959.

Fruttiger, M., L. Karlsson, A.C. Hall, A. Abramsson, A.R. Calver, H. Bostrom, K. Willetts, C.H. Bertold, J.K. Heath, C. Betsholtz, and W.D. Richardson. 1999. Defective oligodendrocyte development and severe hypomyelination in PDGF-A knockout mice. Development 126: 457–467.

Gallo, V., J.M. Zhou, C.J. McBain, P. Wright, P.L. Knutson, and R.C. Armstrong. 1996. Oligodendrocyte progenitor cell proliferation and lineage progression are regulated by glutamate receptor-mediated K+ channel block. J Neurosci 16: 2659–2670.

Giri, S., M. Khan, R. Rattan, I. Singh, and A.K. Singh. 2006. Krabbe disease: psychosine-mediated activation of phospholipase A2 in oligodendrocyte cell death. J Lipid Res 47: 1478–1492.

Gow, A., V.L. Friedrich, and R.A. Lazzarini. 1994. Many naturally-occurring mutations of myelin proteolipid protein impair its intracellular-transport. J Neurosci Res 37: 574–583.

Haq, E., S. Giri, I. Singh, and A.K. Singh. 2003. Molecular mechanism of psychosine-induced cell death in human oligodendrocyte cell line. J Neurochem 86: 1428–1440.

Hayakawa, K., L.D. Pham, A.T. Som, B.J. Lee, S. Guo, E.H. Lo, and K. Arai. 2011. Vascular endothelial growth factor regulates the migration of oligodendrocyte precursor cells. J Neurosci 31: 10666–10670.

He, W., C. Ingraham, L. Rising, S. Goderie, and S. Temple. 2001. Multipotent stem cells from the mouse basal forebrain contribute GABAergic neurons and oligodendrocytes to the cerebral cortex during embryogenesis. J Neurosci 21: 8854–8862.

Heiman, M., A. Schaefer, S. Gong, J.D. Peterson, M. Day, K.E. Ramsey, M. Suarez-Farinas, C. Schwarz, D.A. Stephan, D.J. Surmeier, P. Greengard, and N. Heintz. 2008. A translational profiling approach for the molecular characterization of CNS cell types. Cell 135: 738–748.

Hughes, E.G., S.H. Kang, M. Fukaya, and D.E. Bergles. 2013. Oligodendrocyte progenitors balance growth with self-repulsion to achieve homeostasis in the adult brain. Nat Neurosci 16: 668–676.

Ishibashi, T., K.A. Dakin, B. Stevens, P.R. Lee, S.V. Kozlov, C.L. Stewart, and R.D. Fields. 2006. Astrocytes promote myelination in response to electrical impulses. Neuron 49: 823–832.

Jevtovic-Todorovic, V., R.E. Hartman, Y. Izumi, N.D. Benshoff, K. Dikranian, C.F. Zorumski, J.W. Olney, and D.F. Wozniak. 2003. Early exposure to common anesthetic agents causes widespread neurodegeneration in the developing rat brain and persistent learning deficits. J Neurosci 23: 876–882.

Juliet, P.A., E.E. Frost, J. Balasubramaniam, and M.R. Del Bigio. 2009. Toxic effect of blood components on perinatal rat subventricular zone cells and oligodendrocyte precursor cell proliferation, differentiation and migration in culture. J Neurochem 109: 1285–1299.

Kakita, A., and J.E. Goldman. 1999. Patterns and dynamics of SVZ cell migration in the postnatal forebrain: monitoring living progenitors in slice preparations. Neuron 23: 461–472.

Kessaris, N., M. Fogarty, P. Iannarelli, M. Grist, M. Wegner, and W.D. Richardson. 2006. Competing waves of oligodendrocytes in the forebrain and postnatal elimination of an embryonic lineage. Nat Neurosci 9: 173–179.

Kiernan, B.W., B. Gotz, A. Faissner, and C. ffrench-Constant. 1996. Tenascin-C inhibits oligodendrocyte precursor cell migration by both adhesion-dependent and adhesion-independent mechanisms. Mol Cell Neurosci 7: 322–335.

Kim, Y., and M. Nirenberg. 1989. Drosophila NK-homeobox genes. Proc Natl Acad Sci U S A 86: 7716–7720.

Ladiwala, U., H. Li, J.P. Antel, and J. Nalbantoglu. 1999. p53 induction by tumor necrosis factor-alpha and involvement of p53 in cell death of human oligodendrocytes. Journal of Neurochemistry 73: 605–611.

Levine, J.M., R. Reynolds, and J.W. Fawcett. 2001a. The oligodendrocyte precursor cell in health and disease. Trends Neurosci 24: 39–47.

Levine, J.M., R. Reynolds, and J.W. Fawcett. 2001b. The oligodendrocyte precursor cell in health and disease. Trends Neurosci 24: 39–47.

Lin, S.C., and D.E. Bergles. 2002. Physiological characteristics of NG2-expressing glial cells. J Neurocytol 31: 537–549.

Lu, Q.R., D. Yuk, J.A. Alberta, Z. Zhu, I. Pawlitzky, J. Chan, A.P. McMahon, C.D. Stiles, and D.H. Rowitch. 2000. Sonic hedgehog—regulated oligodendrocyte lineage genes encoding bHLH proteins in the mammalian central nervous system. Neuron 25: 317–329.

Lu, Q.R., T. Sun, Z. Zhu, N. Ma, M. Garcia, C.D. Stiles, and D.H. Rowitch. 2002. Common developmental requirement for Olig function indicates a motor neuron/oligodendrocyte connection. Cell 109: 75–86.

Ludwin, S.K. 1997. The pathobiology of the oligodendrocyte. J Neuropathol Exp Neurol 56: 111–124.

Lytle, J.M., R. Chittajallu, J.R. Wrathall, and V. Gallo. 2009. NG2 cell response in the CNP-EGFP mouse after contusive spinal cord injury. Glia 57: 270–285.

Mabie, P.C., M.F. Mehler, R. Marmur, A. Papavasiliou, Q. Song, and J.A. Kessler. 1997. Bone morphogenetic proteins induce astroglial differentiation of oligodendroglial-astroglial progenitor cells. J Neurosci 17: 4112–4120.

Matute, C., I. Torre, F. Perez-Cerda, A. Perez-Samartin, E. Alberdi, E. Etxebarria, A.M. Arranz, R. Ravid, A. Rodriguez-Antigueedad, M.V. Sanchez-Gomez, and M. Domercq. 2007. P2X(7) receptor blockade prevents ATP excitotoxicity in oligodendrocytes and ameliorates experimental autoimmune encephalomyelitis. J Neurosci 27: 9525–9533.

McKinnon, R.D., T. Matsui, M. Aranda, and M. Dubois-Dalcq. 1991. A role for fibroblast growth factor in oligodendrocyte development. Ann N Y Acad Sci 638: 378–386.

McTigue, D.M., P.J. Horner, B.T. Stokes, and F.H. Gage. 1998. Neurotrophin-3 and brain-derived neurotrophic factor induce oligodendrocyte proliferation and myelination of regenerating axons in the contused adult rat spinal cord. J Neurosci 18: 5354–5365.

McTigue, D.M., and R.B. Tripathi. 2008. The life, death, and replacement of oligodendrocytes in the adult CNS. J Neurochem 107: 1–19.

Mignot, C., O. Boespflug-Tanguy, A. Gelot, A. Dautigny, D. Pham-Dinh, and D. Rodriguez. 2004. Alexander disease: putative mechanisms of an astrocytic encephalopathy. Cell Mol Life Sci 61: 369–385.

Miller, R.H., and K. Ono. 1998. Morphological analysis of the early stages of oligodendrocyte development in the vertebrate central nervous system. Microsc Res Tech 41: 441–453.

Miller, R.H. 2002. Regulation of oligodendrocyte development in the vertebrate CNS. Prog Neurobiol 67: 451–467.

Moser, H., P. Dubey, and A. Fatemi. 2004. Progress in x-linked adrenoleukodystrophy. Curr Opin Neurol 17: 263–269.

Muller, D., C. Wang, G. Skibo, N. Toni, H. Cremer, V. Calaora, G. Rougon, and J.Z. Kiss. 1996. PSA-NCAM is required for activity-induced synaptic plasticity. Neuron 17: 413–422.

Nakatsuji, Y., and R.H. Miller. 2001. Control of oligodendrocyte precursor proliferation mediated by density-dependent cell cycle protein expression. Dev Neurosci 23: 356–363.

Nguyen, L., J.M. Rigo, V. Rocher, S. Belachew, B. Malgrange, B. Rogister, P. Leprince, and G. Moonen. 2001. Neurotransmitters as early signals for central nervous system development. Cell Tissue Res 305: 187–202.

Nusse, R. 2005. Wnt signaling in disease and in development. Cell Res 15: 28–32.

Ono, K., R. Bansal, J. Payne, U. Rutishauser, and R.H. Miller. 1995. Early development and dispersal of oligodendrocyte precursors in the embryonic chick spinal cord. Development 121: 1743–1754.

Ono, K., Y. Yasui, U. Rutishauser, and R.H. Miller. 1997. Focal ventricular origin and migration of oligodendrocyte precursors into the chick optic nerve. Neuron 19: 283–292.

Orentas, D.M., and R.H. Miller. 1996. The origin of spinal cord oligodendrocytes is dependent on local influences from the notochord. Dev Biol 177: 43–53.

Orentas, D.M., and R.H. Miller. 1998. Regulation of oligodendrocyte development. Mol Neurobiol 18: 247–259.

Ortega, M.C., A. Bribian, S. Peregrin, M.T. Gil, O. Marin, and F. de Castro. 2012. Neuregulin-1/ErbB4 signaling controls the migration of oligodendrocyte precursor cells during development. Exp Neurol 235: 610–620.

Paez, P.M., D. Fulton, C.S. Colwell, and A.T. Campagnoni. 2009a. Voltage-operated Ca(2+) and Na(+) channels in the oligodendrocyte lineage. J Neurosci Res 87: 3259–3266.

Paez, P.M., D.J. Fulton, V. Spreuer, V. Handley, C.W. Campagnoni, W.B. Macklin, C. Colwell, and A.T. Campagnoni. 2009b. Golli myelin basic proteins regulate oligodendroglial progenitor cell migration through voltage-gated Ca2+ influx. J Neurosci 29: 6663–6676.

Paez, P.M., D.J. Fulton, V. Spreur, V. Handley, and A.T. Campagnoni. 2010. Multiple kinase pathways regulate voltage-dependent Ca2+ influx and migration in oligodendrocyte precursor cells. J Neurosci 30: 6422–6433.

Prestoz, L., E. Chatzopoulou, G. Lemkine, N. Spassky, B. Lebras, T. Kagawa, K. Ikenaka, B. Zalc, and J.L. Thomas. 2004. Control of axonophilic migration of oligodendrocyte precursor cells by Eph-ephrin interaction. Neuron Glia Biol 1: 73–83.

Raff, M.C., B. Durand, and F.B. Gao. 1998. Cell number control and timing in animal development: the oligodendrocyte cell lineage. The Int J Dev Biol 42: 263–267.

Richardson, W.D., N. Pringle, M.J. Mosley, B. Westermark, and M. Dubois-Dalcq. 1988. A role for platelet-derived growth factor in normal gliogenesis in the central nervous system. Cell 53: 309–319.

Robinson, S., M. Tani, R.M. Strieter, R.M. Ransohoff, and R.H. Miller. 1998. The chemokine growth-regulated oncogene-alpha promotes spinal cord oligodendrocyte precursor proliferation. J Neurosci 18: 10457–10463.

Rogister, B., S. Belachew, and G. Moonen. 1999. Oligodendrocytes: from development to demyelinated lesion repair. Acta Neurol Belg 99: 32–39.

Rowitch, D.H., Q.R. Lu, N. Kessaris, and W.D. Richardson. 2002. An 'oligarchy' rules neural development. Trends Neurosci 25: 417–422.

Sakry, D., A. Neitz, J. Singh, R. Frischknecht, D. Marongiu, F. Biname, S.S. Perera, K. Endres, B. Lutz, K. Radyushkin, J. Trotter, and T. Mittmann. 2014. Oligodendrocyte precursor cells modulate the neuronal network by activity-dependent ectodomain cleavage of glial NG2. PLoS Biol 12: e1001993.

Sakry, D., and J. Trotter. 2016. The role of the NG2 proteoglycan in OPC and CNS network function. Brain Res 1638(Pt B): 161–166.

Salter, M.G., and R. Fern. 2005. NMDA receptors are expressed in developing oligodendrocyte processes and mediate injury. Nature 438: 1167–1171.

Shen, S., J. Li, and P. Casaccia-Bonnefil. 2005. Histone modifications affect timing of oligodendrocyte progenitor differentiation in the developing rat brain. J Cell Biol 169: 577–589.

Simons, M., and K. Trajkovic. 2006. Neuron-glia communication in the control of oligodendrocyte function and myelin biogenesis. J Cell Sci 119: 4381–4389.

Slikker, W., Jr., X. Zou, C.E. Hotchkiss, R.L. Divine, N. Sadovova, N.C. Twaddle, D.R. Doerge, A.C. Scallet, T.A. Patterson, J.P. Hanig, M.G. Paule, and C. Wang. 2007. Ketamine-induced neuronal cell death in the perinatal rhesus monkey. Toxicol Sci 98: 145–158.

Small, R.K., P. Riddle, and M. Noble. 1987. Evidence for migration of oligodendrocyte—type-2 astrocyte progenitor cells into the developing rat optic nerve. Nature 328: 155–157.

Smith, J., E. Ladi, M. Mayer-Proschel, and M. Noble. 2000. Redox state is a central modulator of the balance between self-renewal and differentiation in a dividing glial precursor cell. Proc Natl Acad Sci U S A 97: 10032–10037.

Spassky, N., F. de Castro, B. Le Bras, K. Heydon, F. Queraud-LeSaux, E. Bloch-Gallego, A. Chedotal, B. Zalc, and J.L. Thomas. 2002. Directional guidance of oligodendroglial migration by class 3 semaphorins and netrin-1. J Neurosci 22: 5992–6004.

Stevens, B., S. Porta, L.L. Haak, V. Gallo, and R.D. Fields. 2002. Adenosine: a neuron-glial transmitter promoting myelination in the CNS in response to action potentials. Neuron 36: 855–868.

Stidworthy, M.F., S. Genoud, W.W. Li, D.P. Leone, N. Mantei, U. Suter, and R.J. Franklin. 2004. Notch1 and Jagged1 are expressed after CNS demyelination, but are not a major rate-determining factor during remyelination. Brain 127: 1928–1941.

Sugimoto, Y., M. Taniguchi, T. Yagi, Y. Akagi, Y. Nojyo, and N. Tamamaki. 2001. Guidance of glial precursor cell migration by secreted cues in the developing optic nerve. Development 128: 3321–3330.

Sussman, C.R., K.L. Dyer, M. Marchionni, and R.H. Miller. 2000. Local control of oligodendrocyte development in isolated dorsal mouse spinal cord. J Neurosci Res 59: 413–420.

Tanaka, H., S.Y. Grooms, M.V.L. Bennett, and R.S. Zukin. 2000. The AMPAR subunit GluR2: still front and center-stage. Brain Res 886: 190–207.

Tekki-Kessaris, N., R. Woodruff, A.C. Hall, W. Gaffield, S. Kimura, C.D. Stiles, D.H. Rowitch, and W.D. Richardson. 2001. Hedgehog-dependent oligodendrocyte lineage specification in the telencephalon. Development 128: 2545–2554.

Temple, S., and M.C. Raff. 1986. Clonal analysis of oligodendrocyte development in culture: evidence for a developmental clock that counts cell divisions. Cell 44: 773–779.

Trotter, J., K. Karram, and A. Nishiyama. 2010. NG2 cells: Properties, progeny and origin. Brain Res Rev 63: 72–82.

Tsai, H.H., E. Frost, V. To, S. Robinson, C. Ffrench-Constant, R. Geertman, R.M. Ransohoff, and R.H. Miller. 2002. The chemokine receptor CXCR2 controls positioning of oligodendrocyte precursors in developing spinal cord by arresting their migration. Cell 110: 373–383.

Tsai, H.H., M. Tessier-Lavigne, and R.H. Miller. 2003. Netrin 1 mediates spinal cord oligodendrocyte precursor dispersal. Development 130: 2095–2105.

Vallstedt, A., J.M. Klos, and J. Ericson. 2005. Multiple dorsoventral origins of oligodendrocyte generation in the spinal cord and hindbrain. Neuron 45: 55–67.

Wang, C., G. Rougon, and J.Z. Kiss. 1994. Requirement of polysialic acid for the migration of the O-2A glial progenitor cell from neurohypophyseal explants. J Neurosci 14: 4446–4457.

Wang, S., A.D. Sdrulla, G. diSibio, G. Bush, D. Nofziger, C. Hicks, G. Weinmaster, and B.A. Barres. 1998. Notch receptor activation inhibits oligodendrocyte differentiation. Neuron 21: 63–75.

Wegner, M. 2008. A matter of identity: transcriptional control in oligodendrocytes. J Mol Neurosci 35: 3–12.

Wolswijk, G., and M. Noble. 1989. Identification of an adult-specific glial progenitor cell. Development 105: 387–400.

Wolswijk, G., and M. Noble. 1992. Cooperation between PDGF and FGF converts slowly dividing o-2a(adult) progenitor cells to rapidly dividing cells with characteristics of o-2a(perinatal) progenitor cells. J Cell Biol 118: 889–900.

Wu, L., and J.G. Belasco. 2008. Let me count the ways: mechanisms of gene regulation by miRNAs and siRNAs. Mol Cell 29: 1–7.

Ye, F., Y. Chen, T. Hoang, R.L. Montgomery, X.H. Zhao, H. Bu, T. Hu, M.M. Taketo, J.H. van Es, H. Clevers, J. Hsieh, R. Bassel-Duby, E.N. Olson, and Q.R. Lu. 2009. HDAC1 and HDAC2 regulate oligodendrocyte differentiation by disrupting the beta-catenin-TCF interaction. Nat Neurosci 12: 829–838.

Young, K.M., K. Psachoulia, R.B. Tripathi, S.J. Dunn, L. Cossell, D. Attwell, K. Tohyama, and W.D. Richardson. 2013. Oligodendrocyte dynamics in the healthy adult CNS: evidence for myelin remodeling. Neuron 77: 873–885.

Yue, T., K. Xian, E. Hurlock, M. Xin, S.G. Kernie, L.F. Parada, and Q.R. Lu. 2006. A critical role for dorsal progenitors in cortical myelination. J Neurosci 26: 1275–1280.

Zhao, X., X. He, X. Han, Y. Yu, F. Ye, Y. Chen, T. Hoang, X. Xu, Q.S. Mi, M. Xin, F. Wang, B. Appel, and Q.R. Lu. 2010. MicroRNA-mediated control of oligodendrocyte differentiation. Neuron 65: 612–626.

Zhou, Q., G. Choi, and D.J. Anderson. 2001. The bHLH transcription factor Olig2 promotes oligodendrocyte differentiation in collaboration with Nkx2.2. Neuron 31: 791–807.

Zou, X., N. Sadovova, T.A. Patterson, R.L. Divine, C.E. Hotchkiss, S.F. Ali, J.P. Hanig, M.G. Paule, W. Slikker, Jr., and C. Wang. 2008. The effects of L-carnitine on the combination of, inhalation anesthetic-induced developmental, neuronal apoptosis in the rat frontal cortex. Neuroscience 151: 1053–1065.

4

Schwann Cells

Xuan Zhang,[1,a,*] *Tucker A. Patterson,*[2] *Merle G. Paule,* [1,b]
Cheng Wang[1,c] *and William Slikker, Jr.*[3]

Introduction

In the nervous system there are two main classes of cells: nerve cells (neurons) and glial cells. In the vertebrate nervous system, glial cells include microglia, oligodendrocytes, astrocytes and Schwann cells. As principal glial cells in the peripheral nervous system (PNS), Schwann cells, so named in honor of the German physiologist Theodor Schwann, carry out the important jobs of nurturing the development, maturation, regulation and regeneration of neurons. In addition, Schwann cells provide insulation for axons and, thus, facilitate axonal transduction of nervous impulses. They also provide neurotrophic support for neurons and guide the regrowth of damaged PNS axons and modulation of neuromuscular synaptic activity. Impairment of hemostasis in the PNS caused by abnormalities in Schwann cells and their interactions with neurons may lead to various neuropathies such as Charcot-Marie-Tooth disease (CMT), Guillain-Barre syndrome (GBS), amyloid polyneuropathy and immune-mediated neuropathy.

[1] Division of Neurotoxicology, HFT-132, National Center for Toxicological Research (NCTR), Food and Drug Administration (FDA), 3900 NCTR Rd, Jefferson, AR 72079.
[a] E-mail: Xuan.Zhang@fda.hhs.gov
[b] E-mail: merle.paule@fda.hhs.gov
[c] E-mail: Cheng.Wang@fda.hhs.gov
[2] National Center for Toxicological Research (NCTR), Food and Drug Administration (FDA), 3900 NCTR Rd, Jefferson, AR 72079.
   E-mail: tucker.patterson@fda.hhs.gov
[3] Office of the Director, National Center for Toxicological Research (NCTR), Food and Drug Administration (FDA), 3900 NCTR Rd, Jefferson, AR 72079.
   E-mail: william.slikker@fda.hhs.gov
* Corresponding author

Schwann Cells, their Development and Maturation

Schwann cells (SCs) are the most common supporting cells in the PNS. During development, most Schwann cells are derived from the neural crest. In the mature nervous system, Schwann cells can be further divided into four classes based on their morphology, biochemical contents, and associated neuronal types: myelinating Schwann cells, non-myelinating Schwann cells, perisynaptic Schwann cells and satellite cells (Armati and Mathey, 2013; Arroyo, 2007; Corfas *et al.*, 2004; Griffin and Thompson, 2008; Ko, 2007; Yamazaki *et al.*, 2011; Zhou *et al.*, 1999). Prior to their development and maturation into Schwann cells from the neural crest, these cells exist as two intermediate cell types: the Schwann-cell precursor (SCP), that is usually present in rat nerves at embryonic-days 14 (E14) and 15, and the immature Schwann cell, which can be found from E17 to the time of birth. Therefore, the development of Schwann cells can be divided into three stages: the transition of crest cells to SCPs, the transition of SCPs to immature Schwann cells and the reversible formation of mature Schwann cells (Jessen and Mirsky, 1999b; Jessen and Mirsky, 1999c; Jessen and Mirsky, 2005; Mirsky and Jessen, 1999; Mirsky *et al.*, 2002).

The neural crest and SCPs

The neural crest is a group of multipotent cells that originates during development from the dorsal surface of the neural tube. The crest cells migrate in a ventral direction to give rise to peripheral glia including Schwann cells (Jessen and Mirsky, 1999b; Jessen and Mirsky, 2005). During this transition, signaling mechanisms instruct the highly migratory neural crest cells (NCCs) to become Schwann cells. *In vitro* studies using cultured rat and mouse NCCs have demonstrated that neuron-derived neuregulin-1 (NRG-1) may guide the transformation of multipotent NCCs towards a glial fate (Lawson and Biscoe, 1979; Lobsiger *et al.*, 2002; Shah *et al.*, 1994). SCP survival is mainly regulated by axonal survival signals, such as β neuregulin. At the transcriptional level, transcription factor Sox 10 plays a key role in the generation of SCP (Britsch, 2001; Jessen *et al.*, 2008; Kidd, 2013; Lobsiger *et al.*, 2002; Mirsky *et al.*, 2008; Paratore *et al.*, 2001). Sox 10 also regulates the express of ErbB3 in NCCs and may crosstalk with the NRG signaling pathway (Britsch, 2001; Lobsiger *et al.*, 2002; Paratore *et al.*, 2001).

Using rats or mice as examples, a population of proliferating neural crest cells migrates from the neural tubes to ganglia during the middle to later stages of embryogenesis. Two days after the migration of neural crest cells (E14-15 for the rat, E12-13 for the mouse), SCPs will start to proliferate while migrating toward the periphery and associating with axons that are growing towards their target tissues (Dong *et al.*, 1999; Jessen *et al.*, 1994; Jessen and Mirsky, 1991; Jessen and Mirsky, 1999b; Lobsiger *et al.*, 2002; Miyamoto and Yamauchi, 2014). During this migration, SCPs serve to separate the axon bundles into smaller units, form normal nerve fascicules and build up Schwann cell-axon units (Garratt *et al.*, 2000a; Jessen and Mirsky, 2005; Weinstein, 1999). As intermediate precursors between neural crest cells and Schwann cells, SCPs play essential roles in the trophic support of neurons (Garratt *et al.*, 2000a;

Jessen and Mirsky, 2005). In addition to Schwann cells, some fibroblasts that are present in peripheral nerves and some neurons in the CNS may derive from SCPs as well (Doetsch, 2003; Gotz, 2003; Jessen and Mirsky, 2005; Joseph *et al.*, 2004).

Similar to NCCs, SCPs have low affinity for the neurotrophin receptor p75, the neural adhesion protein L1 and the intermediate filament nestin (Jessen *et al.*, 1994; Lobsiger *et al.*, 2000; Lobsiger *et al.*, 2002). However, SCPs do locate along the growing sensory and motor axons in peripheral nerves (Dong *et al.*, 1999; Jessen and Mirsky, 1991; Lobsiger *et al.*, 2002) and are distinguished from NCCs by different molecular expressions, morphologies, and cell-cell interactions. Compared to NCCs, SCPs have a higher expression of the growth associated protein 43 (GAP43) and a lower expression of the myelin gene protein zero (P0), the peripheral myelin protein 22 (PMP22) and Oct-6 (Blanchard *et al.*, 1996; Jessen *et al.*, 1994; Jessen and Mirsky, 1999b; Lobsiger *et al.*, 2002). SCPs present an epithelial type morphology, are greatly motile and have extensive cell-cell contacts with nearby cells (Jessen *et al.*, 1994; Kidd, 2013; Lobsiger *et al.*, 2002).

Early Schwann cells

Around E17 in rats (E15 in mice), early Schwann cells are generated from SCPs through rapid differentiation (Jessen and Mirsky, 1999c; Jessen and Mirsky, 2005; Lobsiger *et al.*, 2002; Mirsky and Jessen, 1999). The generation of early Schwann cells follows a rostrocaudal and mediolateral trajectory starting in the spinal roots and emanating to the ends of the nerves. Compared with SCPs, early Schwann cells present with increased levels of the sulfated galactocerebroside O4 antigen and have high levels of expression of the small calcium binding protein S100 and the intermediate filament protein glial fibrillary acidic protein (GFAP) (Jessen *et al.*, 1994; Jessen and Mirsky, 1999b; Jessen *et al.*, 1990; Lobsiger *et al.*, 2002; Mirsky *et al.*, 1990; Mirsky and Jessen, 1999). Studies have also shown that transcription factors, such as Oct-6 and Krox20 are up-regulated in early Schwann cells whereas Krox24 is down-regulated (Arroyo *et al.*, 1998; Lobsiger *et al.*, 2002). The transition from SCPs to early Schwann cells is believed to be irreversible, whereas the other two stages of development are thought to be plastic and readily reversible (Jessen and Mirsky, 2005). β neuregulin supports conversion of SCPs to early Schwann cells and also acts as an axonal mitogen and survival factor for early Schwann cells (Grinspan *et al.*, 1996; Jessen and Mirsky, 1999a; Mirsky and Jessen, 1999; Morrissey *et al.*, 1995; Syroid *et al.*, 1996; Trachtenberg and Thompson, 1996).

During differentiation, the proliferation of early Schwann cells increases and reaches a peak around E20 in the rat, declining quickly after birth as final differentiations are finished (Lobsiger *et al.*, 2002; Stewart *et al.*, 1993). The differentiation of early Schwann cells into mature myelinating or non-myelinating Schwann cells usually takes place from E15 until the end of the first postnatal week. During this period, the number of Schwann cells is limited by target-derived trophic support and the apoptotic death of Schwann cells in nerves can be detected by the first postnatal week (Grinspan *et al.*, 1996; Jessen and Mirsky, 1999a; Lobsiger *et al.*, 2002).

Myelinating Schwann Cells (MSCs)

Shortly after birth, MSCs and non-myelinating Schwann Cells (NMSCs) can be differentiated in peripheral nerves. MSCs wrap around individual large-diameter axons of all motor neurons and some sensory neurons, forming a myelin sheath. The myelin sheath insulates axons and significantly increases the velocity of nerve conduction. MSCs are also critical for the formation of the nodes of Ranvier, thin areas of myelin along the axon that support the saltatory conduction of action potentials. Additionally, the myelin sheath helps to protect axons from various physical stresses, infections and inappropriate immune reactions (Corfas *et al.*, 2004; Jessen and Mirsky, 1999a; Lobsiger *et al.*, 2002; Miyamoto and Yamauchi, 2014). Compared with MSCs, NMSCs associate with the smaller-diameter axons of C-fibers originating from most sensory and all postganglionic sympathetic neurons. Remak bundles consist of several sensory axons that are engulfed by one NMSC: the individual axons in the bundle are separated by thin extensions of the cytoplasm of NMSC (Corfas *et al.*, 2004; Lobsiger *et al.*, 2002). The diameter of associated axons determines the genetic separation of the Schwann cells into the two mature cell types, MSCs and NMSCs. Studies have demonstrated that regeneration of large-diameter axons can induce myelination by NMSCs (Aguayo *et al.*, 1976a; Aguayo *et al.*, 1976b; Lobsiger *et al.*, 2002; Weinberg *et al.*, 1975). These differentiation processes are strictly regulated by only a couple of transcription factors, such as Oct-6, Krox20 and Sox10 (Wegner, 2000a; Wegner, 2000b).

MSCs that form myelin sheaths are very large cells that express myelin proteins including myelin basic protein (MBP), peripheral myelin protein 22 (PMP22), myelin gene protein zero (P0), myelin-associated glycoprotein (MAG) and myelin and lymphocyte protein (MAL). These myelin proteins are essential for the formation and function of myelin sheaths which are critical for rapid nerve conduction. The myelination processes can be divided into three phases: cell migration; cell elongation along axons, and myelination. The pre-myelinating process includes MSC migration and elongation (Miyamoto and Yamauchi, 2014) and these processes are modulated by specific transcription factors and protein factors. Among them, neuregulin-1 (Nrg 1), provides pivotal axonal signals that regulate SC development (Corfas *et al.*, 2004; Garratt *et al.*, 2000a; Lemke, 1996; Lemke, 2006; Lobsiger *et al.*, 2002).

As a growth factor, Nrg 1 is also known as a potent mitogen for SCs that regulates SC proliferation, migration and myelination through binding to its receptors, the Erb B2/3 receptors. During development, SCs elongate along and keep continuous contact with their axons. The axon-SC interaction through Nrg1 and ErbB binding not only promotes and induces the glial differentiation of neural crest cells while decreasing neuronal differentiation, but also promotes the maturation, survival and migration of SCPs (Jessen and Mirsky, 2005; Lai, 2005; Riethmacher *et al.*, 1997; Woldeyesus *et al.*, 1999). Additionally, Nrg 1 serves as a key modulator of the physiological properties of SCs by inducing the expression of sodium channels and upregulating gap junction communication between adjacent SCs (Aquino *et al.*, 2006; Chandross *et al.*, 1996; Corfas *et al.*, 2004; Dong *et al.*, 1999; Mahanthappa *et al.*, 1996; Newbern and Birchmeier, 2010; Shah *et al.*, 1994; Syroid *et al.*, 1996; Taveggia *et al.*, 2005; Wilson and Chiu, 1993).

As triggers to induce myelination, Nrg1 and Erb B signaling determine whether SCs will myelinate and how much myelin is produced (Leimeroth *et al.*, 2002; Lemke, 2006; Newbern and Birchmeier, 2010; Taveggia *et al.*, 2005). A variety of transcription factors that control myelination process are modulated by Nrg1. For example, the expression of Oct6/SCIP and Egr2/Krox-20 are upregulated by Nrg1 (Leimeroth *et al.*, 2002; Lemke, 2006; Murphy *et al.*, 1996; Newbern and Birchmeier, 2010). Knock-outs of NRG1 type III resulted in poorly ensheathed sensory neurons in mice and the failure of myelination. It is suggested that axonal NRG1 controls the initiation of the myelination process (Lemke, 2006; Taveggia *et al.*, 2005). In a study using a mouse model, reduced expression of ErbB2 leads to a widespread peripheral neuropathy characterized by abnormally thin myelin sheaths and fewer myelin layers (Garratt, 2000b). Another study using mutant and transgenic mice demonstrated that reduced Nrg1 expression causes hypomyelination and decreased nerve conduction velocity, whereas the overexpression of Nrg1 induces hypermyelination (Michailov *et al.*, 2004).

Non-Myelinating Schwann Cells (NMSCs)

Small axons with 0.5–1.5 μm diameters, such as the C fiber nociceptive neurons, the postganglionic sympathetic fibers and some preganglionic sympathetic and parasympathetic fibers, are defined as Remak fibers and are surrounded by non-myelinating Schwann cells. In peripheral nerves, the majority of nerve fibers are unmyelinated with an approximately four-fold larger number of unmyelinated axons than myelinated axons (Armati and Mathey, 2013; Griffin and Thompson, 2008). Each NMSC ensheaths nerve fiber by embedding the axon within their plasma membrane grooves without forming compact myelin. Compared to MSCs, NMSCs have a high expression of glial fibrillary acidic protein (GFAP), express low–affinity neurotrophin receptor p75 and the cell adhesion molecule L1, no expression of myelin basic protein (MBP) (Corfas *et al.*, 2004; Faissner *et al.*, 1984; Jessen *et al.*, 1990), and have much smaller intercellular spaces along individual axons (Armati and Mathey, 2013; Griffin and Thompson, 2008).

Nerve fibers can be only partially myelinated. The beginnings and ends of axons are usually unmyelinated, such as in the initial and the terminal parts of Aδ and Aβ sensory nerves and motor fibers (Griffin and Thompson, 2008). According to a study in monkeys, Aδ nociceptors in skin lack myelination about 10 cm from the termini in the skin and continue un-myelinated distally (Griffin and Thompson, 2008; Peng *et al.*, 1999). Therefore, the distinction between myelinated and unmyelinated fibers is not absolute and appropriate neuronal signals can initiate myelination by all of NMSCs (Griffin and Thompson, 2008; Taveggia *et al.*, 2005).

Perisynaptic Schwann Cells (PSCs)

In 1960, Birks *et al.* visualized the structures of neuromuscular junctions (NMJ) using electron microscopy (Birks *et al.*, 1960). Compared with MSCs, which wrap around individual axons to form myelin sheaths, the somata of PSCs extend processes that cover the terminal of the presynaptic motor nerve without wrapping, and the postsynaptic acetylcholine receptors (AchRs) are aligned in turn (Corfas *et al.*, 2004; Griffin and Thompson, 2008).

As a model for evaluating the function of PSCs at NMJs, frog NMJs have served well. Markers that have been useful in the investigation of PSC behavior include peanut agglutinin (PNA), which marks the extracellular matrix around the frog PSCs, and the monoclonal antibody (mAb)2A12, which labels the surface of the PSC membrane (Astrow *et al.*, 1998; Corfas *et al.*, 2004; Griffin and Thompson, 2008; Ko, 1987). Although PSCs are not required for the initial formation of NMJs, they present after the formation of nerve-muscle contact and extend their processes onto the nerve terminals. The excessive sprouting of PSCs seems to enhance nerve terminal growth: ablation of PSC with mAb 2A12 causes reduction of nerve synapse growth, retraction of existing synapses and decreases in nerve-evoked muscle tension (Reddy *et al.*, 2003).

PSCs play important roles not only in synaptogenesis but also in synaptic transmission and connectivity. Possessing neurotransmitter receptors, such as purinergic and muscarinic receptors, PSCs can be stimulated by impulses from motor nerves. The binding of purines and acetylcholine to their receptors leads to the release of Ca^{2+} from intracellular stores. These Ca^{2+} transients are specific to the synapse, having been found only in the PSCs at NMJs and not in MSCs along axons (Jahromi *et al.*, 1992; Reist and Smith, 1992). Todd and colleagues have shown that the calcium response can be modulated by the neurotrophins BDNF and NT3 when applied to the synapse (Todd *et al.*, 2007). These results indicate that synaptic transmissions can be detected by PSCs and that a Ca^{2+} dependent signal cascade can be triggered. Activation of muscarinic receptors on PSCs down-regulates the expression of GFAP via a cAMP pathway, whereas the blockade of this signaling enhances the level of GFAP and promotes PSC sprouting (Georgiou and Charlton, 1999; Georgiou *et al.*, 1994). In response to neuronal and synaptic activities, PSCs can feed back to the nerve terminal to modulate neuronal activities as well. Blockade of G protein signaling in PSCs with GDPbetaS, reduced synaptic depression induced by high frequency neuronal stimuli, and activation of G proteins in PSCs enhanced the depression in neurotransmitter release (Robitaille, 1998). Using immunohistochemical labelling, Pinard and colleagues showed that glutamate-aspartate transporters (GLASTs) are widely expressed on PSCs and the feedback from PSCs caused release of glutamate from nerve terminals (Pinard *et al.*, 2003).

Satellite Glial Cells (SGC)

Satellite cells completely wrap around the cell bodies of most sensory ganglia, such as the dorsal root ganglia (DRG) and sympathetic and parasympathetic ganglia, without forming myelin (Armati and Mathey, 2013; Hanani, 2005). As a functional unit, each sensory neuron has its own SGC sheath, which usually includes several SGCs. SGCs express low levels of GFAP, which significantly increases after axonal damage. SGCs also express S100 proteins and glutamine synthetase (GS) is the most useful marker for SGCs (Gonzalez-Martinez *et al.*, 2003; Ichikawa *et al.*, 1997; Sandelin *et al.*, 2004; Vega *et al.*, 1991). SGCs around sensory neurons are connected by gap junctions and exhibit transporters for glutamate and GABA. It has been postulated that SGCs support neurons by supplying nutrients and modulating extracellular ion and neurotransmitter levels (Hanani, 2005; Ren and Dubner, 2010; Scholz, 1974; Schon, 1974).

Ren and colleagues reviewed the modulation of sensory nerve activity in DRG by SGCs through paracrine-type signaling (Ren, 2010; Ren and Dubner, 2010). The injection of complete Freund's adjuvant (CFA) into the hind paw of mice and CFA-induced sciatic nerve neuritis lead to increases in the number of gap junctions between SGCs, a phenomenon that is consistent with peripheral sensitization (Dublin and Hanani, 2007; Ledda *et al.*, 2009). Thus, during chronic inflammatory pain, noxious stimuli can enhance the communication between SGCs and increase neuronal excitability. Using *in situ* hybridization and immunohistochemical techniques, Zhou and colleagues demonstrated that after nerve injury, the expression of nerve growth factor (NGF) and neurotrophin-3 (NT3) was upregulated in SGCs surrounding neurons in damaged DRG. The presence of p75-immunoreactive satellite cells indicated that the noradrenergic terminals sprouting around the lesioned DRG were enhanced by NT3 (Zhou *et al.*, 1999). SGCs may also regulate the excitability of sensory ganglion by buffering extracellular potassium (K^+) concentrations. In a rat model, alteration in the expression of inward rectifying K^+ channel Kir4.1 in SGCs around trigeminal ganglion caused neuropathic pain (Tang *et al.*, 2010; Vit *et al.*, 2008). Feedback between SGCs and neurons is also involved in pain hypersensitivity. Neuronal release of calcitonin gene-related peptide (CGRP) may initiate the production of IL-1β in SGCs and, in turn, induce the production of prostaglandin E_2 (PGE_2) through the cyclooxygenase-2 (COX2) pathway. Release of PGE_2 from SGCs may potentiate the production of CGRP in trigeminal ganglion neurons (Capuano *et al.*, 2009).

On the other hand, SGCs can also down-regulate chronic pain by releasing ATP, which will bind to P2Y receptors on neurons. Neuronal P2X receptors can be downregulated by the ATP effect and, in turn, reduce nociceptor signaling (Chen *et al.*, 2012; Chen *et al.*, 2008).

Schwann Cells in Peripheral Nerve Repair and Regeneration

Unlike nerves in the CNS, injured peripheral nerves can regenerate and re-innervate their appropriate targets. SCs within the PNS promote nerve regeneration. Peripheral nerve injury causes neuron damage and myelin breakdown. After injury, SCs switch to a distinct phenotype and provide help by clearing the myelin debris, recruiting macrophages into the lesions, protecting damaged neurons and initiating axonal regeneration and re-myelination (Susuki, 2014). Activated SCs proliferate and phagocytize the myelin debris and damaged axons along with macrophages and stimulate axonal regrowth (Chen *et al.*, 2007; Susuki, 2014).

Following injury, damaged axons can regenerate and the affected neurons will respond to enhance axonal regeneration (Chen *et al.*, 2007; Susuki, 2014). Usually, the regrowth of the axon starts within a couple of weeks. During the repair process, SCs make contact with the regrowing axons and initiate re-myelination. Growth-related genes, such as those encoding intrinsic neurotrophic factors and key transcription factors are up-regulated by SCs (Rodrigues *et al.*, 2012). In a mouse model, in which the sciatic nerve was crushed, increased axonal excitability was present during peripheral nerve regeneration. This increased axonal excitability resulted from increases in the number and conductance of Na^+ channels. Initially, the regenerated myelin segment was thin and short, after which it became thicker. These regenerated segments lead to a reduced velocity of nerve conduction (Nakata *et al.*, 2008). At the same time, Nodes of Ranvier need to be reformed and the clustering of Na (+) channels at the nodes of Ranvier is required for saltatory conduction. During the process of myelin regeneration and Nodes of Ranvier reformation, SCs play an important role in the accumulation of these channels along the axon (Eshed *et al.*, 2005).

In demyelinating neuropathies and peripheral nerve injury, in order to provide support for axonal regeneration, SCs need to transit to the repair phenotype, proliferate and re-differentiate. Multiple players are involved in this process, such as transcription factors, extracellular matrix proteins, neurotrophic factors and hormones (Chen *et al.*, 2007; Hoke *et al.*, 2002; Jessen and Mirsky, 2008; Rahmatullah *et al.*, 1998; Susuki, 2014; You *et al.*, 1997). SC plasticity is modulated by negative and positive transcriptional regulators of myelination. Among these factors, the transcription factor c-Jun is a key molecule that is up-regulated after nerve injury and it triggers SC dedifferentiation. In c-Jun mutant mice, the regenerative ability is lost in SCs, axonal outgrowth and target re-innervation fail and neuronal death is increased (Arthur-Farraj *et al.*, 2012; Jessen and Mirsky, 2008; Scheib and Hoke, 2013). These results indicate that SCs promote neuronal survival and axon outgrowth after injury.

Remyelination along regrown axons is required for functional recovery. SCs encapsulate regenerated axons with a new myelin sheath. During this process, the signaling cascade modulated by axonal NRG1 plays a crucial role. During development, axonally released NRG1 molecules activate ErbB2 and ErbB3 receptors that are expressed on SCs and modulate SC differentiation, proliferation, migration, axon encapsulation and myelination (Fricker *et al.*, 2011). Following sciatic nerve transection in rats, the expression of NRG1 and erbB2 and erbB3 receptors in SCs significantly increases and the expression of phosphorylated ErbB2 and ErbB3 increases acutely and transiently (Fricker and Bennett, 2011; Guertin *et al.*, 2005; Kwon *et al.*, 1997). The increased expression of ErbB2 and ErbB3 presents about 3 days after peripheral nerve injury, a time at which SCs begin injury-induced proliferation. In sciatic nerve crush injury in *Nrg1* mutant mice, remyelination of axons was significantly reduced (Fricker *et al.*, 2011). In mice that overexpress NRG1 in neurons, remyelination was more apparent than that in wild type. Target ablation of NRG1 in SCs severely impaired re-myelination after crush injury of sciatic nerve (Stassart *et al.*, 2013).

Re-myelination after peripheral nerve injury requires close contact between axons and SCs. The molecular mechanisms underlying this process include the involvement of cell adhesion molecules (CAMs) from the nectin and nectin-like (Necl, also known as SynCAM or Cadm) family (Maurel *et al.*, 2007; Spiegel *et al.*, 2007; Susuki *et al.*, 2011; Zelano *et al.*, 2009). During re-myelination, Necl4 plays a crucial role in Schwann cell–axon interaction. Necl4 is highly expressed on myelinated Schwann cells and is mainly located along the internodes in direct contact with Necl1, which is localized mainly on axons. Therefore, the interaction between these two CAMs mediates Schwann cell adhesion. The disruption of the interaction between Necl1 and Necl4 will interfere with axonal re-myelination (Spiegel *et al.*, 2007). In an *in vivo* study, 7–10 mm segments from sciatic nerves that were resected from rats were investigated. *In situ* hybridization (ISH) demonstrated decreases in necl-1 mRNA signal in the injured nerve and increases in the signals for necl-4 and necl-5 mRNA. Alterations in necl immunoreactivity show close relationships between axon and SC markers (Spiegel *et al.*, 2007; Zelano *et al.*, 2009).

Summary

Schwann cells are recognized as principle glial cells in the peripheral nervous system and play important role in development and maturity. The abnormalities of Schwann cells and their signaling with surrounded neurons result in various peripheral nerve disorders. Schwann cells are primarily discussed in the context of their ability in axonal myelination, nerve regeneration and remyelination after injury and various neuropathies. However, they are further categorized into four groups and all these cells are derived from neural crest cells. Based on the rapid progress of molecular biological techniques and extend research results, these Schwann cell subsets have key functions in keeping homeostasis of normal nervous system, synapse formation, perineuronal organization and immune modulation.

Disclaimer

This document has been reviewed in accordance with United States Food and Drug Administration (FDA) policy and approved for publication. Approval does not signify that the contents necessarily reflect the position or opinions of the FDA. The findings and conclusions in this report are those of the author and do not necessarily represent the views of the FDA.

References

Aguayo, A.J., L. Charron, and G.M. Bray. 1976a. Potential of Schwann cells from unmyelinated nerves to produce myelin: a quantitative ultrastructural and radiographic study. J Neurocytol. 5: 565–573.

Aguayo, A.J., J. Epps, L. Charron, and G.M. Bray. 1976b. Multipotentiality of Schwann cells in cross-anastomosed and grafted myelinated and unmyelinated nerves: quantitative microscopy and radioautography. Brain Res 104: 1–20.

Aquino, J.B., J. Hjerling-Leffler, M. Koltzenburg, T. Edlund, M.J. Villar, and P. Ernfors. 2006. *In vitro* and *in vivo* differentiation of boundary cap neural crest stem cells into mature Schwann cells. Experimental Neurology 198: 438–449.

Armati, P.J., and E.K. Mathey. 2013. An update on Schwann cell biology—immunomodulation, neural regulation and other surprises. Journal of the Neurological Sciences 333: 68–72.

Arroyo, E.J., J.R. Bermingham, Jr., M.G. Rosenfeld, and S.S. Scherer. 1998. Promyelinating Schwann cells express Tst-1/SCIP/Oct-6. The Journal of Neuroscience : The Official Journal of the Society for Neuroscience 18: 7891–7902.

Arroyo E.J., S.S. 2007. The molecular organization of myelinating Schwann cells. pp. 37–54. *In*: A. P.J., (ed.). The Biology of Schwann Cells. Cambridge University Press, New York, NY, USA.

Arthur-Farraj, P.J., M. Latouche, D.K. Wilton, S. Quintes, E. Chabrol, A. Banerjee, A. Woodhoo, B. Jenkins, M. Rahman, M. Turmaine, G.K. Wicher, R. Mitter, L. Greensmith, A. Behrens, G. Raivich, R. Mirsky, and K.R. Jessen. 2012. c-Jun reprograms Schwann cells of injured nerves to generate a repair cell essential for regeneration. Neuron 75: 633–647.

Astrow, S.H., H. Qiang, and C.P. Ko. 1998. Perisynaptic Schwann cells at neuromuscular junctions revealed by a novel monoclonal antibody. Journal of Neurocytology 27: 667–681.

Birks, R., B. Katz, and R. Miledi. 1960. Physiological and structural changes at the amphibian myoneural junction, in the course of nerve degeneration. The Journal of Physiology 150: 145–168.

Blanchard, A.D., A. Sinanan, E. Parmantier, R. Zwart, L. Broos, D. Meijer, C. Meier, K.R. Jessen, and R. Mirsky. 1996. Oct-6 (SCIP/Tst-1) is expressed in Schwann cell precursors, embryonic Schwann cells, and postnatal myelinating Schwann cells: comparison with Oct-1, Krox-20, and Pax-3. Journal of Neuroscience Research 46: 630–640.

Britsch, S., D.E. Goerich, D. Riethmacher, R.I. Peirano, M. Rossner, K.A. Nave, C. Birchmeier, and M. Wegner. 2001. The transcription factor Sox10 is a key regulator of peripheral glial development. Genes and Development 15: 66–78.

Capuano, A., A. De Corato, L. Lisi, G. Tringali, P. Navarra, and C. Dello Russo. 2009. Proinflammatory-activated trigeminal satellite cells promote neuronal sensitization: relevance for migraine pathology. Molecular Pain 5: 43.

Chandross, K.J., D.C. Spray, R.I. Cohen, N.M. Kumar, M. Kremer, R. Dermietzel, and J.A. Kessler. 1996. TNF alpha inhibits Schwann cell proliferation, connexin46 expression, and gap junctional communication. Molecular and Cellular Neurosciences 7: 479–500.

Chen, Z.L., W.M. Yu, and S. Strickland. 2007. Peripheral regeneration. Annual Review of Neuroscience 30: 209–233.

Chen, Y., X. Zhang, C. Wang, G. Li, Y. Gu, and L.Y. Huang. 2008. Activation of P2X7 receptors in glial satellite cells reduces pain through downregulation of P2X3 receptors in nociceptive neurons. Proceedings of the National Academy of Sciences of the United States of America 105: 16773–16778.

Chen, Y., G. Li, and L.Y. Huang. 2012. P2X7 receptors in satellite glial cells mediate high functional expression of P2X3 receptors in immature dorsal root ganglion neurons. Molecular Pain 8: 9.

Corfas, G., M.O. Velardez, C.P. Ko, N. Ratner, and E. Peles. 2004. Mechanisms and roles of axon-Schwann cell interactions. The Journal of Neuroscience : The Official Journal of the Society for Neuroscience 24: 9250–9260.

Doetsch, F. 2003. The glial identity of neural stem cells. Nature Neuroscience 6: 1127–1134.

Dong, Z., A. Sinanan, D. Parkinson, E. Parmantier, R. Mirsky, and K.R. Jessen. 1999. Schwann cell development in embryonic mouse nerves. Journal of Neuroscience Research 56: 334–348.

Dublin, P., and M. Hanani. 2007. Satellite glial cells in sensory ganglia: their possible contribution to inflammatory pain. Brain, Behavior, and Immunity 21: 592–598.

Eshed, Y., K. Feinberg, S. Poliak, H. Sabanay, O. Sarig-Nadir, I. Spiegel, J.R. Bermingham, Jr., and E. Peles. 2005. Gliomedin mediates Schwann cell-axon interaction and the molecular assembly of the nodes of Ranvier. Neuron 47: 215–229.

Faissner, A., J. Kruse, J. Nieke, and M. Schachner. 1984. Expression of neural cell adhesion molecule L1 during development, in neurological mutants and in the peripheral nervous system. Brain Research 317: 69–82.

Fricker, F.R., and D.L. Bennett. 2011. The role of neuregulin-1 in the response to nerve injury. Future Neurology 6: 809–822.

Fricker, F.R., N. Lago, S. Balarajah, C. Tsantoulas, S. Tanna, N. Zhu, S.K. Fageiry, M. Jenkins, A.N. Garratt, C. Birchmeier, and D.L. Bennett. 2011. Axonally derived neuregulin-1 is required for remyelination and regeneration after nerve injury in adulthood. The Journal of Neuroscience : The Official Journal of the Society for Neuroscience 31: 3225–3233.

Garratt, A.N., S. Britsch, and C. Birchmeier. 2000a. Neuregulin, a factor with many functions in the life of a schwann cell. BioEssays : News and Reviews in Molecular, Cellular and Developmental Biology 22: 987–996.

Garratt, A.N., O. Voiculescu, P. Topilko, P. Charnay, and C. Birchmeier. 2000b. A dual role of erbB2 in myelination and in expansion of the schwann cell precursor pool. The Journal of Cell Biology 148: 1035–1046.

Georgiou, J., R. Robitaille, W.S. Trimble, and M.P. Charlton. 1994. Synaptic regulation of glial protein expression *in vivo*. Neuron 12: 443–455.

Georgiou, J., and M.P. Charlton. 1999. Non-myelin-forming perisynaptic schwann cells express protein zero and myelin-associated glycoprotein. Glia 27: 101–109.

Gonzalez-Martinez, T., P. Perez-Pinera, B. Diaz-Esnal, and J.A. Vega. 2003. S-100 proteins in the human peripheral nervous system. Microscopy Research and Technique 60: 633–638.

Gotz, M. 2003. Glial cells generate neurons—master control within CNS regions: developmental perspectives on neural stem cells. Neuroscientist 9: 379–397.

Griffin, J.W., and W.J. Thompson. 2008. Biology and pathology of nonmyelinating Schwann cells. Glia 56: 1518–1531.

Grinspan, J.B., M.A. Marchionni, M. Reeves, M. Coulaloglou, and S.S. Scherer. 1996. Axonal interactions regulate Schwann cell apoptosis in developing peripheral nerve: neuregulin receptors and the role of neuregulins. The Journal of Neuroscience : The Official Journal of the Society for Neuroscience 16: 6107–6118.

Guertin, A.D., D.P. Zhang, K.S. Mak, J.A. Alberta, and H.A. Kim. 2005. Microanatomy of axon/glial signaling during Wallerian degeneration. The Journal of Neuroscience : The Official Journal of the Society for Neuroscience 25: 3478–3487.

Hanani, M. 2005. Satellite glial cells in sensory ganglia: from form to function. Brain Research. Brain Research Reviews 48: 457–476.

Hoke, A., T. Gordon, D.W. Zochodne, and O.A. Sulaiman. 2002. A decline in glial cell-line-derived neurotrophic factor expression is associated with impaired regeneration after long-term Schwann cell denervation. Experimental Neurology 173: 77–85.

Ichikawa, H., D.M. Jacobowitz, and T. Sugimoto. 1997. S100 protein-immunoreactive primary sensory neurons in the trigeminal and dorsal root ganglia of the rat. Brain Research 748: 253–257.

Jahromi, B.S., R. Robitaille, and M.P. Charlton. 1992. Transmitter release increases intracellular calcium in perisynaptic Schwann cells *in situ*. Neuron 8: 1069–1077.

Jessen, K.R., L. Morgan, H.J. Stewart, and R. Mirsky. 1990. Three markers of adult non-myelin-forming Schwann cells, 217c(Ran-1), A5E3 and GFAP: development and regulation by neuron-Schwann cell interactions. Development 109: 91–103.

Jessen, K.R., and R. Mirsky. 1991. Schwann cell precursors and their development. Glia 4: 185–194.

Jessen, K.R., A. Brennan, L. Morgan, R. Mirsky, A. Kent, Y. Hashimoto, and J. Gavrilovic. 1994. The Schwann cell precursor and its fate: a study of cell death and differentiation during gliogenesis in rat embryonic nerves. Neuron 12: 509–527.

Jessen, K.R., and R. Mirsky. 1999a. Developmental regulation in the Schwann cell lineage. Advances in Experimental Medicine and Biology 468: 3–12.

Jessen, K.R., and R. Mirsky. 1999b. Schwann cells and their precursors emerge as major regulators of nerve development. Trends in Neurosciences 22: 402–410.

Jessen, K.R., and R. Mirsky. 1999c. Why do Schwann cells survive in the absence of axons? Annals of the New York Academy of Sciences 883: 109–115.

Jessen, K.R., and R. Mirsky. 2005. The origin and development of glial cells in peripheral nerves. Nature Reviews. Neuroscience 6: 671–682.

Jessen, K.R., and R. Mirsky. 2008. Negative regulation of myelination: relevance for development, injury, and demyelinating disease. Glia 56: 1552–1565.

Jessen, K.R., R. Mirsky, and J. Salzer. 2008. Introduction. Schwann cell biology. Glia 56: 1479–1480.

Joseph, N.M., Y.S. Mukouyama, J.T. Mosher, M. Jaegle, S.A. Crone, E.L. Dormand, K.F. Lee, D. Meijer, D.J. Anderson, and S.J. Morrison. 2004. Neural crest stem cells undergo multilineage differentiation in developing peripheral nerves to generate endoneurial fibroblasts in addition to Schwann cells. Development 131: 5599–5612.

Kidd, G.J., N. Ohno, and B.D. Trapp. 2013. Biology of Schwann cells. pp. 55–79. *In*: G. Said, and C. Krarup (eds.). Pheripheral Nerve Disorders. Scotland, Elsevier.

Ko, C.P., Y. Sugiura, and Z. Feng. 2007. The biology of perisynaptic (terminal) Schwann cells. pp. 72–99. *In*: P.J. Armati (ed.). The Biology of Schwann Cells. Cambridge University Press, New York, NY, USA.

Ko, C.P. 1987. A lectin, peanut agglutinin, as a probe for the extracellular matrix in living neuromuscular junctions. Journal of Neurocytology 16: 567–576.

Kwon, Y.K., A. Bhattacharyya, J.A. Alberta, W.V. Giannobile, K. Cheon, C.D. Stiles, and S.L. Pomeroy. 1997. Activation of ErbB2 during wallerian degeneration of sciatic nerve. The Journal of Neuroscience : The Official Journal of the Society for Neuroscience 17: 8293–8299.

Lai, C. 2005. Peripheral glia: Schwann cells in motion. Current Biology : CB 15: R332–334.

Lawson, S.N., and T.J. Biscoe. 1979. Development of mouse dorsal root ganglia: an autoradiographic and quantitative study. Journal of Neurocytology 8: 265–274.

Ledda, M., E. Blum, S. De Palo, and M. Hanani. 2009. Augmentation in gap junction-mediated cell coupling in dorsal root ganglia following sciatic nerve neuritis in the mouse. Neuroscience 164: 1538–1545.

Leimeroth, R., C. Lobsiger, A. Lussi, V. Taylor, U. Suter, and L. Sommer. 2002. Membrane-bound neuregulin1 type III actively promotes Schwann cell differentiation of multipotent Progenitor cells. Developmental Biology 246: 245–258.

Lemke, G. 1996. Neuregulins in development. Molecular and Cellular Neurosciences 7: 247–262.

Lemke, G. 2006. Neuregulin-1 and myelination. Science's STKE : Signal Transduction Knowledge Environment. 2006: pe11.

Lobsiger, C.S., B. Schweitzer, V. Taylor, and U. Suter. 2000. Platelet-derived growth factor-BB supports the survival of cultured rat Schwann cell precursors in synergy with neurotrophin-3. Glia 30: 290–300.

Lobsiger, C.S., V. Taylor, and U. Suter. 2002. The early life of a Schwann cell. Biological Chemistry 383: 245–253.

Mahanthappa, N.K., E.S. Anton, and W.D. Matthew. 1996. Glial growth factor 2, a soluble neuregulin, directly increases Schwann cell motility and indirectly promotes neurite outgrowth. The Journal of Neuroscience : The Official Journal of the Society for Neuroscience 16: 4673–4683.

Maurel, P., S. Einheber, J. Galinska, P. Thaker, I. Lam, M.B. Rubin, S.S. Scherer, Y. Murakami, D.H. Gutmann, and J.L. Salzer. 2007. Nectin-like proteins mediate axon Schwann cell interactions along the internode and are essential for myelination. J Cell Biol 178: 861–874.

Michailov, G.V., M.W. Sereda, B.G. Brinkmann, T.M. Fischer, B. Haug, C. Birchmeier, L. Role, C. Lai, M.H. Schwab, and K.A. Nave. 2004. Axonal neuregulin-1 regulates myelin sheath thickness. Science 304: 700–703.

Mirsky, R., C. Dubois, L. Morgan, and K.R. Jessen. 1990. O4 and A007-sulfatide antibodies bind to embryonic Schwann cells prior to the appearance of galactocerebroside; regulation of the antigen by axon-Schwann cell signals and cyclic AMP. Development 109: 105–116.

Mirsky, R., and K.R. Jessen. 1999. The neurobiology of Schwann cells. Brain Pathology 9: 293–311.

Mirsky, R., K.R. Jessen, A. Brennan, D. Parkinson, Z. Dong, C. Meier, E. Parmantier, and D. Lawson. 2002. Schwann cells as regulators of nerve development. Journal of Physiology, Paris 96: 17–24.

Mirsky, R., A. Woodhoo, D.B. Parkinson, P. Arthur-Farraj, A. Bhaskaran, and K.R. Jessen. 2008. Novel signals controlling embryonic Schwann cell development, myelination and dedifferentiation. Journal of the Peripheral Nervous System : JPNS 13: 122–135.

Miyamoto, Y., and J. Yamauchi. 2014. Recent insights into molecular mechanisms that control growth factor receptor-mediated Schwann cell morphological changes during development. pp. 5–27. *In*: Y.J., Sango K., (ed.). Schwann Cell Development and Pathology. Tokyo, Springer.

Morrissey, T.K., A.D. Levi, A. Nuijens, M.X. Sliwkowski, and R.P. Bunge. 1995. Axon-induced mitogenesis of human Schwann cells involves heregulin and p185erbB2. Proceedings of the National Academy of Sciences of the United States of America 92: 1431–1435.

Murphy, P., P. Topilko, S. Schneider-Maunoury, T. Seitanidou, A. Baron-Van Evercooren, and P. Charnay. 1996. The regulation of Krox-20 expression reveals important steps in the control of peripheral glial cell development. Development 122: 2847–2857.

Nakata, M., H. Baba, K. Kanai, T. Hoshi, S. Sawai, T. Hattori, and S. Kuwabara. 2008. Changes in Na(+) channel expression and nodal persistent Na(+) currents associated with peripheral nerve regeneration in mice. Muscle & Nerve 37: 721–730.

Newbern, J., and C. Birchmeier. 2010. Nrg1/ErbB signaling networks in Schwann cell development and myelination. Seminars in cell & Developmental Biology 21: 922–928.

Paratore, C., D.E. Goerich, U. Suter, M. Wegner, and L. Sommer. 2001. Survival and glial fate acquisition of neural crest cells are regulated by an interplay between the transcription factor Sox10 and extrinsic combinatorial signaling. Development 128: 3949–3961.

Peng, Y.B., M. Ringkamp, J.N. Campbell, and R.A. Meyer. 1999. Electrophysiological assessment of the cutaneous arborization of Adelta-fiber nociceptors. Journal of Neurophysiology 82: 1164–1177.

Pinard, A., S. Levesque, J. Vallee, and R. Robitaille. 2003. Glutamatergic modulation of synaptic plasticity at a PNS vertebrate cholinergic synapse. The European Journal of Neuroscience 18: 3241–3250.

Rahmatullah, M., A. Schroering, K. Rothblum, R.C. Stahl, B. Urban, and D.J. Carey. 1998. Synergistic regulation of Schwann cell proliferation by heregulin and forskolin. Molecular and Cellular Biology 18: 6245–6252.

Reddy, L.V., S. Koirala, Y. Sugiura, A.A. Herrera, and C.P. Ko. 2003. Glial cells maintain synaptic structure and function and promote development of the neuromuscular junction *in vivo*. Neuron 40: 563–580.

Reist, N.E., and S.J. Smith. 1992. Neurally evoked calcium transients in terminal Schwann cells at the neuromuscular junction. Proceedings of the National Academy of Sciences of the United States of America 89: 7625–7629.

Ren, K. 2010. Emerging role of astroglia in pain hypersensitivity. The Japanese Dental Science Review 46: 86.

Ren, K., and R. Dubner. 2010. Interactions between the immune and nervous systems in pain. Nature Medicine 16: 1267–1276.

Riethmacher, D., E. Sonnenberg-Riethmacher, V. Brinkmann, T. Yamaai, G.R. Lewin, and C. Birchmeier. 1997. Severe neuropathies in mice with targeted mutations in the ErbB3 receptor. Nature 389: 725–730.

Robitaille, R. 1998. Modulation of synaptic efficacy and synaptic depression by glial cells at the frog neuromuscular junction. Neuron 21: 847–855.

Rodrigues, M.C., A.A. Rodrigues, Jr., L.E. Glover, J. Voltarelli, and C.V. Borlongan. 2012. Peripheral nerve repair with cultured schwann cells: getting closer to the clinics. The Scientific World Journal 2012: 413091.

Sandelin, M., S. Zabihi, L. Liu, G. Wicher, and E.N. Kozlova. 2004. Metastasis-associated S100A4 (Mts1) protein is expressed in subpopulations of sensory and autonomic neurons and in Schwann cells of the adult rat. The Journal of Comparative Neurology 473: 233–243.

Scheib, J., and A. Hoke. 2013. Advances in peripheral nerve regeneration. Nature Reviews. Neurology 9: 668–676.

Scholz, S., and J.S. Kelly. 1974. The characterization of 3H-GABA uptake into the satellite glial cells of rat sensory ganglia. Brain Res 66.

Schon, F., and J.S. Kelly. 1974. Autoradiographic localization of 3H-glutamate and 3H-GABA over satellite glial cells. Brain Res 66: 275–288.

Shah, N.M., M.A. Marchionni, I. Isaacs, P. Stroobant, and D.J. Anderson. 1994. Glial growth factor restricts mammalian neural crest stem cells to a glial fate. Cell 77: 349–360.

Spiegel, I., K. Adamsky, Y. Eshed, R. Milo, H. Sabanay, O. Sarig-Nadir, I. Horresh, S.S. Scherer, M.N. Rasband, and E. Peles. 2007. A central role for Necl4 (SynCAM4) in Schwann cell-axon interaction and myelination. Nature Neuroscience 10: 861–869.

Stassart, R.M., R. Fledrich, V. Velanac, B.G. Brinkmann, M.H. Schwab, D. Meijer, M.W. Sereda, and K.A. Nave. 2013. A role for Schwann cell-derived neuregulin-1 in remyelination. Nature Neuroscience 16: 48–54.

Stewart, H.J., L. Morgan, K.R. Jessen, and R. Mirsky. 1993. Changes in DNA synthesis rate in the Schwann cell lineage *in vivo* are correlated with the precursor—Schwann cell transition and myelination. The European Journal of Neuroscience 5: 1136–1144.

Susuki, K., A.R. Raphael, Y. Ogawa, M.C. Stankewich, E. Peles, W.S. Talbot, and M.N. Rasband. 2011. Schwann cell spectrins modulate peripheral nerve myelination. Proceedings of the National Academy of Sciences of the United States of America 108: 8009–8014.

Susuki, K. 2014. Schwann cell-dependent regulation of peripheral nerve injury and repair. pp. 69–79. *In*: Y.J., Sango K. (ed.). Schwann Cell Development and Pathology. Tokyo, Springer.

Syroid, D.E., P.R. Maycox, P.G. Burrola, N. Liu, D. Wen, K.F. Lee, G. Lemke, and T.J. Kilpatrick. 1996. Cell death in the Schwann cell lineage and its regulation by neuregulin. Proceedings of the National Academy of Sciences of the United States of America 93: 9229–9234.

Tang, X., T.M. Schmidt, C.E. Perez-Leighton, and P. Kofuji. 2010. Inwardly rectifying potassium channel Kir4.1 is responsible for the native inward potassium conductance of satellite glial cells in sensory ganglia. Neuroscience 166: 397–407.

Taveggia, C., G. Zanazzi, A. Petrylak, H. Yano, J. Rosenbluth, S. Einheber, X. Xu, R.M. Esper, J.A. Loeb, P. Shrager, M.V. Chao, D.L. Falls, L. Role, and J.L. Salzer. 2005. Neuregulin-1 type III determines the ensheathment fate of axons. Neuron 47: 681–694.

Todd, K.J., D.S. Auld, and R. Robitaille. 2007. Neurotrophins modulate neuron-glia interactions at a vertebrate synapse. The European Journal of Neuroscience 25: 1287–1296.

Trachtenberg, J.T., and W.J. Thompson. 1996. Schwann cell apoptosis at developing neuromuscular junctions is regulated by glial growth factor. Nature 379: 174–177.

Vega, J.A., M.E. del Valle-Soto, B. Calzada, and J.C. Alvarez-Mendez. 1991. Immunohistochemical localization of S-100 protein subunits (alpha and beta) in dorsal root ganglia of the rat. Cellular and Molecular Biology 37: 173–181.

Vit, J.P., P.T. Ohara, A. Bhargava, K. Kelley, and L. Jasmin. 2008. Silencing the Kir4.1 potassium channel subunit in satellite glial cells of the rat trigeminal ganglion results in pain-like behavior in the absence of nerve injury. The Journal of Neuroscience : The Official Journal of the Society for Neuroscience 28: 4161–4171.

Wegner, M. 2000a. Transcriptional control in myelinating glia: flavors and spices. Glia 31: 1–14.

Wegner, M. 2000b. Transcriptional control in myelinating glia: the basic recipe. Glia 29: 118–123.

Weinberg, H.J., P.S. Spencer, and C.S. Raine. 1975. Aberrant PNS development in dystrophic mice. Brain Research 88: 532–537.

Weinstein, D.E. 1999. The role of Schwann cells in neural regeneration. Neuroscientist 5: 208–216.

Wilson, G.F., and S.Y. Chiu. 1993. Mitogenic factors regulate ion channels in Schwann cells cultured from newborn rat sciatic nerve. The Journal of Physiology 470: 501–520.

Woldeyesus, M.T., S. Britsch, D. Riethmacher, L. Xu, E. Sonnenberg-Riethmacher, F. Abou-Rebyeh, R. Harvey, P. Caroni, and C. Birchmeier. 1999. Peripheral nervous system defects in erbB2 mutants following genetic rescue of heart development. Genes & Development 13: 2538–2548.

Yamazaki, S., H. Ema, G. Karlsson, T. Yamaguchi, H. Miyoshi, S. Shioda, M.M. Taketo, S. Karlsson, A. Iwama, and H. Nakauchi. 2011. Nonmyelinating Schwann cells maintain hematopoietic stem cell hibernation in the bone marrow niche. Cell 147: 1146–1158.

You, S., T. Petrov, P.H. Chung, and T. Gordon. 1997. The expression of the low affinity nerve growth factor receptor in long-term denervated Schwann cells. Glia 20: 87–100.

Zelano, J., S. Plantman, N.P. Hailer, and S. Cullheim. 2009. Altered expression of nectin-like adhesion molecules in the peripheral nerve after sciatic nerve transection. Neuroscience Letters 449: 28–33.

Zhou, X.F., Y.S. Deng, E. Chie, Q. Xue, J.H. Zhong, E.M. McLachlan, R.A. Rush, and C.J. Xian. 1999. Satellite-cell-derived nerve growth factor and neurotrophin-3 are involved in noradrenergic sprouting in the dorsal root ganglia following peripheral nerve injury in the rat. The European Journal of Neuroscience 11: 1711–1722.

5

Neuroendocrine Cells

Zhen He,[1,a,*] *Sherry A. Ferguson,*[1,b] *Tucker A. Patterson*[1,c]
and *Merle G. Paule*[2]

Introduction

The concept of "neuroendocrine cells"

A new era of prominence for neuroendocrine cells (NECs) is beginning: heretofore, the concept or systematic acknowledgement of NECs has not received much attention. It is challenging within a single chapter to describe a coherent nomenclature for NECs in the absence of consistency in the literature. Terms such as "neurosecretory cells" and "NECs" are used to describe the same cell types and there is growing acknowledgment that these cells can be "healthy", "precancerous" or "in various pathological/pathophysiological states" (see later discussions). There currently appears to be a dichotomous or dual-standard in play when defining NECs (see also later comments). Thus, it seems worthy to highlight the state of knowledge concerning NECs and to propose terminology with the hope that it will become more commonly used in discussions of NECs.

---

[1] National Center for Toxicological Research (NCTR), Food and Drug Administration (FDA), 3900 NCTR Rd, Jefferson, AR 72079.
[a] E-mail: Zhen.He@fda.hhs.gov
[b] E-mail: Sherry.Ferguson@fda.hhs.gov
[c] E-mail: tucker.patterson@fda.hhs.gov
[2] Division of Neurotoxicology, HFT-132, National Center for Toxicological Research (NCTR), Food and Drug Administration (FDA), 3900 NCTR Rd, Jefferson, AR 72079.
  E-mail: merle.paule@fda.hhs.gov
* Corresponding author

Here, NECs are defined as those cells that interact with the nervous system and hormones of the endocrine glands or that release hormones into blood in response to a neural stimulus [http://medical-dictionary.thefreedictionary.com/ Neuroendocrine+cell]. The term NEC can thus cover a wide variety of cells. Developmentally, neurons and endocrine cells are derived from different embryonic tissues: the former arising from the ectoderm/neuroectoderm, and which receive and directly transmit signals to neurons or other types of cells to affect specific biological functions; the latter arising from nonectodermal tissue, and indirectly and diffusely controlling their specific biological functions by releasing hormones into the blood stream. For the present discussion those conventional endocrine cells will not be considered as NECs. Here, we consider cells to be NECs in a somewhat arbitrary fashion. Cells, wherever they may reside, that are in physical contact with neurons or their projections and release hormones are taken to be NECs. In addition, there is an increasing trend in the literature to call cells bearing both neuronal and endocrine markers—such as those identified using immunohistochemical approaches—NECs (see later discussion), even if it cannot be verified that they are physically connected to neuronal cells or processes. Accordingly, both of the above classifications of NECs will be utilized in this chapter.

NECs can be located throughout the body both outside and within the central nervous system (CNS) and we will identify them as such. NECs release hormones in response to signals sent by other cells, usually neurons, and the hormones are typically released into the blood, at least locally, where they generally act to maintain homeostasis.

Like other cell types, NECs may malfunction due to various pathological and/or pathophysiological states, such as disease, trauma, infection and intoxication. NECs are, thus, clinically relevant since they can be involved in pathogenic processes and participate in the etiology in a variety of diseases. Abnormal responses by adrenal NECs, for example, are related to the many clinical events. Insulin shock can be elicited by the interruption of splanchnic nerve-adrenal synaptic connections or functions (Hamelink *et al.*, 2002). Sporadic hypoxic events can occur during episodes of sleep apnea and lead to the occurrence of hypertension due to inappropriate adrenal catecholamine secretion (Fletcher, 1997; Dick *et al.*, 2007; EC, 1997; Prabhakar and Peng, 2004). Factors involved with diabetes mellitus and metabolic syndrome symptoms are thought to be attributable to improper adrenal epinephrine secretion in response to changes in blood glucose levels (Ziegler *et al.*, 2012). Neuroendocrine tumors (NETs) may develop from NECs, and it has been proposed that pheochromocytoma, a rare catecholamine-producing neuroendocrine tumor, is derived from the chromaffin NECs in the adrenal medulla. A gradual accumulation of pro-oxidants due to metabolic oxidative stress is thought to lead to proto-oncogene activation, tumor suppressor gene inactivation, DNA damage, and genomic instability leading to pheochromocytoma (Pacak, 2011).

Neuroendocrine system and NECs

In general, the conventional concept of a neuroendocrine system combines aspects of the nervous system with aspects of the endocrine system, highlighting the integration and cooperation between the two. In a broad sense, reference to the neuroendocrine system implies that its fundamental components are the endocrine organs, tissue and cells that respond to nervous system input or activity. The hypothalamus, or at least some subsets of it, and the pineal gland (see below) can be thought of as neuroendocrine organs, while NECs are present throughout the body. As shown in Figure 5.1, NECs both outside and within the CNS should be considered conceptually distinct from the other major types of nerve and endocrine cells.

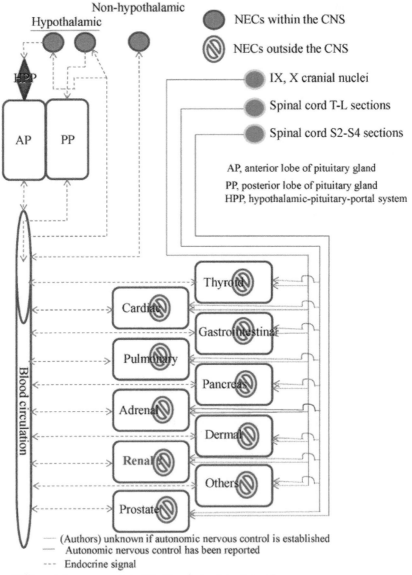

Figure 5.1. Neuroendocrine system & neuroendocrine cells (NECs). The red highlighted text indicates that no NEC has been currently identified within the human kidney.

Cells in the anterior lobe of the pituitary gland that produce thyroid stimulating hormone (TSH), follicle-stimulating hormone (FSH), luteinizing hormone (LH), prolactin (PRL), growth hormone (GH) and adrenocorticotropic hormone (ACTH) are not considered NECs when they are not in physical contact with any neurons in the CNS. Parvocellular neurosecretory cells (NECs) in the hypothalamus control endocrine cells of the anterior pituitary by releasing regulatory hormones via a vascular route called the hypothalamo-hypophyseal portal system (the hypophyseal portal circulation): the hypothalamic capillaries leading to infundibular blood vessels drain into a second capillary bed in the anterior pituitary. The hypothalamic releasing hormones govern the release of pituitary hormones by binding to anterior pituitary endocrine cells after diffusing out of the second capillary bed.

On the other hand, cells in the innermost (medullary) part of the adrenal gland can be referred to as NECs because direct connections exist between sympathetic nerve fibers and those cells. Sympathetic neurons of the autonomic nervous system send descending nerve fibers directly to NECs in the adrenal medullae and stimulate those cells to release adrenal medullary hormones via synaptic input. In this way, the two systems function together, effectively integrating the nervous system and endocrine activation.

Hypothalamic NECs play a pivotal role in balancing the body's hormonal status. As an example, NECs in the hypothalamus produce and release corticotropin-releasing hormone (CRH), which in turn stimulates the anterior lobe of the pituitary gland to produce and secrete ACTH. Thereafter, ACTH activates the adrenal cortex to yield and discharge cortisol. As a negative feedback mechanism, cortisol inhibits the release of both ACTH and CRH. Accordingly, the NECs receive the inhibitory signal from cortisol then act to decrease the subsequent release of hormone from the pituitary endocrine cells. In reality, most control mechanisms are far more complicated. In response to a given stimulus, many types of NECs over a wide area can be simultaneously activated. For example, a meal can result in the simultaneous increase in serum insulin, testosterone, androstenedione and dehydroepiandrosterone, along with reductions in cortisol and 17 alpha hydroxyprogesterone (Parra *et al.*, 1995), resulting in the activation of multiple NECs as well as multiple hypothalamic-pituitary-endocrine glands. In addition, a given NEC can receive control signals from multiple sources. For example, with respect to the lactotrophic axis, prolactin induces electric current in two separate components in the hypothalamic dopamine-NECs, each of which appears to be mediated through distinct signaling pathways (Lyons *et al.*, 2012). To control lactation and other biological functions such as sexual drive, fertility and body weight, dopamine is released from NECs to inhibit prolactin production when serum prolactin increases.

Defining NECs

As mentioned in Section The concept of "neuroendocrine cells", NECs can be defined by their biological functions and/or their expression of both neuronal and endocrine markers, in addition to their physical location and intercellular connections. NECs vary considerably in size, morphology and fine structure, even if they are the same type of cells synthesizing identical hormones (Ekengren *et al.*, 1978). Similar characteristics of these NECs can be classified as discussed below in terms of common cellular and subcellular features.

Morphology of the neuron-NEC connection

Hormone-immunoreactive (-ir) NECs may have rough or smooth surfaces, with or without somatic and dendritic appendages. Both dendrites and perikarya of NECs may form synaptic connections with axons that do not exhibit hormone immunoreactivity.

Electrical properties of NECs

NECs appear to have electrical membrane properties that are similar to those of typical neurons (Kandel, 1964). Analysis of postsynaptic potentials indicates the presence of recurrent collaterals and thus, the concept of the NEC as a cell type whose axons only form contacts with blood vessels (and not neurons or other effector cells) is not supported (Kandel, 1964): the physiological and morphological evidence show that NECs can have multiple axonal branches (Hayward, 1974). There is also a differential expression of AMPA receptors (AMPARs) and NMDA receptors in different types of NECs, and these differences can contribute to differential firing patterns of the NECs in response to physiological stimuli (Stern *et al.*, 1999).

Subcellular structures

One of the essential biological functions of NECs is to manufacture, transport and release cell-specific hormone(s). Accordingly, the organelles involved in the synthesis, packaging and transportation of a specific hormone are generally well-developed in NECs. Critical and pathway-specific molecules involved in the synthesis, packaging and transportation of cell-specific hormones are associated with cellular organelles and can be visualized immunohistochemically or morphologically using light or electron microscopy. Typical ultrastructural signals may include immunolabeling of the synthesized hormone in association with rough endoplasmic reticulum, free ribosomes and hormone-secretory granules. Prohormones are typically synthesized in the endoplasmic reticulum (ER), a type of subcellular organelle characterized by flattened, membrane-enclosed sacs or tubes or cisternae. These ER products move to the Golgi complex (GC), another subcellular organelle consisting of stacks of cisternae connected by tubules and attached to a variety of transporting vehicles. In the GC, prohormones are sorted and packed into secretory granules, which are then sent to their destination. Modern technology makes possible 3-dimensional viewing of these structures (Ladinsky *et al.*, 1999; Rambourg *et al.*, 1974), as living images (Polishchuk *et al.*, 2000; van Rijnsoever *et al.*, 2008) and allows for a determination of the precise location of key molecular components and their association with the subcellular organelles (Rabouille and Klumperman, 2005; Zeuschner *et al.*, 2006). In parallel with morphological approaches, biochemical and genetic analyses have described in detail the relevant molecular machineries that form the secretory and endocytic pathways (Martinez-Alonso *et al.*, 2013).

Immunohistochemical characteristics

Double or triple immunohistochemical labeling approaches can be used to define aspects of the physical connections between presynaptic nerves and hormone-synthesizing NECs. For example, it has been shown by visualizing dopamine-beta-hydroxylase (DBH)- or phenylethanolamine-N-methyltransferase (PNMT)-ir axons that catecholaminergic fibers establish contacts with the dendrites and cell bodies of TRH-ir cells (Liposits *et al.*, 1987).

Pitfalls in defining neuroendocrine cells

In reality, many hormones released by NECs can serve as neurotransmitters. Dopamine is a well-known neurotransmitter that can be secreted into the blood stream where it exerts a hormonal biological function. Conversely, thyrotropin-releasing hormone (TRH), by definition as a hormone, may also function as a neurotransmitter: it has been shown that TRH-ir cells form en passant type or terminal synapses on central epinephrine neurons (Liposits *et al.*, 1987). Thus, distinguishing NECs from neurons may depend upon whether the released molecules enter the circulation and how those molecules affect down-stream effectors. In addition, the time-course over which the biological effects are induced by the released molecules can help defining the cell type: immediate effects suggest neuron-like qualities whereas slow or delayed effects suggest NEC-like qualities.

NECs Outside the CNS

The concept of "NECs outside the CNS" is proposed here in order to distinguish them from "NECs within the CNS", highlighting some substantial differences between them which are still under investigation. Relative to NECs within the CNS, NECs outside the CNS typically display the following features. First, evidence of direct physical connections (synapses) between peripheral nerves and NECs outside the CNS can be elusive. Thus, a NEC outside the CNS may be defined exclusively by its characteristic neuroendocrine marker(s) [see later for citation/evidence]. Second, the embryonic tissue from which NECs outside the CNS derive may be unknown or unidentifiable. Third, NECs outside the CNS, as distinguished from neuroendocrine tumor cells (NETCs), may also reside within pathophysiological sites (Christensen *et al.*, 1988; Miller and Muller, 1995; Oba *et al.*, 2013), highlighting their potential clinical significance in some diseases.

APUD cell series

The acronym "APUD" derives from the words Amine Precursor Uptake and Decarboxylase. In the early 1960s, scientist A. G. E. Pearse coined the term "APUD" to describe a group of cells that share the common function of secreting a low molecular weight polypeptide hormone (Carvalheira *et al.*, 1968; Pearse, 1969) and then expanded the term to include central (hypothalamic and pineal) NECs and pituitary endocrine cells (Pearse, 1975). Nevertheless, some cells in the APUD series can be categorized as NECs. While being primarily utilized in the clinic to detect NETCs, current positron emission tomography (PET) approaches allow for the molecular imaging of cells with APUD features: ^{18}F-dihydroxyphenylalanine (^{18}F-DOPA) and ^{11}C-5-hydroxytroptophan (^{11}C-5-HTP) are PET tracers that can be used to define cells with APUD features because they can identify the uptake of these monoamine precursors (Oberg, 2012; Sundin *et al.*, 2007). In addition, it has been shown that the following peptides with hormone or hormone-like effects are expressed in specific cells in the gastrointestinal tract: somatostatin, pancreatic polypeptide (PP), peptide YY, glucagon, secretin, vasoactive intestinal peptide, gastrin, cholecystokinin (CCK), neurotensin, motilin, gastrin-releasing peptide, substance P, enkephalin and serotonin (Rawdon and Andrew, 1999). Neuron specific enolase is also a common marker for both gastrointestinal endocrine cells and enteric nerves (Bishop *et al.*, 1982), suggesting that at least some of these cells can be categorized as NECs as defined earlier [see Section Introduction]. To follow the logic that well-known hormone-producing cells should not be re-classified and also to embrace the idea that cells meeting the definition of "NECs" should be re-scrutinized, Table 5.1 lists the individual hormones and hormone-like peptides for which specific cell types can be screened for possible categorization as peripheral NECs. Alternatively, a cell bearing a hormone-like peptide (serving as a biomarker) which also expresses a neuron-specific protein may be "exclusively" coined as a "NEC" after ruling out the possibility of NETCs.

Thyroid NECs

Parafollicular cells in the thyroid (also called C cells) are NECs that secrete calcitonin. They reside in the connective tissue adjacent to the thyroid follicles. While they are not homogenously distributed (Rink *et al.*, 2001), they are typically situated basally in the epithelium with no direct contact to the follicular lumen. These cells are large and stain much more weakly than the follicular cells or colloid in conventional histological preparations. There are a limited number of those cells in the thyroid (Rink *et al.*, 2001) and embryologically, they are derived from neural crest cells. Defining these cells using conventional methods involves double labeling of calcitonin- and chromogranin-immunoreactive deposits and their location. These parafollicular cells can also secrete serotonin and several neuroendocrine peptides such as somatostatin and CGRP, suggesting their neuroectoderm origin. Because these cells also express TRH, they may also have a paracrine role in regulating thyroid hormone production. The parafollicular cells may become cancerous, leading to medullary carcinoma of the thyroid.

Pulmonary NECs

Pulmonary NECs are regarded to be components of the pulmonary neuroendocrine system that consist of a specific group of airway epithelial cells (Van Lommel *et al.*, 1999). Pulmonary NECs can be solitary or clustered to form neuroepithelial bodies (NEBs). Having oxygen sensitive chemoreceptors, they serve local and reflex-mediated regulatory functions as well as regulate airway growth and development. Pulmonary NECs have been heavily investigated for over half a century and their existence has been well accepted (Van Lommel *et al.*, 1999). All pulmonary NECs, whether solitary or clustered, are defined by their characteristic core granules that contain one or more bioactive amines and/or neuropeptides (Van Lommel *et al.*, 1999). Identified peptides include, but are not limited to: cholecystokinin, endothelin, peptide YY, helodermin and pituitary adenylate cyclase-activating protein, bombesin (also known as neuromedin B and gastrin-releasing peptide), synaptophysin, chromogranin A, and calcitonin (Van Lommel *et al.*, 1999). Accordingly, the morphological and histochemical characteristics of the pulmonary NECs include immunoreactivities to those molecules in association with the cellular organelles. It is worth noting that the biological functions of pulmonary NECs include serving as

Table 5.1. Regular hormones and hormone-like peptides found in NECs located outside of the CNS.

A) Regular hormones

Hormone	Where produced	Primary/known biological role
Thyroid Stimulating Hormone (TSH)	Pituitary gland (anterior lobe)	Stimulate production of thyroxine (T4) and triiodothyronine (T3)
Follicle-Stimulating Hormone (FSH)	Pituitary gland (anterior lobe)	Regulate development and pubertal maturation
Luteinizing Hormone (LH)	Pituitary gland (anterior lobe)	Trigger ovulation and development of the corpus luteum in females or production of testosterone in males
Prolactin (PRL)	Pituitary gland (anterior lobe)	Stimulate milk production in female mammals
Growth Hormone (GH)	Pituitary gland (anterior lobe)	Stimulate growth, cell reproduction and regeneration
Adrenocorticotropic hormone (ACTH)	Pituitary gland (anterior lobe)	Increase production and release of corticosteroids
Melatonin	Pineal gland	Govern biological and sleep rhythms
Thyroxine	Thyroid gland	Regulate metabolic rate
*Calcitonin	Thyroid gland	Reduce blood calcium
Parathyroid hormone (PTH)	Parathyroid glands	Increase blood calcium
Insulin	Pancreas	Control of blood sugar
Cortisol, Norepinephrine, Epinephrine, Adrenaline	Adrenal glands	Regulate biological functions in response to emotional changes and stress
Androgens	Testes, ovaries and adrenal glands	Control male secondary sexual features, sexual arousal in males and females
Estrogens	Testes, ovaries and adrenal glands	Involved in breast development and menarche in females
Progesterone	Testes, ovaries and adrenal glands	Involved in preparation of uterus for implantation of fertilized egg

*Calcitonin is highlighted indicating that this is considered as a regular hormone but it may serve as a biomarker for labeling a NEC.

Table 5.1. contd....

Table 5.1. contd.

B) Hormone-like peptides

Hormone-like peptides	Where produced	Primary/known biological role
Islet-associated protein-2 (IA-2) and IA-2β	GI NECs	Neuroendocrine-specific protein tyrosine phosphatases (PTPs)/insulin secretion-relevant
Pancreatic polypeptide	Pancreas and GI NECs	Regulate pancreatic secretion
Serotonin	GI-NECs	Movement of food across the GI tract
Cholecystokinin/gastrin-1 (CCK)	GI-NECs	Stimulate the digestion of fat and protein
Somatostatin	GI-NECs	Regulate gastric acid and pepsin secretion
Secretin	GI-NECs	Regulate water homeostasis
Gastric inhibitory polypeptide	GI-NECs	Induce insulin secretion and regulate GI motility and acid secretion
Glucagon-like peptide-1	GI-NECs	Regulate acid secretion and gastric motility
Glucagon	GI-NECs	Maintain blood glucose homeostasis
Gastrin-releasing peptide (GRE)	GI-NECs	Stimulate the release of Gastrin
Gastrin	GI-NECs/GI, pancreas	Stimulate acid secretion and gastric motility
Motilin	GI-NECs	Control inter-digestive migrating contractions
Vasoactive intestinal peptide	GI-NECs/pancreas	Regulate cardiovascular functions and modulate GI motility

GI, gastrointestinal; NEC, neuroendocrine cells

chemoreceptors that react to local/micro-environmental changes. Pulmonary NECs have ultrastructures capable of sensing changes in the airway lumen (Van Lommel *et al.*, 1999) and they also have synaptic structures with surrounding peripheral nervous system projections (Van Lommel *et al.*, 1999). The ultrastructural evidence suggests that pulmonary NECs have the potential to secrete hormones in response to either airway luminal stimulation, neural input or both. Accordingly, the pulmonary NECs, or at least a subset of them, qualify as NECs according to the definition. It has been recommended that the solitary and clustered pulmonary NECs be called "pulmonary NECs" and "NEBs" respectively, to avoid confusion. NEB-associated pulmonary NECs are distinguished from isolated pulmonary NECs by their innervation and by their specific patterns of gene expression. The origin of pulmonary NECs remains unknown but may likely be the neuroectoderm and/or endodermis.

Cardiac NECs

The topic of cardiac NECs was reviewed nearly three decades ago with the literature indicating that these cells likely participate in both local and distant regulation of cardiac metabolism and function (Galoyan, 1986). The term "cardioactive neurohormones of the hypothalamus" was coined in reference to two coronary dilatory compounds from the rat hypothalamus because: (a) when injected intravenously in minute doses they cause a dilation of coronary vessels in cats; (b) they are produced by the neurosecretory cells of the hypothalamus; and (c) the heart is the target organ (Galoyan, 1986). The immunoreactivity of neurohormone "G" (NHG) was found around cardiac ganglion cells and in close topographical contact with coronary vessels and capillaries (Abrahamyan *et al.*, 2002) indicating that there is the potential that NHG may serve as a hormone that could be secreted directly into the local or systemic circulation. In addition, cardiac ganglion cells may be characterized as having both neuronal and neuroendocrine properties (Galoyan, 1986).

Pancreatic NECs

Chromogranin A, a neuroendocrine cell marker, is present in early human fetuses when the number of chromogranin A-immunoreactive cells surpasses that of insulin- and glucagon-containing cells (Krivova *et al.*, 2014). In addition, the appearance of chromogranin A-ir cells precedes the appearance of insulin- and glucagon-producing cells in the pancreatic islets and ducts during all periods of gestation (Krivova *et al.*, 2014). Thus, the timing of the appearance of chromogranin A suggests a role in the development of the endocrine part of the pancreas as well as the pancreatic NECs. In contrast to adrenal NECs that originate from the neuroectoderm, the pancreas arises from a dorsal and a ventral bud from the endoderm which fuse together to form the single organ (Slack, 1995). In reality, pancreatic β-cells have a neuroendocrine function to release insulin and GABA (γ-aminobutyric acid, a well-known inhibitory neurotransmitter) (Cuttitta *et al.*, 2013; L'Amoreaux *et al.*, 2010), suggesting that a subset of the β-cells are candidate NECs. Other pancreatic NECs can be defined using immunoreactivity to ghrelin and somatostatin (Kasacka *et al.*, 2012), both of which function as either neuropeptides or hormones and are considered to be NEC biomarkers. In adult humans, cells expressing ghrelin are very scarce (approximately 1%) (Wierup *et al.*, 2002), and this scarcity is a potential obstacle to their detection and identification using morphological, immunohistochemical and ultrastructural criteria.

Adrenal medullar NECs

In humans, chromaffinoblasts begin to migrate into the cortex of the adrenal in the second embryonic month. *Per* definition, mature chromaffin cells (ECs)—or at least a subset of them in the innermost part of the medulla—fully qualify as NECs. On the basis of embryonic origin and physiological, biochemical and morphological characteristics, the definition of NEC includes neuroectodermal origin, pre-ganglionic sympathetic innervation, and the capability to synthesize, store and release catecholamines (Benedeczky, 1983). The polygonal chromaffin cells form rounded islands as well as shorter or longer cords alongside sinusoidal vessels (Benedeczky, 1983), suggesting a vascular exit for their secretory products. The best and most extensively used method to identify ECs is the "chromate staining" method which produces colored chromaffin reaction-products within the cytoplasm (Benedeczky, 1983). Under the electronic microscope the adreno-medullary cells display chromaffin granules of various sizes (Koval *et al.*, 2001), in addition to a well-developed Golgi apparatus and abundant rough-surfaced endoplasmic reticulum. These cells receive synaptic input from the splanchnic nerve and release epinephrine and norepinephrine into the circulation. In studies using ^{14}C-labeled tyrosine, tyrosine is converted to dopamine, then norepinephrine and thereafter epinephrine (Goodall and Kirshner, 1958). Concentrations of the catecholamines epinephrine and norepinephrine surge in the serum in response to stress (Davis *et al.*, 2000; Gerlo *et al.*, 1991; Gonzalez-Trapaga *et al.*, 2000). The adrenal medulla is a major source for circulating norepinephrine and the sole source for peripheral epinephrine (Goldstein *et al.*, 1983; Vollmer, 1996), highlighting the importance of these NECs.

Pheochromocytoma cell lines are available from both human (Pfragner *et al.*, 1998) and rat adrenal origins (PC-12). Ironically, these cells express all of the necessary biomarkers to qualify as NECs including the presence of neuroendocrine granules, positive immunoreactivity to chromogranin and related peptides, neuron specific enolase and vasoactive intestinal peptide (Pfragner *et al.*, 1998). Furthermore, while some of these NETCs may grow aggressively (poorly differentiated neuroendocrine carcinoma), a great majority of these NETCs are relatively slow growing and remain well differentiated (Grozinsky-Glasberg *et al.*, 2008), suggesting a future research need for the additional proof required to distinguish NECs from NETCs.

Renal NECs

The kidney is an organ representative of those in which there has been no unequivocal demonstration of the presence of intrinsic NECs, even though there have been significant efforts into how renal neuroendocrine tumors develop (Guy *et al.*, 1999). Using an immunohistochemical approach for two NEC biomarkers, chromogranin and serotonin, the presence of NECs in the kidneys of fetuses, infants, children and adults could not be demonstrated within the renal parenchyma or hilum.

Gastrointestinal NECs

Enterochromaffin (EC) cells are candidate NECs that are nearly omnipresent throughout the gastrointestinal epithelium (Sjolund *et al.*, 1983). Other potential NECs cells that have been identified in this location include enteroglucagon/peptide YY (PYY) (EG/L cells), neurotensin (N cells), somatostatin (SST; D cells), gastrin (CCK cells), and pro-melanocyte stimulating hormone (MSH) cells (Sjolund *et al.*, 1983). EC cells can be found at the base of intestinal crypts (Sjolund *et al.*, 1983); these cells have specific identifying characteristics including: immunoreactivity to tryptophan hydroxylase (TPH-ir) and chromogranin A (CgA-ir); and the presence of pleomorphic granules (oblong, ovoid, kidney shaped, triangular, or U shaped) that can be ultrastructurally identified (Kidd *et al.*, 2006). These cells display infrabasal extensions suggesting a neural nature and/or potential innervation by peripheral nerves (Challacombe and Wheeler, 1983).

Prostatic NECs

NECs in the prostate gland are scattered within the grandular epithelium and can be defined by triple immunoreactivity to chromofranin A & B and secretogranin II (Schmid *et al.*, 1994) and their geographical and morphologic features. In humans, differentiation diagnostic tests can separate normal NECs from cancerous cells because prostatic neuroendocrine tumor cells may also express all of the aforementioned proteins NEC markers. Interestingly, expression of chromofranin B is predominant in prostate cancer cells, noticeably in poorly differentiated tumors (Schmid *et al.*, 1994).

Dermal NECs

Attempts to define NECs in skin can be tracked over half a century when McGavran (McGavran, 1964) reported chromaffin cells in the human dermis. Dermal chromaffin cells contain membrane-limited, electron-opaque granules that are similar to those in the medullary cells of the human adrenal that are concentrated at one pole of the cell and adjacent to this pole there are three unmyelinated axons within a Schwann cell, suggesting direct innervation from peripheral nerves.

Other peripheral NECs

Using the key words "neuroendocrine cell" and the names of specific organs, hundreds of publications can be found, most related to neoplasms. One of the reasons for this is that neoplastic neuroendocrine cells can develop in almost all tissues and organs, indicating their clinical significance and prevalence. On the other hand, most peripheral NECs are very sparsely distributed throughout a host organ, highlighting the difficulty in identifying them using the aforementioned qualification criteria. Nevertheless, given the continuous advances being made in science and technology, it is expected that an ever increasing number of peripheral NECs will be identified.

NECs within the CNS

Hypothalamic NECs

The hypothalamus and the pituitary gland are the major centers of neuroendocrine integration in the body. Hypothalamic NECs release hormones into the blood to control the secretion of pituitary hormones. These hormones include thyrotropin-releasing hormone (TRH), gonadotropin-releasing hormone (GnRH), growth hormone-releasing hormone (GHRH), corticotropin-releasing hormone (CRH), somatostatin and dopamine. All are released into capillaries and travel immediately—via portal veins —to a second capillary bed in the anterior lobe of the pituitary where they exert their biological effects (Table 5.2). Instead of being released into the hypothalamic portal vein, two other hypothalamic hormones (vasopressin and oxytocin) are transported in neurons to the posterior lobe of the pituitary where they are released into the circulation. Hypothetically, these NECs are vulnerable to axonal damage that originates from the vasculature such as that caused by chemical toxicants as there is no effective blood-brain barrier in the areas where the NECs secrete their hormones (McKinley *et al.*, 1983; Simpson, 1981). For example, TRH-ir NECs sustain extensive axonal damage following development of experimental allergic encephalomyelitis, the damage being attributable to mechanical or chemical factors in association with inflammatory foci (White *et al.*, 1990).

Table 5.2. Hormones produced by NECs within the CNS.

Hormone	Where produced	Primary biological role
TRH	Hypothalamic PVN	Stimulate release of TSH
GnRH	The medial preoptic area of hypothalamus	Stimulate release of FSH & LH
GHRH	Hypothalamic ARN	Stimulate release of GH
CRH	Hypothalamic PVN and SCN	Stimulate release of ACTH
Somatostatin	Hypothalamic ARN	Inhibit the release of GH & TSH
Dopamine	Hypothalamic ARN	Inhibit the release of PRL
Vasopressin	Hypothalamic PVN and SON	Antidiuretic
Oxytocin	Hypothalamic PVN and SON	Stimulate contractions of the uterus and the release of milk
Melanocyte-stimulating hormone (MSH)	Hypothalamic ARN and intermediate lobe of the pituitary gland	Stimulate the production and release of melanin (melanogenesis)
Melatonin	Pineal body or gland	Control the daily night-day cycle

PVN, paraventricular nucleus; SON, supraoptic nucleus; SCN, suprachiasmatic nuclei; ARN, arcuate nucleus.

Neuroendocrine cells producing GHRH, somatostatin and dopamine reside in the arcuate nucleus (ARN) and control the secretion of the pituitary hormones, GH, TSH and PRL, respectively (Table 5.2). TRH- and CRH-releasing NECs, in controlling pituitary TSH or ACTH secretion, are located in the hypothalamic paraventricular nucleus (PVN) (Aizawa and Greer, 1981; Jackson and Reichlin, 1979; Lechan and Jackson, 1982; Lechan *et al.*, 1983; Lennard *et al.*, 1993; Martin and Reichlin, 1972; Vale *et al.*, 1981). The NECs synthesizing GnRH for the control of pituitary FSH and LH release exist in the hypothalamic medial preoptic area and septal region (Merchenthaler *et al.*, 1980). While these cells originate from different locations in the hypothalamus, all of their axons terminate at the median eminence and secrete their hormones (GHRH, somatostatin and dopamine, TRH, CRH and GnRH) into the portal capillary system through which secretion of pituitary hormones is controlled.

There are other two hypothalamic NECs, the oxytocin- and vasopressin-NECs, that exhibit different characteristics both in their anatomy (axons' ending in the posterior pituitary lobe other than the median eminence) and their down-stream molecules. Unlike the other hypothalamic hormones that target an endocrine organ (i.e., pituitary gland) and stimulate the production of pituitary hormones, oxytocin and vasopressin directly serve as the circulatory hormones to affect their targeted organs, such as the uterus and the kidney. Finally, the NECs producing melanocyte-stimulating hormone (MSH) remain ill-defined but acknowledged (see later).

NECs producing TRH

These cells may have either rough or smooth surfaces with or without somatic and dendritic appendages (Liposits *et al.*, 1987). Ultrastructural analysis reveals the following features: TRH-immunolabel associated with rough endoplasmic reticulum, free ribosomes and neurosecretory granules as well as non-labeled axonal synaptic connections onto both cellular dendrites and perikarya (Liposits *et al.*, 1987). Localization of dopamine-beta-hydroxylase (DBH)- or phenylethanolamine-N-methyltransferase (PNMT)-ir axons and TRH-synthesizing cells demonstrates that catecholaminergic fibers establish contacts with the dendrites and cell bodies of the TRH-ir cells.

These cells reside in the paraventricular nucleus (PVN) and the suprachiasmatic nucleus. They are responsible for the endocrine control of hypophyseal ACTH release. Using an immunofluorescent approach, NECs producing CRH varied in size by less than 25 microns in diameter in both mice and rats (Smith *et al.*, 2014; Wittmann *et al.*, 2005). Cocaine- and amphetamine-regulated transcript (CART) is a neuropeptide widely expressed in the brain and CART-containing axons heavily innervate the PVN and form synaptic connections with the surface of hypophysiotropic NECs that produce CRH (Sarkar *et al.*, 2004). CART increases CRH mRNA levels and the stereotaxic central administration of CART into the PVN causes a rapid increase in plasma ACTH and corticosterone levels through induction of CRH release (Smith *et al.*, 2004). Anatomical and functional studies suggest that CART peptides may have a role in the regulation of neuroendocrine and autonomic responses during stress by their involvement in the control of the hypothalamic-pituitary-adrenal axis (Balkan *et al.*, 2012).

NECs producing GnRH

Cells synthesizing gonadotropin-releasing hormone [GnRH also known as luteinizing-hormone-releasing hormone (LHRH) and luliberin] are primarily located in the medial preoptic area but are small in number (Ajika, 1979). These cells vary between 10–13 microns in diameter and are oval shaped with beaded string-like processes toward the organum vasculosum of the lamina terminalis (OVLT). LHRH-positive terminals are densely localized in the OVLT with the beaded string-like axons terminating most often close to blood vessels (Ajika, 1979). Ultramicroscopically, the buttons of the axons contain many LHRH-positive granules with diameters of about 100 nm. Some of these terminals terminate directly on the basement membrane of the pericapillary space, suggesting a biological function to influence release of pituitary FSH and/or LH via secretion of GnRH into the hypothalamus-pituitary portal circulation. In terms of clinical relevance, both GnRH agonists and antagonists have been developed for the treatment of hormone-dependent diseases via their effects on the hypothalamic-pituitary-gonadal axis (Nakata *et al.*, 2014).

NECs producing somatostatin

These cells reside in the ventromedial nucleus of the hypothalamus and project to the median eminence where somatostatin is released via axons into the hypophyseal portal circulation as mentioned previously. A negative feedback mechanism is in operation such that these cells respond to high circulating concentrations of growth hormone and somatomedins by increasing the release of somatostatin to reduce the rate of secretion of growth hormone. Somatostatin may also serve as a neurotransmitter and/or modulator. There are populations of somatostatin neurons in the arcuate and periventricular nuclei (Krulich *et al.*, 1977), the periventricular areas of the preoptic area and anterior hypothalamus (Hoffman and Hayes, 1979), the hippocampus, cortex and striatum (Billova *et al.*, 2007), and the brainstem nucleus of the solitary tract (Maley and Elde, 1982), suggesting a role for somatostatin and somatostatin-producing neuroendocrine neuronal cells in multiple biological functions in addition to the regulation of growth hormone.

NECs producing somatostatin, in addition to NECs producing GHRH, control the ultradian rhythm of GH secretion via sophisticated patterns of release of somatostatin into the hypothalamic-pituitary portal circulation. In addition, the hypothalamic NECs producing somatostatin integrate information from noradrenergic projections from the ventrolateral nucleus of the solitary tract, locus coeruleus (LC) and the parabrachial nucleus (PBN) (Iqbal *et al.*, 2005).

NECs producing dopamine

Dopamine-producing NECs responsible for governing release of the pituitary PRL reside in the PVN and in the dorsomedial portion of the middle ARN in rats (Freeman *et al.*, 2000; Lerant and Freeman, 1998). However, "NECs" are named such only when their characteristics meet the previous definition criteria. It is very difficult to differentiate individual dopaminergic neurons that may play a role in neuroendocrine regulation from the NECs that are biologically and morphologically distinguishable. For example, bilateral electrolytic lesions of the A8, A9 and A10 regions of the substantia nigra reduce dopamine levels in the hypothalamic median eminence by 40% (Kizer *et al.*, 1976), implying that many dopamine-producing cells other than those residing within the hypothalamus may also be NECs. Recall that the median eminence is the area from which NECs secrete hormones into the portal vessels to control the pituitary endocrine cells. This also suggests that a subset of dopamine-containing axons in the hypothalamic median eminence originate from neurons rather than from NECs. It appears that the dopamine-producing cells located mostly in the dorsomedial part of the arcuate nucleus are NECs because they project to the external zone of the median eminence (Kawano and Daikoku, 1987) from where dopamine is released into the hypothalamic portal circulation. The dopamine-producing cells in the periventricular nucleus and in the rostral arcuate nucleus, on the other hand, project to both the intermediate and neural lobe of the pituitary gland (Holzbauer and Racke, 1985). Accordingly, determining whether these cells are NECs remains elusive.

NECs producing MSH

Prudently, only subsets of cells that produce MSH can be called NECs. The following evidence supports the notion that some MSH-producing cells are neuroendocrine rather than endocrine cells: (a) many are located in the hypothalamus (Kawai *et al.*, 1984), a part of the central nervous system; (b) in addition to being MSH-immunoreactive, these cells also express neuron-specific marker(s) such as beta-endorphin (Micevych and Elde, 1982) and gamma-MSH is also present at the synapse-like structure of the nerve terminal where it appears to function as a neurotransmitter or a neuromodulator (Osamura *et al.*, 1982); (c) these cells are under the direct control of central neurons and it appears that release of MSH by these cells is under the control of neurotransmitter(s) (Tomiko *et al.*, 1983); and most importantly (d) MSH-containing axons project to the median eminence (Daikoku *et al.*, 1985) suggesting that MSH is released into blood stream via the hypothalamic pituitary portal circulation. Alternatively, the following observations argue against the statement above, suggesting that at least some MSH-producing cells should be classified as endocrine cells: first, MSH (or intermedins) is also produced by cells in the intermediate lobe of the pituitary gland, a well-accepted endocrine gland; second, MSH belongs to a group of peptides called the melanocortins which include ACTH, α-MSH, β-MSH and γ-MSH and ACTH, an obvious hypophyseal hormone, and they can be co-localized with MSH within the same cell (Osamura *et al.*, 1982); and finally, these peptides are all cleavage products of a larger precursor peptide called pro-opiomelanocortin (POMC).

NECs producing oxytocin

These cells are located primarily in the paraventricular (PVN) and supraoptic nuclei (SON) of the hypothalamus (Vandesande and Dierickx, 1975) and they project axons down the infundibulum to terminals in the posterior pituitary. The perikarya of these cells do not appear to be morphologically distinct from NECs producing vasopressin. Cell diameter varies from 15 ~ 25 microns in rats as determined using an immunofluorescent approach (Richards *et al.*, 2005). Differences in immunoreactivity to oxytocin and vasopressin and electrophysiological characteristics distinguish these cell types from each other (Stern *et al.*, 1999). Interestingly, oxytocin is one of the few hormones employing a positive feedback mechanism: uterine contractions stimulate the release of oxytocin from the posterior pituitary, which in turn, increases uterine contractions. This positive feedback loop continues throughout labor.

As previously mentioned, the cell bodies of these cells can be located in the PVN and the SON of the hypothalamus (Vandesande and Dierickx, 1975) with axons projecting to the posterior pituitary. Morphologic and geometric characteristics of oxytocin and vasopressin-NECs are similar, yet that exhibit fundamental differences in their biology, biochemistry and electrophysiology.

Non-hypothalamic NECs

Pinealocytes. Mammalian pinealocytes are NECs, at least in concept: in addition to residing in the pineal gland in the center of the brain, they produce hormones—

including melatonin—that are released into the circulation, and they also receive direct neuronal input that controls their functions. Ironically, the structure has two well accepted names: pineal body and pineal gland. The former addresses its anatomical features (pine cone shape) while the latter highlights its characteristic of producing several important hormones. In fact, the pineal body or gland is an organ located in the CNS, but independent from isolation via the blood-brain-barrier (BBB). It is probably appropriate at this time to now classify the pineal body or gland as a neuroendocrine organ because its major components are neuroendocrine cells, pinealocytes. In addition to the hormone-producing characteristics as part of defining NECs, neuronal control of pinealocytes can be traced to the suprachiasmatic nuclei (SCN) of the hypothalamus through a complex pathway including the superior cervical ganglion (SCG), intermediolateral nucleus of the upper thoracic spinal cord (IML), and parvicellular subdivisions of the hypothalamic paraventricular nucleus (PVN) (Larsen *et al.*, 1998). Immunoreactivity of dopamine-beta-hydroxylase (DBH), essential for the formation of the melatonin synthesis-regulating substance noradrenaline (NA), is located in pineal nerve fibers coming from the superior cervical ganglia (Schroder and Vollrath, 1985). It is indeed intriguing that a "central" neuroendocrine organ, the pineal body or gland, is directly controlled by peripheral nerves via the SCG and IML. Simonneaux and Ribelayga (2003) have formulated a scheme of multiple controls over the pineal gland including central inputs from the lateral hypothalamus and the habenular nucleus, peripheral inputs from the trigeminal ganglia and the pterygopalatine ganglia, and hormonal inputs from the circulation as well as paracrine inputs in addition to the major input initiated by visual (light) information (Simonneaux and Ribelayga, 2003). In addition, immunohistochemical characteristics of the pineal gland include enriched deposits of phosphorylated neurofilaments (neuron-specific) and glial fibrillary acidic protein (GFAP, astrocyte-specific) [http://www.microscopyu. com/staticgallery/fluorescence/ratbrain29.html], demonstrating the complexity of the pineal neuroendocrine organ in not only how it is constructed, but also in how it is regulated and functions.

Endocrine Disruptors and Toxicity to Neuroendocrine Cells

Concept and outline of the field

There is accumulating evidence that endocrine disruptors may pathophysiologically affect NECs peripherally and/or centrally, directly or indirectly, and lead to subclinical or clinical consequences. For example, bisphenol A (BPA), an endocrine disruptor at high doses, has been linked to the development of cancer (Jenkins *et al.*, 2009; Prins *et al.*, 2008a; Prins *et al.*, 2008b), obesity (Newbold *et al.*, 2009b; Newbold *et al.*, 2007), abnormalities in reproductive organ functions (Newbold *et al.*, 2009a), and neurological impairment (Leranth *et al.*, 2008; Zhou *et al.*, 2009). NECs are the primary neuronal targets of endocrine disruptors and they seem to be most susceptible to disruption during prenatal and early postnatal developmental periods (El Majdoubi, 2011). In the periphery, endocrine disruptors have been associated with the pathogenesis of insulin resistance (Polyzos *et al.*, 2012). Centrally, endocrine disrupting chemicals can alter the neuroendocrine GnRH regulatory network at all hierarchical levels: at sufficient doses, genistein and BPA can suppress inhibitory components and activate stimulatory components of the GnRH network; dioxins can inhibit the GnRH network (Mueller and Heger, 2014); prenatal exposure to polychlorinated biphenyls (PCBs), other endocrine disrupting chemicals, can alter gene expression in the hypothalamic preoptic area and affect brain-derived neurotrophic factor, GABA(B) receptors-1 and -2, IGF-1, kisspeptin receptor, NMDA receptor subunits NR2b and NR2c, prodynorphin, and TGFα (Dickerson *et al.*, 2011). Ethinyl estradiol (EE2), a potent synthetic estrogen, increases the number of GnRH-ir cells and GnRH fibers, decreases the size of GnRH-ir cells and modifies their migration pattern (Vosges *et al.*, 2010), highlighting that estrogens, estrogen-like compounds, and endocrine disruptors can affect central NECs. Nevertheless, there is still a knowledge gap as to how endocrine disruptors can affect NECs to promote the development of disease. In the following sections we describe methods for assessing the potential neuroendocrine toxicities associated with exposure to BPA.

Anatomy of the preoptic area

Endocrine active compounds are found extensively in the environment and could, with sufficient exposure, interfere with normal development and/or function. Within the central nervous system are several sexually dimorphic structures that are critically dependent on developmentally appropriate hormonal concentrations. Alterations of those concentrations, via exposure to an endocrine active compound or compounds, could alter the ultimate expression of sexual dimorphism. One of the best known and characterized of the sexually dimorphic brain structures is the sexually dimorphic nucleus of the preoptic area (SDN-POA), a cluster of cells located in the hypothalamic preoptic area of rats. Larger in males, the SDN-POA (or a similarly located structure) is present in many other mammals, including sheep and ferrets (Baum *et al.*, 1996; Roselli and Stormshak, 2010), and humans (Baum *et al.*, 1996; Garcia-Falgueras *et al.*, 2011; Roselli and Stormshak, 2010; Swaab and Fliers, 1985).

Initial measurements of the SDN-POA were conducted using the Nissl stain method which delineated the SDN-POA boundaries using the negatively charged RNA blue dye with thionin or cresyl violet. As demonstrated in Figure 5.2, the SDN-POA can be visualized and the borders defined using a dense blue stain (highlighted in yellow) and its surrounding structures that serve as anatomical land marks. These include the anterior commissure, the 3rd ventricle (3V), optic chiasm (OX) and superchiasmatic nucleus (SN). In addition, there is increasing evidence that the calbindin-D28K (CB28) immunoreactivity-delineated nucleus-like structure located in the preoptic area can be defined at the SDN-POA and used to determine volume. This is partially due to the fact that CB28 immunoreactivity provides a clearer boundary that is more easily distinguishable from the surrounding CB28 immunoreactivity-negative structures (He *et al.*, 2013a; He *et al.*, 2013b; He *et al.*, 2012). Also, the SDN-POA-like area in mice cannot be delineated using Nissl stain (Orikasa and Sakuma, 2010; Young, 1982), but is distinguishable using CB28 immunoreactivity (Bodo and Rissman, 2008; Budefeld *et al.*, 2008; Edelmann *et al.*, 2007; Orikasa and Sakuma,

2010). Initially, the CB28-delineated SDN-POA was considered a subdivision of the SDN-POA as determined using the Nissl method (Sickel and McCarthy, 2000), and our findings are consistent with that interpretation. Using the Nissl approach the longitudinal axis of the male SDN-POA extends 540–630 μm, whereas in the female it has a length of 180–270 μm. Using the CB28 immunoreactivity approach a clear calbindin-D28K-positive cell mass is recognizable over a length of 90–360 μm, with a range of 180–360 μm in males and a range of 90–270 μm in females (He *et al.*, 2013b).

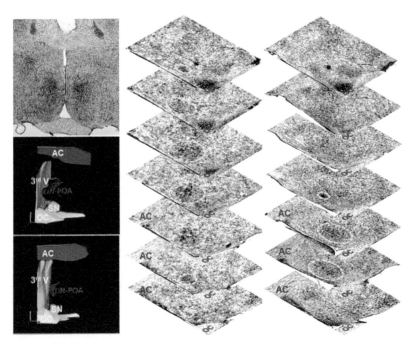

Figure 5.2. Serial brain sections and three-dimensional view of the male (middle image in far left column and middle series) and female (last image in far left column and right series) rat sexually dimorphic nucleus of the preoptic area (SDN-POA). Landmark structures that are also found throughout serial sections containing the SDN-POA are labeled including the 3rd ventricle (3V), anterior commissure (AC), suprachiasmatic nucleus (SON) and the optic chiasm (OX). These structures were traced and their images used in constructing the three-dimensional images. The yellow dotted circle highlights the SDN-POA. The Stereo Investigation System (MBF Bioscience, Williston, VT) was used to reconstruct these views of the SDN-POA and its surrounding anatomical structures from seven sequential slices stained with thionin.

Sex hormones and the effects of estrogen-like compounds on the SDN-POA

The size of the SDN-POA is sensitive to exogenous sex hormones and estrogen-like compounds during development. Estrogen agonists like diethylstilbestrol increase the sexually dimorphic nucleus of the preoptic area volume in female rats (Faber *et al.*, 1993). On the other hand, estrogen antagonists like tamoxifen decrease the volume of the male SDN-POA (Dohler *et al.*, 1984; Vancutsem and Roessler, 1997). Lifetime dietary exposure to genistein (5–500 ppm) or nonylphenol (25–750 ppm) increased SDN-POA volume in adult male, but not female, rats (Scallet *et al.*, 2004). A study in our laboratory demonstrated that gestational treatment of pregnant dams followed by direct treatment of the pups after birth with low doses of the putative estrogen-like compound bisphenol A significantly increased the SDN-POA volume of postnatal day 21 male rats, but had no effect in same-age females (He *et al.*, 2012). As expected, the reference estrogen, ethinyl estradiol (EE2 at 5 µg/kg/day), increased the volume of the SDN-POA of postnatal day (PND) 21 females and the higher ethinyl estradiol dose of 10 µg/kg/day also increased the volume of the SDN-POA of PND 21 males (He *et al.*, 2012).

At the cellular level, several developmental mechanisms that shape the SDN-POA have been proposed including neurogenesis, migration and apoptosis. Estrogen and estrogen-disruptors may affect the development of the SDN-POA through these cellular mechanisms. The neurogenesis mechanism is thought to involve the activities of stem cells residing within a stem cell reservoir such as the 3rd ventricle stem cell niche or residing sporadically within the brain parenchyma. The migration mechanism implies that changes in cellular migration speed and direction affect the size of the SDN-POA, as well as the number of cells found therein. For example, progenitor cells arising from the stem cell niche or surrounding area move toward the SDN-POA, leading to an increase in cell number and size. Alternatively, cells residing within the SDN-POA, such as calbindin immunoreactive cells, moving radially away from the center of the SDN-POA may expand its volume. The apoptosis mechanism implies that female rats exhibit more apoptotic cells within the SDN-POA than males leading to a smaller size of the nucleus. As a corollary of that mechanism, inhibition or disruption of apoptosis in females would lead to an enlarged or masculinized SDN-POA.

In our recent studies, we have identified a stem cell reservoir at the rostral end of the 3rd ventricle, very close to two sexually-dimorphic brain structures (Figure 5.3) (He *et al.*, 2013a): the anteroventral periventricular nucleus of the hypothalamus (AVPV) and the SDN-POA. We have demonstrated that the activities of stem cells in the hypothalamic region including the SDN-POA at PND 21 are robust and that importantly, neural stem cells in this area maintain their capacity to proliferate (Figure 5.4) (He *et al.*, 2013a). These findings indicate that exogenous estrogen treatment, and potentially exposure to estrogen-disruptors, may increase hypothalamic neurogenesis, which in turn reshapes the SDN-POA.

Future studies of estrogen's effect on central NECs

Important questions remain as to how estrogens and estrogen-like compounds can affect central NECs like the GnRH-ir positive cells in the hypothalamus. In rodents, prenatal or postnatal exposure to BPA, a compound that at relatively high doses can have estrogen-like effects, can increase the number of kisspeptin positive cells in the female AVPV and the calbindin-positive cells in the male SDN-POA (Naule *et al.*, 2014; Patisaul *et al.*, 2007), indicating that neural stem cell activities may play a role. BPA can increase the lordosis quotient in naive females, elevate systemic estradiol levels, and exacerbate responses normally induced by ovarian estradiol during the

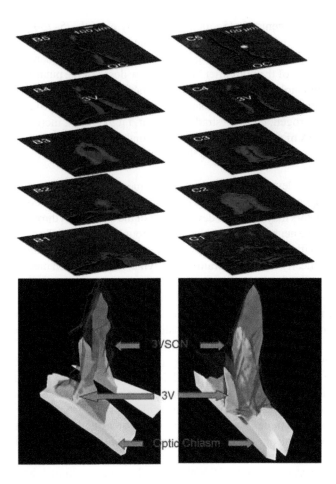

Figure 5.3. Mapping and reconstructing the 3rd Ventricle Stem Cell Niche, weanling vs. adult. Images in the left (upper) series were acquired from a PND21 male rat and images in the right (upper) series were taken from an adult (PND110) male rat, demonstrating the nestin-positive region/cells (red label) that extend into the parenchyma of the hypothalamus (~ 0.1 mm or more from the ventricular wall) at the rostral end of the 3rd ventricle, namely the 3rd ventricle stem cell niche (3VSCN). Distance between the 2 adjacent brain slices is 90 μm. These sequential images were employed to reconstruct 3-dimensional images of the 3VSCN (bottom left and right panels) using the Stereo Investigation software from MBF Bioscience. SVZ, subventricular zone; OC, optic chiasm; 3V, the 3rd ventricle; DAPI-staining, blue labelings.

postnatal/prepubertal period (Naule *et al.*, 2014). On the other hand, neither the number of GnRH NECs nor GnRH NEC activation is altered by BPA (Naule *et al.*, 2014; Patisaul *et al.*, 2007). Accordingly, BPA at the doses (0.25 mg injection x 4 over postnatal days 1–2 or perinatal exposure at 0.05 ~ 5 mg/kg/day) used in those studies does not appear to have a direct impact on GnRH NECs at the cellular level. Future studies may benefit by focusing on the connections between the NECs (GnRH-ir cells) and the up-stream neurons that regulate them. Of considerable interest are the molecular signaling pathways underlying the effects of BPA and similar compounds.

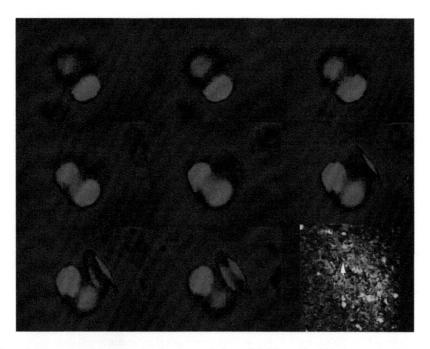

Figure 5.4. Representative images showing the morphology of dividing cells within the sexually dimorphic nucleus of the preoptic area. Lower right panel: This relatively low power image illustrates the CB28-immunoreactivity (green fluorescence) delineation of a unilateral adult male SDN-POA within which the red fluorescence (Ki67-immunoreactivity) is located. The series images (n = 8) demonstrate the cell at the telophase of cell division defined in the image located at low right panel. The series images were acquired with a sequential order at 1 μm intervals along the Z axis.

Disclaimer

This document has been reviewed in accordance with United States Food and Drug Administration (FDA) policy and approved for publication. Approval does not signify that the contents necessarily reflect the position or opinions of the FDA nor does mention of trade names or commercial products constitute endorsement or recommendation for use. The findings and conclusions in this report are those of the authors and do not necessarily represent the views of the FDA.

Contributions

Z.H. made the initial draft. M.G.P., S.A.F. and T.A.P. significantly edited the manuscript.

Acknowledgements

The authors wish to thank Li Cui, M.D., Ph.D. for her reorganization and optimization of the figures.

References

Abrahamyan, S.S., M. Fodor, A.A. Galoyan, and M. Palkovits. 2002. Distribution of the hypothalamic cardioactive hormone "G"-protein complex (PCG) in neuronal elements of the heart in intact and vagotomized rats. Neurochem Res 27: 381–388.

Ajika, K. 1979. Simultaneous localization of LHRH and catecholamines in rat hypothalamus. J Anat 128: 331–347.

Aizawa, T., and M.A. Greer. 1981. Delineation of the hypothalamic area controlling thyrotropin secretion in the rat. Endocrinology 109: 1731–8.

Balkan, B., A. Keser, O. Gozen, E.O. Koylu, T. Dagci, M.J. Kuhar, and S. Pogun. 2012. Forced swim stress elicits region-specific changes in CART expression in the stress axis and stress regulatory brain areas. Brain Res 1432: 56–65.

Baum, M.J., S.A. Tobet, J.A. Cherry, and R.G. Paredes. 1996. Estrogenic control of preoptic area development in a carnivore, the ferret. Cell Mol Neurobiol 16: 117–128.

Benedeczky, I. 1983. The functional morphology of chromaffin cells. Acta Biol Hung 34: 137–154.

Billova, S., A.S. Galanopoulou, N.G. Seidah, X. Qiu, and U. Kumar. 2007. Immunohistochemical expression and colocalization of somatostatin, carboxypeptidase-E and prohormone convertases 1 and 2 in rat brain. Neuroscience 147: 403–418.

Bishop, A.E., J.M. Polak, P. Facer, G.L. Ferri, P.J. Marangos, and A.G. Pearse. 1982. Neuron specific enolase: a common marker for the endocrine cells and innervation of the gut and pancreas. Gastroenterology 83: 902–915.

Bodo, C., and E.F. Rissman. 2008. The androgen receptor is selectively involved in organization of sexually dimorphic social behaviors in mice. Endocrinology 149: 4142–4150.

Budefeld, T., N. Grgurevic, S.A. Tobet, and G. Majdic. 2008. Sex differences in brain developing in the presence or absence of gonads. Dev Neurobiol 68: 981–995.

Carvalheira, A.F., U. Welsch, and A.G. Pearse. 1968. Cytochemical and ultrastructural observations on the argentaffin and argyrophil cells of the gastro-intestinal tract in mammals, and their place in the APUD series of polypeptide-secreting cells. Histochemie 14: 33–46.

Challacombe, D.N., and E.E. Wheeler. 1983. Possible neural projections from enterochromaffin cells. Lancet 2: 1502.

Christensen, W.N., E.W. Strong, M.S. Bains, and J.M. Woodruff. 1988. Neuroendocrine differentiation in the glandular peripheral nerve sheath tumor. Pathologic distinction from the biphasic synovial sarcoma with glands. Am J Surg Pathol 12: 417–426.

Cuttitta, C.M., S.R. Guariglia, A.E. Idrissi, and W.J. L'Amoreaux. 2013. Taurine's effects on the neuroendocrine functions of pancreatic beta cells. Adv Exp Med Biol 775: 299–310.

Daikoku, S., Y. Okamura, H. Kawano, Y. Tsuruo, M. Maegawa, and T. Shibasaki. 1985. CRF-containing neurons of the rat hypothalamus. Cell Tissue Res 240: 575–584.

Davis, S.N., P. Galassetti, D.H. Wasserman, and D. Tate. 2000. Effects of gender on neuroendocrine and metabolic counterregulatory responses to exercise in normal man. J Clin Endocrinol Metab 85: 224–230.

Dick, T.E., Y.H. Hsieh, N. Wang, and N. Prabhakar. 2007. Acute intermittent hypoxia increases both phrenic and sympathetic nerve activities in the rat. Exp Physiol 92: 87–97.

Dickerson, S.M., S.L. Cunningham, and A.C. Gore. 2011. Prenatal PCBs disrupt early neuroendocrine development of the rat hypothalamus. Toxicol Appl Pharmacol 252: 36–46.

Dohler, K.D., S.S. Srivastava, J.E. Shryne, B. Jarzab, A. Sipos, and R.A. Gorski. 1984. Differentiation of the sexually dimorphic nucleus in the preoptic area of the rat brain is inhibited by postnatal treatment with an estrogen antagonist. Neuroendocrinology 38: 297–301.

Edelmann, M., C. Wolfe, E.M. Scordalakes, E.F. Rissman, and S. Tobet. 2007. Neuronal nitric oxide synthase and calbindin delineate sex differences in the developing hypothalamus and preoptic area. Dev Neurobiol 67: 1371–1381.

Ekengren, B., J. Peute, and G. Fridberg. 1978. Gonadotropic cells in the Atlantic salmon, Salmo salar. An experimental immunocytological, electron microscopical study. Cell Tissue Res 191: 187–203.

El Majdoubi, M. 2011. Stem cell-derived *in vitro* models for investigating the effects of endocrine disruptors on developing neurons and neuroendocrine cells. J Toxicol Environ Health B Crit Rev 14: 292–299.

Faber, K.A., L. Ayyash, S. Dixon, and C.L. Hughes, Jr. 1993. Effect of neonatal diethylstilbestrol exposure on volume of the sexually dimorphic nucleus of the preoptic area of the hypothalamus and pituitary responsiveness to gonadotropin-releasing hormone in female rats of known anogenital distance at birth. Biol Reprod 48: 947–951.

Fletcher, E.C. 1997. Sympathetic activity and blood pressure in the sleep apnea syndrome, Respiration 64 (Suppl. 1): 22–28.

Freeman, M.E., B. Kanyicska, A. Lerant, and G. Nagy. 2000. Prolactin: structure, function, and regulation of secretion. Physiol Rev 80: 1523–1631.

Galoyan, A. 1986. Neuroendocrine heart and hypothalamus. Neurochem Res 11: 769–787.

Garcia-Falgueras, A., L. Ligtenberg, F.P. Kruijver, and D.F. Swaab. 2011. Galanin neurons in the intermediate nucleus (InM) of the human hypothalamus in relation to sex, age, and gender identity. J Comp Neurol 519: 3061–3084.

Gerlo, E.A., D.F. Schoors, and A.G. Dupont. 1991. Age- and sex-related differences for the urinary excretion of norepinephrine, epinephrine, and dopamine in adults. Clin Chem 37: 875–878.

Goldstein, D.S., R. McCarty, R.J. Polinsky, and I.J. Kopin. 1983. Relationship between plasma norepinephrine and sympathetic neural activity. Hypertension 5: 552–559.

Gonzalez-Trapaga, J.L., R.A. Nelesen, J.E. Dimsdale, P.J. Mills, B. Kennedy, R.J. Parmer, and M.G. Ziegler. 2000. Plasma epinephrine levels in hypertension and across gender and ethnicity. Life Sci 66: 2383–2392.

Goodall, M., and N. Kirshner. 1958. Biosynthesis of epinephrine and norepinephrine by sympathetic nerves and ganglia. Circulation 17: 366–371.

Grozinsky-Glasberg, S., I. Shimon, M. Korbonits, and A.B. Grossman. 2008. Somatostatin analogues in the control of neuroendocrine tumours: efficacy and mechanisms. Endocr Relat Cancer 15: 701–720.

Guy, L., L.R. Begin, L.L. Oligny, G.B. Brock, S. Chevalier, and A.G. Aprikian. 1999. Searching for an intrinsic neuroendocrine cell in the kidney. An immunohistochemical study of the fetal, infantile and adult kidney. Pathol Res Pract 195: 25–30.

Hamelink, C., O. Tjurmina, R. Damadzic, W.S. Young, E. Weihe, H.W. Lee, and L.E. Eiden. 2002. Pituitary adenylate cyclase-activating polypeptide is a sympathoadrenal neurotransmitter involved in catecholamine regulation and glucohomeostasis. Proc Natl Acad Sci U S A 99: 461–466.

Hayward, J.N. 1974. Physiological and morphological identification of hypothalamic magnocellular neuroendocrine cells in goldfish preoptic nucleus. J Physiol 239: 103–124.

He, Z., M.G. Paule, and S.A. Ferguson. 2012. Low oral doses of bisphenol A increase volume of the sexually dimorphic nucleus of the preoptic area in male, but not female, rats at postnatal day 21. Neurotoxicol Teratol 34: 331–337.

He, Z., S.A. Ferguson, L. Cui, L.J. Greenfield, Jr., and M.G. Paule. 2013a. Role of neural stem cell activity in postweaning development of the sexually dimorphic nucleus of the preoptic area in rats. PLoS One 8: e54927.

He, Z., S.A. Ferguson, L. Cui, L.J. Greenfield, and M.G. Paule. 2013b. Development of the sexually dimorphic nucleus of the preoptic area and the influence of estrogen-like compounds. Neural Regen Res 8: 2763–2774.

Hoffman, G.E., and T.A. Hayes. 1979. Somatostatin neurons and their projections in dog diencephalon. J Comp Neurol 186: 371–91.

Holzbauer, M., and K. Racke. 1985. The dopaminergic innervation of the intermediate lobe and of the neural lobe of the pituitary gland. Med Biol 63: 97–116.

Iqbal, J., T.R. Manley, Q. Yue, M.R. Namavar, and I.J. Clarke. 2005. Noradrenergic regulation of hypothalamic cells that produce growth hormone-releasing hormone and somatostatin and the effect of altered adiposity in sheep. J Neuroendocrinol 17: 341–352.

Jackson, I.M., and S. Reichlin. 1979. Thyrotropin-releasing hormone in the blood of the frog, Rana pipiens: its nature and possible derivation from regional locations in the skin. Endocrinology 104: 1814–21.

Jenkins, S., N. Raghuraman, I. Eltoum, M. Carpenter, J. Russo, and C.A. Lamartiniere. 2009. Oral exposure to bisphenol a increases dimethylbenzanthracene-induced mammary cancer in rats. Environ Health Perspect 117: 910–915.

Kandel, E.R. 1964. Electrical properties of hypothalamic neuroendocrine cells. J Gen Physiol 47: 691–717.

Kasacka, I., E. Arciszewska, M.M. Winnicka, and A. Lewandowska. 2012. Effects of CP 55,940—agonist of CB1 cannabinoid receptors on ghrelin and somatostatin producing cells in the rat pancreas. Folia Histochem Cytobiol 50: 111–117.

Kawai, Y., S. Inagaki, S. Shiosaka, T. Shibasaki, N. Ling, M. Tohyama, and Y. Shiotani. 1984. The distribution and projection of gamma-melanocyte stimulating hormone in the rat brain: an immunohistochemical analysis. Brain Res 297: 21–32.

Kawano, H., and S. Daikoku. 1987. Functional topography of the rat hypothalamic dopamine neuron systems: retrograde tracing and immunohistochemical study. J Comp Neurol 265: 242–253.

Kidd, M., I.M. Modlin, G.N. Eick, and M.C. Champaneria. 2006. Isolation, functional characterization, and transcriptome of Mastomys ileal enterochromaffin cells. Am J Physiol Gastrointest Liver Physiol 291: G778–791.

Kizer, J.S., M. Palkovits, and M.J. Brownstein. 1976. The projections of the A8, A9 and A10 dopaminergic cell bodies: evidence for a nigral-hypothalamic-median eminence dopaminergic pathway. Brain Res 108: 363–370.

Koval, L.M., E.N. Yavorskaya, and E.A. Lukyanetz. 2001. Electron microscopic evidence for multiple types of secretory vesicles in bovine chromaffin cells. Gen Comp Endocrinol 121: 261–277.

Krivova, Y.S., V.M. Barabanov, A.E. Proshchina, and S.V. Savel'ev. 2014. Distribution of chromogranin A in human fetal pancreas. Bull Exp Biol Med 156: 865–868.

Krulich, L., M. Quijada, J.E. Wheaton, P. Illner, and S.M. McCann. 1977. Localization of hypophysiotropic neurohormones by assay of sections from various brain areas. Fed Proc 36: 1953–1959.

L'Amoreaux, W.J., C. Cuttitta, A. Santora, J.F. Blaize, J. Tachjadi, and A. El Idrissi. 2010. Taurine regulates insulin release from pancreatic beta cell lines. J Biomed Sci 17 Suppl 1: S11.

Ladinsky, M.S., D.N. Mastronarde, J.R. McIntosh, K.E. Howell, and L.A. Staehelin. 1999. Golgi structure in three dimensions: functional insights from the normal rat kidney cell. J Cell Biol 144: 1135–1149.

Larsen, P.J., L.W. Enquist, and J.P. Card. 1998. Characterization of the multisynaptic neuronal control of the rat pineal gland using viral transneuronal tracing. Eur J Neurosci 10: 128–145.

Lechan, R.M., and I.M. Jackson. 1982. Immunohistochemical localization of thyrotropin-releasing hormone in the rat hypothalamus and pituitary. Endocrinology 111: 55–65.

Lechan, R.M., M.E. Molitch and I.M. Jackson. 1983. Distribution of immunoreactive human growth hormone-like material and thyrotropin-releasing hormone in the rat central nervous system: evidence for their coexistence in the same neurons. Endocrinology 112: 877–84.

Lerant, A., and M.E. Freeman. 1998. Ovarian steroids differentially regulate the expression of PRL-R in neuroendocrine dopaminergic neuron populations: a double label confocal microscopic study. Brain Res 802: 141–154.

Leranth, C., T. Hajszan, K. Szigeti-Buck, J. Bober, and N.J. MacLusky. 2008. Bisphenol A prevents the synaptogenic response to estradiol in hippocampus and prefrontal cortex of ovariectomized nonhuman primates. Proc Natl Acad Sci U S A 105: 14187–14191.

Liposits, Z., W.K. Paull, P. Wu, I.M. Jackson, and R.M. Lechan. 1987. Hypophysiotrophic thyrotropin releasing hormone (TRH) synthesizing neurons. Ultrastructure, adrenergic innervation and putative transmitter action. Histochemistry 88: 1–10.

Lyons, D.J., A. Hellysaz, and C. Broberger. 2012. Prolactin regulates tuberoinfundibular dopamine neuron discharge pattern: novel feedback control mechanisms in the lactotrophic axis. J Neurosci 32: 8074–8083.

Maley, B., and R. Elde. 1982. Immunohistochemical localization of putative neurotransmitters within the feline nucleus tractus solitarii. Neuroscience 7: 2469–2490.

Martin, J.B., and S. Reichlin. 1972. Plasma thyrotropin (TSH) response to hypothalamic electrical stimulation and to injection of synthetic thyrotropin releasing hormone (TRH). Endocrinology 90: 1079–85.

Martinez-Alonso, E., M. Tomas, and J.A. Martinez-Menarguez. 2013. Morpho-functional architecture of the Golgi complex of neuroendocrine cells. Front Endocrinol (Lausanne) 4: 41.

McGavran, M.H. 1964. "Chromaffin" Cell: Electron Microscopic Identification in the Human Dermis. Science 145: 275–276.

McKinley, M.J., D.A. Denton, M. Leventer, J. Penschow, R.S. Weisinger, and R.D. Wright. 1983. Morphology of the organum vasculosum of the lamina terminalis (OVLT) of the sheep. Brain Res Bull 11: 649–657.

Merchenthaler, I., I. Lengvari, J. Horvath, and G. Setalo. 1980. Immunohistochemical study of the LHRH-synthesizing neuron system of aged female rats. Cell Tissue Res 209: 499–503.

Micevych, P.E., and R.P. Elde. 1982. Neurons containing alpha-melanocyte stimulating hormone and beta-endorphin immunoreactivity in the cat hypothalamus. Peptides 3: 655–662.

Miller, R.R., and N.L. Muller. 1995. Neuroendocrine cell hyperplasia and obliterative bronchiolitis in patients with peripheral carcinoid tumors. Am J Surg Pathol 19: 653–658.

Mueller, J.K., and S. Heger. 2014. Endocrine disrupting chemicals affect the gonadotropin releasing hormone neuronal network. Reprod Toxicol 44: 73–84.

Nakata, D., T. Masaki, A. Tanaka, M. Yoshimatsu, Y. Akinaga, M. Asada, R. Sasada, M. Takeyama, K. Miwa, T. Watanabe, and M. Kusaka. 2014. Suppression of the hypothalamic-pituitary-gonadal axis by TAK-385 (relugolix), a novel, investigational, orally active, small molecule gonadotropin-releasing hormone (GnRH) antagonist: studies in human GnRH receptor knock-in mice. Eur J Pharmacol 723: 167–174.

Naule, L., M. Picot, M. Martini, C. Parmentier, H. Hardin-Pouzet, M. Keller, I. Franceschini, and S. Mhaouty-Kodja. 2014. Neuroendocrine and behavioral effects of maternal exposure to oral bisphenol A in female mice. J Endocrinol 220: 375–388.

Newbold, R.R., E. Padilla-Banks, R.J. Snyder, and W.N. Jefferson. 2007. Perinatal exposure to environmental estrogens and the development of obesity. Mol Nutr Food Res 51: 912–917.

Newbold, R.R., W.N. Jefferson, and E. Padilla-Banks. 2009a. Prenatal exposure to bisphenol a at environmentally relevant doses adversely affects the murine female reproductive tract later in life. Environ Health Perspect 117: 879–885.

Newbold, R.R., E. Padilla-Banks, and W.N. Jefferson. 2009b. Environmental estrogens and obesity. Mol Cell Endocrinol 304: 84–89.

Oba, H., K. Nishida, S. Takeuchi, H. Akiyama, K. Muramatsu, M. Kurosumi, and T. Kameya. 2013. Diffuse idiopathic pulmonary neuroendocrine cell hyperplasia with a central and peripheral carcinoid and multiple tumorlets: a case report emphasizing the role of neuropeptide hormones and human gonadotropin-alpha. Endocr Pathol 24: 220–228.

Oberg, K. 2012. Molecular imaging radiotherapy: theranostics for personalized patient management of neuroendocrine tumors (NETs). Theranostics 2: 448–458.

Orikasa, C., and Y. Sakuma. 2010. Estrogen configures sexual dimorphism in the preoptic area of C57BL/6J and ddN strains of mice. J Comp Neurol 518: 3618–3629.

Osamura, R.Y., N. Komatsu, K. Watanabe, Y. Nakai, I. Tanaka, and H. Imura. 1982. Immunohistochemical and immunocytochemical localization of gamma-melanocyte stimulating hormone (gamma-MSH)-like immunoreactivity in human and rat hypothalamus. Peptides 3: 781–787.

Pacak, K. 2011. Phaeochromocytoma: a catecholamine and oxidative stress disorder. Endocr Regul 45: 65–90.

Parra, A., J. Barron, M. Mota-Gonzalez, V. Ibarra, and A. Espinosa de los Monteros. 1995. Serum androgen changes during meal-induced hyperinsulinemia and after acute sequential blockade and hyperstimulation of insulin release in women with polycystic ovary syndrome. Arch Med Res 26 Spec No: S209–217.

Patisaul, H.B., A.E. Fortino, and E.K. Polston. 2007. Differential disruption of nuclear volume and neuronal phenotype in the preoptic area by neonatal exposure to genistein and bisphenol-A. Neurotoxicology 28: 1–12.

Pearse, A.G. 1969. The cytochemistry and ultrastructure of polypeptide hormone-producing cells of the APUD series and the embryologic, physiologic and pathologic implications of the concept. J Histochem Cytochem 17: 303–313.

Pearse, A.G. 1975. Neurocristopathy, neuroendocrine pathology and the APUD concept. Z Krebsforsch Klin Onkol Cancer Res Clin Oncol 84: 1–18.

Pfragner, R., A. Behmel, D.P. Smith, B.A. Ponder, G. Wirnsberger, I. Rinner, S. Porta, T. Henn, and B. Niederle. 1998. First continuous human pheochromocytoma cell line: KNA. Biological, cytogenetic and molecular characterization of KNA cells. J Neurocytol 27: 175–186.

Polishchuk, R.S., E.V. Polishchuk, P. Marra, S. Alberti, R. Buccione, A. Luini, and A.A. Mironov. 2000. Correlative light-electron microscopy reveals the tubular-saccular ultrastructure of carriers operating between Golgi apparatus and plasma membrane. J Cell Biol 148: 45–58.

Polyzos, S.A., J. Kountouras, G. Deretzi, C. Zavos, and C.S. Mantzoros. 2012. The emerging role of endocrine disruptors in pathogenesis of insulin resistance: a concept implicating nonalcoholic fatty liver disease. Curr Mol Med 12: 68–82.

Prabhakar, N.R., and Y.J. Peng. 2004. Peripheral chemoreceptors in health and disease. J Appl Physiol (1985) 96: 359–366.

Prins, G.S., W.Y. Tang, J. Belmonte, and S.M. Ho. 2008a. Developmental exposure to bisphenol A increases prostate cancer susceptibility in adult rats: epigenetic mode of action is implicated. Fertil Steril 89: e41.

Prins, G.S., W.Y. Tang, J. Belmonte, and S.M. Ho. 2008b. Perinatal exposure to oestradiol and bisphenol A alters the prostate epigenome and increases susceptibility to carcinogenesis. Basic Clin Pharmacol Toxicol 102: 134–138.

Rabouille, C., and J. Klumperman. 2005. Opinion: The maturing role of COPI vesicles in intra-Golgi transport. Nat Rev Mol Cell Biol 6: 812–7.

Rambourg, A., Y. Clermont, and A. Marraud. 1974. Three-dimensional structure of the osmium-impregnated Golgi apparatus as seen in the high voltage electron microscope. Am J Anat 140: 27–45.

Rawdon, B.B., and A. Andrew. 1999. Gut endocrine cells in birds: an overview, with particular reference to the chemistry of gut peptides and the distribution, ontogeny, embryonic origin and differentiation of the endocrine cells. Prog Histochem Cytochem 34: 3–82.

Richards, D.S., R.M. Villalba, F.J. Alvarez, and J.E. Stern. 2005. Expression of GABAB receptors in magnocellular neurosecretory cells of male, virgin female and lactating rats. J Neuroendocrinol 17: 413–423.

Rink, T., H. Fitz, H.J. Schroth, and S. Braun. 2001. Development of the parafollicular cells in recurrent goiter. Eur J Endocrinol 144: 485–489.

Roselli, C.E., and F. Stormshak. 2010. The ovine sexually dimorphic nucleus, aromatase, and sexual partner preferences in sheep. J Steroid Biochem Mol Biol 118: 252–256.

Sarkar, S., G. Wittmann, C. Fekete, and R.M. Lechan. 2004. Central administration of cocaine- and amphetamine-regulated transcript increases phosphorylation of cAMP response element binding protein in corticotropin-releasing hormone-producing neurons but not in prothyrotropin-releasing hormone-producing neurons in the hypothalamic paraventricular nucleus. Brain Res 999: 181–192.

Scallet, A.C., R.L. Divine, R.R. Newbold, and K.B. Delclos. 2004. Increased volume of the calbindin D28k-labeled sexually dimorphic hypothalamus in genistein and nonylphenol-treated male rats. Toxicol Sci 82: 570–576.

Schmid, K.W., B. Helpap, M. Totsch, R. Kirchmair, B. Dockhorn-Dworniczak, W. Bocker, and R. Fischer-Colbrie. 1994. Immunohistochemical localization of chromogranins A and B and secretogranin II in normal, hyperplastic and neoplastic prostate. Histopathology 24: 233–239.

Schroder, H., and L. Vollrath. 1985. Distribution of dopamine-beta-hydroxylase-like immunoreactivity in the rat pineal organ. Histochemistry 83: 375–380.

Sickel, M.J., and M.M. McCarthy. 2000. Calbindin-D28k immunoreactivity is a marker for a subdivision of the sexually dimorphic nucleus of the preoptic area of the rat: developmental profile and gonadal steroid modulation. J Neuroendocrinol 12: 397–402.

Simonneaux, V., and C. Ribelayga. 2003. Generation of the melatonin endocrine message in mammals: a review of the complex regulation of melatonin synthesis by norepinephrine, peptides, and other pineal transmitters. Pharmacol Rev 55: 325–395.

Simpson, J.B. 1981. The circumventricular organs and the central actions of angiotensin. Neuroendocrinology 32: 248–256.

Sjolund, K., G. Sanden, R. Hakanson, and F. Sundler. 1983. Endocrine cells in human intestine: an immunocytochemical study. Gastroenterology 85: 1120–1130.

Slack, J.M. 1995. Developmental biology of the pancreas. Development 121: 1569–1580.

Smith, J.A., L. Wang, H. Hiller, C.T. Taylor, A.D. de Kloet, and E.G. Krause. 2014. Acute hypernatremia promotes anxiolysis and attenuates stress-induced activation of the hypothalamic-pituitary-adrenal axis in male mice. Physiol Behav 136: 91–96.

Smith, S.M., J.M. Vaughan, C.J. Donaldson, J. Rivier, C. Li, A. Chen, and W.W. Vale. 2004. Cocaine- and amphetamine-regulated transcript activates the hypothalamic-pituitary-adrenal axis through a corticotropin-releasing factor receptor-dependent mechanism. Endocrinology 145: 5202–5209.

Stern, J.E., M. Galarreta, R.C. Foehring, S. Hestrin, and W.E. Armstrong. 1999. Differences in the properties of ionotropic glutamate synaptic currents in oxytocin and vasopressin neuroendocrine neurons. J Neurosci 19: 3367–3375.

Sundin, A., U. Garske, and H. Orlefors. 2007. Nuclear imaging of neuroendocrine tumours. Best Pract Res Clin Endocrinol Metab 21: 69–85.

Swaab, D.F., and E. Fliers. 1985. A sexually dimorphic nucleus in the human brain. Science 228: 1112–1115.

Tomiko, S.A., P.S. Taraskevich, and W.W. Douglas. 1983. GABA acts directly on cells of pituitary pars intermedia to alter hormone output. Nature 301: 706–707.

Van Lommel, A., T. Bolle, W. Fannes, and J.M. Lauweryns. 1999. The pulmonary neuroendocrine system: the past decade. Arch Histol Cytol 62: 1–16.

van Rijnsoever, C., V. Oorschot, and J. Klumperman. 2008. Correlative light-electron microscopy (CLEM) combining live-cell imaging and immunolabeling of ultrathin cryosections. Nat Methods 5: 973–80.

Vancutsem, P.M., and M.L. Roessler. 1997. Neonatal treatment with tamoxifen causes immediate alterations of the sexually dimorphic nucleus of the preoptic area and medial preoptic area in male rats. Teratology 56: 220–228.

Vandesande, F., and K. Dierickx. 1975. Identification of the vasopressin producing and of the oxytocin producing neurons in the hypothalamic magnocellular neurosecretory system of the rat. Cell Tissue Res 164: 153–162.

Vollmer, R.R. 1996. Selective neural regulation of epinephrine and norepinephrine cells in the adrenal medulla—cardiovascular implications. Clin Exp Hypertens 18: 731–751.

Vosges, M., Y. Le Page, B.C. Chung, Y. Combarnous, J.M. Porcher, O. Kah, and F. Brion. 2010. 17alpha-ethinylestradiol disrupts the ontogeny of the forebrain GnRH system and the expression of brain aromatase during early development of zebrafish. Aquat Toxicol 99: 479–491.

White, S.R., G.K. Samathanam, R.M. Bowker, and M.W. Wessendorf. 1990. Damage to bulbospinal serotonin-, tyrosine hydroxylase-, and TRH-containing axons occurs early in the development of experimental allergic encephalomyelitis in rats. J Neurosci Res 27: 89–98.

Wierup, N., H. Svensson, H. Mulder, and F. Sundler. 2002. The ghrelin cell: a novel developmentally regulated islet cell in the human pancreas. Regul Pept 107: 63–69.

Wittmann, G., Z. Liposits, R.M. Lechan, and C. Fekete. 2005. Origin of cocaine- and amphetamine-regulated transcript-containing axons innervating hypophysiotropic corticotropin-releasing hormone-synthesizing neurons in the rat. Endocrinology 146: 2985–2991.

Young, J.K. 1982. A comparison of hypothalami of rats and mice: lack of gross sexual dimorphism in the mouse. Brain Res 239: 233–239.

Zeuschner, D., W.J. Geerts, E. van Donselaar, B.M. Humbel, J.W. Slot, A.J. Koster, and J. Klumperman. 2006. Immuno-electron tomography of ER exit sites reveals the existence of free COPII-coated transport carriers. Nat Cell Biol 8: 377–83.

Zhou, R., Z. Zhang, Y. Zhu, L. Chen, M. Sokabe, and L. Chen. 2009. Deficits in development of synaptic plasticity in rat dorsal striatum following prenatal and neonatal exposure to low-dose bisphenol A. Neuroscience 159: 161–171.

Ziegler, M.G., H. Elayan, M. Milic, P. Sun, and M. Gharaibeh. 2012. Epinephrine and the metabolic syndrome. Curr Hypertens Rep 14: 1–7.

6

Neural Stem Cells in the Adult Brain: Altered Activity in Aging, Alzheimer's Disease, Mood Disorders and Epilepsy

Ashok K. Shetty[1],* and *Bharathi Hattiangady*[2]

Introduction

Neural stem/progenitor cells (NSCs) are seen in both the developing and adult central nervous system (CNS). These are multipotent cells, which exhibit self-renewal as well as produce a vast majority of cells in the developing CNS (Gage and Temple, 2013). The development of cerebral cortex commences in the anterior neural tube and is specified by homeobox proteins encoded by Distal-less (Dlx) family of genes DLX1 and DLX2 and NK2 homeobox 1 gene (NKX2.1) (Rubenstein and Rakic, 1999). A subset of radial glial cells (RGCs) that extend from their cell body located in the ventricular zone to the pial surface in the primordial CNS are considered the primary NSCs (Noctor *et al.*, 2001). These multipotent NSCs express Sox-2 (sex determining region Y [SRY]-box 2, a transcription factor), nestin (a primitive neurofilament protein)

1 Professor and Associate Director, Institute for Regenerative Medicine Texas A&M Health Science Center College of Medicine, Temple, TX 76502.
   E-mail: Shetty@medicine.tamhsc.edu
2 Assistant Professor, Department of Molecular and Cellular Medicine, Texas A&M Health Science Center College of Medicine at Scott & White, 5701 Airport Road, Module C, Temple, TX 76501.
   E-mail: hattiangady@medicine.tamhsc.edu
* Corresponding author

and glial fibrillary acidic protein (GFAP, an intermediate filament protein). Through proliferation, they repeatedly produce more restricted neuronal and glial progenitors, which are referred to as transit amplifying (or intermediate) progenitors (Englund *et al.*, 2005). These progenitors display limited self-renewal and proliferative activity than NSCs and hence end up in producing differentiated progeny (Davis and Temple, 1994; Gage and Temple, 2013). However, the multipotent property of neuronal progenitors varies considerably during development. This is evident from observations that progenitors generated early during development can give rise to multiple types of neurons but progenitors generated later in development cannot produce the earlier fates (Frantz and McConnell, 1996; McConnell and Kaznowski, 1991).

During development of the CNS, NSCs and progenitors appear to be committed to the generation of specific types of neural cells (Taverna *et al.*, 2014). The precise mechanisms by which CNS progenitors are temporally restricted or committed to certain lineages are still unknown. However, results of many studies suggest that production of gliogenic cues such as cardiotrophin, transcription factor sequences, DNA methylation changes and chromatin modifications underlie the temporal control of progenitor cell output (Barnabe-Heider *et al.*, 2005; Gage and Temple, 2013; Pereira *et al.*, 2010). While the CNS development and patterning occurs in a modular fashion with specific CNS regions acting as organizers (Gage and Temple, 2013; Martinez *et al.*, 1999; Puelles and Rubenstein, 2003), neural progenitor and precursor cells retain ability to cross regional boundaries during development. A classic example is the migration of GABA-ergic progenitors during development. These progenitors originate in the ganglionic eminences of the ventral forebrain but later migrate tangentially into the overlying dorsal cortex where they mature and get incorporated into the circuitry as GABA-ergic interneurons (Anderson *et al.*, 1997).

Once the CNS is developed with the essential numbers and varieties of progenitors and differentiated cells, active NSCs in most regions undergo depletion in number, either through terminal differentiation or through attainment of a quiescent state. The production of neurons from NSCs and their progenitors typically occurs from the early embryonic development period to early postnatal periods whereas the production of glia during development occurs from the late embryonic development period to postnatal periods (Gotz and Huttner, 2005; Paridaen and Huttner, 2014). The production of glia (from NSCs or glial progenitors derived from NSCs) however continues at reduced levels in all regions of the brain thereafter (Gallo and Deneen, 2014; Guerout *et al.*, 2014; Rowitch and Kriegstein, 2010; Urban and Guillemot, 2014). Thus, in the postnatal and adult periods, NSCs or their progenitors capable of generating neurons are restricted to only a few regions of the CNS known as neurogenic regions, which mainly include the dentate gyrus (DG) of the hippocampal formation and the subventricular zone (SVZ) lining the walls of ventricles in the forebrain. In these regions, NSCs proliferate on a daily basis and generate neurons and glia throughout life in most mammalian brains. This phenomenon, now referred to as adult neurogenesis, has come to limelight through extensive research during the last 2 decades. This finding has overthrown one of the foremost dogmas pertaining to the mammalian brain that neurons were generated in entirety during development and hence are not reinstated when they die due to disease or injury. While the idea of cell division and neuronal differentiation occurring in the adult brain was emerged as early as 1960's via an observation that new cells are generated in the olfactory system and the DG of the adult brain (Altman, 1969; Altman and Das, 1965), lack of definitive markers to detect neurons made this finding mostly go unnoticed. The subsequent reports demonstrating newly born neurons in the adult DG and the olfactory bulb through tritiatedthymidine labeling (Kaplan and Hinds, 1977) and the involvement of newly born neurons in learning of new songs in adult canaries (Goldman and Nottebohm, 1983) gave additional evidence but did not provide enough impetus for exploring the concept of adult neurogenesis in detail.

An inspiring discovery in early nineties that both neurons and astrocytes can be produced in culture from cells isolated from the adult mammalian brain (Reynolds and Weiss, 1992) however prompted a colossal curiosity for probing NSCs and neurogenesis in the adult brain. Eriksson and colleagues in 1998 offered the first proof for the presence of NSCs in the adult human hippocampus (Eriksson *et al.*, 1998) through demonstration of newly born neurons in the autopsied hippocampus of patients who received 5'-bromodeoxyuridine (BrdU) for examining cancer cell kinetics prior to their death. Since then, there has been an immense attention for learning the behavior of adult NSCs and neurogenesis under normal and disease conditions. A recent study, by providing a strong proof for the existence of NSCs in the DG of humans, has further stimulated this field (Spalding *et al.*, 2013). This study showed that significant neurogenesis from NSCs persists even during old age in the human hippocampus. Since NSC activity and neurogenesis in the adult hippocampus is now confirmed in virtually all mammalian species including humans, in this chapter, we mainly focus on discussing the present knowledge on the activity of NSCs and neurogenesis in the adult mammalian hippocampus under standard and disease settings. Additionally, we confer how deviant or declined NSC activity and neurogenesis in aging and neurological disorders can interfere with cognitive and mood function.

NSCs in Neurogenic and Non-neurogenic Regions of the Adult Brain

Numerous studies have now clearly established the occurrence of proliferating NSCs and neurogenesis throughout life in two distinct regions of the adult mammalian brain: the subgranular zone (SGZ) of the DG in the hippocampal formation and the subventricular zone (SVZ) covering the walls of forebrain lateral ventricles (Ihrie and Alvarez-Buylla, 2011; Ming and Song, 2011; Yao *et al.*, 2012). As seen in the developing CNS, NSCs in the adult brain exhibit ability for widespread proliferation, self-renewal and generating all three major CNS cell types (Ihrie and Alvarez-Buylla, 2011; Ming and Song, 2011; Yao *et al.*, 2012). In the SGZ of the DG, proliferation of NSCs and intermediate progenitors produce mostly new neurons and some glia. Newly generated neurons then migrate radially into the granule cell layer (GCL), where they mature as dentate granule cells with dendrites projecting into the molecular layer and axons projecting into the stratum lucidum of the CA3. These new granule cells then establish afferent connectivity with axons from the entorhinal cortex, and efferent connectivity with the CA3 pyramidal neurons (Ming and Song, 2011; Yao *et al.*, 2012). In the SVZ, daily division of NSCs and/or intermediate progenitors generates cohorts of young neurons (neuroblasts), which journey into the olfactory bulb through a conduit named as rostral migratory stream (RMS) (Ihrie and Alvarez-Buylla, 2011). Upon arrival, they undergo differentiation and maturation into olfactory interneurons. Neural stem cells also persist in the SGZ and SVZ of the adult human brain. While NSCs in the human SGZ generates new granule cells for the DG as in rodents (Eriksson *et al.*, 1998; Spalding *et al.*, 2013), NSCs in the human SVZ are believed to contribute new neurons to the striatum (Ernst *et al.*, 2014). This is in sharp contrast to the SVZ of rodents generating neurons that under normal conditions migrate only to the olfactory bulb. Nonetheless, NSCs in the two neurogenic regions share several key characteristics (Urban and Guillemot, 2014). These include expression of markers such as Sox-2, nestin and GFAP, low level of proliferation under normal conditions, their location adjacent to the microvasculature and restricted potential. In the rodent brain, the SGZ-NSCs typically produce dentate granule cells and astrocytes whereas the SVZ-NSCs classically produce neurons of the olfactory bulb and oligodendrocytes (Urban and Guillemot, 2014).

Limited amount of neurogenesis probably also occurs in "non-neurogenic" regions of the adult brain in certain circumstances (Gould, 2007). The regions of adult brain displaying some NSC activity comprise the CA1 subfield of the hippocampus (Rietze *et al.*, 2000), substantia nigra of the midbrain (Zhao *et al.*, 2003), neocortex (Cameron and Dayer, 2008), striatum (Bedard *et al.*, 2006), amygdaloid nuclei (Fowler *et al.*, 2008), piriform cortex (Shapiro *et al.*, 2007), and hypothalamus (Yuan and Arias-Carrion, 2011). However, it is still unclear whether putative NSCs in these regions are capable of generating a significant number of neurons in physiological conditions. It appears that these cells either sustain the properties of quiescent NSCs or intermittently generate glial cells such as astrocytes and oligodendrocytes (Gage and Temple, 2013). Yet, when separated and harvested in a culture dish containing FGF-2 and/or other factors, these NSCs undergo differentiation into neurons. This suggests that, stimulation of NSCs or their progeny for neuronal differentiation in non-neurogenic regions requires extrinsic factors that promote NSC proliferation and differentiation (Palmer *et al.*, 1999). Indeed, several studies have suggested that administration of certain neuroprotective compounds and drugs or lifestyle changes (such as voluntary exercise) can activate NSCs (Pieper *et al.*, 2010; Walker *et al.*, 2015). Examples include increased neurogenesis in the hypothalamus of spontaneously hypertensive rats with voluntary exercise (Niwa *et al.*, 2016), enhanced neurogenesis in hypothalamic and habenular regions following chronic fluoxetine treatment (Sachs and Caron, 2015). However, it remains to be verified whether sub-chronic or chronic administration of neurogenic drugs would be adequate for reversing deficits associated with neurological disorders.

Properties of NSC Niche in the Subgranular Zone and the Subventricular Zone

The NSC niche comprises varied types of cells and structures, which include astrocytes, neurons, axon projections and blood vessels. The primary function of this niche is to offer an apt milieu that maintains the bulk of NSCs quiescent and undifferentiated

(Morrison and Spradling, 2008; Urban and Guillemot, 2014). Furthermore, the niche presents a diversity of signals that regulate the behavior of NSCs and/or their progenitors to adjust the production of new cells to the needs of the tissue (Faigle and Song, 2013; Urban and Guillemot, 2014).

In the SGZ of the hippocampus, a subclass of slowly dividing cells that are positive for GFAP and present a single radial process that extends into the molecular layer through the GCL are thought to be NSCs or radial glial cells (Type 1 cells) (Ihrie and Alvarez-Buylla, 2008; Kempermann *et al.*, 2004; Seri *et al.*, 2001). Thus, type 1 cells are a subset of astrocytes but express several NSC markers, which include Sox-2, nestin, vimentin and musashi-1 (Filippov *et al.*, 2003; Hattiangady and Shetty, 2008a; Lagace *et al.*, 2007; Okano *et al.*, 2005). Lugert and colleagues suggest that cells having short, horizontal processes in the SGZ are also Type 1 cells (Lugert *et al.*, 2010). Both cell types (radial and non-radial type 1 cells) seem to change their behavior from active to quiescent states and vice versa (Rolando and Taylor, 2014). Inactivity of NSCs between their proliferative states is believed to prevent their exhaustion in normal conditions as well as reduce chromosomal aberrations resulting from frequent division (Rolando and Taylor, 2014). Type 1 cells do not divide often. When they do, they mostly undergo asymmetric division, which not only maintains self-renewal but also generates cohorts of transit amplifying cells (TAPs, type 2 cells or intermediate progenitors). However, the type of cues determines whether radial or non-radial type 1 NSCs get activated following injury or pathology in the adult DG. A good example for this phenomenon is observations that radial NSCs proliferate in response to physical exercise whereas non-radial NSCs proliferate in conditions such as seizures (Lugert *et al.*, 2010). Type 2 progenitor cells are smaller in size, occur in clusters, express Sox-2 and mammalian achaete-scute homolog 1 (Mash1) and display short tangential processes. These cells typically proliferate briskly and produce a pool of double cortin (DCX) expressing neuroblasts (Type 3 cells) and glia (Encinas *et al.*, 2011; Hsieh, 2012). Neuroblasts mature to form full-fledged granule cells expressing markers such as the neuron specific nuclear antigen (NeuN, a marker of mature neurons), Prox-1 (a transcription factor specific to dentate granule cells) and calbindin (a calcium binding protein typically expressed in mature granule cells). Newly born granule cells also establish afferent connectivity with the axons coming from the entorhinal cortex (perforant path fibers) and efferent connectivity with the CA3 pyramidal neurons. All of these events occur within 6–8 weeks after birth of neurons in the rodent hippocampus (Hsieh, 2012; Zhao *et al.*, 2006). In the SVZ, stemness of primary NSCs (also referred to as type B cells) is maintained by Notch signaling (Basak *et al.*, 2012). Like most type 1 cells in the SGZ, B cells in the SVZ display radial glial cell morphology, express Sox-2, GFAP, nestin and vimentin, and undergo division occasionally to produce Mash1 and nestin expressing type C cells or TAPs. Type C cells divide rapidly and produce type A cells or neuroblasts expressing DCX and polysialylated neural cell adhesion molecule (PSA-NCAM). These neuroblasts migrate along the RMS to the olfactory bulb where they undergo maturation and integration into the circuitry (Lois *et al.*, 1996). Studies have shown that subsets of adult SVZ-NSCs are restricted and varied pertaining to generation of distinct types of neurons (Merkle *et al.*, 2007).

Activity of NSCs can be influenced by a multitude of signals arising from the local milieu. These include signals from different cell types residing in and around

NSC niches, which include astrocytes, oligodendrocyte progenitors, endothelial cells and pericytes in blood vessels, microglia, mature granule cells and GABA-ergic interneurons (Gemma and Bachstetter, 2013; Goldman and Chen, 2011; Licht and Keshet, 2015; Palmer *et al.*, 2000). Moreover, non-cellular factors such as secreted molecules and the extracellular matrix proteins can also influence self-renewal and differentiation properties of NSCs (Hsieh, 2012; Morrens *et al.*, 2012; Song *et al.*, 2002; Tozuka *et al.*, 2005). Particularly, astrocytes have been demonstrated to be a source of many paracrine factors having vital roles in neurogenesis. These comprise Notch, Sonic hedgehog (Shh), bone morphogenetic proteins (BMPs) and Wingless-type MMTV integration site family (Wnt) proteins (Ables *et al.*, 2010; Ahn and Joyner, 2005; Hsieh, 2012; Lie *et al.*, 2005). In the SVZ, it has been shown that notch regulates NSC identity and self-renewal, whereas epidermal growth factor receptor (EGFR)on NSCs specifically affects their proliferation and migration (Aguirre *et al.*, 2010). Moreover, a recent study showed that NSC activity is also influenced by self-derived neurotrophic factors such as vascular endothelial growth factor (VEGF) (Kirby *et al.*, 2015). Thus, NSC division and self-renewal, as well as differentiation of NSC offspring into mature neurons or glia are regulated by signals from other cell types residing in a multicellular niche as well as NSCs themselves.

Factors Regulating of NSC Activity and Neurogenesis in the Adult Hippocampus

A number of transcription factors and signaling pathways are involved in the preservation of NSCs and differentiation of NSC derived cells in the SGZ. First, notch signaling is involved in the preservation of both radial and non-radial type 1 cells with the expression of transcription factors Sox-2 and Hes5 (hairy and enhancer of split 5, encodes a member of a family of basic helix–loop–helix [bHLH] transcription repressors) (Ables *et al.*, 2010; Hsieh, 2012; Lugert *et al.*, 2010; Lugert *et al.*, 2012). Notch signaling negatively regulates cell cycle exit and suppresses the expression of pro-neural transcription factors and thereby blocks differentiation of NSCs into neurons (Rolando and Taylor, 2014). Several other transcription factors also believed to play roles in the maintenance of NSCs, which comprise Pax 6 (paired domain and homeodomain-containing transcription factor), Ascl1 (another bHLH transcription factor), Fox03 (forkhead domain transcription factor), REST (repressor element 1 silencing transcription factor), and TLX (nuclear hormone transcription factor) (Gao *et al.*, 2011; Hsieh, 2012; Jessberger *et al.*, 2008; Renault *et al.*, 2009). Second, the generation of type 2 cells (TAPs) and type 3 cells (neuroblasts) from types 1 and 2 cells respectively, encompasses a successive up-regulation of transcription factors Neurogenin 2 (abHLH transcription factor involved in neural fate-choice decision) and Tbr2 (T-brain gene 2, involved in conversion of radial glia into intermediate progenitors during cortex development). Third, differentiation of neuroblasts into mature neurons involves the expression of NeuroD1 (Neuronal differentiation 1) (Kuwabara *et al.*, 2009; Seib *et al.*, 2013), and Sox-3, Sox-11 and FoxG1 (Forkhead box G1, a Winged Helix transcriptional repressor) (Haslinger *et al.*, 2009; Shen *et al.*, 2006; Wang *et al.*, 2006). Fourth, inhibition of Wnt activity results in reduced neurogenesis whereas its activation via Wnt3a protein increases neurogenesis (Lie *et al.*, 2005). Interestingly, removal of a negative regulator of the Wnt/ß-catenin pathway such as Dickkopf1 (Dkk1) improves neurogenesis (Seib *et al.*, 2013). Fifth, the survival and maturation of newly generated neurons is influenced by Prox-1 (*prospero*-related homeobox gene) (Jessberger *et al.*, 2008; Karalay *et al.*, 2011) and CREB (cyclic AMP response element-binding protein) (Jagasia *et al.*, 2009).

The extent of adult hippocampal neurogenesis is also regulated by neurotransmitters such as the gamma-amino butyric acid (GABA), serotonin (5-hyroxytryptamine, 5-HT), acetylcholine (ACh) and dopamine. Pertaining to GABA, it is fascinating that GABA released by a subpopulation of interneurons expressing the calcium binding protein parvalbumin (PV) in the DG regulates both stem cell quiescence and neuronal cell fate decision by newly born cells. Non-synaptic GABA released by PV+ interneurons represses the activation of quiescent NSCs through activation of γ2-containing $GABA_A$ receptors expressed on NSCs, which averts the exhaustion of primary NSCs by rapid proliferation in normal conditions (Song *et al.*, 2012). Another recent study shows that pharmacological inhibition of $GABA_B$ receptors stimulates NSC proliferation and genetic deletion of $GABA_{B1}$ receptor subunits increases NSC proliferation and differentiation of neuroblasts *in vivo* (Giachino *et al.*, 2014). Thus, by signaling through both $GABA_A$ and $GABA_{B1}$ receptors, GABA regulates the extent of neurogenesis. On the other hand, axons of PV+ interneurons establish immature synapses on newly born neurons to promote the survival of dividing DCX+ immature neurons (Song *et al.*, 2013). Studies on links between 5-HT and hippocampal neurogenesis using increased levels of 5-HT, reduction of serotonergic neurons or stimulation of 5-HT receptors have demonstrated that 5-HT can increase the generation of new neurons in the DG (Doze and Perez, 2012). Besides, a recent study has shown that a serotonergic deficit in the hippocampus as defined by reduced levels of serotonin (5-HT) 1B receptor, decreased 5-HT neurotransmitter levels, and a loss of serotonergic nerve terminals innervating the DG/CA3 subfield (but not associated with changes in the number of serotonergic neurons in the raphe nuclei) can greatly diminish hippocampal neurogenesis (Kohl *et al.*, 2016). Regarding the role of ACh, studies have shown that hippocampal neurogenesis declines with ablation of cholinergic neurons in the medial septum but increases with cholinergic agonist treatment (Mohapel *et al.*, 2005). Additionally, increased expression of vesicular acetylcholine transporter (VAChT) in the hippocampus increases the dendritic complexity of adult born hippocampal neurons and improves acquisition of spatial memory in aging (Nagy and Aubert, 2015). With regard to the influence of dopamine on neurogenesis, studies have demonstrated that dopamine increases hippocampal neurogenesis through activation of D1-like receptor (Takamura *et al.*, 2014) and lesion of dopaminergic neurons impairs adult hippocampal neurogenesis (Schlachetzki *et al.*, 2016).

NSCs and Neurogenesis in the Adult Human Brain

The DG of the human brain has also been validated as a neurogenic region. The first evidence came from postmortem hippocampal tissues of five cancer patients who received intravenous infusion of 5'-bromodeoxyuridine (BrdU) for diagnostic purposes (Eriksson *et al.*, 1998). Investigation of BrdU+ cells in the DG using

BrdU-NeuN and BrdU-GFAP dual immunofluorescence in these tissues revealed ~22% neurons and ~18% astrocytes among newly born cells. The residual BrdU+ cells were smaller in dimensions and seemed to be quiescent undifferentiated NSCs. Cells that expressed BrdU and calbindin were also seen, which presented additional confirmation for the existence of newly born granule cells in the adult human DG. Notwithstanding this finding, skepticism predominated in the field as some studies have implied that neurogenesis in nonhuman primates is relatively scarcer than in rodents (Amrein *et al.*, 2011; Kheirbek and Hen, 2013). However, an investigation using a unique birth-dating methodology centered on the peak of carbon isotope 14 (^{14}C) that was released into the atmosphere during the above ground nuclear bomb tests between 1945 and 1963, presented a strong confirmation for extensive neurogenesis in the DG of humans throughout adulthood (Kempermann, 2013; Spalding *et al.*, 2013). This approach banked on the concept that, eating plants that have soaked up ^{14}C from the atmosphere or consuming meat of animals that ate plants absorbing atmospheric ^{14}C would cause incorporation of ^{14}C into the DNA of dividing human cells (Welberg, 2013). As ^{14}C is merged into the DNA during cell division, the ^{14}C material of a cell would thus portray ^{14}C levels in the atmosphere at the time of birth of the cell. Because atomic bomb testing in 1950s and 1960s triggered a spike in atmospheric ^{14}C levels and intensities weakened after 1963 (because of limited test ban treaty), measurement of the quantity of ^{14}C in cellular DNA offered information on the birth date of cells (Welberg, 2013). Spalding and colleagues sorted neurons and glial cells in hippocampal samples obtained from post-mortem brains of people who were born in various years in the 20th century. Next, they purified neuronal DNA, measured ^{14}C levels and determined the age of granule cells by comparing the amount of incorporated ^{14}C in granule cells with the atmospheric ^{14}C that existed in different calendar years (Spalding *et al.*, 2013). This study uncovered that: (i) hippocampus is the most conspicuous region for adult neurogenesis in the human brain; (ii) ~80% of granule cells in the DG undergo renewal in adulthood and ~700 new neurons are added per day (Kempermann, 2013); (iii) men and women displayed quite similar turnover rate of DG granule cells; (iv) hippocampal neurogenesis in humans undergoes only a modest decrease with aging, unlike in rodents where a dramatic age-related decline in neurogenesis is observed (Hattiangady and Shetty, 2008a; Kuhn *et al.*, 1996; Rao *et al.*, 2005; Rao *et al.*, 2006). Collectively, the findings in this investigation provided a strong evidence for the occurrence of NSCs as well as robust neurogenesis in the adult human hippocampus.

Apart from the hippocampus, studies have examined possible neurogenesis in the neocortex and olfactory bulbs of the human brain. These regions did not reveal significant neurogenesis however (Bergmann *et al.*, 2012; Bhardwaj *et al.*, 2006). Pertaining to the SVZ, an earlier study demonstrated that the infant human SVZ and rostral migratory stream contain an extensive corridor of migrating immature neurons before 18 months of age but this germinal activity subsides in older children and is nearly extinct by adulthood (Sanai *et al.*, 2011).

However, a recent study by Ernst and colleagues, by employing histological and ^{14}C birth-dating approach, suggests that neurons derived from the human SVZ form interneurons in the adjacent striatum (Ernst *et al.*, 2014). This study also suggested that the neuronal turnover in the striatum appears restricted to interneurons, and striatal neurons generated during postnatal periods are preferentially depleted in patients with Huntington's disease, a disorder in which striatal interneurons degenerate. This contrasts with the SVZ of rodents generating neurons that, under normal conditions, migrate only to the olfactory bulb and differentiate into olfactory interneurons (Lim and Alvarez-Buylla, 2014).

Functional Significance of NSC Activity and Neurogenesis in the Adult Hippocampus

Since newly generated neurons (granule cells) integrate well into the existing hippocampal circuitry, questions about their function have received great interest and scrutiny (van Praag *et al.*, 2002). These neurons display increased amenability for long-term potentiation (a substrate underlying learning and memory) for about a month after their birth (Ming and Song, 2011; Schmidt-Hieber *et al.*, 2004; Wang *et al.*, 2000). Thus, persistent production of new dentate granule cells maintains a pool of neurons in the DG with special properties, which also meets some of the unique computational needs of the hippocampus (Appleby *et al.*, 2011; Kempermann, 2013). Furthermore, adult DG neurogenesis is believed to be vital for hippocampus-dependent learning and memory function (Braun and Jessberger, 2014). At first, multiple studies connected the amount of hippocampal neurogenesis with spatial learning and memory functioning and proposed a relationship between spatial memory ability and the quantity of newly born neurons. Supporting evidence comprised observations that: (i) increased hippocampus-dependent memory function and enhanced DG neurogenesis are seen in animals housed in enriched environmental settings, such as larger cages comprising enriching objects/toys or having access to built-in running wheels (Kempermann *et al.*, 1997; van Praag *et al.*, 1999); (ii) aged rats displaying capability for memory retrieval in a water maze probe test have higher levels of DG neurogenesis than aged rats exhibiting inability for retrieval of memory (Drapeau *et al.*, 2003); (iii) co-existence of impaired hippocampus-dependent memory and greatly declined DG neurogenesis in animals that underwent stress (Gould *et al.*, 1999); and (iv) impairments in certain types of hippocampus-dependent memory when DG neurogenesis is ablated using approaches such as whole brain radiation or administration of drugs that target dividing NSCs and their progeny (Shors *et al.*, 2001; Snyder *et al.*, 2005). Despite these observations, opposing outcomes in other studies and misgivings pertaining to the efficiency of methods employed to selectively ablate the entire DG neurogenesis without altering the existing hippocampal cytoarchitecture or circuitry prompted some doubts on the stated function of DG neurogenesis in spatial learning and memory (Braun and Jessberger, 2014; Deng *et al.*, 2010).

A series of subsequent studies however provided stronger verification that populations of new granule cells that survive are those that are successfully integrated into the learning and memory circuitry. First, it has been shown that spatial learning cleverly adds or rejects newly born granule cells contingent upon their stage of the maturation and functional status (Deng *et al.*, 2010; Tronel *et al.*, 2010). This was evidenced through observation that memory retrieval task after the water maze training preferentially activates new granule cells that were ~4–6 weeks old at the time of learning the task and ~10 weeks old at the time of memory retrieval test in mice (Kee *et al.*, 2007). Second, computational models suggested a role for newly born granule cells in the encoding of temporal information as well as in the maintenance of old memories during the encoding of new information (Aimone *et al.*, 2010; Aimone *et al.*, 2011; Deng *et al.*, 2010). Third, elimination of hippocampal neurogenesis through advanced genetic ablation methods caused spatial learning deficits (Dupret *et al.*, 2008; Jessberger *et al.*, 2009; Zhang *et al.*, 2008). Fourth, silencing of ~4-week-old newly born granule cells after water maze learning substantially disrupted the retrieval of spatial memory (Gu *et al.*, 2012). Fifth, many investigations via selective examination of DG circuitry have showed that newly generated neurons influence pattern separation function (Clelland *et al.*, 2009; Kheirbek *et al.*, 2012; Nakashiba *et al.*, 2012; Sahay *et al.*, 2011a; Sahay *et al.*, 2011b). Pattern separation is ability for discerning analogous experiences by storing similar representations in a non-overlapping way (Leutgeb *et al.*, 2007; Yassa and Stark, 2011). Studies have implied that scarceness of activation in the dentate granule cell layer (resulting in sparse functional connectivity between DG and CA3) helps pattern separation function as it enhances global remapping, which is a process by which similar representations are encoded by non-overlapping neuronal ensembles (O'Reilly and McClelland, 1994). As *per* this concept, a recent study suggests that adult born neurons modulate DG excitability via excitation of GABA-ergic interneurons, which in turn increases sparse coding in the granule cell layer and thereby positively influences pattern separation function (Ikrar *et al.*, 2013).

Yet, one of the questions being conferred is, whether the extent of neurogenesis taking place in adulthood is sufficient for affecting hippocampus function. Multiple levels of observations support an affirmative answer to this query. To begin with, the amount of neurogenesis going on after the initial development of the DG is considerable, as ~9,000 new neurons are born every day in the DG of a young adult rat, and the number of new granule cells generated each month is equivalent to ~6% of the total size of the granule cell population (Cameron and McKay, 2001). It is ~700 new neurons *per* day for an annual turnover rate of 1.75% in the human brain (Kempermann, 2013). Yet, the volume of DG does not grow as a function of age because a sizable fraction of newly generated neurons die over a period of time after their birth (Dayer *et al.*, 2003; Gould *et al.*, 1999). Therefore, even with continual neurogenesis, the total number of granule cells in the DG remains nearly stable all through life in normal conditions. Furthermore, though newly born granule cells have been demonstrated to replace some granule cells that were born during early development, they frequently appear to restitute granule cells that were generated in the adulthood (Dayer *et al.*, 2003; Kempermann, 2012). Moreover, newly born granule cells have a lower threshold for eliciting long-term potentiation (LTP), which is an indication of increased synaptic plasticity by these neurons (Schmidt-Hieber *et al.*, 2004; Snyder *et al.*, 2001). Additionally, studies have shown that newly born granule cells constitute a major component of granule cells that contribute to LTP in the DG because the preexisting network is inhibited from being activated (Garthe *et al.*, 2009; Saxe *et al.*, 2006). These observations support the idea that newly born granule cells are more plastic than older granule cells and are more likely to generate action potentials in response to an incoming stimulus. Accordingly, it is likely that young granule cells react to a larger diversity of synaptic input, in comparison to older (preexisting) granule cells that seem to respond to a more specific input. This discrepancy in reaction between young and old granule cells is thought to be advantageous because young granule cells being tuned to a broad variety of inputs can be good integrators and older cells displaying high input specificity can be better separators (Kempermann, 2012; Marin-Burgin *et al.*, 2012). These characteristics seem crucial for the DG to accomplish its functional roles, which encompass enabling pattern separation and circumventing catastrophic interference, giving contextual and affective annotations, and contribute to cognitive flexibility in conditions where fresh information needs to be combined to previously formed representations (Kempermann, 2012; Marin-Burgin *et al.*, 2012). From the above, it appears that incessant production of new granule cells is a necessity for maintaining normal hippocampus function, as it offers a pool of highly plastic immature neurons to enable intricate computational demands (Kempermann, 2012). It is thought that a DG network dominated by young granule cells is predisposed towards construing similar but not identical inputs as separate, whereas older granule cells are prejudiced towards interpreting similar but not identical inputs as alike (Gage and Temple, 2013). Psychiatric disorders are also associated with declined DG neurogenesis (Bergmann and Frisen, 2013). It is believed that fewer newly born neurons impairing pattern separation function also impair the ability for distinguishing threats from similar, but safe, situations and contribute to a generalized perception of fear and anxiety in posttraumatic stress disorder (Bergmann and Frisen, 2013; Kheirbek *et al.*, 2012). Additionally, the extent of increase in hippocampus neurogenesis determines some of the positive effects of antidepressants in animal models (Santarelli *et al.*, 2003). Overall, there is adequate evidence to support the idea that adult hippocampal neurogenesis is one of the important substrates in the brain for sustaining normal cognitive, memory and mood function.

Age-related Changes in NSC Activity and Neurogenesis in the Hippocampus

Neurogenesis declines considerably with age in the hippocampus of rats and mice, the two species most extensively quantified for neurogenesis (Drapeau and Nora Abrous, 2008). A multitude of studies have examined different stages of neurogenesis such as proliferation of NSCs and their progenitors, extent of production of newly born cells *per* day or over a period of several days, short-term and long-term survival of newly born cells, migration of newly born cells into the GCL and differentiation of newly born cells into neurons. Investigation of the proliferation of putative NSCs and the extent of neurogenesis revealed considerable declines between the young adult and middle age (Figure 6.1) (Cameron and McKay, 1999; Kempermann *et al.*, 1998; Kuhn *et al.*, 1996; Lemaire *et al.*, 2006; Rao *et al.*, 2005; Rao *et al.*, 2006). Some studies have also suggested further declines in putative NSC proliferation and neurogenesis between the middle age and old age (Bizon and Gallagher, 2003; Bondolfi *et al.*, 2004; Rao *et al.*, 2006). However, from most studies, it is apparent that most of the decline in neurogenesis during adulthood occurs between young adult and middle age (Driscoll *et al.*, 2006; Nacher *et al.*, 2003; Rao *et al.*, 2005; Seki and Arai, 1995).

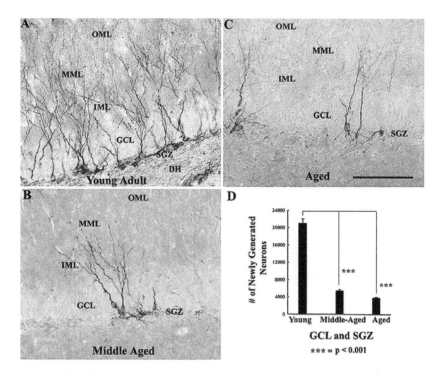

Figure 6.1. Arrangement of newly generated neurons in the dentate gyrus (DG) of young adult (A), middle aged (B), aged (C) rats visualized with doublecortin (DCX) immunostaining. Note that the density of DCX positive neurons is much greater in the young adult DG. DCX positive neurons in the young adult are present throughout the DG along the subgranular zone (SGZ) and the granule cell layer (GCL). Whereas, DCX positive neurons in the middle-aged and aged DG are in clusters of 2–4 cells separated by much larger gaps along the SGZ and GCL. Figure D compares the total number of DCX positive neurons between the young adult, middle aged and aged DG. Note that the total number is reduced dramatically as early as middle age but exhibits no significant change between the middle age and the old age. DH, dentate hilus; IML, inner molecular layer; MML, middle molecular layer; OML, outer molecular layer. Scale bar, A–C = 50 μm. *Figure reproduced from Rao et al., Eur J Neurosci, 21: 464–476, 2005.*

Notwithstanding a drastic decline in the proliferation of NSCs and/or their progenitors, aging does not adversely affect the short-term survival of newly born cells (Kempermann *et al.*, 1998; Lichtenwalner *et al.*, 2001; McDonald and Wojtowicz, 2005; Rao *et al.*, 2005). One of the studies using F344 rats has also demonstrated that the long-term (5-months) survival of adult born neurons does not vary between young adult, middle-aged and aged rats (Rao *et al.*, 2005). Aging however slows down the radial migration of newly born cells from the SGZ to the GCL in the first four weeks after birth (Heine *et al.*, 2004), possibly because of a decline in the expression of PSA-NCAM (Drapeau and Nora Abrous, 2008; Seki and Arai, 1995). Nonetheless, by 5 months after their birth, newly born cells in the young adult, middle-aged and aged animals take up comparable positions in the GCL (Rao *et al.*, 2005). Pertaining to the differentiation of newly born cells into neurons, several studies in mice and a few studies in rats have suggested that neuronal differentiation of newly born cells declines with aging (Bizon *et al.*, 2004; Bondolfi *et al.*, 2004; Heine *et al.*,

2004; Kempermann *et al.*, 1998; Lichtenwalner *et al.*, 2001; Molofsky *et al.*, 2006; Nacher *et al.*, 2003). However, studies in F344 rats did not see significant difference in neuronal fate-choice decision of newly born cells between young, adult, middle-aged and aged rats (Figure 6.2) (Rao *et al.*, 2005; Rao *et al.*, 2006). Some of the discrepancies between these studies may be due to differences in the age of animals examined and/or

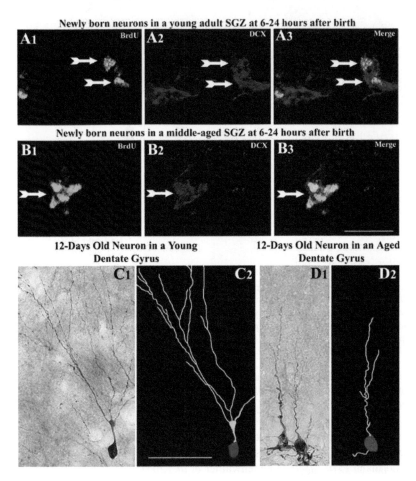

Figure 6.2. A1–B3, Neuronal differentiation of newly born cells at 6–24 hours after birth in the subgranular zone of a young dentate gyrus (A1–A3) and a middle-aged dentate gyrus (B1-B3) visualized using doublecortin (DCX) and 5'-bromodeoxyuridine (BrdU) dual immunofluorescence and confocal microscopy. Note that a great majority of newly born cells express the early neuronal marker DCX in both age groups. Arrows indicate cells that are immunopositive for both BrdU and DCX. Scale bar, 25 μm. C1–D2, Comparison of the growth of 12-days old neurons from a young dentate gyrus (C1) and an aged dentate gyrus (D1) visualized through doublecortin (DCX) and 5'-bromodeoxyuridine (BrdU) dual immunostaining of hippocampal sections at 12 days after four BrdU injections over an 18-hour period. Note BrdU immunoreactivity in the nucleus and DCX immunoreactivity in the soma and dendrites of these cells. C2 and D2 illustrate the dendritic growth of neurons in C1 and D1 traced with Neurolucida. Note that the overall dendritic growth at 12 days after birth is much greater in the neuron from the young dentate gyrus (C2) than the neuron from the aged dentate gyrus (D2). Scale bar, 50 μm. *Figure reproduced from Rao et al., Aging Cell, 5: 545–558, 2006.*

differential survival periods between BrdU administration and analyses (Drapeau and Nora Abrous, 2008). Measurement of dendritic growth of newly born neurons during the DCX expression phase (at ~12 days after birth) revealed retarded maturation of dendrites in middle aged and aged animals (Figure 6.2) (Rao *et al.*, 2006). Collectively, the above studies demonstrated that age-related decline in neurogenesis in rodents is predominantly because of reduced proliferation of NSCs and their progenitors.

Studies have examined the potential mechanisms underlying age-related changes in neurogenesis. Particularly, changes in intrinsic properties of NSCs and their progenitors and alterations in the microenvironment have received considerable interest. Quantification using BrdU labeling and endogenous markers of cell proliferation suggested declines in the size of dividing NSCs and their progenitors by 3–4 times in the SGZ between young adult and middle age (Olariu *et al.*, 2007). Some of the earlier studies have also suggested that the number of putative NSCs (i.e., GFAP, nestin, vimentin expressing radial glia) decline with aging in the GCL (Alonso, 2001; Nacher *et al.*, 2003). However, a study utilizing Sox-2 as a putative marker of NSCs and their progenitors and stereological counts of Sox-2+ cells in the SGZ demonstrated that the overall number of Sox-2+ cells in the SGZ remains constant between young adult, middle-aged, and aged F344 rats though the percentage of Sox-2+ cells expressing proliferation markers such as BrdU and Ki67 declines drastically with age (Figure 6.3) (Hattiangady and Shetty, 2008a). This study implied that, aging is not associated with a decline in the total number of putative NSCs and progenitors in the SGZ but rather associated with a decline in the proportion of NSCs and their progenitors that proliferate (active NSCs). This is a result of increased quiescence of NSCs with aging, perhaps because of age-related changes in the NSC niche and the microenvironment (Drapeau and Nora Abrous, 2008; Hattiangady and Shetty, 2008a). Furthermore, comparison of the percentages of Sox-2+ cells expressing BrdU or Ki-67 in the SGZ, examined after four injections of BrdU (once every 6 hours over a period of 18 hours) and 6 hours of survival period after the last BrdU injection suggested a lengthening of the cell cycle of NSCs between 4-months and 12-months old F344 rats as well as between 12-months and 24-months old F344 rats (Hattiangady and Shetty, 2008a). However, studies in younger animals (young adult and 10-months old Sprague-Dawley rats), using a protocol of multiple BrdU and tritiated thymidine injections coupled with endogenous proliferation markers, did not show such lengthening of the cell cycle (Olariu *et al.*, 2007). The discrepancy is likely due to differences in the age and/or species of animals examined in the two studies. Collectively, the above results demonstrated that, aging is not associated with NSC loss in the SGZ of hippocampus. This conclusion is also corroborated by another rat study, which suggested that increased numbers of quiescent NSCs in the SGZ in aging is due to reductions in Notch-dependent active NSCs (Lugert *et al.*, 2010). However, a study using a mouse model has suggested that NSC pool gets exhausted in aging because of differentiation of NSCs into astrocytes (Encinas *et al.*, 2011), implying species-specific differences regarding the persistence of SGZ-NSCs in aging. It is likely that higher order animals maintain NSCs in greater numbers during old age. This notion is supported by observations that only moderate decrease in hippocampal neurogenesis occurs during aging in humans (Spalding *et al.*, 2013).

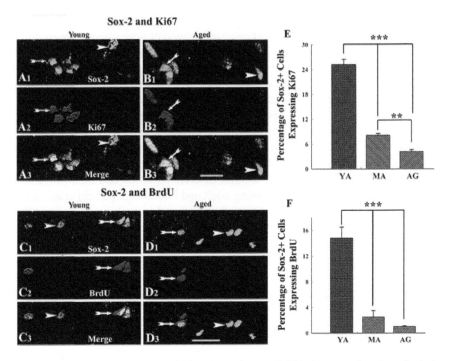

Figure 6.3. Sox-2 immunoreactive cells in the subgranular zone (SGZ) of young and aged rats displaying Ki67 (A1–B3) or BrdU (C1–D3), visualized by dual immunofluorescence and Z-section analyses using confocal microscopy. Note that all proliferating cells (as identified by the expression of Ki67 or BrdU) are positive for Sox-2 (arrows). Arrowheads denote Sox-2 immunoreactive cells lacking Ki67 or BrdU. Scale bar, 25 μm. Figures E and F illustrate the percentages of Sox-2 immunoreactive cells expressing Ki67 (E) and BrdU (F). Note that the percentages of Sox-2 immunoreactive cells expressing Ki67 or BrdU decline considerably during the course of aging. ** = $p < 0.01$; *** = $p < 0.001$. *Figure reproduced from Hattiangady and Shetty, Neurobiol Aging, 29: 129–147, 2008.*

Studies have also examined the potential reasons for decreased proliferation of NSCs with aging. A quantitative real time RT-PCR study showed that age-related decrease in hippocampal neurogenesis is not associated with a decline in the expression of multiple genes important for NSC proliferation and neurogenesis in the DG (Shetty *et al.*, 2013). Most genes important for cell cycle arrest, regulation of cell differentiation, growth factors and cytokine levels, synaptic functions, apoptosis, cell adhesion and cell signaling, and regulation of transcription displayed stable expression in the DG with aging. The exceptions included increased expression of genes important for NSC proliferation and neurogenesis (Stat3 and Shh), DNA damage response and NF-kappaB signaling (Cdk5rap3), neuromodulation (Adora1), and decreased expression of a gene important for neuronal differentiation (HeyL). However, several age-related changes in the microenvironment can suppress NSC proliferation. First, increased levels of stress hormone corticosterone during aging (Drapeau and Nora Abrous, 2008; Sapolsky, 1992) is likely involved in reducing neurogenesis in aging, as adrenalectomy in senescent rats reduces corticosterone levels in association with increased proliferation of NSCs and enhanced neurogenesis (Cameron and McKay, 1999; Montaron *et al.*, 2006; Montaron *et al.*, 1999). Second, increased glutamatergic neurotransmission or hyperexcitability in the aged DG likely contributes to decreased neurogenesis in aging, as injections of NMDA receptor antagonist enhance NSC proliferation in the SGZ (Nacher *et al.*, 2003). Third, age-related decreases in the concentration of NSC mitogenic factors such as epidermal growth factor (EGF), fibroblast growth factor-2 (FGF-2), insulin-like growth factor-1 (IGF-1), brain-derived neurotrophic factor (BDNF), vascular endothelial growth factor (VEGF) and/or some of their receptors may also underlie decreased NSC proliferation and neurogenesis in aging (Figure 6.4) (Chadashvili and Peterson, 2006; Enwere *et al.*, 2004; Lai *et al.*, 2000; Shetty *et al.*, 2005; Sonntag *et al.*, 1999). Indeed, administration of

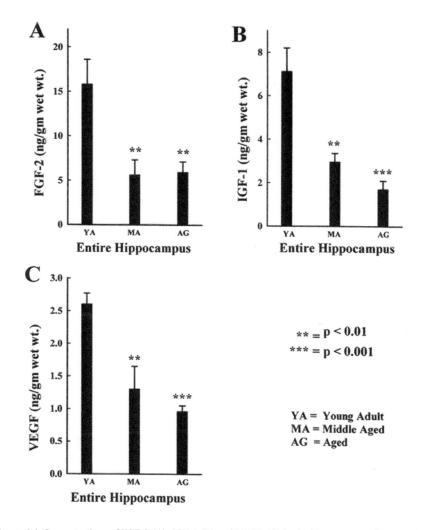

Figure 6.4. Concentrations of FGF-2 (A), IGF-1 (B) and VEGF (C) in the hippocampus of young adult (YA) middle-aged (MA) and aged (AG) F344 rats. Note that the concentrations of FGF-2, IGF-1 and VEGF decrease considerably between young age and middle age but exhibit no significant decrease between middle age and old age. *Figure reproduced from Shetty et al., Glia, 51: 173–186, 2005.*

some of these neurotrophic factors (EGF, FGF-2, IGF-1) increases NSC proliferation and neurogenesis in the aged DG (Figure 6.5) (Enwere *et al.*, 2004; Jin *et al.*, 2003; Lichtenwalner *et al.*, 2001; Rai *et al.*, 2007). Fourth, age-related reductions in microvasculature, decreased cerebral blood flow, low-grade inflammation and increased distance between NSCs and capillaries in the hippocampus (Hattiangady

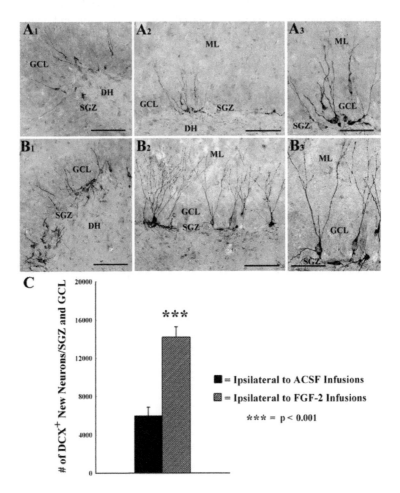

Figure 6.5. Distribution of newly born neurons in the subgranular zone (SGZ) and the granule cell layer (GCL) at 8 days following two-week infusions of ACSF or FGF-2 into the posterior lateral ventricle, visualized by doublecortin (DCX) immunostaining. A1 and A2 show DCX⁺ neurons in representative regions of the crest and upper blade of the SGZ and GCL ipsilateral to ACSF infusions. B1 and B2 illustrate DCX⁺ neurons in representative regions of the crest and upper blade of the SGZ and GCL ipsilateral to FGF-2 infusions. A3 and B3 are magnified views of a region from A2 and B2 demonstrating the morphology and extent of dendrites in individual neurons. ML, molecular layer; DH, dentate hilus. Scale bar, A1, A2, B1 and B2 = 100 μm; A3 and B3 = 50 μm. Bar chart in C compares the status of neurogenesis (based on the total number of DCX⁺ newly born neurons) between SGZ and GCL ipsilateral to ACSF infusions and SGZ and GCL ipsilateral to FGF-2 infusions. Note that the amount of neurogenesis is 138% greater in the SGZ and GCL ipsilateral to FGF-2 infusions than SGZ and GCL ipsilateral to ACSF infusions. *Figure reproduced from Rai et al., Eur J Neurosci, 26: 1765–1779, 2007.*

and Shetty, 2008a; Kodali *et al.*, 2015; Sonntag *et al.*, 1997) may also be contributing to reduced neurogenesis in aging, as strategies that improve microvasculature density and diminish inflammation enhance neurogenesis in the aged hippocampus (Kodali *et al.*, 2015). Thus, a multitude of age-related changes in the hippocampus contribute to decreased NSC activity and neurogenesis in aging.

An interesting issue is whether age-related decrease in DG neurogenesis contributes to memory dysfunction in aging. In normal conditions, animals with preserved spatial memory (aged unimpaired) display a higher level of NSC proliferation and increased survival of newly born cells and neurons, in comparison to animals displaying spatial memory impairments (aged impaired), which suggested that neurogenesis participates in learning and memory (Drapeau *et al.*, 2003; Drapeau and Nora Abrous, 2008; Driscoll *et al.*, 2006). However, a few other studies failed to demonstrate a link between decreased SGZ cell proliferation and spatial memory (Bizon and Gallagher, 2003; Merrill *et al.*, 2003), likely due to differences in strain, gender and number of animals employed as well as timing of BrdU injection and analyses. Additional studies showed that, in aged unimpaired rats, learning increases the survival of cells generated ~1 week before the learning episode but decreases the survival of cells produced during the early phase of learning (Drapeau *et al.*, 2007), implying that spatial memory abilities critically depend on the dynamic regulation of adult-born neuron survival. Furthermore, strategies that increase DG neurogenesis also improve memory function in aged animals. For instance, both short-term and long-term exposures to enriched environment enhanced DG neurogenesis in aged mice along with improved memory function (Kempermann *et al.*, 2002; Kempermann *et al.*, 1998). Furthermore, in middle-aged and aged animals, physical exercise enhanced neurogenesis along with improved spatial learning ability (Kronenberg *et al.*, 2006; Lugert *et al.*, 2010; van Praag *et al.*, 2005). Likewise, prevention of memory dysfunction observed in aged F344 rats following resveratrol treatment in late middle age was found to be associated with more than 2 fold increase in hippocampal neurogenesis (Kodali *et al.*, 2015). Thus, several correlative studies support the notion that declined hippocampal neurogenesis is one of the factors contributing to age-related memory dysfunction. Interestingly, the overall decline in DG neurogenesis during adult life is relatively modest in normal humans (Spalding *et al.*, 2013), in comparison to a massive decline seen in rodents (Rao *et al.*, 2005; Rao *et al.*, 2006). This may suggest that humans with intact memory function likely maintain higher levels of neurogenesis and individuals afflicted with age-related memory dysfunction display reduced levels of neurogenesis because of increased quiescence of NSCs. Stimulation of quiescent NSCs in such individuals may improve memory function. However, better understanding of these issues will require development of advanced *in vivo* imaging techniques that quantify NSC activity and the extent of neurogenesis in the human brain.

Activity of Hippocampal NSCs and Neurogenesis in Alzheimer's Disease

Alzheimer's disease (AD) affects ~5 million Americans and ~45 million people worldwide. Difficulties for learning, reasoning, judgment and communication, depression and personality changes are features seen in individuals afflicted with AD (Goedert and Spillantini, 2006; Mu and Gage, 2011). A multitude of factors can lead to AD, which include genetics (e.g., presence of a risk gene apolipoprotein E ε4), and environmental and lifestyle factors such as heart disease, stroke, high blood pressure, diabetes and obesity (Reitz and Mayeux, 2014). Progression of AD symptoms is typically associated with the accumulation of amyloid-beta (Aβ) in the extracellular space (amyloid plaques) and tau inside neurons (neurofibrillary tangles) (Mu and Gage, 2011; Reitz and Mayeux, 2014). Release Aβ proteins occur when amyloid precursor protein (APP) is sequentially cleaved by β- and γ-secretases. As plaques and tangles accumulate, other changes such as impaired transfer of information at synapses, loss of synapses and neurodegeneration become evident (Selkoe, 2001).

Neuropathology in AD can affect multiple brain areas but the hippocampus is one of the most vulnerable brain regions in AD (Mu and Gage, 2011; Ohm, 2007; Scheff and Price, 2006). It has been suggested that, in the early stages of AD, alterations in NSC activity and hippocampal neurogenesis underlie memory and mood impairments (Lazarov *et al.*, 2010). It appears that augmented NSC activity and hippocampal neurogenesis generally associate with the severity of the disease in senile AD (Jin *et al.*, 2004b). Yet, investigations on younger (pre-senile) AD brains imply either no changes or diminished neurogenesis with increased levels of BMP (Boekhoorn *et al.*, 2006; Crews *et al.*, 2010). Investigations of NSC activity and neurogenesis in animal models of AD have also provided conflicting findings. In a transgenic mouse model of AD with the Swedish and Indiana mutations under the platelet derived growth factor promoter, increased neurogenesis was seen (Jin *et al.*, 2004a; Lopez-Toledano and Shelanski, 2007), which is consistent with the findings in senile AD brains (Jin *et al.*, 2004a). Analyses in other models of AD have mostly revealed decreased neurogenesis however. These comprise the presinilin1 (PS1) conditional knock-out mouse (Feng *et al.*, 2001), mouse overexpressing a mutant form of APP (the Swedish mutation APP695) (Haughey *et al.*, 2002), the mutant PS1 knock-in mouse (Wang *et al.*, 2004), the mutant P117L transgenic mouse (Wen *et al.*, 2004), the homozygous PDAPP mouse (Donovan *et al.*, 2006), the "double knock-in" mouse carrying targeted mutations in both APP and PS-1 (Zhang *et al.*, 2007), the mouse overexpressing the Swedish variant of APP and the exon 9-deleted variant of human PS1 (Verret *et al.*, 2007), the triple transgenic mouse 3xTg-AD that harbors three mutant genes (APPswe, PS1M146V and tauP301L) (Rodriguez *et al.*, 2008), and Tg2576 transgenic mouse overexpressing the human APP isoform harboring the Swedish double mutation (Krezymon *et al.*, 2013). Disagreements in findings between different investigations likely echo dissimilarities in the stage of disease at the time of examination, drug treatment given to patients and the approaches used for analyzing proliferating cells (Jin *et al.*, 2006). Systematic investigation of larger cohorts of AD brains at different stages of the disease will be required to fully understand the links between AD and changes in neurogenesis. In addition, issues such as the age of patient at the time of AD onset, the type and doses of drugs taken, and the duration of drug therapy in different AD patient populations need to be considered. Diminished neurogenesis observed in some mouse models of AD may reflect the evolved stage of the disease at the time of investigation because intensified unfavorable alterations are probable in the milieu with time after disease onset (Chen *et al.*, 2008; Perry *et al.*, 2012).

It is currently unknown whether therapies that normalize the activity of NSCs and neurogenesis would slow down the evolution of AD when such interventions were applied at the early stage of disease. Animal studies performed so far are mostly supportive of this possibility however. A multitude of approaches that augment neurogenesis have been shown to enhance cognitive functioning in several transgenic mouse models of AD. The stratagems comprise physical exercise or exposure to enriched environment (Costa *et al.*, 2007; Nichol *et al.*, 2009; Nichol *et al.*, 2007; Tapia-Rojas *et al.*, 2016), inhibition of microglial activation (Biscaro *et al.*, 2012), choline supplementation (Velazquez *et al.*, 2013), neural precursor cell grafting (Ben-Menachem-Zidon *et al.*, 2014), curcumin treatment (Tiwari *et al.*, 2014) electro acupuncture (Li *et al.*, 2014) and directed expression of Neurod1 in cycling hippocampal progenitors (Richetin *et al.*, 2015). Considering these, examining neurogenesis in patients with mild cognitive impairment is important because normal aging is not associated with significant decreases in neurogenesis in humans (Spalding *et al.*, 2013) and impaired neurogenesis likely precedes the deposition of Aβ plaques (Chuang, 2010) and accumulation of hyperphosphorylated microtubule associated protein tau. Indeed, a recent study using a mouse model of tauopathy showed that neurogenesis declines prior to a robust aggregation of hyperphosphorylated tau (Komuro *et al.*, 2015). This necessitates development of more advanced techniques for accurately quantifying neurogenesis *in vivo* in healthy and disease conditions (Ho *et al.*, 2013), as currently available techniques such as the measurement of blood flow in the DG or the detection of a spectroscopic biomarker by magnetic resonance spectroscopy can only provide an indirect measure of neurogenesis in human brains (Manganas *et al.*, 2007; Pereira *et al.*, 2007).

Links between NSC Activity and Psychiatric Disorders

Aberrant NSC activity may contribute to several psychiatric disorders such as schizophrenia, autism spectrum disorders (ASDs) and major depressive disorder (MDD). First, mutations in Disrupted-in-schizophrenia (DISC1) cause an extremely rare monogenic form of schizophrenia (Ladran *et al.*, 2013). Consistent with this, dominant negative DISC1 expression during mouse cortical development causes cellular and behavioral phenotypes similar to that seen in schizophrenia (Clapcote *et al.*, 2007; Kamiya *et al.*, 2005), and silencing of DISC1 in NSCs causes depolarizing GABA signaling, which leads to hypertrophy of soma, accelerated dendritic growth of newly born neurons, abnormal neural organization and aberrant behavior (Kim *et al.*, 2012). Second, a deletion at 22q11 enhances the genetic risk of schizophrenia and ASDs (Ladran *et al.*, 2013). Accordingly, diminished dosage of 22q11 genes in a mouse model results in reduced proliferation of NSCs (particularly the intermediate progenitors), and aberrant migration and altered connectivity of neural progenitors during development (Meechan *et al.*, 2009). Furthermore, loss of an autism risk gene (mef2c) in NSCs has been shown to cause cellular and behavioral phenotypes comparable to that seen in ASDs (Ladran *et al.*, 2013).

Major depressive disorders are also associated with altered NSC activity and hippocampal neurogenesis. Nearly 15% of the population is afflicted with depression and considerable morbidity and disability is seen in people affected with MDDs (Pandya *et al.*, 2012). One of the hallmarks of MDDs is a reduced size of the hippocampus, which likely reflects declined DG neurogenesis for extended periods (Lee *et al.*, 2013; Masi and Brovedani, 2011). However, contribution of other changes such as neurodegeneration, reduced branching of axons and dendrites and diminished synapses cannot be ruled (Duman and Aghajanian, 2012; Masi and Brovedani, 2011; Sapolsky, 2001). Several changes in the hippocampus likely contribute to altered NSC activity in depression, which may include decreased serotonergic neurotransmission, increased levels of stress and cortisol and reduced levels of p-CREB and BDNF (D'Sa and Duman, 2002). This premise is based on observations that, selective serotonin reuptake inhibitors (SSRIs) such as fluoxetine augment neurogenesis via stimulation of the $5HT_{1A}$ receptors (Santarelli *et al.*, 2003), stress enhances cortisol levels and decreases neurogenesis (Gould *et al.*, 1997; Sapolsky, 2004), and lower concentration of p-CREB and BDNF diminish neurogenesis (Lee *et al.*, 2002; Nakagawa *et al.*, 2002). While the precise mechanisms through which neurogenesis affects mood are remain to be identified, investigations in animal prototypes have offered support for a link between mood and hippocampal neurogenesis. For instance, experimental approaches that provoke depressive-like behavior (e.g., maternal separation or social hierarchy related stress) also decrease neurogenesis (Lajud *et al.*, 2012; Mirescu *et al.*, 2004). Additionally, enhanced neurogenesis with improved cognitive and mood function is consistently seen in animals exposed to antidepressant activities such as physical exercise, enriched environment or treated with antidepressant drugs (Lee *et al.*, 2013; Sahay and Hen, 2007; Santarelli *et al.*, 2003; van Praag *et al.*, 1999).

Since not all drugs with antidepressant effects require adequate neurogenesis to ease depression, enhanced neurogenesis outcome after certain antidepressant treatments was once regarded as an epiphenomenon (David *et al.*, 2009). Furthermore, genetic alterations that increase hippocampal neurogenesis in normal animals enhance pattern separation function but do not seem to improve mood function (Sahay *et al.*, 2011a). However, several other studies corroborate the idea that positive effects of many antidepressants are mediated through improved DG neurogenesis. This is substantiated through following observations. The time of onset of increased neurogenesis correlates with the beneficial effect of antidepressant treatment. This was clear from the failure of fluoxetine and imipramine to have an antidepressant effect when proliferating NSCs in the hippocampus were ablated (Santarelli *et al.*, 2003). Studies also suggest that newly generated neurons serve as a buffer for stress response through their effects on the hypothalamic–pituitary–adrenal (HPA) axis (Snyder *et al.*, 2011). Furthermore, therapeutic outcomes from antidepressants occur only with a chronic treatment schedule, which suggest that antidepressants do not have an acute effect on serotonergic neurotransmission. These observations rather point to a requirement for antidepressant-induced improvements in neurogenesis, which likely takes several weeks before drug-induced neurons can become functionally integrated (Braun and Jessberger, 2014). Moreover, diminished anxiety and depression-related behaviors were observed when adult neurogenesis was enhanced via removal of pro-apoptotic gene Bax in NSCs, in mice treated chronically with corticosterone (Hill *et al.*, 2015).

Also, increased depressive-like phenotype is seen in adulthood when genetic variations affect postnatal neurogenesis (such as deletion of huntingtin-associated protein 1 expression) (Xiang *et al.*, 2015). From the above, it is apparent that modified activity of NSCs and neurogenesis is one of the conspicuous characteristics of MDD. Some investigations also support the idea that decreased neurogenesis leads to depression. Thus, drugs or compounds capable of augmenting NSC activity and neurogenesis may alleviate cognitive and mood dysfunction in people afflicted with MDD. Such drugs may also sustain positive mood in normal individuals.

NSC Activity and Neurogenesis in Status Epilepticus and Temporal Lobe Epilepsy

Epilepsy, the third most prevalent neurological disorder in the USA after Alzheimer's disease and stroke, affects > 2 million Americans and ~65 million people worldwide. It is an ailment characterized by spontaneous recurrent seizures (SRS) and comorbidities such as cognitive dysfunction and depression. Temporal lobe epilepsy (TLE) is the most common type of epilepsy among partial epilepsies and accounts for ~60% of patients with epilepsy. It is frequently resistant to antiepileptic drugs (Engel, 2001; Strine *et al.*, 2005). While the etiology of TLE is unknown in most cases, an initial precipitating injury (IPI) such as status epilepticus (SE), head injury with loss of consciousness, injuries/concussions during early childhood, malformations in brain development, infections such as meningitis or encephalitis or childhood febrile seizures leads to TLE in others (Lewis, 2005; Pitkanen and Sutula, 2002). Characteristically, a dormant period of several years is observed between an IPI and the emergence of chronic TLE illustrated by SRS originating from the temporal lobe, and impairments in cognition and mood (Bell *et al.*, 2011; Butler and Zeman, 2008; Devinsky, 2004). In addition, TLE is often accompanying with hippocampal sclerosis due to a sizeable loss of neurons in the dentate hilus, and CA1 and CA3 subfields. Studies performed in various models of TLE in the last fifteen years have pointed out that altered activity of NSCs and abnormal integration of newly born neurons are some of the most conspicuous aberrant alterations observed in TLE (Hester and Danzer, 2014; Kuruba *et al.*, 2009; Parent *et al.*, 1997; Scharfman and Gray, 2007).

Numerous animal model studies have demonstrated greatly enhanced NSC proliferation and neurogenesis in the early phase of TLE (for 2–3 weeks following acute seizures or SE) (Bengzon *et al.*, 1997; Hattiangady *et al.*, 2004; Parent *et al.*, 1997). Nonetheless, acute seizures do not modify the division of type 1 NSCs in the SGZ but promote considerable proliferation of TAPs (type 2 cells) and immature neurons (Jessberger *et al.*, 2005). Indeed, the hippocampus from young TLE patients (< 2–4 years old) exhibits increased proliferation of NSC-like cells (Blumcke *et al.*, 2001), indicating that the commencement of TLE in pediatric patients is likely accompanied by augmented DG neurogenesis. Although the precise processes prompting the seizure-induced surge in DG neurogenesis are unknown, several possibilities have been proposed. For instance, proliferation of types 2 and 3 cells as well as the facilitation of neuronal differentiation of newly born cells are thought to be mediated by multiple neurotrophic factors that are believed to be released from dying neurons, deafferented neurons and reactive glia after seizures (Banerjee *et al.*, 2005; Bugra *et al.*, 1994; Croll *et al.*, 2004; Lowenstein *et al.*, 1993; Shetty *et al.*, 2004; Shetty *et al.*, 2003). Moreover, elevated concentrations of gamma-amino butyric acid (GABA) and neuropeptide Y (NPY) in the DG during the early post-seizure period are believed to play roles, as both GABA and NPY can considerably influence DG neurogenesis (Ge *et al.*, 2007; Masiulis *et al.*, 2011; Scharfman and Gray, 2006). Additionally, it has been suggested that repressor element 1-silencing transcription factor (REST, a transcriptional repressor that functions to limit transcription of target genes) plays a role in seizure-induced neurogenesis surge (Jessberger *et al.*, 2007). Increased proliferation of NSCs due to amplified neuronal activity during and after seizures cannot be ruled out, as increased excitatory stimuli is known to influence NSC proliferation and the production of new neurons (Deisseroth *et al.*, 2004).

Increased DG neurogenesis in the early phase of TLE is also associated with aberrant migration of a significant fraction of newly born granule cells into the hilus and the molecular layer (Figure 6.6) (Hattiangady *et al.*, 2004; Rao *et al.*, 2008; Hester and Danzer, 2014; Parent *et al.*, 1997; Scharfman *et al.*, 2000; Scharfman and Gray, 2007; Shetty and Hattiangady, 2007). One of the likely causes of such abnormal migration is the loss or reduced levels of reelin (a migration guidance cue released from a subclass of GABA-ergic interneurons in the dentate hilus) after acute seizures (Gong *et al.*, 2007). Because of their abnormal incorporation into the CA3 network, the ectopic newly born granule cells likely become seizure inducing "hub cells". They seem to facilitate the development of pro-epileptogenic circuitry, as they display activation when epileptic rats are experiencing SRS and react with a longer latency to onset of evoked responses after perforant path fiber stimulation (Scharfman *et al.*, 2003a; Scharfman *et al.*, 2003b; Scharfman *et al.*, 2002). It has also been shown that, these cells display basal dendrites and establish afferent connectivity with mossy fiber terminals (axons of other granule cells) (Pierce *et al.*, 2005; Shapiro *et al.*, 2005; Walter *et al.*, 2007), sprout axons (mossy fibers) abnormally into the dentate inner molecular layer (Hester and Danzer, 2014), exhibit spontaneous bursts of action potentials and contribute to SRS in chronically epileptic animals (Jung *et al.*, 2004; McCloskey *et al.*, 2006; Scharfman *et al.*, 2000). Hippocampal tissues obtained from patients with TLE also comprise similarly displaced granule cells (Houser, 1990).

While studies in some animal models of TLE had raised doubts about the role of ectopically placed newly born granule cells in the development of chronic TLE (Pekcec *et al.*, 2011; Pekcec *et al.*, 2007; Raedt *et al.*, 2007), findings in other animal prototypes of TLE consistently supported the involvement of these cells as one of the pathophysiological changes contributing to SRS. These include observations that the frequency of SRS declines when seizure-induced neurogenesis is reduced through antimitotic agent treatment (Jung *et al.*, 2004) and animals with fewer ectopic newly born neurons display reduced numbers of seizures than animals exhibiting large numbers of ectopic newly born neurons (McCloskey *et al.*, 2006). Suppression of this aberrant neurogenesis may also minimize co-morbidities of TLE such as cognitive and mood dysfunction because animal studies have shown that blockage of abnormal DG neurogenesis after acute seizures can prevent seizure-induced cognitive impairments (Jessberger *et al.*, 2007; Pekcec *et al.*, 2008). Nonetheless, a recent study has introduced some fresh doubts about the role of seizure-induced ectopic granule cells in promoting

Intact Young DG

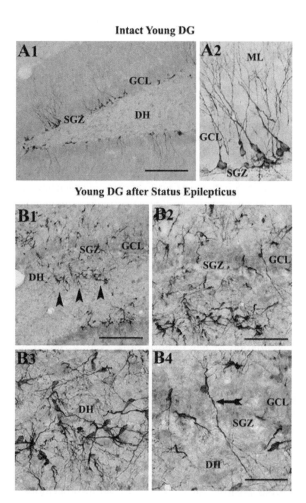

Young DG after Status Epilepticus

Figure 6.6. Distribution of newly born neurons in the subgranular zone-granule cell layer (SGZ-GCL) of different groups, visualized through doublecortin (DCX) immunostaining. The groups include DG of a naïve young adult (A1, A2) and a young adult rat at 13 days after status epilepticus (B1, B2). A2 and B2 are magnified views of regions from A1 and B1. Note that the density of newly born neurons increases after SE (B1, B2), in comparison to the age-matched naïve rat (A1, A2). Furthermore, in the rat that underwent SE, there is an increased density of ectopically migrated newly born neurons in the dentate hilus (indicated by arrowheads in B1 and magnified views in B2 and B3) and molecular layer (B1, B2). Figure B4 shows some newly born neurons with long basal dendrites (arrows). Scale bars, A1, B1 = 200 μm; B2 = 100 μm A2, B3, B4 = 50 μm. *Figure reproduced from Rao et al., Hippocampus, 18: 931–944, 2008.*

SRS. In this study, genetic ablation of neurogenesis that occurred in the four weeks prior to the induction of acute seizures (in nestin-thymidine kinase [nestin-TK] EGFP mice using ganciclovir treatment) resulted in reduction of chronic seizure frequency for prolonged periods and prevented epilepsy-associated cognitive dysfunction (Cho *et al.*, 2015). Interestingly, ablation of neurogenesis that occurred in the weeks before and the weeks after acute seizure insult did not decrease the frequency of SRS. Some of this discrepancy may be attributable to the loss of proliferating nestin-TK+ reactive astrocytes in the hippocampus when seizure-generated new nestin+ cells (precursors of new dentate granule cells) in the DG were ablated (Cho *et al.*, 2015). Another issue is that, not all seizure-generated new granule cells take up ectopic positions. Significant numbers of seizure-generated new granule cells migrate appropriately into the granule cell layer as in the normal brain, display hypoexcitability (unlike ectopically placed new granule cells) and likely serve as an inhibitory gate to prevent the propagation of seizure activity entering the hippocampus (Iyengar *et al.*, 2015; Murphy *et al.*, 2011). Ablation of such normal granule cells that are born after acute seizures may therefore cause increased epileptogenesis and exacerbate the frequency and/or intensity of spontaneous seizures. Thus, despite a large number of studies, the precise role of post-seizure neurogenesis in epileptogenesis is still unclear. Advanced techniques that selectively remove/silence ectopically placed newly born granule cells in the dentate hilus and the molecular layer but spare appropriately integrated newly born granule cells in the GCL will be needed to address this issue further.

The chronic phase of TLE is associated with greatly waned neurogenesis (Hattiangady *et al.*, 2004; Hattiangady and Shetty, 2010). An inverse relationship has been observed between the frequency of SRS and the extent of neurogenesis (Hattiangady *et al.*, 2004). However, chronic epilepsy does not affect the overall addition or long-term survival of newly born cells in the GCL but greatly reduces the neuronal differentiation of newly born cells. Only 4–5% of newly born cells differentiate into neurons in the chronic phase, in comparison to 73–80% of such cells exhibiting neuronal differentiation in the intact hippocampus (Hattiangady and Shetty, 2010). Thus, decreased neurogenesis in chronic epilepsy is a result of dramatic declines in the neuronal fate-choice decision of newly born cells. Apart from a major decrease, neurogenesis in the chronic phase of epilepsy maintains abnormal features seen in the early phase, which include abnormal migration of some newly born neurons into the hilus and occurrences of basal dendrites from newly born neurons that incorporate into the GCL (Hattiangady and Shetty, 2008b). Decreased density of newly born neurons positive for PSA-NCAM has also been observed in the DG samples of children exhibiting frequent SRS and adult TLE patients (Crespel *et al.*, 2005; Fahrner *et al.*, 2007; Mathern *et al.*, 2002; Pirttila *et al.*, 2005). Greatly waned neurogenesis in chronic epilepsy appears to be due to multiple factors. These include modified microenvironment of NSC niches in the SGZ including reduced levels of neurotrophic factors (such as BDNF, FGF-2 and IGF-1) vital for NSC activity and neuronal differentiation (Hattiangady *et al.*, 2004; Shetty *et al.*, 2003), diminution in the numbers of multipotent NSCs that can generate neurons (Hattiangady and Shetty, 2008b) and modifications in signaling pathways initiating reduced neuronal differentiation of NSC progeny, such as an enhanced concentration of Dickkopf-1 (an inhibitor of neurogenesis promoting protein Wnt) (Busceti *et al.*, 2007; Lie *et al.*, 2005). Persistence of abnormal migration of newly born neurons in the chronic phase of epilepsy likely reflects reduced concentration of reelin (a protein that guides the migration of newly born neurons into the GCL) due to considerable loss of reelin+ interneurons in the dentate hilus (Gong *et al.*, 2007). Considering the functions of hippocampal neurogenesis discussed earlier, altered neurogenesis is likely one of the major causes of memory and mood impairments in TLE patients. This association is supported by the finding that the overall granule cell density is a significant predictor accounting for the total memory capacity in an individual TLE patient (Pauli *et al.*, 2006; Siebzehnrubl and Blumcke, 2008) and increased production of new neurons in the hippocampus is essential for effective action of antidepressants (Perera *et al.*, 2007; Sahay and Hen, 2007; Santarelli *et al.*, 2003). While application of strategies that increase neurogenesis seem appealing for easing memory and mood impairments in chronic epilepsy (Barkas *et al.*, 2012; Shetty, 2011), the overall benefit from increasing neurogenesis in chronic epilepsy will however be contingent upon whether newly born granule cells take up positions in the GCL and establish normal connectivity with the residual CA3 pyramidal neurons. Strategies capable of not only enhancing the proliferation of NSCs and neuronal differentiation of NSC progeny but also increasing reelin levels in the hippocampus may be required to normalize neurogenesis in chronic epilepsy conditions.

Conclusions

Neural stem cells that endure in most regions of the adult brain exist in a state of quiescence whereas NSCs in the SGZ of the hippocampus and the SVZ lining the lateral ventricles maintain significant proliferative activity and generate new neurons throughout life. In the human brain, the hippocampus neurogenesis is widespread and adds nearly 700 neurons per day whereas the SVZ neurogenesis is limited to contributing some medium sized spiny neurons to the striatum. Furthermore, from a functional perspective, NSCs in the adult hippocampus have received greater attention because neurons generated from these cells continually get incorporated into the adult hippocampus circuitry and influence functions such as memory and mood. Proliferation of NSCs and neurogenesis in the adult hippocampus are regulated by multiple physiological stimuli and signaling pathways. The process of neurogenesis by NSCs and the behavior of newly born neurons are also sensitive to pathological environment. Altered behaviors of NSCs and/or newly born neurons have been shown to contribute to functional impairments in aging and several neurological disorders. Neural stem cells in the adult hippocampus appear suitable as therapeutic targets for the treatment of hippocampus dysfunction in aging, early stages of the AD, depressive disorders and chronic epilepsy because their stimulation and the resulting increased neurogenesis would likely be beneficial for improving both memory and mood function.

Acknowledgements

This work was supported by grants from the State of Texas (Emerging Technology Funds to A.K.S.), the Department of Defense (Peer Reviewed Medical Research Program Grant to A.K.S.), and the Department of Veterans Affairs (Research Career Scientist and VA Merit Review Awards to A.K.S.). Authors thank Drs. M.S. Rao, R. Kuruba, B. Shuai, G. Shetty, V. Parihar, M. Kodali and B. Waldau for their excellent contributions to SHETTY LAB neural stem cell and neurogenesis studies discussed in this chapter.

Departments of Defense and Veterans Affairs and United States Government Disclaimer

The contents of this article suggest the views of authors and do not represent the views of the U.S. Department of Defense, U.S. Department of Veterans Affairs or the United States Government.

References

Ables, J.L., N.A. Decarolis, M.A. Johnson, P.D. Rivera, Z. Gao, D.C. Cooper, F. Radtke, J. Hsieh, and A.J. Eisch. 2010. Notch1 is required for maintenance of the reservoir of adult hippocampal stem cells. J Neurosci 30: 10484–10492.

Aguirre, A., M.E. Rubio, and V. Gallo. 2010. Notch and EGFR pathway interaction regulates neural stem cell number and self-renewal. Nature 467: 323–327.

Ahn, S., and A.L. Joyner. 2005. *In vivo* analysis of quiescent adult neural stem cells responding to Sonic hedgehog. Nature 437: 894–897.

Aimone, J.B., W. Deng, and F.H. Gage. 2010. Adult neurogenesis: integrating theories and separating functions. Trends Cogn Sci 14: 325–337.

Aimone, J.B., W. Deng, and F.H. Gage. 2011. Resolving new memories: a critical look at the dentate gyrus, adult neurogenesis, and pattern separation. Neuron 70: 589–596.

Alonso, G. 2001. Proliferation of progenitor cells in the adult rat brain correlates with the presence of vimentin-expressing astrocytes. Glia 34: 253–266.

Altman, J., and G.D. Das. 1965. Autoradiographic and histological evidence of postnatal hippocampal neurogenesis in rats. J Comp Neurol 124: 319–335.

Altman, J. 1969. Autoradiographic and histological studies of postnatal neurogenesis. IV. Cell proliferation and migration in the anterior forebrain, with special reference to persisting neurogenesis in the olfactory bulb. J Comp Neurol 137: 433–457.

Amrein, I., K. Isler, and H.P. Lipp. 2011. Comparing adult hippocampal neurogenesis in mammalian species and orders: influence of chronological age and life history stage. Eur J Neurosci 34: 978–987.

Anderson, S.A., D.D. Eisenstat, L. Shi, and J.L. Rubenstein. 1997. Interneuron migration from basal forebrain to neocortex: dependence on Dlx genes. Science 278: 474–476.

Appleby, P.A., G. Kempermann, and L. Wiskott. 2011. The role of additive neurogenesis and synaptic plasticity in a hippocampal memory model with grid-cell like input. PLoS Comput Biol 7: e1001063.

Banerjee, S.B., R. Rajendran, B.G. Dias, U. Ladiwala, S. Tole, and V.A. Vaidya. 2005. Recruitment of the Sonic hedgehog signalling cascade in electroconvulsive seizure-mediated regulation of adult rat hippocampal neurogenesis. Eur J Neurosci 22: 1570–1580.

Barkas, L., E. Redhead, M. Taylor, A. Shtaya, D.A. Hamilton, and W.P. Gray. 2012. Fluoxetine restores spatial learning but not accelerated forgetting in mesial temporal lobe epilepsy. Brain 135: 2358–2374.

Barnabe-Heider, F., J.A. Wasylnka, K.J. Fernandes, C. Porsche, M. Sendtner, D.R. Kaplan, and F.D. Miller. 2005. Evidence that embryonic neurons regulate the onset of cortical gliogenesis via cardiotrophin-1. Neuron 48: 253–265.

Basak, O., C. Giachino, E. Fiorini, H.R. Macdonald, and V. Taylor. 2012. Neurogenic subventricular zone stem/progenitor cells are Notch1-dependent in their active but not quiescent state. J Neurosci 32: 5654–5666.

Bedard, A., C. Gravel, and A. Parent. 2006. Chemical characterization of newly generated neurons in the striatum of adult primates. Exp Brain Res 170: 501–512.

Bell, B., J.J. Lin, M. Seidenberg, and B. Hermann. 2011. The neurobiology of cognitive disorders in temporal lobe epilepsy. Nat Rev Neurol 7: 154–164.

Ben-Menachem-Zidon, O., Y. Ben-Menahem, T. Ben-Hur, and R. Yirmiya. 2014. Intra-hippocampal transplantation of neural precursor cells with transgenic over-expression of IL-1 receptor antagonist rescues memory and neurogenesis impairments in an Alzheimer's disease model. Neuropsychopharmacology 39: 401–414.

Bengzon, J., Z. Kokaia, E. Elmer, A. Nanobashvili, M. Kokaia, and O. Lindvall. 1997. Apoptosis and proliferation of dentate gyrus neurons after single and intermittent limbic seizures. Proc Natl Acad Sci U S A 94: 10432–10437.

Bergmann, O., and J. Frisen. 2013. Neuroscience. Why adults need new brain cells. Science 340: 695–696.

Bergmann, O., J. Liebl, S. Bernard, K. Alkass, M.S. Yeung, P. Steier, W. Kutschera, L. Johnson, M. Landen, H. Druid, K.L. Spalding, and J. Frisen. 2012. The age of olfactory bulb neurons in humans. Neuron 74: 634–639.

Bhardwaj, R.D., M.A. Curtis, K.L. Spalding, B.A. Buchholz, D. Fink, T. Bjork-Eriksson, C. Nordborg, F.H. Gage, H. Druid, P.S. Eriksson, and J. Frisen. 2006. Neocortical neurogenesis in humans is restricted to development. Proc Natl Acad Sci U S A 103: 12564–12568.

Biscaro, B., O. Lindvall, G. Tesco, C.T. Ekdahl, and R.M. Nitsch. 2012. Inhibition of microglial activation protects hippocampal neurogenesis and improves cognitive deficits in a transgenic mouse model for Alzheimer's disease. Neurodegener Dis 9: 187–198.

Bizon, J.L., and M. Gallagher. 2003. Production of new cells in the rat dentate gyrus over the lifespan: relation to cognitive decline. Eur J Neurosci 18: 215–219.

Bizon, J.L., H.J. Lee, and M. Gallagher. 2004. Neurogenesis in a rat model of age-related cognitive decline. Aging Cell 3: 227–234.

Blumcke, I., J.C. Schewe, S. Normann, O. Brustle, J. Schramm, C.E. Elger, and O.D. Wiestler. 2001. Increase of nestin-immunoreactive neural precursor cells in the dentate gyrus of pediatric patients with early-onset temporal lobe epilepsy. Hippocampus 11: 311–321.

Boekhoorn, K., M. Joels, and P.J. Lucassen. 2006. Increased proliferation reflects glial and vascular-associated changes, but not neurogenesis in the presenile Alzheimer hippocampus. Neurobiol Dis 24: 1–14.

Bondolfi, L., F. Ermini, J.M. Long, D.K. Ingram, and M. Jucker. 2004. Impact of age and caloric restriction on neurogenesis in the dentate gyrus of C57BL/6 mice. Neurobiol Aging 25: 333–340.

Braun, S.M., and S. Jessberger. 2014. Adult neurogenesis and its role in neuropsychiatric disease, brain repair and normal brain function. Neuropathol Appl Neurobiol 40: 3–12.

Bugra, K., H. Pollard, G. Charton, J. Moreau, Y. Ben-Ari, and M. Khrestchatisky. 1994. aFGF, bFGF and flg mRNAs show distinct patterns of induction in the hippocampus following kainate-induced seizures. Eur J Neurosci 6: 58–66.

Busceti, C.L., F. Biagioni, E. Aronica, B. Riozzi, M. Storto, G. Battaglia, F.S. Giorgi, R. Gradini, F. Fornai, A. Caricasole, F. Nicoletti, and V. Bruno. 2007. Induction of the Wnt inhibitor, Dickkopf-1, is associated with neurodegeneration related to temporal lobe epilepsy. Epilepsia 48: 694–705.

Butler, C.R., and A.Z. Zeman. 2008. Recent insights into the impairment of memory in epilepsy: transient epileptic amnesia, accelerated long-term forgetting and remote memory impairment. Brain 131: 2243–2263.

Cameron, H.A., and R.D. McKay. 1999. Restoring production of hippocampal neurons in old age. Nat Neurosci 2: 894–897.

Cameron, H.A., and R.D. McKay. 2001. Adult neurogenesis produces a large pool of new granule cells in the dentate gyrus. J Comp Neurol 435: 406–417.

Cameron, H.A., and A.G. Dayer. 2008. New interneurons in the adult neocortex: small, sparse, but significant? Biol Psychiatry 63: 650–655.

Chadashvili, T., and D.A. Peterson. 2006. Cytoarchitecture of fibroblast growth factor receptor 2 (FGFR-2) immunoreactivity in astrocytes of neurogenic and non-neurogenic regions of the young adult and aged rat brain. J Comp Neurol 498: 1–15.

Chen, Q., A. Nakajima, S.H. Choi, X. Xiong, S.S. Sisodia, and Y.P. Tang. 2008. Adult neurogenesis is functionally associated with AD-like neurodegeneration. Neurobiol Dis 29: 316–326.

Cho, K.O., Z.R. Lybrand, N. Ito, R. Brulet, F. Tafacory, L. Zhang, L. Good, K. Ure, S.G. Kernie, S.G. Birnbaum, H.E. Scharfman, A.J. Eisch, and J. Hsieh. 2015. Aberrant hippocampal neurogenesis contributes to epilepsy and associated cognitive decline. Nat Commun 6: 6606.

Chuang, T.T. 2010. Neurogenesis in mouse models of Alzheimer's disease. Biochim Biophys Acta 1802: 872–880.

Clapcote, S.J., T.V. Lipina, J.K. Millar, S. Mackie, S. Christie, F. Ogawa, J.P. Lerch, K. Trimble, M. Uchiyama, Y. Sakuraba, H. Kaneda, T. Shiroishi, M.D. Houslay, R.M. Henkelman, J.G. Sled, Y. Gondo, D.J. Porteous, and J.C. Roder. 2007. Behavioral phenotypes of Disc1 missense mutations in mice. Neuron 54: 387–402.

Clelland, C.D., M. Choi, C. Romberg, G.D. Clemenson, Jr., A. Fragniere, P. Tyers, S. Jessberger, L.M. Saksida, R.A. Barker, F.H. Gage, and T.J. Bussey. 2009. A functional role for adult hippocampal neurogenesis in spatial pattern separation. Science 325: 210–213.

Costa, D.A., J.R. Cracchiolo, A.D. Bachstetter, T.F. Hughes, K.R. Bales, S.M. Paul, R.F. Mervis, G.W. Arendash, and H. Potter. 2007. Enrichment improves cognition in AD mice by amyloid-related and unrelated mechanisms. Neurobiol Aging 28: 831–844.

Crespel, A., V. Rigau, P. Coubes, M.C. Rousset, F. de Bock, H. Okano, M. Baldy-Moulinier, J. Bockaert, and M. Lerner-Natoli. 2005. Increased number of neural progenitors in human temporal lobe epilepsy. Neurobiol Dis 19: 436–450.

Crews, L., A. Adame, C. Patrick, A. Delaney, E. Pham, E. Rockenstein, L. Hansen, and E. Masliah. 2010. Increased BMP6 levels in the brains of Alzheimer's disease patients and APP transgenic mice are accompanied by impaired neurogenesis. J Neurosci 30: 12252–12262.

Croll, S.D., J.H. Goodman, and H.E. Scharfman. 2004. Vascular endothelial growth factor (VEGF) in seizures: a double-edged sword. Adv Exp Med Biol 548: 57–68.

D'Sa, C., and R.S. Duman. 2002. Antidepressants and neuroplasticity. Bipolar Disord 4: 183–194.

David, D.J., B.A. Samuels, Q. Rainer, J.W. Wang, D. Marsteller, I. Mendez, M. Drew, D.A. Craig, B.P. Guiard, J.P. Guilloux, R.P. Artymyshyn, A.M. Gardier, C. Gerald, I.A. Antonijevic, E.D. Leonardo, and R. Hen. 2009. Neurogenesis-dependent and -independent effects of fluoxetine in an animal model of anxiety/depression. Neuron 62: 479–493.

Davis, A.A., and S. Temple. 1994. A self-renewing multipotential stem cell in embryonic rat cerebral cortex. Nature 372: 263–266.

Dayer, A.G., A.A. Ford, K.M. Cleaver, M. Yassaee, and H.A. Cameron. 2003. Short-term and long-term survival of new neurons in the rat dentate gyrus. J Comp Neurol 460: 563–572.

Deisseroth, K., S. Singla, H. Toda, M. Monje, T.D. Palmer, and R.C. Malenka. 2004. Excitation-neurogenesis coupling in adult neural stem/progenitor cells. Neuron 42: 535–552.

Deng, W., J.B. Aimone, and F.H. Gage. 2010. New neurons and new memories: how does adult hippocampal neurogenesis affect learning and memory? Nat Rev Neurosci 11: 339–350.

Devinsky, O. 2004. Therapy for neurobehavioral disorders in epilepsy. Epilepsia 45 Suppl 2: 34–40.

Donovan, M.H., U. Yazdani, R.D. Norris, D. Games, D.C. German, and A.J. Eisch. 2006. Decreased adult hippocampal neurogenesis in the PDAPP mouse model of Alzheimer's disease. J Comp Neurol 495: 70–83.

Doze, V.A., and D.M. Perez. 2012. G-protein-coupled receptors in adult neurogenesis. Pharmacol Rev 64: 645–675.

Drapeau, E., W. Mayo, C. Aurousseau, M. Le Moal, P.V. Piazza, and D.N. Abrous. 2003. Spatial memory performances of aged rats in the water maze predict levels of hippocampal neurogenesis. Proc Natl Acad Sci U S A 100: 14385–14390.

Drapeau, E., M.F. Montaron, S. Aguerre, and D.N. Abrous. 2007. Learning-induced survival of new neurons depends on the cognitive status of aged rats. J Neurosci 27: 6037–6044.

Drapeau, E., and D. Nora Abrous. 2008. Stem cell review series: role of neurogenesis in age-related memory disorders. Aging Cell 7: 569–589.

Driscoll, I., S.R. Howard, J.C. Stone, M.H. Monfils, B. Tomanek, W.M. Brooks, and R.J. Sutherland. 2006. The aging hippocampus: a multi-level analysis in the rat. Neuroscience 139: 1173–1185.

Duman, R.S., and G.K. Aghajanian. 2012. Synaptic dysfunction in depression: potential therapeutic targets. Science 338: 68–72.

Dupret, D., J.M. Revest, M. Koehl, F. Ichas, F. De Giorgi, P. Costet, D.N. Abrous, and P.V. Piazza. 2008. Spatial relational memory requires hippocampal adult neurogenesis. PLoS One 3: e1959.

Encinas, J.M., T.V. Michurina, N. Peunova, J.H. Park, J. Tordo, D.A. Peterson, G. Fishell, A. Koulakov, and G. Enikolopov. 2011. Division-coupled astrocytic differentiation and age-related depletion of neural stem cells in the adult hippocampus. Cell Stem Cell 8: 566–579.

Engel, J., Jr. 2001. Mesial temporal lobe epilepsy: what have we learned? Neuroscientist 7: 340–352.

Englund, C., A. Fink, C. Lau, D. Pham, R.A. Daza, A. Bulfone, T. Kowalczyk, and R.F. Hevner. 2005. Pax6, Tbr2, and Tbr1 are expressed sequentially by radial glia, intermediate progenitor cells, and postmitotic neurons in developing neocortex. J Neurosci 25: 247–251.

Enwere, E., T. Shingo, C. Gregg, H. Fujikawa, S. Ohta, and S. Weiss. 2004. Aging results in reduced epidermal growth factor receptor signaling, diminished olfactory neurogenesis, and deficits in fine olfactory discrimination. J Neurosci 24: 8354–8365.

Eriksson, P.S., E. Perfilieva, T. Bjork-Eriksson, A.M. Alborn, C. Nordborg, D.A. Peterson, and F.H. Gage. 1998. Neurogenesis in the adult human hippocampus. Nat Med 4: 1313–1317.

Ernst, A., K. Alkass, S. Bernard, M. Salehpour, S. Perl, J. Tisdale, G. Possnert, H. Druid, and J. Frisen. 2014. Neurogenesis in the striatum of the adult human brain. Cell 156: 1072–1083.

Fahrner, A., G. Kann, A. Flubacher, C. Heinrich, T.M. Freiman, J. Zentner, M. Frotscher, and C.A. Haas. 2007. Granule cell dispersion is not accompanied by enhanced neurogenesis in temporal lobe epilepsy patients. Exp Neurol 203: 320–332.

Faigle, R., and H. Song. 2013. Signaling mechanisms regulating adult neural stem cells and neurogenesis. Biochim Biophys Acta 1830: 2435–2448.

Feng, R., C. Rampon, Y.P. Tang, D. Shrom, J. Jin, M. Kyin, B. Sopher, M.W. Miller, C.B. Ware, G.M. Martin, S.H. Kim, R.B. Langdon, S.S. Sisodia, and J.Z. Tsien. 2001. Deficient neurogenesis in forebrain-specific presenilin-1 knockout mice is associated with reduced clearance of hippocampal memory traces. Neuron 32: 911–926.

Filippov, V., G. Kronenberg, T. Pivneva, K. Reuter, B. Steiner, L.P. Wang, M. Yamaguchi, H. Kettenmann, and G. Kempermann. 2003. Subpopulation of nestin-expressing progenitor cells in the adult murine hippocampus shows electrophysiological and morphological characteristics of astrocytes. Mol Cell Neurosci 23: 373–382.

Fowler, C.D., Y. Liu, and Z. Wang. 2008. Estrogen and adult neurogenesis in the amygdala and hypothalamus. Brain Res Rev 57: 342–351.

Frantz, G.D., and S.K. McConnell. 1996. Restriction of late cerebral cortical progenitors to an upper-layer fate. Neuron 17: 55–61.

Gage, F.H., and S. Temple. 2013. Neural stem cells: generating and regenerating the brain. Neuron 80: 588–601.

Gallo, V., and B. Deneen. 2014. Glial development: the crossroads of regeneration and repair in the CNS. Neuron 83: 283–308.

Gao, Z., K. Ure, P. Ding, M. Nashaat, L. Yuan, J. Ma, R.E. Hammer, and J. Hsieh. 2011. The master negative regulator REST/NRSF controls adult neurogenesis by restraining the neurogenic program in quiescent stem cells. J Neurosci 31: 9772–9786.

Garthe, A., J. Behr, and G. Kempermann. 2009. Adult-generated hippocampal neurons allow the flexible use of spatially precise learning strategies. PLoS One 4: e5464.

Ge, S., D.A. Pradhan, G.L. Ming, and H. Song. 2007. GABA sets the tempo for activity-dependent adult neurogenesis. Trends Neurosci 30: 1–8.

Gemma, C., and A.D. Bachstetter. 2013. The role of microglia in adult hippocampal neurogenesis. Front Cell Neurosci 7: 229.

Giachino, C., M. Barz, J.S. Tchorz, M. Tome, M. Gassmann, J. Bischofberger, B. Bettler, and V. Taylor. 2014. GABA suppresses neurogenesis in the adult hippocampus through GABAB receptors. Development 141: 83–90.

Goedert, M., and M.G. Spillantini. 2006. A century of Alzheimer's disease. Science 314: 777–781.

Goldman, S.A., and Z. Chen. 2011. Perivascular instruction of cell genesis and fate in the adult brain. Nat Neurosci 14: 1382–1389.

Goldman, S.A., and F. Nottebohm. 1983. Neuronal production, migration, and differentiation in a vocal control nucleus of the adult female canary brain. Proc Natl Acad Sci U S A 80: 2390–2394.

Gong, C., T.W. Wang, H.S. Huang, and J.M. Parent. 2007. Reelin regulates neuronal progenitor migration in intact and epileptic hippocampus. J Neurosci 27: 1803–1811.

Gotz, M., and W.B. Huttner. 2005. The cell biology of neurogenesis. Nat Rev Mol Cell Biol 6: 777–788.

Gould, E. 2007. How widespread is adult neurogenesis in mammals? Nat Rev Neurosci 8: 481–488.

Gould, E., B.S. McEwen, P. Tanapat, L.A. Galea, and E. Fuchs. 1997. Neurogenesis in the dentate gyrus of the adult tree shrew is regulated by psychosocial stress and NMDA receptor activation. J Neurosci 17: 2492–2498.

Gould, E., P. Tanapat, N.B. Hastings, and T.J. Shors. 1999. Neurogenesis in adulthood: a possible role in learning. Trends Cogn Sci 3: 186–192.

Gu, Y., M. Arruda-Carvalho, J. Wang, S.R. Janoschka, S.A. Josselyn, P.W. Frankland, and S. Ge. 2012. Optical controlling reveals time-dependent roles for adult-born dentate granule cells. Nat Neurosci 15: 1700–1706.

Guerout, N., X. Li, and F. Barnabe-Heider. 2014. Cell fate control in the developing central nervous system. Exp Cell Res 321: 77–83.

Haslinger, A., T.J. Schwarz, M. Covic, and D.C. Lie. 2009. Expression of Sox11 in adult neurogenic niches suggests a stage-specific role in adult neurogenesis. Eur J Neurosci 29: 2103–2114.

Hattiangady, B., M.S. Rao, and A.K. Shetty. 2004. Chronic temporal lobe epilepsy is associated with severely declined dentate neurogenesis in the adult hippocampus. Neurobiol Dis 17: 473–490.

Hattiangady, B., and A.K. Shetty. 2008a. Aging does not alter the number or phenotype of putative stem/progenitor cells in the neurogenic region of the hippocampus. Neurobiol Aging 29: 129–147.

Hattiangady, B., and A.K. Shetty. 2008b. Implications of decreased hippocampal neurogenesis in chronic temporal lobe epilepsy. Epilepsia 49 Suppl 5: 26–41.

Hattiangady, B., and A.K. Shetty. 2010. Decreased neuronal differentiation of newly generated cells underlies reduced hippocampal neurogenesis in chronic temporal lobe epilepsy. Hippocampus 20: 97–112.

Haughey, N.J., A. Nath, S.L. Chan, A.C. Borchard, M.S. Rao, and M.P. Mattson. 2002. Disruption of neurogenesis by amyloid beta-peptide, and perturbed neural progenitor cell homeostasis, in models of Alzheimer's disease. J Neurochem 83: 1509–1524.

Heine, V.M., S. Maslam, M. Joels, and P.J. Lucassen. 2004. Prominent decline of newborn cell proliferation, differentiation, and apoptosis in the aging dentate gyrus, in absence of an age-related hypothalamus-pituitary-adrenal axis activation. Neurobiol Aging 25: 361–375.

Hester, M.S., and S.C. Danzer. 2014. Hippocampal granule cell pathology in epilepsy—a possible structural basis for comorbidities of epilepsy? Epilepsy Behav 38: 105–116.

Hill, A.S., A. Sahay, and R. Hen. 2015. Increasing adult hippocampal neurogenesis is sufficient to reduce anxiety and depression-like behaviors. Neuropsychopharmacology 40: 2368–2378.

Ho, N.F., J.M. Hooker, A. Sahay, D.J. Holt, and J.L. Roffman. 2013. *In vivo* imaging of adult human hippocampal neurogenesis: progress, pitfalls and promise. Mol Psychiatry 18: 404–416.

Houser, C.R. 1990. Granule cell dispersion in the dentate gyrus of humans with temporal lobe epilepsy. Brain Res 535: 195–204.

Hsieh, J. 2012. Orchestrating transcriptional control of adult neurogenesis. Genes Dev 26: 1010–1021.

Ihrie, R.A., and A. Alvarez-Buylla. 2008. Cells in the astroglial lineage are neural stem cells. Cell Tissue Res 331: 179–191.

Ihrie, R.A., and A. Alvarez-Buylla. 2011. Lake-front property: a unique germinal niche by the lateral ventricles of the adult brain. Neuron 70: 674–686.

Ikrar, T., N. Guo, K. He, A. Besnard, S. Levinson, A. Hill, H.K. Lee, R. Hen, X. Xu, and A. Sahay. 2013. Adult neurogenesis modifies excitability of the dentate gyrus. Front Neural Circuits 7: 204.

Iyengar, S.S., J.J. LaFrancois, D. Friedman, L.J. Drew, C.A. Denny, N.S. Burghardt, M.V. Wu, J. Hsieh, R. Hen, and H.E. Scharfman. 2015. Suppression of adult neurogenesis increases the acute effects of kainic acid. Exp Neurol 264: 135–149.

Jagasia, R., K. Steib, E. Englberger, S. Herold, T. Faus-Kessler, M. Saxe, F.H. Gage, H. Song, and D.C. Lie. 2009. GABA-cAMP response element-binding protein signaling regulates maturation and survival of newly generated neurons in the adult hippocampus. J Neurosci 29: 7966–7977.

Jessberger, S., B. Romer, H. Babu, and G. Kempermann. 2005. Seizures induce proliferation and dispersion of doublecortin-positive hippocampal progenitor cells. Exp Neurol 196: 342–351.

Jessberger, S., K. Nakashima, G.D. Clemenson, Jr., E. Mejia, E. Mathews, K. Ure, S. Ogawa, C.M. Sinton, F.H. Gage, and J. Hsieh. 2007. Epigenetic modulation of seizure-induced neurogenesis and cognitive decline. J Neurosci 27: 5967–5975.

Jessberger, S., N. Toni, G.D. Clemenson, Jr., J. Ray, and F.H. Gage. 2008. Directed differentiation of hippocampal stem/progenitor cells in the adult brain. Nat Neurosci 11: 888–893.

Jessberger, S., R.E. Clark, N.J. Broadbent, G.D. Clemenson, Jr., A. Consiglio, D.C. Lie, L.R. Squire, and F.H. Gage. 2009. Dentate gyrus-specific knockdown of adult neurogenesis impairs spatial and object recognition memory in adult rats. Learn Mem 16: 147–154.

Jin, K., Y. Sun, L. Xie, S. Batteur, X.O. Mao, C. Smelick, A. Logvinova, and D.A. Greenberg. 2003. Neurogenesis and aging: FGF-2 and HB-EGF restore neurogenesis in hippocampus and subventricular zone of aged mice. Aging Cell 2: 175–183.

Jin, K., V. Galvan, L. Xie, X.O. Mao, O.F. Gorostiza, D.E. Bredesen, and D.A. Greenberg. 2004a. Enhanced neurogenesis in Alzheimer's disease transgenic (PDGF-APPSw, Ind) mice. Proc Natl Acad Sci U S A 101: 13363–13367.

Jin, K., A.L. Peel, X.O. Mao, L. Xie, B.A. Cottrell, D.C. Henshall, and D.A. Greenberg. 2004b. Increased hippocampal neurogenesis in Alzheimer's disease. Proc Natl Acad Sci U S A 101: 343–347.

Jin, K., L. Xie, X.O. Mao, and D.A. Greenberg. 2006. Alzheimer's disease drugs promote neurogenesis. Brain Res 1085: 183–188.

Jung, K.H., K. Chu, M. Kim, S.W. Jeong, Y.M. Song, S.T. Lee, J.Y. Kim, S.K. Lee, and J.K. Roh. 2004. Continuous cytosine-b-D-arabinofuranoside infusion reduces ectopic granule cells in adult rat hippocampus with attenuation of spontaneous recurrent seizures following pilocarpine-induced status epilepticus. Eur J Neurosci 19: 3219–3226.

Kamiya, A., K. Kubo, T. Tomoda, M. Takaki, R. Youn, Y. Ozeki, N. Sawamura, U. Park, C. Kudo, M. Okawa, C.A. Ross, M.E. Hatten, K. Nakajima, and A. Sawa. 2005. A schizophrenia-associated mutation of DISC1 perturbs cerebral cortex development. Nat Cell Biol 7: 1167–1178.

Kaplan, M.S., and J.W. Hinds. 1977. Neurogenesis in the adult rat: electron microscopic analysis of light radioautographs. Science 197: 1092–1094.

Karalay, O., K. Doberauer, K.C. Vadodaria, M. Knobloch, L. Berti, A. Miquelajauregui, M. Schwark, R. Jagasia, M.M. Taketo, V. Tarabykin, D.C. Lie, and S. Jessberger. 2011. Prospero-related homeobox 1 gene (Prox1) is regulated by canonical Wnt signaling and has a stage-specific role in adult hippocampal neurogenesis. Proc Natl Acad Sci U S A 108: 5807–5812.

Kee, N., C.M. Teixeira, A.H. Wang, and P.W. Frankland. 2007. Preferential incorporation of adult-generated granule cells into spatial memory networks in the dentate gyrus. Nat Neurosci 10: 355–362.

Kempermann, G., H.G. Kuhn, and F.H. Gage. 1997. More hippocampal neurons in adult mice living in an enriched environment. Nature 386: 493–495.

Kempermann, G., H.G. Kuhn, and F.H. Gage. 1998. Experience-induced neurogenesis in the senescent dentate gyrus. J Neurosci 18: 3206–3212.

Kempermann, G., D. Gast, and F.H. Gage. 2002. Neuroplasticity in old age: sustained fivefold induction of hippocampal neurogenesis by long-term environmental enrichment. Ann Neurol 52: 135–143.

Kempermann, G., S. Jessberger, B. Steiner, and G. Kronenberg. 2004. Milestones of neuronal development in the adult hippocampus. Trends Neurosci 27: 447–452.

Kempermann, G. 2012. Neuroscience. Youth culture in the adult brain. Science 335: 1175–1176.

Kempermann, G. 2013. Neuroscience. What the bomb said about the brain. Science 340: 1180–1181.

Kheirbek, M.A., K.C. Klemenhagen, A. Sahay, and R. Hen. 2012. Neurogenesis and generalization: a new approach to stratify and treat anxiety disorders. Nat Neurosci 15: 1613–1620.

Kheirbek, M.A., and R. Hen. 2013. (Radio)active neurogenesis in the human hippocampus. Cell 153: 1183–1184.

Kim, J.Y., C.Y. Liu, F. Zhang, X. Duan, Z. Wen, J. Song, E. Feighery, B. Lu, D. Rujescu, D. St Clair, K. Christian, J.H. Callicott, D.R. Weinberger, H. Song, and G.L. Ming. 2012. Interplay between DISC1 and GABA signaling regulates neurogenesis in mice and risk for schizophrenia. Cell 148: 1051–1064.

Kirby, E.D., A.A. Kuwahara, R.L. Messer, and T. Wyss-Coray. 2015. Adult hippocampal neural stem and progenitor cells regulate the neurogenic niche by secreting VEGF. Proc Natl Acad Sci U S A 112: 4128–4133.

Kodali, M., V.K. Parihar, B. Hattiangady, V. Mishra, B. Shuai, and A.K. Shetty. 2015. Resveratrol prevents age-related memory and mood dysfunction with increased hippocampal neurogenesis and microvasculature, and reduced glial activation. Sci Rep 5: 8075.

Kohl, Z., N. Ben Abdallah, J. Vogelgsang, L. Tischer, J. Deusser, D. Amato, S. Anderson, C.P. Muller, O. Riess, E. Masliah, S. Nuber, and J. Winkler. 2016. Severely impaired hippocampal neurogenesis associates with an early serotonergic deficit in a BAC alpha-synuclein transgenic rat model of Parkinson's disease. Neurobiol Dis 85: 206–217.

Komuro, Y., G. Xu, K. Bhaskar, and B.T. Lamb. 2015. Human tau expression reduces adult neurogenesis in a mouse model of tauopathy. Neurobiol Aging 36: 2034–2042.

Krezymon, A., K. Richetin, H. Halley, L. Roybon, J.M. Lassalle, B. Frances, L. Verret, and C. Rampon. 2013. Modifications of hippocampal circuits and early disruption of adult neurogenesis in the tg2576 mouse model of Alzheimer's disease. PLoS One 8: e76497.

Kronenberg, G., A. Bick-Sander, E. Bunk, C. Wolf, D. Ehninger, and G. Kempermann. 2006. Physical exercise prevents age-related decline in precursor cell activity in the mouse dentate gyrus. Neurobiol Aging 27: 1505–1513.

Kuhn, H.G., H. Dickinson-Anson, and F.H. Gage. 1996. Neurogenesis in the dentate gyrus of the adult rat: age-related decrease of neuronal progenitor proliferation. J Neurosci 16: 2027–2033.

Kuruba, R., B. Hattiangady, and A.K. Shetty. 2009. Hippocampal neurogenesis and neural stem cells in temporal lobe epilepsy. Epilepsy Behav 14 Suppl 1: 65–73.

Kuwabara, T., J. Hsieh, A. Muotri, G. Yeo, M. Warashina, D.C. Lie, L. Moore, K. Nakashima, M. Asashima, and F.H. Gage. 2009. Wnt-mediated activation of NeuroD1 and retro-elements during adult neurogenesis. Nat Neurosci 12: 1097–1105.

Ladran, I., N. Tran, A. Topol, and K.J. Brennand. 2013. Neural stem and progenitor cells in health and disease. Wiley Interdiscip Rev Syst Biol Med 5: 701–715.

Lagace, D.C., M.C. Whitman, M.A. Noonan, J.L. Ables, N.A. DeCarolis, A.A. Arguello, M.H. Donovan, S.J. Fischer, L.A. Farnbauch, R.D. Beech, R.J. DiLeone, C.A. Greer, C.D. Mandyam, and A.J. Eisch. 2007. Dynamic contribution of nestin-expressing stem cells to adult neurogenesis. J Neurosci 27: 12623–12629.

Lai, M., C.J. Hibberd, P.D. Gluckman, and J.R. Seckl. 2000. Reduced expression of insulin-like growth factor 1 messenger RNA in the hippocampus of aged rats. Neurosci Lett 288: 66–70.

Lajud, N., A. Roque, M. Cajero, G. Gutierrez-Ospina, and L. Torner. 2012. Periodic maternal separation decreases hippocampal neurogenesis without affecting basal corticosterone during the stress hyporesponsive period, but alters HPA axis and coping behavior in adulthood. Psychoneuroendocrinology 37: 410–420.

Lazarov, O., M.P. Mattson, D.A. Peterson, S.W. Pimplikar, and H. van Praag. 2010. When neurogenesis encounters aging and disease. Trends Neurosci 33: 569–579.

Lee, J., W. Duan, and M.P. Mattson. 2002. Evidence that brain-derived neurotrophic factor is required for basal neurogenesis and mediates, in part, the enhancement of neurogenesis by dietary restriction in the hippocampus of adult mice. J Neurochem 82: 1367–1375.

Lee, M.M., A. Reif, and A.G. Schmitt. 2013. Major depression: a role for hippocampal neurogenesis? Curr Top Behav Neurosci 14: 153–179.

Lemaire, V., S. Lamarque, M. Le Moal, P.V. Piazza, and D.N. Abrous. 2006. Postnatal stimulation of the pups counteracts prenatal stress-induced deficits in hippocampal neurogenesis. Biol Psychiatry 59: 786–792.

Leutgeb, J.K., S. Leutgeb, M.B. Moser, and E.I. Moser. 2007. Pattern separation in the dentate gyrus and CA3 of the hippocampus. Science 315: 961–966.

Lewis, D.V. 2005. Losing neurons: selective vulnerability and mesial temporal sclerosis. Epilepsia 46 Suppl 7: 39–44.

Li, X., F. Guo, Q. Zhang, T. Huo, L. Liu, H. Wei, L. Xiong, and Q. Wang. 2014. Electroacupuncture decreases cognitive impairment and promotes neurogenesis in the APP/PS1 transgenic mice. BMC Complement Altern Med 14: 37.

Licht, T., and E. Keshet. 2015. The vascular niche in adult neurogenesis. Mech Dev 138 Pt 1: 56–62.

Lichtenwalner, R.J., M.E. Forbes, S.A. Bennett, C.D. Lynch, W.E. Sonntag, and D.R. Riddle. 2001. Intracerebroventricular infusion of insulin-like growth factor-I ameliorates the age-related decline in hippocampal neurogenesis. Neuroscience 107: 603–613.

Lie, D.C., S.A. Colamarino, H.J. Song, L. Desire, H. Mira, A. Consiglio, E.S. Lein, S. Jessberger, H. Lansford, A.R. Dearie, and F.H. Gage. 2005. Wnt signalling regulates adult hippocampal neurogenesis. Nature 437: 1370–1375.

Lim, D.A., and A. Alvarez-Buylla. 2014. Adult neural stem cells stake their ground. Trends Neurosci 37: 563–571.

Lois, C., J.M. Garcia-Verdugo, and A. Alvarez-Buylla. 1996. Chain migration of neuronal precursors. Science 271: 978–981.

Lopez-Toledano, M.A., and M.L. Shelanski. 2007. Increased neurogenesis in young transgenic mice overexpressing human APP(Sw, Ind). J Alzheimers Dis 12: 229–240.

Lowenstein, D.H., M.S. Seren, and F.M. Longo. 1993. Prolonged increases in neurotrophic activity associated with kainate-induced hippocampal synaptic reorganization. Neuroscience 56: 597–604.

Lugert, S., O. Basak, P. Knuckles, U. Haussler, K. Fabel, M. Gotz, C.A. Haas, G. Kempermann, V. Taylor, and C. Giachino. 2010. Quiescent and active hippocampal neural stem cells with distinct morphologies respond selectively to physiological and pathological stimuli and aging. Cell Stem Cell 6: 445–456.

Lugert, S., M. Vogt, J.S. Tchorz, M. Muller, C. Giachino, and V. Taylor. 2012. Homeostatic neurogenesis in the adult hippocampus does not involve amplification of Ascl1(high) intermediate progenitors. Nat Commun 3: 670.

Manganas, L.N., X. Zhang, Y. Li, R.D. Hazel, S.D. Smith, M.E. Wagshul, F. Henn, H. Benveniste, P.M. Djuric, G. Enikolopov, and M. Maletic-Savatic. 2007. Magnetic resonance spectroscopy identifies neural progenitor cells in the live human brain. Science 318: 980–985.

Marin-Burgin, A., L.A. Mongiat, M.B. Pardi, and A.F. Schinder. 2012. Unique processing during a period of high excitation/inhibition balance in adult-born neurons. Science 335: 1238–1242.

Martinez, S., P.H. Crossley, I. Cobos, J.L. Rubenstein, and G.R. Martin. 1999. FGF8 induces formation of an ectopic isthmic organizer and isthmocerebellar development via a repressive effect on Otx2 expression. Development 126: 1189–1200.

Masi, G., and P. Brovedani. 2011. The hippocampus, neurotrophic factors and depression: possible implications for the pharmacotherapy of depression. CNS Drugs 25: 913–931.

Masiulis, I., S. Yun, and A.J. Eisch. 2011. The interesting interplay between interneurons and adult hippocampal neurogenesis. Mol Neurobiol 44: 287–302.

Mathern, G.W., J.L. Leiphart, A. De Vera, P.D. Adelson, T. Seki, L. Neder, and J.P. Leite. 2002. Seizures decrease postnatal neurogenesis and granule cell development in the human fascia dentata. Epilepsia 43 Suppl 5: 68–73.

McCloskey, D.P., T.M. Hintz, J.P. Pierce, and H.E. Scharfman. 2006. Stereological methods reveal the robust size and stability of ectopic hilar granule cells after pilocarpine-induced status epilepticus in the adult rat. Eur J Neurosci 24: 2203–2210.

McConnell, S.K., and C.E. Kaznowski. 1991. Cell cycle dependence of laminar determination in developing neocortex. Science 254: 282–285.

McDonald, H.Y., and J.M. Wojtowicz. 2005. Dynamics of neurogenesis in the dentate gyrus of adult rats. Neurosci Lett 385: 70–75.

Meechan, D.W., E.S. Tucker, T.M. Maynard, and A.S. LaMantia. 2009. Diminished dosage of 22q11 genes disrupts neurogenesis and cortical development in a mouse model of 22q11 deletion/DiGeorge syndrome. Proc Natl Acad Sci U S A 106: 16434–16445.

Merkle, F.T., Z. Mirzadeh, and A. Alvarez-Buylla. 2007. Mosaic organization of neural stem cells in the adult brain. Science 317: 381–384.

Merrill, D.A., R. Karim, M. Darraq, A.A. Chiba, and M.H. Tuszynski. 2003. Hippocampal cell genesis does not correlate with spatial learning ability in aged rats. J Comp Neurol 459: 201–207.

Ming, G.L., and H. Song. 2011. Adult neurogenesis in the mammalian brain: significant answers and significant questions. Neuron 70: 687–702.

Mirescu, C., J.D. Peters, and E. Gould. 2004. Early life experience alters response of adult neurogenesis to stress. Nat Neurosci 7: 841–846.

Mohapel, P., G. Leanza, M. Kokaia, and O. Lindvall. 2005. Forebrain acetylcholine regulates adult hippocampal neurogenesis and learning. Neurobiol Aging 26: 939–946.

Molofsky, A.V., S.G. Slutsky, N.M. Joseph, S. He, R. Pardal, J. Krishnamurthy, N.E. Sharpless, and S.J. Morrison. 2006. Increasing p16INK4a expression decreases forebrain progenitors and neurogenesis during ageing. Nature 443: 448–452.

Montaron, M.F., K.G. Petry, J.J. Rodriguez, M. Marinelli, C. Aurousseau, G. Rougon, M. Le Moal, and D.N. Abrous. 1999. Adrenalectomy increases neurogenesis but not PSA-NCAM expression in aged dentate gyrus. Eur J Neurosci 11: 1479–1485.

Montaron, M.F., E. Drapeau, D. Dupret, P. Kitchener, C. Aurousseau, M. Le Moal, P.V. Piazza, and D.N. Abrous. 2006. Lifelong corticosterone level determines age-related decline in neurogenesis and memory. Neurobiol Aging 27: 645–654.

Morrens, J., W. Van Den Broeck, and G. Kempermann. 2012. Glial cells in adult neurogenesis. Glia 60: 159–174.

Morrison, S.J., and A.C. Spradling. 2008. Stem cells and niches: mechanisms that promote stem cell maintenance throughout life. Cell 132: 598–611.

Mu, Y., and F.H. Gage. 2011. Adult hippocampal neurogenesis and its role in Alzheimer's disease. Mol Neurodegener 6: 85.

Murphy, B.L., R.Y. Pun, H. Yin, C.R. Faulkner, A.W. Loepke, and S.C. Danzer. 2011. Heterogeneous integration of adult-generated granule cells into the epileptic brain. J Neurosci 31: 105–117.

Nacher, J., G. Alonso-Llosa, D.R. Rosell, and B.S. McEwen. 2003. NMDA receptor antagonist treatment increases the production of new neurons in the aged rat hippocampus. Neurobiol Aging 24: 273–284.

Nagy, P.M., and I. Aubert. 2015. Overexpression of the vesicular acetylcholine transporter enhances dendritic complexity of adult-born hippocampal neurons and improves acquisition of spatial memory during aging. Neurobiol Aging 36: 1881–1889.

Nakagawa, S., J.E. Kim, R. Lee, J. Chen, T. Fujioka, J. Malberg, S. Tsuji, and R.S. Duman. 2002. Localization of phosphorylated cAMP response element-binding protein in immature neurons of adult hippocampus. J Neurosci 22: 9868–9876.

Nakashiba, T., J.D. Cushman, K.A. Pelkey, S. Renaudineau, D.L. Buhl, T.J. McHugh, V. Rodriguez Barrera, R. Chittajallu, K.S. Iwamoto, C.J. McBain, M.S. Fanselow, and S. Tonegawa. 2012. Young dentate granule cells mediate pattern separation, whereas old granule cells facilitate pattern completion. Cell 149: 188–201.

Nichol, K.E., A.I. Parachikova, and C.W. Cotman. 2007. Three weeks of running wheel exposure improves cognitive performance in the aged Tg2576 mouse. Behav Brain Res 184: 124–132.

Nichol, K., S.P. Deeny, J. Seif, K. Camaclang, and C.W. Cotman. 2009. Exercise improves cognition and hippocampal plasticity in APOE epsilon4 mice. Alzheimers Dement 5: 287–294.

Niwa, A., M. Nishibori, S. Hamasaki, T. Kobori, K. Liu, H. Wake, S. Mori, T. Yoshino, and H. Takahashi. 2016. Voluntary exercise induces neurogenesis in the hypothalamus and ependymal lining of the third ventricle. Brain Struct Funct 221: 1653–1666.

Noctor, S.C., A.C. Flint, T.A. Weissman, R.S. Dammerman, and A.R. Kriegstein. 2001. Neurons derived from radial glial cells establish radial units in neocortex. Nature 409: 714–720.

O'Reilly, R.C., and J.L. McClelland. 1994. Hippocampal conjunctive encoding, storage, and recall: avoiding a trade-off. Hippocampus 4: 661–682.

Ohm, T.G. 2007. The dentate gyrus in Alzheimer's disease. Prog Brain Res 163: 723–740.

Okano, H., H. Kawahara, M. Toriya, K. Nakao, S. Shibata, and T. Imai. 2005. Function of RNA-binding protein Musashi-1 in stem cells. Exp Cell Res 306: 349–356.

Olariu, A., K.M. Cleaver, and H.A. Cameron. 2007. Decreased neurogenesis in aged rats results from loss of granule cell precursors without lengthening of the cell cycle. J Comp Neurol 501: 659–667.

Palmer, T.D., E.A. Markakis, A.R. Willhoite, F. Safar, and F.H. Gage. 1999. Fibroblast growth factor-2 activates a latent neurogenic program in neural stem cells from diverse regions of the adult CNS. J Neurosci 19: 8487–8497.

Palmer, T.D., A.R. Willhoite, and F.H. Gage. 2000. Vascular niche for adult hippocampal neurogenesis. J Comp Neurol 425: 479–494.

Pandya, M., M. Altinay, D.A. Malone, Jr., and A. Anand. 2012. Where in the brain is depression? Curr Psychiatry Rep 14: 634–642.

Parent, J.M., T.W. Yu, R.T. Leibowitz, D.H. Geschwind, R.S. Sloviter, and D.H. Lowenstein. 1997. Dentate granule cell neurogenesis is increased by seizures and contributes to aberrant network reorganization in the adult rat hippocampus. J Neurosci 17: 3727–3738.

Paridaen, J.T., and W.B. Huttner. 2014. Neurogenesis during development of the vertebrate central nervous system. EMBO Rep 15: 351–364.

Pauli, E., M. Hildebrandt, J. Romstock, H. Stefan, and I. Blumcke. 2006. Deficient memory acquisition in temporal lobe epilepsy is predicted by hippocampal granule cell loss. Neurology 67: 1383–1389.

Pekcec, A., M. Muhlenhoff, R. Gerardy-Schahn, and H. Potschka. 2007. Impact of the PSA-NCAM system on pathophysiology in a chronic rodent model of temporal lobe epilepsy. Neurobiol Dis 27: 54–66.

Pekcec, A., C. Fuest, M. Muhlenhoff, R. Gerardy-Schahn, and H. Potschka. 2008. Targeting epileptogenesis-associated induction of neurogenesis by enzymatic depolysialylation of NCAM counteracts spatial learning dysfunction but fails to impact epilepsy development. J Neurochem 105: 389–400.

Pekcec, A., M. Lupke, R. Baumann, H. Seifert, and H. Potschka. 2011. Modulation of neurogenesis by targeted hippocampal irradiation fails to affect kindling progression. Hippocampus 21: 866–876.

Pereira, A.C., D.E. Huddleston, A.M. Brickman, A.A. Sosunov, R. Hen, G.M. McKhann, R. Sloan, F.H. Gage, T.R. Brown, and S.A. Small. 2007. An *in vivo* correlate of exercise-induced neurogenesis in the adult dentate gyrus. Proc Natl Acad Sci U S A 104: 5638–5643.

Pereira, J.D., S.N. Sansom, J. Smith, M.W. Dobenecker, A. Tarakhovsky, and F.J. Livesey. 2010. Ezh2, the histone methyltransferase of PRC2, regulates the balance between self-renewal and differentiation in the cerebral cortex. Proc Natl Acad Sci U S A 107: 15957–15962.

Perera, T.D., J.D. Coplan, S.H. Lisanby, C.M. Lipira, M. Arif, C. Carpio, G. Spitzer, L. Santarelli, B. Scharf, R. Hen, G. Rosoklija, H.A. Sackeim, and A.J. Dwork. 2007. Antidepressant-induced neurogenesis in the hippocampus of adult nonhuman primates. J Neurosci 27: 4894–4901.

Perry, E.K., M. Johnson, A. Ekonomou, R.H. Perry, C. Ballard, and J. Attems. 2012. Neurogenic abnormalities in Alzheimer's disease differ between stages of neurogenesis and are partly related to cholinergic pathology. Neurobiol Dis 47: 155–162.

Pieper, A.A., S. Xie, E. Capota, S.J. Estill, J. Zhong, J.M. Long, G.L. Becker, P. Huntington, S.E. Goldman, C.H. Shen, M. Capota, J.K. Britt, T. Kotti, K. Ure, D.J. Brat, N.S. Williams, K.S. MacMillan, J. Naidoo, L. Melito, J. Hsieh, J. De Brabander, J.M. Ready, and S.L. McKnight. 2010. Discovery of a proneurogenic, neuroprotective chemical. Cell 142: 39–51.

Pierce, J.P., J. Melton, M. Punsoni, D.P. McCloskey, and H.E. Scharfman. 2005. Mossy fibers are the primary source of afferent input to ectopic granule cells that are born after pilocarpine-induced seizures. Exp Neurol 196: 316–331.

Pirttila, T.J., A. Manninen, L. Jutila, J. Nissinen, R. Kalviainen, M. Vapalahti, A. Immonen, L. Paljarvi, K. Karkola, I. Alafuzoff, E. Mervaala, and A. Pitkanen. 2005. Cystatin C expression is associated with granule cell dispersion in epilepsy. Ann Neurol 58: 211–223.

Pitkanen, A., and T.P. Sutula. 2002. Is epilepsy a progressive disorder? Prospects for new therapeutic approaches in temporal-lobe epilepsy. Lancet Neurol 1: 173–181.

Puelles, L., and J.L. Rubenstein. 2003. Forebrain gene expression domains and the evolving prosomeric model. Trends Neurosci 26: 469–476.

Raedt, R., P. Boon, A. Persson, A.M. Alborn, T. Boterberg, A. Van Dycke, B. Linder, T. De Smedt, W.J. Wadman, E. Ben-Menachem, and P.S. Eriksson. 2007. Radiation of the rat brain suppresses seizure-induced neurogenesis and transiently enhances excitability during kindling acquisition. Epilepsia 48: 1952–1963.

Rai, K.S., B. Hattiangady, and A.K. Shetty. 2007. Enhanced production and dendritic growth of new dentate granule cells in the middle-aged hippocampus following intracerebroventricular FGF-2 infusions. Eur J Neurosci 26: 1765–1779.

Rao, M.S., B. Hattiangady, A. Abdel-Rahman, D.P. Stanley, and A.K. Shetty. 2005. Newly born cells in the ageing dentate gyrus display normal migration, survival and neuronal fate choice but endure retarded early maturation. Eur J Neurosci 21: 464–476.

Rao, M.S., B. Hattiangady, and A.K. Shetty. 2006. The window and mechanisms of major age-related decline in the production of new neurons within the dentate gyrus of the hippocampus. Aging Cell 5: 545–558.

Rao, M.S., B. Hattiangady, and A.K. Shetty. 2008. Status epilepticus during old age is not associated with enhanced hippocampal neurogenesis. Hippocampus 18: 931–944.

Reitz, C., and R. Mayeux. 2014. Alzheimer disease: epidemiology, diagnostic criteria, risk factors and biomarkers. Biochem Pharmacol 88: 640–651.

Renault, V.M., V.A. Rafalski, A.A. Morgan, D.A. Salih, J.O. Brett, A.E. Webb, S.A. Villeda, P.U. Thekkat, C. Guillerey, N.C. Denko, T.D. Palmer, A.J. Butte, and A. Brunet. 2009. FoxO3 regulates neural stem cell homeostasis. Cell Stem Cell 5: 527–539.

Reynolds, B.A., and S. Weiss. 1992. Generation of neurons and astrocytes from isolated cells of the adult mammalian central nervous system. Science 255: 1707–1710.

Richetin, K., C. Leclerc, N. Toni, T. Gallopin, S. Pech, L. Roybon, and C. Rampon. 2015. Genetic manipulation of adult-born hippocampal neurons rescues memory in a mouse model of Alzheimer's disease. Brain 138: 440–455.

Rietze, R., P. Poulin, and S. Weiss. 2000. Mitotically active cells that generate neurons and astrocytes are present in multiple regions of the adult mouse hippocampus. J Comp Neurol 424: 397–408.

Rodriguez, J.J., V.C. Jones, M. Tabuchi, S.M. Allan, E.M. Knight, F.M. LaFerla, S. Oddo, and A. Verkhratsky. 2008. Impaired adult neurogenesis in the dentate gyrus of a triple transgenic mouse model of Alzheimer's disease. PLoS One 3: e2935.

Rolando, C., and V. Taylor. 2014. Neural stem cell of the hippocampus: development, physiology regulation, and dysfunction in disease. Curr Top Dev Biol 107: 183–206.

Rowitch, D.H., and A.R. Kriegstein. 2010. Developmental genetics of vertebrate glial-cell specification. Nature 468: 214–222.

Rubenstein, J.L., and P. Rakic. 1999. Genetic control of cortical development. Cereb Cortex 9: 521–523.

Sachs, B.D., and M.G. Caron. 2015. Chronic fluoxetine increases extra-hippocampal neurogenesis in adult mice. Int J Neuropsychopharmacol 18.

Sahay, A., and R. Hen. 2007. Adult hippocampal neurogenesis in depression. Nat Neurosci 10: 1110–1115.

Sahay, A., K.N. Scobie, A.S. Hill, C.M. O'Carroll, M.A. Kheirbek, N.S. Burghardt, A.A. Fenton, A. Dranovsky, and R. Hen. 2011a. Increasing adult hippocampal neurogenesis is sufficient to improve pattern separation. Nature 472: 466–470.

Sahay, A., D.A. Wilson, and R. Hen. 2011b. Pattern separation: a common function for new neurons in hippocampus and olfactory bulb. Neuron 70: 582–588.

Sanai, N., T. Nguyen, R.A. Ihrie, Z. Mirzadeh, H.H. Tsai, M. Wong, N. Gupta, M.S. Berger, E. Huang, J.M. Garcia-Verdugo, D.H. Rowitch, and A. Alvarez-Buylla. 2011. Corridors of migrating neurons in the human brain and their decline during infancy. Nature 478: 382–386.

Santarelli, L., M. Saxe, C. Gross, A. Surget, F. Battaglia, S. Dulawa, N. Weisstaub, J. Lee, R. Duman, O. Arancio, C. Belzung, and R. Hen. 2003. Requirement of hippocampal neurogenesis for the behavioral effects of antidepressants. Science 301: 805–809.

Sapolsky, R.M. 1992. Do glucocorticoid concentrations rise with age in the rat? Neurobiol Aging 13: 171–174.

Sapolsky, R.M. 2001. Depression, antidepressants, and the shrinking hippocampus. Proc Natl Acad Sci U S A 98: 12320–12322.

Sapolsky, R.M. 2004. Is impaired neurogenesis relevant to the affective symptoms of depression? Biol Psychiatry 56: 137–139.

Saxe, M.D., F. Battaglia, J.W. Wang, G. Malleret, D.J. David, J.E. Monckton, A.D. Garcia, M.V. Sofroniew, E.R. Kandel, L. Santarelli, R. Hen, and M.R. Drew. 2006. Ablation of hippocampal neurogenesis impairs contextual fear conditioning and synaptic plasticity in the dentate gyrus. Proc Natl Acad Sci U S A 103: 17501–17506.

Scharfman, H.E., J.H. Goodman, and A.L. Sollas. 2000. Granule-like neurons at the hilar/CA3 border after status epilepticus and their synchrony with area CA3 pyramidal cells: functional implications of seizure-induced neurogenesis. J Neurosci 20: 6144–6158.

Scharfman, H.E., A.L. Sollas, and J.H. Goodman. 2002. Spontaneous recurrent seizures after pilocarpine-induced status epilepticus activate calbindin-immunoreactive hilar cells of the rat dentate gyrus. Neuroscience 111: 71–81.

Scharfman, H.E., A.E. Sollas, R.E. Berger, J.H. Goodman, and J.P. Pierce. 2003a. Perforant path activation of ectopic granule cells that are born after pilocarpine-induced seizures. Neuroscience 121: 1017–1029.

Scharfman, H.E., A.L. Sollas, R.E. Berger, and J.H. Goodman. 2003b. Electrophysiological evidence of monosynaptic excitatory transmission between granule cells after seizure-induced mossy fiber sprouting. J Neurophysiol 90: 2536–2547.

Scharfman, H.E., and W.P. Gray. 2006. Plasticity of neuropeptide Y in the dentate gyrus after seizures, and its relevance to seizure-induced neurogenesis. EXS 95: 193–211.

Scharfman, H.E., and W.P. Gray. 2007. Relevance of seizure-induced neurogenesis in animal models of epilepsy to the etiology of temporal lobe epilepsy. Epilepsia 48 Suppl 2: 33–41.

Scheff, S.W., and D.A. Price. 2006. Alzheimer's disease-related alterations in synaptic density: neocortex and hippocampus. J Alzheimers Dis 9: 101–115.

Schlachetzki, J.C., T. Grimm, Z. Schlachetzki, N.M. Ben Abdallah, B. Ettle, P. Vohringer, B. Ferger, B. Winner, S. Nuber, and J. Winkler. 2016. Dopaminergic lesioning impairs adult hippocampal neurogenesis by distinct modification of alpha-synuclein. J Neurosci Res 94: 62–73.

Schmidt-Hieber, C., P. Jonas, and J. Bischofberger. 2004. Enhanced synaptic plasticity in newly generated granule cells of the adult hippocampus. Nature 429: 184–187.

Seib, D.R., N.S. Corsini, K. Ellwanger, C. Plaas, A. Mateos, C. Pitzer, C. Niehrs, T. Celikel, and A. Martin-Villalba. 2013. Loss of Dickkopf-1 restores neurogenesis in old age and counteracts cognitive decline. Cell Stem Cell 12: 204–214.

Seki, T., and Y. Arai. 1995. Age-related production of new granule cells in the adult dentate gyrus. Neuroreport 6: 2479–2482.

Selkoe, D.J. 2001. Presenilin, Notch, and the genesis and treatment of Alzheimer's disease. Proc Natl Acad Sci U S A 98: 11039–11041.

Seri, B., J.M. Garcia-Verdugo, B.S. McEwen, and A. Alvarez-Buylla. 2001. Astrocytes give rise to new neurons in the adult mammalian hippocampus. J Neurosci 21: 7153–7160.

Shapiro, L.A., M.J. Korn, and C.E. Ribak. 2005. Newly generated dentate granule cells from epileptic rats exhibit elongated hilar basal dendrites that align along GFAP-immunolabeled processes. Neuroscience 136: 823–831.

Shapiro, L.A., K.L. Ng, R. Kinyamu, P. Whitaker-Azmitia, E.E. Geisert, M. Blurton-Jones, Q.Y. Zhou, and C.E. Ribak. 2007. Origin, migration and fate of newly generated neurons in the adult rodent piriform cortex. Brain Struct Funct 212: 133–148.

Shen, L., H.S. Nam, P. Song, H. Moore, and S.A. Anderson. 2006. FoxG1 haploinsufficiency results in impaired neurogenesis in the postnatal hippocampus and contextual memory deficits. Hippocampus 16: 875–890.

Shetty, A.K., V. Zaman, and G.A. Shetty. 2003. Hippocampal neurotrophin levels in a kainate model of temporal lobe epilepsy: a lack of correlation between brain-derived neurotrophic factor content and progression of aberrant dentate mossy fiber sprouting. J Neurochem 87: 147–159.

Shetty, A.K., M.S. Rao, B. Hattiangady, V. Zaman, and G.A. Shetty. 2004. Hippocampal neurotrophin levels after injury: Relationship to the age of the hippocampus at the time of injury. J Neurosci Res 78: 520–532.

Shetty, A.K., B. Hattiangady, and G.A. Shetty. 2005. Stem/progenitor cell proliferation factors FGF-2, IGF-1, and VEGF exhibit early decline during the course of aging in the hippocampus: role of astrocytes. Glia 51: 173–186.

Shetty, A.K., and B. Hattiangady. 2007. Concise review: prospects of stem cell therapy for temporal lobe epilepsy. Stem Cells 25: 2396–2407.

Shetty, A.K. 2011. Progress in cell grafting therapy for temporal lobe epilepsy. Neurotherapeutics 8: 721–735.

Shetty, G.A., B. Hattiangady, and A.K. Shetty. 2013. Neural stem cell- and neurogenesis-related gene expression profiles in the young and aged dentate gyrus. Age (Dordr) 35: 2165–2176.

Shors, T.J., G. Miesegaes, A. Beylin, M. Zhao, T. Rydel, and E. Gould. 2001. Neurogenesis in the adult is involved in the formation of trace memories. Nature 410: 372–376.

Siebzehnrubl, F.A., and I. Blumcke. 2008. Neurogenesis in the human hippocampus and its relevance to temporal lobe epilepsies. Epilepsia 49 Suppl 5: 55–65.

Snyder, J.S., N. Kee, and J.M. Wojtowicz. 2001. Effects of adult neurogenesis on synaptic plasticity in the rat dentate gyrus. J Neurophysiol 85: 2423–2431.

Snyder, J.S., N.S. Hong, R.J. McDonald, and J.M. Wojtowicz. 2005. A role for adult neurogenesis in spatial long-term memory. Neuroscience 130: 843–852.

Snyder, J.S., A. Soumier, M. Brewer, J. Pickel, and H.A. Cameron. 2011. Adult hippocampal neurogenesis buffers stress responses and depressive behaviour. Nature 476: 458–461.

Song, H., C.F. Stevens, and F.H. Gage. 2002. Astroglia induce neurogenesis from adult neural stem cells. Nature 417: 39–44.

Song, J., C. Zhong, M.A. Bonaguidi, G.J. Sun, D. Hsu, Y. Gu, K. Meletis, Z.J. Huang, S. Ge, G. Enikolopov, K. Deisseroth, B. Luscher, K.M. Christian, G.L. Ming, and H. Song. 2012. Neuronal circuitry mechanism regulating adult quiescent neural stem-cell fate decision. Nature 489: 150–154.

Song, J., J. Sun, J. Moss, Z. Wen, G.J. Sun, D. Hsu, C. Zhong, H. Davoudi, K.M. Christian, N. Toni, G.L. Ming, and H. Song. 2013. Parvalbumin interneurons mediate neuronal circuitry-neurogenesis coupling in the adult hippocampus. Nat Neurosci 16: 1728–1730.

Sonntag, W.E., C.D. Lynch, P.T. Cooney, and P.M. Hutchins. 1997. Decreases in cerebral microvasculature with age are associated with the decline in growth hormone and insulin-like growth factor 1. Endocrinology 138: 3515–3520.

Sonntag, W.E., C.D. Lynch, S.A. Bennett, A.S. Khan, P.L. Thornton, P.T. Cooney, R.L. Ingram, T. McShane, and J.K. Brunso-Bechtold. 1999. Alterations in insulin-like growth factor-1 gene and protein expression and type 1 insulin-like growth factor receptors in the brains of ageing rats. Neuroscience 88: 269–279.

Spalding, K.L., O. Bergmann, K. Alkass, S. Bernard, M. Salehpour, H.B. Huttner, E. Bostrom, I. Westerlund, C. Vial, B.A. Buchholz, G. Possnert, D.C. Mash, H. Druid, and J. Frisen. 2013. Dynamics of hippocampal neurogenesis in adult humans. Cell 153: 1219–1227.

Strine, T.W., R. Kobau, D.P. Chapman, D.J. Thurman, P. Price, and L.S. Balluz. 2005. Psychological distress, comorbidities, and health behaviors among U.S. adults with seizures: results from the 2002 National Health Interview Survey. Epilepsia 46: 1133–1139.

Takamura, N., S. Nakagawa, T. Masuda, S. Boku, A. Kato, N. Song, Y. An, Y. Kitaichi, T. Inoue, T. Koyama, and I. Kusumi. 2014. The effect of dopamine on adult hippocampal neurogenesis. Prog Neuropsychopharmacol Biol Psychiatry 50: 116–124.

Tapia-Rojas, C., F. Aranguiz, L. Varela-Nallar, and N.C. Inestrosa. 2016. Voluntary Running Attenuates Memory Loss, Decreases Neuropathological Changes and Induces Neurogenesis in a Mouse Model of Alzheimer's Disease. Brain Pathol 26: 62–74.

Taverna, E., M. Gotz, and W.B. Huttner. 2014. The cell biology of neurogenesis: toward an understanding of the development and evolution of the neocortex. Annu Rev Cell Dev Biol 30: 465–502.

Tiwari, S.K., S. Agarwal, B. Seth, A. Yadav, S. Nair, P. Bhatnagar, M. Karmakar, M. Kumari, L.K. Chauhan, D.K. Patel, V. Srivastava, D. Singh, S.K. Gupta, A. Tripathi, R.K. Chaturvedi, and K.C. Gupta. 2014. Curcumin-loaded nanoparticles potently induce adult neurogenesis and reverse cognitive deficits in Alzheimer's disease model via canonical Wnt/beta-catenin pathway. ACS Nano 8: 76–103.

Tozuka, Y., S. Fukuda, T. Namba, T. Seki, and T. Hisatsune. 2005. GABAergic excitation promotes neuronal differentiation in adult hippocampal progenitor cells. Neuron 47: 803–815.

Tronel, S., A. Fabre, V. Charrier, S.H. Oliet, F.H. Gage, and D.N. Abrous. 2010. Spatial learning sculpts the dendritic arbor of adult-born hippocampal neurons. Proc Natl Acad Sci U S A 107: 7963–7968.

Urban, N., and F. Guillemot. 2014. Neurogenesis in the embryonic and adult brain: same regulators, different roles. Front Cell Neurosci 8: 396.

van Praag, H., B.R. Christie, T.J. Sejnowski, and F.H. Gage. 1999. Running enhances neurogenesis, learning, and long-term potentiation in mice. Proc Natl Acad Sci U S A 96: 13427–13431.

van Praag, H., A.F. Schinder, B.R. Christie, N. Toni, T.D. Palmer, and F.H. Gage. 2002. Functional neurogenesis in the adult hippocampus. Nature 415: 1030–1034.

van Praag, H., T. Shubert, C. Zhao, and F.H. Gage 2005. Exercise enhances learning and hippocampal neurogenesis in aged mice. J Neurosci 25: 8680–8685.

Velazquez, R., J.A. Ash, B.E. Powers, C.M. Kelley, M. Strawderman, Z.I. Luscher, S.D. Ginsberg, E.J. Mufson, and B.J. Strupp. 2013. Maternal choline supplementation improves spatial learning and adult hippocampal neurogenesis in the Ts65Dn mouse model of Down syndrome. Neurobiol Dis 58: 92–101.

Verret, L., J.L. Jankowsky, G.M. Xu, D.R. Borchelt, and C. Rampon. 2007. Alzheimer's-type amyloidosis in transgenic mice impairs survival of newborn neurons derived from adult hippocampal neurogenesis. J Neurosci 27: 6771–6780.

Walker, A.K., P.D. Rivera, Q. Wang, J.C. Chuang, S. Tran, S. Osborne-Lawrence, S.J. Estill, R. Starwalt, P. Huntington, L. Morlock, J. Naidoo, N.S. Williams, J.M. Ready, A.J. Eisch, A.A. Pieper, and J.M. Zigman. 2015. The P7C3 class of neuroprotective compounds exerts antidepressant efficacy in mice by increasing hippocampal neurogenesis. Mol Psychiatry 20: 500–508.

Walter, C., B.L. Murphy, R.Y. Pun, A.L. Spieles-Engemann, and S.C. Danzer. 2007. Pilocarpine-induced seizures cause selective time-dependent changes to adult-generated hippocampal dentate granule cells. J Neurosci 27: 7541–7552.

Wang, S., B.W. Scott, and J.M. Wojtowicz. 2000. Heterogenous properties of dentate granule neurons in the adult rat. J Neurobiol 42: 248–257.

Wang, R., K.T. Dineley, J.D. Sweatt, and H. Zheng. 2004. Presenilin 1 familial Alzheimer's disease mutation leads to defective associative learning and impaired adult neurogenesis. Neuroscience 126: 305–312.

Wang, Y., S. Ristevski, and V.R. Harley. 2006. SOX13 exhibits a distinct spatial and temporal expression pattern during chondrogenesis, neurogenesis, and limb development. J Histochem Cytochem 54: 1327–1333.

Welberg, L. 2013. Neurogenesis: a bombshell of a finding. Nat Rev Neurosci 14: 522.

Wen, P.H., P.R. Hof, X. Chen, K. Gluck, G. Austin, S.G. Younkin, L.H. Younkin, R. DeGasperi, M.A. Gama Sosa, N.K. Robakis, V. Haroutunian, and G.A. Elder. 2004. The presenilin-1 familial Alzheimer disease mutant P117L impairs neurogenesis in the hippocampus of adult mice. Exp Neurol 188: 224–237.

Xiang, J., S. Yan, S.H. Li, and X.J. Li. 2015. Postnatal loss of hap1 reduces hippocampal neurogenesis and causes adult depressive-like behavior in mice. PLoS Genet 11: e1005175.

Yao, J., Y. Mu, and F.H. Gage. 2012. Neural stem cells: mechanisms and modeling. Protein Cell 3: 251–261.

Yassa, M.A., and C.E. Stark. 2011. Pattern separation in the hippocampus. Trends Neurosci 34: 515–525.

Yuan, T.F., and O. Arias-Carrion. 2011. Adult neurogenesis in the hypothalamus: evidence, functions, and implications. CNS Neurol Disord Drug Targets 10: 433–439.

Zhang, C., E. McNeil, L. Dressler, and R. Siman. 2007. Long-lasting impairment in hippocampal neurogenesis associated with amyloid deposition in a knock-in mouse model of familial Alzheimer's disease. Exp Neurol 204: 77–87.

Zhang, C.L., Y. Zou, W. He, F.H. Gage, and R.M. Evans. 2008. A role for adult TLX-positive neural stem cells in learning and behaviour. Nature 451: 1004–1007.

Zhao, M., S. Momma, K. Delfani, M. Carlen, R.M. Cassidy, C.B. Johansson, H. Brismar, O. Shupliakov, J. Frisen, and A.M. Janson. 2003. Evidence for neurogenesis in the adult mammalian substantia nigra. Proc Natl Acad Sci U S A 100: 7925–7930.

Zhao, C., E.M. Teng, R.G. Summers, Jr., G.L. Ming, and F.H. Gage. 2006. Distinct morphological stages of dentate granule neuron maturation in the adult mouse hippocampus. J Neurosci 26: 3–11.

7

Neural Stem Cells and Transplantation

Malathi Srivatsan

Introduction

Neural Stem cells are immature, undifferentiated, pluripotent cells that can multiply and differentiate into specialized cell types (Evans and Kaufman, 1981; Martin, 1981). Transplantation of these stem cells or their derivatives, and mobilization of already endogenous stem cells within the adult brain have both been proposed as future therapies for neurodegenerative diseases. These treatments became an area of intense research globally due to the hope that they can lead to recovery in function for millions of people. Prior to this, researching cures for loss of function due to neuron death or neural degeneration was at best not very promising, as mature neurons do not multiply. Because fully differentiated neurons in higher organisms have unique morphological, physiological and biochemical properties, they are generally post-mitotic and will not divide to replace dead or injured cells (Currais *et al.*, 2009). As a result neuronal loss either from degeneration or injury leads to irreparable functional loss. Recent advances in stem cell research offer a hope of recovery in function for millions of people suffering worldwide. This chapter focusses on the existing need for stem cells, the advancement in techniques used to differentiate stem cells into specific neuron types, the results from transplantation in animal models, human clinical trials and the challenges yet to be overcome.

Professor of Neurobiology, P.O.Box: 599, Department of Biological Sciences, Arkansas State University, Jonesboro, AR 72401.
E-mail: msrivatsan@astate.edu

The Growing Need for Stem Cell-derived Neurons and Glial Cells for Transplantation Therapy

The World Health Organization reports that 250,000 to 500,000 people suffer a spinal cord injury (SCI) each year (WHO Fact sheet N°384). About 276,000 Americans currently suffer from spinal cord injuries and each year 12,500 new patients are added to this list (National SCI Statistical center data 2014). In addition about 300,000 patients are disabled permanently due to traumatic brain injury (TBI, DoD report 2014). Injury and death of neurons in the central nervous system (CNS) lead to permanent loss of function in these otherwise healthy individuals.

As life expectancy is increasing worldwide, the number of patients suffering from age-related neurodegenerative disorders such as Alzheimer's disease (AD), Parkinson's disease (PD), Huntington's disease (HD), Amyotrophic Lateral Sclerosis (ALS) and stroke is also increasing. Approximately 44 million patients worldwide are suffering from AD and associated dementia and this number is expected to triple by the year 2050 (World Alzheimer's report 2014). AD is an irreversible, progressive disease that affects memory and leads to cognitive decline and behavioral and psychiatric disorders for approximately 5 million sufferers in America (U.S. Department of Health and Human Services report 2013). The loss of neurons, predominantly cholinergic, along with the presence of amyloid plaques and neurofibrillary tangles is the hallmarks of AD (NIA report 2014). Like AD, PD is a frightening and irreversible neurodegenerative disorder. It was estimated that the number of individuals over 50 years of age with PD was between 4.1 and 4.6 million in the most populous nations in 2005 and was projected to more than double to between 8.7 and 9.3 million by 2030 (Dorsey *et al.*, 2007). Degeneration of nigrostriatal dopaminergic neurons, the main pathology in PD, causes its trademark impairment in movement. PD symptoms may include tremors, slowed movement (bradykinesia), rigidity of muscles, impaired posture/balance and loss of automatic movements. Yet another traumatic neurodegenerative disease is ALS, also known as Lou Gehrig's disease. It is a progressive motor neuron disease (MND) that results from the degeneration of neurons that control voluntary movement. Patients exhibit muscle weakness and paralysis, as well as impaired speaking, swallowing, and breathing and have a life expectancy of two to five years from the onset of symptoms. In the USA the prevalence of ALD/MND is around four patients per 100,000 people; thus about 12,000 Americans are currently suffering from ALS (Mehta *et al.*, 2014). HD is an autosomal dominant genetic disorder that affects muscle coordination and leads to mental decline and behavioral symptoms. HD results from neuronal destruction in the caudate nucleus, located within the basal ganglia. The disorder is more prevalent in Europe, North America, and Australia than in Asia (Pringsheim *et al.*, 2012). It is estimated that approximately 30,000 people in America suffer from HD and 16% of all cases are estimated to be juvenile HD. Strokes are the fourth leading cause of death in the US and each year about 750,000 Americans suffer from them. Although the risk of a stroke increases with age, strokes can occur at any age (CDC Fact sheet 2014). During a stroke, neurons in affected areas of brain die due to lack of blood supply. This leads to loss of function associated with the affected region of the brain. Apart from these major neurological disorders, degenerative changes in the glial cells can also lead to diseases such as multiple sclerosis (MS). Degeneration of oligodendrocytes impairs the myelin sheath surrounding axons, normally used to help conduct electrical signals through them. Axons that have been demyelinated tend to atrophy and eventually degenerate. MS is the most common demyelinating disease and both genetic and environmental factors appear to contribute to this disease which is generally classified as an autoimmune disorder (Barcellos *et al.*, 2002; Marrie, 2004; Weiner, 2004). Approximately 400,000 people in the US and 2.5 million worldwide suffer from MS (National MS society data). The neurodegeneration that causes these disorders impairs function in many different ways for patients, and negatively affects quality of life for both patients and care givers through symptoms and rising health care costs. This crisis is forcing researchers to quickly find ways to treat functional loss due to death or degeneration of neurons and oligodendroglia. Since each of these disorders requires a specific type of neuron for transplantation therapy, in recent years, a promising strategy has been to try to differentiate stem cells into neurons of specific phenotype or oligodendrocytes so that specific types of dead/injured neurons or oligodendrocytes can be replaced with live ones (Mitrecic *et al.*, 2012).

Differentiating Specific Types of Neurons and Oligodendrocytes

With the availability of genomic data, high-throughput screening methods and embryonic stem (ES) cell lines, molecular mechanisms underlying the differentiation of specific nervous system cells have been identified. Access to human ES cells has allowed the characterization of these cells and is currently driving attempts to differentiate them into specific cell types and populations. While we first came to know about the pluripotent ES cells obtained from mouse embryos in 1981 (Evans and Kaufman, 1981; Martin, 1981) it took another 17 years to see the first published report on human ES cell lines (Thomson *et al.*, 1998). Since the publication of these reports we have learned that embryonic (Cai and Grabel, 2007; Carpenter *et al.*, 2001; Dhara *et al.*, 2008; Robertson *et al.*, 2008) and adult stem cells (Alvarez-Buylla *et al.*, 2008; Alvarez-Buylla and Garcia-Verdugo, 2002; Merkle *et al.*, 2007) have the potential to differentiate into cell types, which may be able to replace injured neurons or promote repair in regions of injury or degeneration in the nervous system. Recent advances in the understanding of the differentiation process have allowed us to generate induced, pluripotent stem cells (iPS cells) by inducing de-differentiation and reprogramming of adult cells into stem cells with the integration of pluripotency-associated genes such as Oct4, Sox2, Klf4, c-myc, Lin28 and Nanog (Takahashi *et al.*, 2007; Takahashi and Yamanaka, 2006); this conveniently avoids the many ethical, religious, and political challenges of retrieving stem cells from human embryos. In addition to replacing dead neurons, stem cells might be able to synthesize and release growth factors that could support the remaining neuronal populations and even stimulate intrinsic regenerative mechanisms within the nervous system. Understanding the development of the nervous system during normal embryonic development has shown that morphogens such as the bone morphogenic protein, sonic hedgehog (Shh), fibroblast growth factors (FGFs) and retinoic acid (RA) induce neuronal specification by interacting in specific time frames (Bally-Cuif and Hammerschmidt, 2003; Briscoe and Ericson, 2001; Jessell, 2000).

Further, tightly regulated temporal and spatial expressions and releases of neurotrophins and the presence of their receptors on differentiating cells appear to promote differentiation to the next step of phenotype determination. These neurotrophins belong to a family of growth factors that include nerve growth factor (NGF), brain-derived neurotrophic factor (BDNF), neurotrophin-3 (NT-3) and neurotrophin-4 (NT-4), which all play crucial roles in the development of the nervous system (Chao, 2003) by binding to and activating receptor tyrosine kinases (Trk) and the p75 neurotrophin receptor (p75NTR). Taking the cues from the findings of studies on normal embryonic development of the nervous system, researchers have used these neurotrophins on human embryonic stem cells (hESCs) and have found that neurotrophic factors can mediate hESCs survival and also modulate their fate (Pyle *et al.*, 2006; Schuldiner *et al.*, 2000).

While several cell biology procedures, e.g., the use of a cocktail of growth factors, are being used to promote neuronal differentiation of stem cells, some challenges in this field remain: getting an adequate number of cells capable of differentiation into a specific cell type through cell division (Armstrong and Syendsen, 2000; Bithell and Williams, 2005) and arresting their cell division to prevent cancer development once transplantation occurs (Molofsky *et al.*, 2003; Rossi *et al.*, 2008). Additional concerns specific to neurons are how to drive the stem cells to differentiate into neurons, as most differentiate into glial cells (Bithell and Williams, 2005) and making sure they will grow axons to the required length, recognize appropriate targets to make synapses, and exhibit optimal electrical activity and conduction for impulse transmission (Armstrong and Syendsen, 2000). Research is progressing at a rapid pace to address these concerns and in the last few years much progress has been made in successfully differentiating stem cells into specific neuron types and also in transplanting them, mostly in animal models.

Cholinergic neurons from stem cells

Cholinergic neurons (CNs) in the basal fore brain, which release acetylcholine as the neurotransmitter and innervate the hippocampus and cortex, are primarily affected in AD. Therefore researchers are interested in generating CNs from stem cells to develop novel therapies for AD including transplantation. Expression of specific homeobox genes and exposure to neurotrophins seem to drive cholinergic neuronal differentiation. Manabe *et al.* reported that the expression of L3/Lhx8, a member of the LIM-homeobox gene family, was necessary for neurons to develop a cholinergic phenotype while other studies revealed the role of neurotrophins in cholinergic differentiation (Manabe *et al.*, 2007). Nilbratt *et al.* found that hESCs expressed neurotrophin receptors (Nilbratt *et al.*, 2010). In their study, stimulation with NT-3, NGF and CNTF neurotrophic factors induced hESCs to differentiate into cholinergic neurons that expressed the cholinergic enzyme choline acetyltransferase, cholinergic $\alpha3$, $\alpha4$ and $\alpha7$ nicotinic acetylcholine receptor subunits (nAChR) and M1, M2 and M3 muscarinic acetylcholine receptor (mAChR) subunits as shown in Figure 7.1. These receptors were functional because exposure to ACh resulted in elevated cytosolic calcium levels which were abolished by the mAChR antagonist scopolamine.

Neurotrophic factors promote cholinergic differentiation in human embryonic stem cell-derived neurons

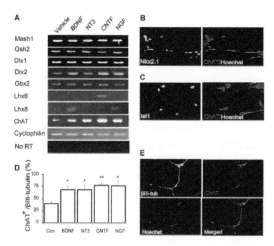

Figure 7.1. (From ref. Nilbratt *et al.*, 2010) Expression of markers for sub-regional identity and for cholinergic neurotransmission in differentiated hES cells. (A) After 18 day of differentiation with BDNF, NT-3, CNTF and NGF, the cells expressed transcripts of several genes expressed the ventral telencephalon, including *Mash1*, *Dlx1*, *Dlx2*, *Gbx2* and *Gsh2*. The LIM-homeobox genes *Lhx6* and *Lhx8*, both confined to the MGE, were up-regulated with CNTF, and with BDNF and NGF treatment, respectively. (B, C) Immunostaining of hES cell-derived cholinergic cells with anti-Nkx2.1 and anti-Islet-1 antibodies 18 days after plating (at 40×). (D) The proportion of neurons expressing ChAT following treatment with neurotrophic factors. All factors tested showed statistically significant effect on the proportion of ChAT$^+$ neurons. $*p <$ 0.05; $**p < 0.001$ as compared to control. Values are expressed as mean $\pm$ S.E.M., $n = 6$. (E) Immunostaining of ChAT (red) in NT-3-differentiated neurons at day 18 (at 20×) (Reproduced with permission from John Wiley and sons).

Using a slightly different strategy, Cristopher *et al.* converted hESCs through exposure to RA into neural progenitor cells which in turn were differentiated into CNs when exposed to Shh, FGF8 and bone morphogenetic protein 9 (BMP9) (Christopher *et al.*, 2011) as BMP9 induces the expression of cholinergic markers in cultured murine septal progenitors (Lopez-Coviella *et al.*, 2005). It should be noted that the effect of BMP9 on cholinergic differentiation during the development of murine brain is both spatially and temporally limited, being effective only in the septal neurogenesis during the E14-E16 period of development (Lopez-Coviella *et al.*, 2000). These researchers also found that nucleofection of the neural progenitor neurosphere cells with DNA, which encodes *Lhx8* and *Gbx1*, also resulted in cholinergic differentiation. In a later study Crompton *et al.* expressed basal forebrain transcription factors NKX2.1 and LHX8, as well as the general forebrain marker FOXG1 while inducing the intrinsic hedgehog signaling in human pluripotent stem cells (hPSCs) using a 3D non-adherent differentiation system (Crompton *et al.*, 2013). Upon differentiation, these self-renewing cells exhibited a cholinergic phenotype expressing the specific markers choline acetyl transferase (ChAT), vesicular acetylcholine transporter (VACht) and LIM homeodomain factor ISL1. These hPSC-derived CNs released ACh and demonstrated a functional cholinergic electrophysiological profile. Most recently Adib *et al.* described a tissue engineering method to first differentiate human bone marrow stromal cells (hBMSCs) into human neural stem cells (hNSCs) and the subsequent differentiation of hNSCs into CNs by sequential replacement of culturing media (Adib *et al.*, 2015). They generated cell aggregates from hBMSCs using basic fibroblast growth factor (bFGF), epidermal growth factor (EGF) and B27 supplement (Invitrogen). The hNSCs were isolated from the generated cell aggregates and differentiated into CNs using culture medium in which bFGF, EGF and B27 were gradually replaced with NGF.

Cholinergic spinal motor neurons from stem cells

There is a growing interest in differentiation of stem cells into motor neurons (MNs) of the spinal cord since this can lead to a cure for paralysis from spinal cord injury or treatment of ALS, a degenerative disorder affecting motor neurons. While studies on embryonic development of spinal motor neurons (MNs) have outlined the pathways of their differentiation (Jessell, 2000; Lee and Pfaff, 2001), RA-mediated differentiation of ES cells from the CCE ES cell line derived from 129/Sv mouse strain (Robertson *et al.*, 1986) led to findings that showed that the combinatorial expression of transcription factors of the LIM family (Islet-1, Islet-2, Lim-3) and homeobox gene 9 (HB-9) characterizes the spinal motor neurons (Renoncourt *et al.*, 1998). Subsequently Harper *et al.* derived motor neurons from mouse ES cells, co-cultured them with C2C12 myoblasts (ATCC) and reported that the ES cell-derived motoneurons grew long axons, established neuromuscular junctions, and induced muscle contractions *in vitro* (Harper *et al.*, 2004). In the following years, Lee *et al.* differentiated hESCs into neural rosettes and subsequently into motor neurons by culturing them in an N2 medium supplemented with ascorbic acid (AA) and BDNF in the presence of RA/ SHH (Lee *et al.*, 2007). These cells were confirmed to be motor neurons based on their morphological, physiological, biochemical and gene expression profiles. To avoid immune rejection, recent research is focusing on deriving patient-specific iPSCs and differentiating them into phenotype-specific neurons for treatment considerations. Dimos *et al.* reprogrammed skin fibroblasts collected from an 82-year-old ALS patient into iPSCs and then successfully differentiated them into motor neurons using the sonic Shh signaling pathway as an agonist and RA (Dimos *et al.*, 2008). These procedures take up to 60 days for generating a functional motor neuron and in order to shorten this length of time, Hester *et al.* developed an adenoviral gene delivery system by encoding transcription factors neurogenin 2 (Ngn2), Isl-1, and LIM/homeobox protein 3 (Lhx3) to hESCs from the HSF-6 hESC cell line (Hester *et al.*, 2011). They also repeated the gene delivery approach with hiPSCs derived from human fibroblasts using retroviral vectors expressing the four Yamanaka factors (OCT3/4, KLF-4, SOX2, and c-MYC). The gene delivery approach induced differentiation into functional MNs with mature electrophysiological properties in both the hESCs and in the hiPSCs, 11 days after gene delivery, with 60–70% differentiation efficiency from hESCs and hiPSCs. Although this approach significantly reduces the time required to generate functional MNs, it is important to determine if the MN identity will be retained after silencing the viral-mediated transcription factor expression upon differentiation. A recent study addresses some of these issues (Amoroso *et al.*, 2013) by providing a non-viral method to produce motor neurons with relatively high yield and rapidity. Amoroso *et al.* also introduced automated procedures for cell identification and colony selection, thus minimizing individual judgment errors (Amoroso *et al.*, 2013). They exposed hESCs and hiPSCs to higher concentration of RA and two Shh pathway activators, purmorphamine and mouse agonist SAG. In three weeks of differentiation a motor neuron yield of 50% was obtained. This protocol used by Amoroso *et al.* is virus-free and therefore may be better suited for translational applications.

Dopaminergic neurons from stem cells

Generating dopaminergic (DA) neurons has great value in cell replacement/ transplantation therapy for PD. Furthermore, these cells will be extremely useful for determining the role of potential therapeutic agents in *in vitro* assays. Functional DA neurons can also help in the *in vitro* studies on understanding the mechanism of action of drugs of abuse. A major constraint for the above needs has been the limited availability of human cells for both basic and therapeutic research. Therefore substantial research efforts have been directed towards successful dopaminergic differentiation. Earlier studies demonstrated that DA neurons could be generated from mouse ES cells when they were transplanted into the rat striatum resulting in proliferation of ES cells into fully differentiated functional DA neurons. ES cell-derived DA neurons caused gradual and sustained behavioral restoration of DA-mediated motor asymmetry (Bjorklund *et al.*, 2002). Researchers probing the mechanisms underlying differentiation of DA neurons showed that the expression or presence of transcription factors Nurr1, Ngn2, FoxA2 and other factors such as Bcl-X_L, Pitx3 were required (Ang, 2006; Besnard *et al.*, 2004; Courtois *et al.*, 2010; Kim *et al.*, 2007; Kim *et al.*, 2003; Lebel *et al.*, 2001; Lee *et al.*, 2010; Sakurada *et al.*, 1999). Schulz and collaborators reported the successful differentiation of hESCs to form neurons expressing markers of midbrain DA neurons adapting a serum-free suspension culture system (Schulz *et al.*, 2004). Their results showed large networks of TH-positive neurons in the aggregates in suspension. In the following year DA neurons were efficiently generated by applying Shh and FGF8 in a specific sequence (Yan *et al.*, 2005). With the growing interest in the use of patients' own iPSCs, several reports showed successful generation of midbrain DA neurons with hiPSC- and PD-derived iPSCs through protocols using FGF8a, WNT1, and a low concentration of RA combined with a use of Shh that belonged to the high activity form (Cooper *et al.*, 2010; Hargus *et al.*, 2010). Boyer *et al.* described a combination of two protocols for differentiating hPS and hES cells into DA neurons (Boyer *et al.*, 2012). The two protocols included first leading the stem cells into rosette formation and then using methods to inhibit alternate differentiation pathways.

More recent research shows that dopaminergic neurons can be generated by treating hESCs with LIM homeobox transcription factor 1 alpha (LMX1A) which is a master protein that regulates the development of the midbrain neurons (Fathi *et al.*, 2015).

Researchers have been exploring stem cell sources other than neural stem cells for generating DA neurons. Once studies established the presence of mesenchymal stem cells in the umbilical cord, interest in storing cord blood has soared resulting in proliferation of cord blood banks. Yan *et al.* reported a strategy to differentiate umbilical

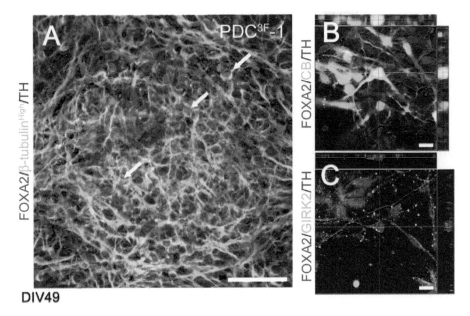

Figure 7.2. (From ref. Cooper *et al.*, 2010) Phenotypic characterization of FOXA2⁺ dopaminergic neurons generated by SHH-C24II and FGF8a from RA-treated human ES/iPS cells. (A) Human PD-iPS cell lines (PDC³ᶠ-1) were competent to generate FOXA2⁺ (red) dopaminergic neurons (TH, blue; β-tubulin, green; arrowheads). (B, C) In cultures differentiated with 500 ng/ml SHH-C24II and FGF8a, cells coexpressed TH (blue), FOXA2 (red) and calbindin (green, B) or weakly GIRK2 (green, C), indicative of an A10 or A9 DA neuron phenotype, respectively. Scale bar A = 50 μm, B, C = 10 μm (Reproduced with permission from Elsevier).

cord-derived mesenchymal stem cells into DA neurons *in vitro* by transfecting Lmx1α and NTN genes through a combination of a recombinant adenovirus with the Lmx1α regulatory factor (Yan *et al.*, 2013). Due to the proximity of embryonic origin and the ease of collection, Chang *et al.* (Chang *et al.*, 2014) used dental pulp stem cells and showed that it would be possible to differentiate dental pulp stem cells into DA neurons.

Stem Cells and Transplantation

In the past twenty to thirty years, cell transplantation as a strategy to treat neurodegenerative disorders or brain and spinal cord injury has gained support due to the lack of a successful long term pharmaceutical or surgical treatment for these conditions. Initially researchers tried transplanting mostly murine ESCs into animal models which was followed by transplanting hESCs into animal models, the success of which has led to a few clinical trials involving transplantation into human CNS.

Transplantation for Parkinson's disease

Although brain stem and cortical areas may also be affected in PD, the neuronal population that is primarily affected is DA neurons of the substantia nigra,

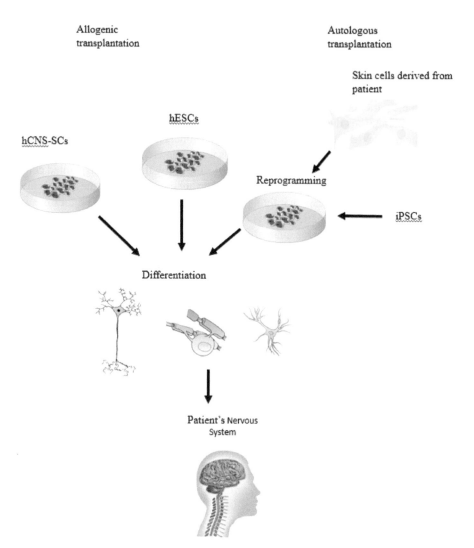

Figure 7.3. Reprogramming skin cells from patients into induced pluripotent stem cells (iPSCs), propagating human CNS-derived stem cells (HCNs-SCs) and human embryonic stem cells for differentiation into neurons, oligodendrocytes and astrocytes for autologous or allogenic transplantation.

so transplantation of DA neurons is considered effective treatment for PD. Following several years of developing protocols for differentiation and transplantation in animal models it appears that the time has come to consider transplantation of hPSC and hESC-derived DA neurons, now headed for clinical trials, as a promising approach in humans. The support for the idea of hPSC-derived DA cell transplantation therapy came from successful studies using animal models as long as thirty years ago. These studies have evaluated the effects of transplantation of fetal mesencephalic tissue containing nigral DA cells into the striatum of rodents and non-human primates (Bjorklund *et al.*, 1980; Perlow *et al.*, 1979; Studer *et al.*, 1998). Behavioral studies showed that following transplantation, recovery with regards to rotational behavior was observed in response to amphetamine in 6-OHDA lesioned animals (Bjorklund *et al.*, 2002) and in unilateral substantia nigra lesioned rats (Studer *et al.*, 1998). The success of these animal studies led to clinical trials in which fetal mesencephalic tissue was transplanted in patients with PD (Lindvall *et al.*, 1994; Ma *et al.*, 2010). Although significant benefit to the patients were reported, questions remain with regards to the efficacy of these transplantations due to the occurrence of dyskinesias (Isacson and Kordower, 2008), thought to result from contaminating serotonergic neurons in the cell graft (Politis *et al.*, 2010). The transplanted fetal cells were reported to survive for four years, yet the motor ratings declined (Ma *et al.*, 2010) suggesting that the transplanted cells might undergo progressive neurodegeneration. Further concerns about the availability of enough fetal tissue indicate that fetal tissue transplantation may not be feasible as a treatment option for PD patients. This indeed emphasizes the need for DA neurons derived from stem cells for transplantation.

Transplantation of stem cell-derived DA neurons has yielded promising results. Kim *et al.* reported that DA neurons derived from mouse ES cells were functional in Parkinsonian rats (Kim *et al.*, 2002). Recently Kim *et al.* converted mouse-tail tip fibroblasts into DA neurons which when transplanted alleviated symptoms in a mouse model of PD (Kim *et al.*, 2011). In a study that followed, Hayashi *et al.* derived dopaminergic neurons from bone marrow stem cells in monkeys and differentiated them into DA neurons (Hayashi *et al.*, 2013). The bone marrow doner macaques were treated with MPTP to induce hemiparkinsonism and then received an autologous transplant of the differentiated DA neurons. Monkeys that received the transplant showed significant improvement in motor function. The company NeuroGeneration Inc., that has trial sites in California, Italy and Estonia cultures biopsied brain tissue from PD patients to expand neural SCs and differentiates them for autologous transplantation. The differentiated mixed neurons and glial cell populations are then implanted at multiple sites in the post-commissural putamen of patients. Patients who were thus treated were reported to show some motor recovery and increased dopamine uptake in the transplanted putamen (Lévesque *et al.*, 2009).

Transplantation for Alzheimer's disease

Early experiments involved transplanting embryonic or fetal cells into animal models of AD. Transplantation of murine septal precursors, porcine cholinergic precursors, and the human embryonic septal/diagonal band region into the hippocampus of rodents resulted in a stable engraftment of the introduced cells, some of which differentiated into cholinergic neurons (LeBlanc *et al.*, 1999; Nilsson *et al.*, 1988). Although this method did not provide enough cells for transplantation, these studies demonstrated the potential for transplanted exogenous basal forebrain CNs to ameliorate memory deficits in adult cortex. As *in vitro* propagation of SCs became possible, researchers began to use SCs for transplantation. Blurton-Jones *et al.* reported that NSCs transplanted in the hippocampus of a mouse model of AD improved memory deficits through a BDNF mediated response (Blurton-Jones *et al.*, 2009). In addition, observations from animal studies have indicated that the transplanted SCs migrated and differentiated into cholinergic neurons, astrocytes, and oligodendrocytes. Kern *et al.* described the implantation of neural SCs into the hippocampus of aged Down syndrome mice (Ts65Dn mouse model) resulting in a significant reduction in the tau/reelin-positive extra cellular granules in the hippocampal area (Kern *et al.*, 2011). To address if differentiated and transplanted basal forebrain CNs would make functional synaptic connections in the host cortex, researchers used FACS-purified populations of basal forebrain CNs differentiated from hESCs and engrafted in cultured entorhinal-hippocampal murine cortical slices. The cells extended long axonal projections and synapsed with host neurons in the cortical slices successfully as revealed by electrophysiology (Christopher *et al.*, 2011). This was followed by *in vivo* experiments. Liu *et al.* differentiated hESCs to a uniform population of homeobox gene NKX2.1 (also known as thyroid transcription factor1)/medial ganglionic eminence (NKX2.1$^+$ MGE)-like progenitor cells (Liu *et al.*, 2013). After transplantation into the hippocampus of mice whose basal forebrain CNs and some GABA neurons in the medial septum had been destroyed by mu P75-saporin, human MGE-like progenitors produced basal forebrain CNs that made synapses with endogenous neurons; mice transplanted with MGE-like cells showed improvements in learning and memory. Menachem-Zidon *et al.* altered the neural SCs to overexpress the IL-1 receptor antagonist (IL-1raTG) and then transplanted them into the hippocampus in a murine model of AD (Tg2576 mice) (Ben-Menachem-Zidon *et al.*, 2014). This resulted in an improvement in memory functioning and neurogenesis in the hippocampus. Transplantation of new SCs can also lead to functional recovery through indirect means. This became obvious when adipose-derived mesenchymal stem cells were transplanted into the hippocampi of APP/PS1 transgenic Alzheimer's disease model mice. The transplantation led to formation of new neuroblasts in the hippocampi of these mice coinciding with reduced cognitive impairment and oxidative stress (Yan *et al.*, 2014).

Transplantation for spinal cord injury and ALS

In October of 2014 there was sensational headlines about 40 year old DarekFidyka from Poland walking again (medical news, 2014) after being paralyzed for four years due to a stab wound that injured his spinal cord in the thoracic region. This recovery happened because surgeons transplanted olfactory ensheathing cells (stem cells from the nose) forming a successful bridge at the injury site, resulting in a promising result for research on stem cell transplantation. In injury cases, often the patient is otherwise a healthy individual without any pathology till the time of injury and this healthy body environment may support SCs ability to bridge. This is a culmination of several years of research and trials beginning with animal models. An early study showed that transplantation of *in vitro*-expanded rat fetal neural progenitor cells led to neurogenesis and motor functional recovery after spinal cord contusion injury in adult rats (Ogawa *et al.*, 2002). In 2004 Harper *et al.* reported that they transplanted motor neuron-committed mouse ES cells (from HB9-GFP-transgenic mouse) into an adult rat's spinal cord and found that the transplanted cells became motor neurons that were able to survive long term (Harper *et al.*, 2004). The axonal growth from these new motor neurons was inhibited by myelin and was overcome by pharmacologic treatment of host animals either with dibutyrylcAMP or with Y27632. Following this, another study (Lee *et al.*, 2007) explored if hESC transplants could also survive long term. Their results presented evidence for *in vivo* survival of hESC-derived motoneurons in adult rats' spinal cords encouraging further research on hESC-based transplantation for motoneuron diseases. Following the success in rats, transplantation was extended to non-human primates. Iwanami *et al.* expanded human neural progenitor cells in culture and transplanted them into the spinal cord of adult marmosets nine days after they had cervical contusion SCIs (Iwanami *et al.*, 2005). Not only did the graft differentiate into neurons and glial cells but it also appeared to improve motor function in the marmosets. A recent paper by Lu *et al.* showed that rat and human neural stem cells injected into a rat's spinal cord injury site were able to engraft into the tissue and integrate into host circuitry (Figure 7.4) to promote functional recovery (Lu *et al.*, 2012).

The breakthrough that olfactory mucosa is an important and readily available source of progenitor cells for neural repair has directed the transplantation of olfactory stem cells to spinal cord injured patients (Lima *et al.*, 2006). Olfactory mucosa autografts were transplanted into lesions at the spinal cord C4–T6 levels in seven patients. All the patients showed improved motor functions. The side effects included a transient pain, relieved with medication, and decrease in sensory perception. In a phase I clinical trial Tabakow *et al.* reported that one year after the transplantation of autologous mucosal olfactory ensheathing cells and olfactory nerve fibroblasts in patients with complete spinal cord injury, both sensory and motor functions improved and there was evidence for spinal cord neural transmission (Tabakow *et al.*, 2013). These results help conclude that intraspinal transplantation of autologous olfactory ensheathing cells is feasible and probably safe.

Neuralstem Inc. (Rockville, MD, USA) has developed NSI-566 cells from human spinal cord progenitor cells for transplantation. In preclinical animal studies, rats with surgically transected spinal cords, rendered permanently and completely paraplegic, were reported to recover significant locomotor function regaining movement in all lower extremity joints after transplantation (van Gorp *et al.*, 2013). Neural stem cells also express nerve growth factors that can help in regeneration. Neural stem started surgeries in a Phase I trial of its NSI-566 neural stem cells for chronic spinal cord injury (cSCI) patients at the University of California, San Diego School of Medicine in September 2014. The success of this trial can lead to more such transplantation treatment options available for patients in USA.

Neural stem cell-based transplantation for ALS

Since it is currently possible to differentiate SCs into motor neurons, naturally researchers want to be able to apply this treatment to ALS. Intrathecal transplantation of human motor neurons derived from neural SCs in the spinal cord of the SOD1G93A mouse ALS model was reported to delay the onset of the disease and extend life span in these animals (Lee, 2014). Similarly intraspinal grafting of human neural precursor cells in ALS rats showed improvements in motor neuron survival and motor function due to the ability of transplanted NPCs to integrate into the spinal cord, differentiate, and form functional synapses with host motor neurons (Xu *et al.*, 2006).

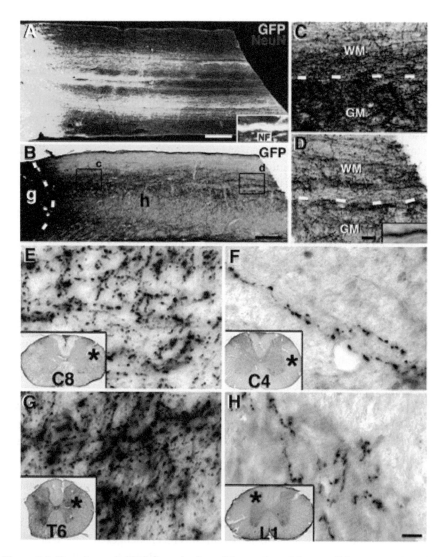

Figure 7.4. (From Lu *et al.*, 2012) Extensive Long-Distance Axonal Outgrowth from Neural Stem Cell Grafts. (A) GFP and NeuN immunolabeling reveals that GFP-expressing neural stem cell grafts robustly extend axons into the host spinal cord rostral and caudal to the T3 complete transection site (caudal shown) over the 12 mm length of the horizontal section. Extensive regions of the host spinal cord contain graft-derived projections in white matter and gray matter. Inset shows that GFP-labeled projections arising from grafts express neurofilament (NF), confirming their identity as axons. (B) Light-level GFP immunolabeling also clearly reveals the density and distribution of axons extending from the lesion site. Boxes are shown at higher magnification in C, D. (C) Dense numbers of GFP-labeled axons are present in host white matter (WM) and gray matter (GM) 2 mm caudal to the lesion, and (D) large numbers of axons remain at the end of the block of tissue, 6 mm caudal to the lesion. (E) A high density of GFP-labeled axons is present in the lateral portion of C8, three spinal segments above the lesion, and (F) remain detectable at C4, 7 spinal segments above the lesion. Asterisk indicates locations of higher magnification view. (G) GFP-labeled axons also extend in dense numbers caudally to T6 white matter and gray matter (shown) and (H) caudally to L1. Overall, axons extend at least 25 mm in each direction. Scale bar: A–B, 600 μm; B–C, 40 μm; D–G, 10 μm (Reproduced with permission from Elsevier).

Several other studies showed similar beneficial effects on animal models of ALS (Lunn *et al.*, 2014). These promising results led to human clinical trials (Feldman *et al.*, 2014; Gordon, 2013). Feldman *et al.* described two critical clinical observations of hSC transplant in ALS patients, (i) repeated intra-spinal injection of hSC into the same patients appeared to be beneficial for treating ALS without causing any severe side effects and (ii) some of the patients without bulbar palsy and early transplant during ALS development showed significant reduction in the development of the disease (Feldman *et al.*, 2014). In a clinical trial in China multiple treatments of olfactory ensheathing cell transplants appeared to promote functional improvement indicating the safety and feasibility of multiple injections to the fragile ALS spine (Chen *et al.*, 2012). Following the success of a phase I clinical trial (Feldman *et al.*, 2014), Neuralstem Inc. has recently launched a phase II clinical trial for transplantation treatment of ALS patients at three centers (Emory University Hospital in Atlanta, Georgia, ALS Clinic at the University of Michigan Health System, in Ann Arbor, Michigan, and Massachusetts General Hospital in Boston—See more details at: http://www.neuralstem.com/cell-therapy-for-als).

Transplantation for Huntington's disease and strokes

Only a few attempts have been made for transplantation for treating Huntington's disease and stroke victims. When striatal grafts were performed, Huntington's patients developed post-graft subdural hematomas, most likely from the graft being in the heavily atrophied basal ganglia (Hauser *et al.*, 2002). A subarachnoidal injection of neural and hematopoietic SCs in stroke patients resulted in significant improvement in function without serious side effects (Bang *et al.*, 2005; Rabinovich *et al.*, 2005). Progressively, neurological deficits have decreased without any adverse side effects in cerebral infracted patients when treated with intravenous mesenchymal stem cell infusion (Rabinovich *et al.*, 2005).

Challenges facing transplantation therapy

Although SCs and, especially iPSCs, are a blessing for patients with incurable neurological disorders, SCs are associated with several challenges in clinical practice. Firstly, stem cell properties such as self-renewal and plasticity are also characteristics of cancer cells and the possibility of patients developing cancer after stem cell transplantation is a dangerous and unacceptable side effect. Dopaminergic neurons made from mouse embryonic stem cells were transplanted into Parkinson's mice and provided some decrease in symptoms, but 20% of the mice receiving the ESCs died from teratoma formation (Nishimura *et al.*, 2003). In all of these rats the cells began to lose their specialization and grew uncontrollably resulting in tumors (Roy *et al.*, 2006). Other concerns specific to iPSCs are chromosomal aberrations and epigenetic modifications which have been reported to occur during reprogramming and may be maintained through differentiation (Lister *et al.*, 2011; Mayshar *et al.*, 2010). Currently we do not know what the effect of these mutations and aberrations will be on the physiology or safety of differentiated cells or whether they will persist after extended culture of an iPSC line (Hussein *et al.*, 2011). In addition, the differentiation potential of iPSC lines may change over time in culture, requiring careful characterization of differentiated cells to ensure consistency and reproducibility (Koehler *et al.*, 2011). A third challenge is immune rejection in case of allogenic SCs grafts (Reddy *et al.*, 2009) requiring immunosuppressive treatment to avoid an immune response against the transplant and the consequent risk of infections. We also have to ask if these new transplanted cells will survive and be functional over time in a patient's brain that has developed the pathology and thus is not a normal internal environment. Current reports show conflicting results from very low to moderate survival rates with some patients showing dramatic improvements for a certain period, whereas others do not (Kim *et al.*, 2011). Yet another area that requires investigation is the effect of neurotransmitters on the newly differentiated transplants. For example, during development as well as in adult neurogenesis, transmitters play a role in cell division and differentiation (Ge *et al.*, 2007). Serotonin has been reported to play a role in modulating neural SC behavior in the hippocampus (Santarelli *et al.*, 2003). Another study showed that fluoxetine, a specific serotonin re-uptake inhibitor commonly used to treat depression, selectively affected the proliferation of immature neuroblasts (Encinas *et al.*, 2006).

Summary

Rapid advances in stem cell research have paved the way for propagation of stem cells in culture, differentiation of these cells into phenotype specific subsets of neurons, and generation of iPSCs for autologous transplantation. This has led to a few clinical trials. Spinal cord and brain injury appears to be a more promising area for transplantation because these patients have had a normal physiology and no neural pathology before their injuries; they have a healthy internal environment for the graft to survive and function. Long term observation and evaluation of transplanted experimental animals as well as patients are required to make sure that the transplantation does not lead to aberrations or tumor formation.

Acknowledgements

Author wishes to thank John Wiley and sons for the permission to use Figure 7.1 from Nilbratt *et al.*, 2010, Elsevier for the permission to use Figure 7.2 from Cooper *et al.*, 2010, and Figure 7.4 from Lu *et al.*, 2012.

References

Adib, S., T. Tiraihi, M. Darvishi, T. Taheri, and H. Kazemi. 2015. Cholinergic differentiation of neural stem cells generated from cell aggregates-derived from human bone marrow stromal cells. Tissue Eng and Reg Med 12: 43–52.

Alvarez-Buylla, A., and J.M. Garcia-Verdugo. 2002. Neurogenesis in adult subventricular zone. J Neurosci 22: 629–634.

Alvarez-Buylla, A., M. Kohwi, T.M. Nguyen, and F.T. Merkle. 2008. The heterogeneity of adult neural stem cells and the emerging complexity of their niche. Cold Spring Harb Symp Quant Biol 73: 357–365.

Amoroso, M.W., G.F. Croft, D.J. Williams, S. O'Keeffe, M.A. Carrasco, A.R. Davis, L. Roybon, D.H. Oakley, T. Maniatis, C.E. Henderson, and H. Wichterle. 2013. Accelerated high-yield generation of limb-innervating motor neurons from human stem cells. J Neurosci 33: 574–586.

Ang, S.L. 2006. Transcriptional control of midbrain dopaminergic neuron development. Development 133: 3499–3506.

Armstrong, R.J.E., and C.N. Syendsen. 2000. Neural stem cells: From cell biology to cell replacement Cell Transplantation 9: 139–152.

Bally-Cuif, L., and M. Hammerschmidt. 2003. Induction and patterning of neuronal development, and its connection to cell cycle control. Curr Opin Neurobiol 13: 16–25.

Bang, O.Y., J.S. Lee, P.H. Lee, and G. Lee. 2005. Autologous mesenchymal stem cell transplantation in stroke patients. Ann Neurol 57: 874–882.

Barcellos, L.F., J.R. Oksenberg, A.J. Green, P. Bucher, J.B. Rimmler, S. Schmidt, M.E. Garcia, R.R. Lincoln, M.A. Pericak-Vance, J.L. Haines, S.L. Hauser, and G. Multiple Sclerosis Genetics. 2002. Genetic basis for clinical expression in multiple sclerosis. Brain 125: 150–158.

Ben-Menachem-Zidon, O., Y. Ben-Menahem, T. Ben-Hur, and R. Yirmiya. 2014. Intra-hippocampal transplantation of neural precursor cells with transgenic over-expression of IL-1 receptor antagonist rescues memory and neurogenesis impairments in an Alzheimer's disease model. Neuropsychopharmacology 39: 401–414.

Besnard, V., S.E. Wert, W.M. Hull, and J.A. Whitsett. 2004. Immunohistochemical localization of Foxa1 and Foxa2 in mouse embryos and adult tissues. Gene Expr Patterns 5: 193–208.

Bithell, A., and B.P. Williams. 2005. Neural stem cells and cell replacement therapy: making the right cells. Clin Sci (Lond) 108: 13–22.

Bjorklund, A., S.B. Dunnett, U. Stenevi, M.E. Lewis, and S.D. Iversen. 1980. Reinnervation of the denervated striatum by substantia nigra transplants: functional consequences as revealed by pharmacological and sensorimotor testing. Brain Res 199: 307–333.

Bjorklund, L.M., R. Sanchez-Pernaute, S. Chung, T. Andersson, I.Y. Chen, K.S. McNaught, A.L. Brownell, B.G. Jenkins, C. Wahlestedt, K.S. Kim, and O. Isacson. 2002. Embryonic stem cells develop into functional dopaminergic neurons after transplantation in a Parkinson rat model. Proc Natl Acad Sci U S A 99: 2344–2349.

Blurton-Jones, M., M. Kitazawa, H. Martinez-Coria, N.A. Castello, F.J. Muller, J.F. Loring, T.R. Yamasaki, W.W. Poon, K.N. Green, and F.M. LaFerla. 2009. Neural stem cells improve cognition via BDNF in a transgenic model of Alzheimer disease. Proc Natl Acad Sci U S A 106: 13594–13599.

Boyer, L.F., B. Campbell, S. Larkin, Y. Mu, and F.H. Gage. 2012. Dopaminergic differentiation of human pluripotent cells. Curr Protoc Stem Cell Biol Chapter 1: Unit1H 6.

Briscoe, J., and J. Ericson. 2001. Specification of neuronal fates in the ventral neural tube. Curr Opin Neurobiol 11: 43–49.

Cai, C., and L. Grabel. 2007. Directing the differentiation of embryonic stem cells to neural stem cells. Dev Dyn 236: 3255–3266.

Carpenter, M.K., M.S. Inokuma, J. Denham, T. Mujtaba, C.P. Chiu, and M.S. Rao. 2001. Enrichment of neurons and neural precursors from human embryonic stem cells. Exp Neurol 172: 383–397.

CDC Fact sheet. 2014. Stroke fact sheet http://www.cdc.gov/stroke/facts.htm (last accessed on 02/08/2015).

Chang, C.C., K.C. Chang, S.J. Tsai, H.H. Chang, and C.P. Lin. 2014. Neurogenic differentiation of dental pulp stem cells to neuron-like cells in dopaminergic and motor neuronal inductive media. J Formos Med Assoc 113: 956–965.

Chao, M.V. 2003. Neurotrophins and their receptors: a convergence point for many signalling pathways. Nat Rev Neurosci 4: 299–309.

Chen, L., D. Chen, H. Xi, Q. Wang, Y. Liu, F. Zhang, H. Wang, Y. Ren, J. Xiao, Y. Wang, and H. Huang. 2012. Olfactory ensheathing cell neurorestorotherapy for amyotrophic lateral sclerosis patients: benefits from multiple transplantations. Cell Transplant 21 Suppl 1: S65–77.

Christopher, J., A. Bissonnette, L. Lyass, B.J. Bhattacharyya, A. Belmadani, R.J. Miller, and J.A. Kessler. 2011. The controlled generation of functional basal forebrain cholinergic neurons from human embryonic stem cells. Stem Cell 29: 802–811.

Cooper, O., G. Hargus, M. Deleidi, A. Blak, T. Osborn, E. Marlow, K. Lee, A. Levy, E. Perez-Torres, A. Yow, and O. Isacson. 2010. Differentiation of human ES and Parkinson's disease iPS cells into ventral midbrain dopaminergic neurons requires a high activity form of SHH, FGF8a and specific regionalization by retinoic acid. Mol Cell Neurosci 45: 258–266.

Courtois, E.T., C.G. Castillo, E.G. Seiz, M. Ramos, C. Bueno, I. Liste, and A. Martinez-Serrano. 2010. *In vitro* and *in vivo* enhanced generation of human A9 dopamine neurons from neural stem cells by Bcl-XL. J Biol Chem 285: 9881–9897.

Crompton, L.A., M.L. Byrne, H. Taylor, T.L. Kerrigan, G. Bru-Mercier, J.L. Badger, P.A. Barbuti, J. Jo, S.J. Tyler, S.J. Allen, T. Kunath, K. Cho, and M.A. Caldwell. 2013. Stepwise, non-adherent differentiation of human pluripotent stem cells to generate basal forebrain cholinergic neurons via hedgehog signaling. Stem Cell Res 11: 1206–1221.

Currais, A., T. Hortobagyi, and S. Soriano. 2009. The neuronal cell cycle as a mechanism of pathogenesis in Alzheimer's disease. Aging (Albany NY) 1: 363–371.

Dhara, S.K., K. Hasneen, D.W. Machacek, N.L. Boyd, R.R. Rao, and S.L. Stice. 2008. Human neural progenitor cells derived from embryonic stem cells in feeder-free cultures. Differentiation 76: 454–464.

Dimos, J.T., K.T. Rodolfa, K.K. Niakan, L.M. Weisenthal, H. Mitsumoto, W. Chung, G.F. Croft, G. Saphier, R. Leibel, R. Goland, H. Wichterle, C.E. Henderson, and K. Eggan. 2008. Induced pluripotent stem cells generated from patients with ALS can be differentiated into motor neurons. Science 321: 1218–1221.

DoD Worldwide Numbers for Traumatic Brain Injury (U.S. Department of Defense Military Health System). 2014. http://dvbic.dcoe.mil/dod-worldwide-numbers-tbi (last accessed on 02/06/2015).

Dorsey, E.R., R. Constantinescu, J.P. Thompson, K.M. Biglan, R.G. Holloway, K. Kieburtz, F.J. Marshall, B.M. Ravina, G. Schifitto, A. Siderowf, and C.M. Tanner. 2007. Projected number of people with Parkinson disease in the most populous nations, 2005 through 2030. Neurology 68: 384–386.

Encinas, J.M., A. Vaahtokari, and G. Enikolopov. 2006. Fluoxetine targets early progenitor cells in the adult brain. Proc Natl Acad Sci U S A 103: 8233–8238.

Evans, M.J., and M.H. Kaufman. 1981. Establishment in culture of pluripotential cells from mouse embryos. Nature 292: 154–156.

Fathi, A., H. Rasouli, M. Yeganeh, G.H. Salekdeh, and H. Baharvand. 2015. Efficient differentiation of human embryonic stem cells toward dopaminergic neurons using recombinant LMX1A factor. Mol Biotechnol 57: 184–194.

Feldman, E.L., N.M. Boulis, J. Hur, K. Johe, S.B. Rutkove, T. Federici, M. Polak, J. Bordeau, S.A. Sakowski, and J.D. Glass. 2014. Intraspinal neural stem cell transplantation in amyotrophic lateral sclerosis: phase 1 trial outcomes. Ann Neurol 75: 363–373.

Ge, S., D.A. Pradhan, G.L. Ming, and H. Song. 2007. GABA sets the tempo for activity-dependent adult neurogenesis. Trends Neurosci 30: 1–8.

Gordon, P.H. 2013. Amyotrophic Lateral Sclerosis: An update for 2013 Clinical Features, Pathophysiology, Management and Therapeutic Trials. Aging Dis 4: 295–310.

Hargus, G., O. Cooper, M. Deleidi, A. Levy, K. Lee, E. Marlow, A. Yow, F. Soldner, D. Hockemeyer, P.J. Hallett, T. Osborn, R. Jaenisch, and O. Isacson. 2010. Differentiated Parkinson patient-derived induced pluripotent stem cells grow in the adult rodent brain and reduce motor asymmetry in Parkinsonian rats. Proc Natl Acad Sci U S A 107: 15921–15926.

Harper, J.M., C. Krishnan, J.S. Darman, D.M. Deshpande, S. Peck, I. Shats, S. Backovic, J.D. Rothstein, and D.A. Kerr. 2004. Axonal growth of embryonic stem cell-derived motoneurons *in vitro* and in motoneuron-injured adult rats. Proc Natl Acad Sci U S A 101: 7123–7128.

Hauser, R.A., P.R. Sandberg, T.B. Freeman, and A.J. Stoessl. 2002. Bilateral human fetal striatal transplantation in Huntington's disease. Neurology 58: 1704; author reply 1704.

Hayashi, T., S. Wakao, M. Kitada, T. Ose, H. Watabe, Y. Kuroda, K. Mitsunaga, D. Matsuse, T. Shigemoto, A. Ito, H. Ikeda, H. Fukuyama, H. Onoe, Y. Tabata, and M. Dezawa. 2013. Autologous mesenchymal stem cell-derived dopaminergic neurons function in parkinsonian macaques. J Clin Invest 123: 272–284.

Hester, M.E., M.J. Murtha, S. Song, M. Rao, C.J. Miranda, K. Meyer, J. Tian, G. Boulting, D.V. Schaffer, M.X. Zhu, S.L. Pfaff, F.H. Gage, and B.K. Kaspar. 2011. Rapid and efficient generation of functional motor neurons from human pluripotent stem cells using gene delivered transcription factor codes. Mol Ther 19: 1905–1912.

Hussein, S.M., N.N. Batada, S. Vuoristo, R.W. Ching, R. Autio, E. Narva, S. Ng, M. Sourour, R. Hamalainen, C. Olsson, K. Lundin, M. Mikkola, R. Trokovic, M. Peitz, O. Brustle, D.P. Bazett-Jones, K. Alitalo, R. Lahesmaa, A. Nagy, and T. Otonkoski. 2011. Copy number variation and selection during reprogramming to pluripotency. Nature 471: 58–62.

Isacson, O., and J.H. Kordower. 2008. Future of cell and gene therapies for Parkinson's disease. Ann Neurol 64 Suppl 2: S122–138.

Iwanami, A., S. Kaneko, M. Nakamura, Y. Kanemura, H. Mori, S. Kobayashi, M. Yamasaki, S. Momoshima, H. Ishii, K. Ando, Y. Tanioka, N. Tamaoki, T. Nomura, Y. Toyama, and H. Okano. 2005. Transplantation of human neural stem cells for spinal cord injury in primates. J Neurosci Res 80: 182–190.

Jessell, T.M. 2000. Neuronal specification in the spinal cord: inductive signals and transcriptional codes. Nat Rev Genet 1: 20–29.

Kern, D.S., K.N. Maclean, H. Jiang, E.Y. Synder, J.R. Sladek, Jr., and K.B. Bjugstad. 2011. Neural stem cells reduce hippocampal tau and reelin accumulation in aged Ts65Dn Down syndrome mice. Cell Transplant 20: 371–379.

Kim, J.H., J.M. Auerbach, J.A. Rodriguez-Gomez, I. Velasco, D. Gavin, N. Lumelsky, S.H. Lee, J. Nguyen, R. Sanchez-Pernaute, K. Bankiewicz, and R. McKay. 2002. Dopamine neurons derived from embryonic stem cells function in an animal model of Parkinson's disease. Nature 418: 50–56.

Kim, J.Y., H.C. Koh, J.Y. Lee, M.Y. Chang, Y.C. Kim, H.Y. Chung, H. Son, Y.S. Lee, L. Studer, R. McKay, and S.H. Lee. 2003. Dopaminergic neuronal differentiation from rat embryonic neural precursors by Nurr1 overexpression. J Neurochem 85: 1443–1454.

Kim, H.J., M. Sugimori, M. Nakafuku, and C.N. Svendsen. 2007. Control of neurogenesis and tyrosine hydroxylase expression in neural progenitor cells through bHLH proteins and Nurr1. Exp Neurol 203: 394–405.

Kim, J., S.C. Su, H. Wang, A.W. Cheng, J.P. Cassady, M.A. Lodato, C.J. Lengner, C.Y. Chung, M.M. Dawlaty, L.H. Tsai, and R. Jaenisch. 2011. Functional integration of dopaminergic neurons directly converted from mouse fibroblasts. Cell Stem Cell 9: 413–419.

Koehler, K.R., P. Tropel, J.W. Theile, T. Kondo, T.R. Cummins, S. Viville, and E. Hashino. 2011. Extended passaging increases the efficiency of neural differentiation from induced pluripotent stem cells. BMC Neurosci 12: 82.

Lebel, M., Y. Gauthier, A. Moreau, and J. Drouin. 2001. Pitx3 activates mouse tyrosine hydroxylase promoter via a high-affinity binding site. J Neurochem 77: 558–567.

LeBlanc, C.J., T.W. Deacon, B.R. Whatley, J. Dinsmore, L. Lin, and O. Isacson. 1999. Morris water maze analysis of 192-IgG-saporin-lesioned rats and porcine cholinergic transplants to the hippocampus. Cell Transplant 8: 131–142.

Lee, S.K., and S.L. Pfaff. 2001. Transcriptional networks regulating neuronal identity in the developing spinal cord. Nat Neurosci 4 Suppl: 1183–1191.

Lee, H., G.A. Shamy, Y. Elkabetz, C.M. Schofield, N.L. Harrsion, G. Panagiotakos, N.D. Socci, V. Tabar, and L. Studer. 2007. Directed differentiation and transplantation of human embryonic stem cell-derived motoneurons. Stem Cells 25: 1931–1939.

Lee, H.S., E.J. Bae, S.H. Yi, J.W. Shim, A.Y. Jo, J.S. Kang, E.H. Yoon, Y.H. Rhee, C.H. Park, H.C. Koh, H.J. Kim, H.S. Choi, J.W. Han, Y.S. Lee, J. Kim, J.Y. Li, P. Brundin, and S.H. Lee. 2010. Foxa2 and Nurr1 synergistically yield A9 nigral dopamine neurons exhibiting improved differentiation, function, and cell survival. Stem Cells 28: 501–512.

Lee, J.K. 2014. Cerebral perfusion pressure: how low can we go? Paediatr Anaesth 24: 647–648.

Lévesque, M.F., N. Toomas, and M. Rezak. 2009. Therapeutic microinjection of autologous adult human neural stem cells and differerntiated neurons for Parkinson's disease: Five-year post-operative outcome. The Open Stem Cell Journal 1: 20–29.

Lima, C., J. Pratas-Vital, P. Escada, A. Hasse-Ferreira, C. Capucho, and J.D. Peduzzi. 2006. Olfactory mucosa autografts in human spinal cord injury: a pilot clinical study. J Spinal Cord Med 29: 191–203.

Lindvall, O., G. Sawle, H. Widner, J.C. Rothwell, A. Bjorklund, D. Brooks, P. Brundin, R. Frackowiak, C.D. Marsden, P. Odin, and S. Rehncrona. 2004. 1994. Evidence for long-term survival and function of dopaminergic grafts in progressive Parkinson's disease. Ann Neurol 35: 172–180.

Lister, R., M. Pelizzola, Y.S. Kida, R.D. Hawkins, J.R. Nery, G. Hon, J. Antosiewicz-Bourget, R. O'Malley, R. Castanon, S. Klugman, M. Downes, R. Yu, R. Stewart, B. Ren, J.A. Thomson, R.M. Evans, and J.R. Ecker. 2011. Hotspots of aberrant epigenomic reprogramming in human induced pluripotent stem cells. Nature 471: 68–73.

Liu, Y., J.P. Weick, H. Liu, R. Krencik, X. Zhang, L. Ma, G.M. Zhou, M. Ayala, and S.C. Zhang. 2013. Medial ganglionic eminence-like cells derived from human embryonic stem cells correct learning and memory deficits. Nat Biotechnol 31: 440–447.

Lopez-Coviella, I., B. Berse, R. Krauss, R.S. Thies, and J.K. Blusztajn. 2000. Induction and maintenance of the neuronal cholinergic phenotype in the central nervous system by BMP-9. Science 289: 313–316.

Lopez-Coviella, I., M.T. Follettie, T.J. Mellott, V.P. Kovacheva, B.E. Slack, V. Diesl, B. Berse, R.S. Thies, and J.K. Blusztajn. 2005. Bone morphogenetic protein 9 induces the transcriptome of basal forebrain cholinergic neurons. Proc Natl Acad Sci U S A 102: 6984–6989.

Lu, P., Y. Wang, L. Graham, K. McHale, M. Gao, D. Wu, J. Brock, A. Blesch, E.S. Rosenzweig, L.A. Havton, B. Zheng, J.M. Conner, M. Marsala, and M.H. Tuszynski. 2012. Long-distance growth and connectivity of neural stem cells after severe spinal cord injury. Cell 150: 1264–1273.

Lunn, J.S., S.A. Sakowski, and E.L. Feldman. 2014. Concise review: Stem cell therapies for amyotrophic lateral sclerosis: recent advances and prospects for the future. Stem Cells 32: 1099–1109.

Ma, Y., C. Tang, T. Chaly, P. Greene, R. Breeze, S. Fahn, C. Freed, V. Dhawan, and D. Eidelberg. 2010. Dopamine cell implantation in Parkinson's disease: long-term clinical and (18)F-FDOPA PET outcomes. J Nucl Med 51: 7–15.

Manabe, T., K. Tatsumi, M. Inoue, M. Makinodan, T. Yamauchi, E. Makinodan, S. Yokoyama, R. Sakumura, and A. Wanaka. 2007. L3/Lhx8 is a pivotal factor for cholinergic differentiation of murine embryonic stem cells. Cell Death Differ 14: 1080–1085.

Marrie, R.A. 2004. Environmental risk factors in multiple sclerosis aetiology. Lancet Neurol 3: 709–718.

Martin, G.R. 1981. Isolation of a pluripotent cell line from early mouse embryos cultured in medium conditioned by teratocarcinoma stem cells. Proc Natl Acad Sci U S A 78: 7634–7638.

Mayshar, Y., U. Ben-David, N. Lavon, J.C. Biancotti, B. Yakir, A.T. Clark, K. Plath, W.E. Lowry, and N. Benvenisty. 2010. Identification and classification of chromosomal aberrations in human induced pluripotent stem cells. Cell Stem Cell 7: 521–531.

Medical News. 2014. Paralyzed man walks again after nose cells repair his spinal cord at http://www.medicalnewstoday.com/articles/284152.php (last accessed on 02/06/2015).

Mehta, P., V. Antao, W. Kaye, M. Sanchez, D. Williamson, L. Bryan, O. Muravov, K. Horton, T. Division of, A.f.T.S. Human Health Sciences, A.G. Disease Registry, C. Centers for Disease, and Prevention. 2014. Prevalence of amyotrophic lateral sclerosis—United States, 2010–2011. MMWR Surveill Summ 63 Suppl 7: 1–14.

Merkle, F.T., Z. Mirzadeh, and A. Alvarez-Buylla. 2007. Mosaic organization of neural stem cells in the adult brain. Science 317: 381–384.

Mitrecic, D., C. Nicaise, L. Klimaschewski, S. Gajovic, D. Bohl, and R. Pochet. 2012. Genetically modified stem cells for the treatment of neurological diseases. Front Biosci (Elite Ed) 4: 1170–1181.

Molofsky, A.V., R. Pardal, T. Iwashita, I.K. Park, M.F. Clarke, and S.J. Morrison. 2003. Bmi-1 dependence distinguishes neural stem cell self-renewal from progenitor proliferation. Nature 425: 962–967.

National Institute of Aging Report. 2014. Alzheimer's Disease: Unraveling the Mystery. http://www.nia.nih.gov/alzheimers/publication/part-2-what-happens-brain-ad/hallmarks-ad (last accessed on 02/06/2015).

National Multiple Sclerosis Society Data at http://www.nationalmssociety.org/About-the-Society/MS-Prevalence (last accessed on 02/10/2015).

National SCI Statistical center data. 2014. Spinal Cord Injury facts and figures at a glance. https://www.nscisc.uab.edu/PublicDocuments/fact_figures_docs/Facts%202014.pdf (last accessed on 02/06/2015).

Nilbratt, M., O. Porras, A. Marutle, O. Hovatta, and A. Nordberg. 2010. Neurotrophic factors promote cholinergic differentiation in human embryonic stem cell-derived neurons. J Cell Mol Med 14: 1476–1484.

Nilsson, O.G., P. Brundin, H. Widner, R.E. Strecker, and A. Bjorklund. 1988. Human fetal basal forebrain neurons grafted to the denervated rat hippocampus produce an organotypic cholinergic reinnervation pattern. Brain Res 456: 193–198.

Nishimura, F., M. Yoshikawa, S. Kanda, M. Nonaka, H. Yokota, A. Shiroi, H. Nakase, H. Hirabayashi, Y. Ouji, J. Birumachi, S. Ishizaka, and T. Sakaki. 2003. Potential use of embryonic stem cells for the treatment of mouse parkinsonian models: improved behavior by transplantation of *in vitro* differentiated dopaminergic neurons from embryonic stem cells. Stem Cells 21: 171–180.

Ogawa, Y., K. Sawamoto, T. Miyata, S. Miyao, M. Watanabe, M. Nakamura, B.S. Bregman, M. Koike, Y. Uchiyama, Y. Toyama, and H. Okano. 2002. Transplantation of *in vitro*-expanded fetal neural progenitor cells results in neurogenesis and functional recovery after spinal cord contusion injury in adult rats. J Neurosci Res 69: 925–933.

Perlow, M.J., W.J. Freed, B.J. Hoffer, A. Seiger, L. Olson, and R.J. Wyatt. 1979. Brain grafts reduce motor abnormalities produced by destruction of nigrostriatal dopamine system. Science 204: 643–647.

Politis, M., K. Wu, C. Loane, N.P. Quinn, D.J. Brooks, S. Rehncrona, A. Bjorklund, O. Lindvall, and P. Piccini. 2010. Serotonergic neurons mediate dyskinesia side effects in Parkinson's patients with neural transplants. Sci Transl Med 2: 38ra46.

Pringsheim, T., K. Wiltshire, L. Day, J. Dykeman, T. Steeves, and N. Jette. 2012. The incidence and prevalence of Huntington's disease: a systematic review and meta-analysis. Mov Disord 27: 1083–1091.

Pyle, A.D., L.F. Lock, and P.J. Donovan. 2006. Neurotrophins mediate human embryonic stem cell survival. Nat Biotechnol 24: 344–350.

Rabinovich, S.S., V.I. Seledtsov, N.V. Banul, O.V. Poveshchenko, V.V. Senyukov, S.V. Astrakov, D.M. Samarin, and V.Y. Taraban. 2005. Cell therapy of brain stroke. Bull Exp Biol Med 139: 126–128.

Reddy, P., M. Arora, M. Guimond, and C.L. Mackall. 2009. GVHD: a continuing barrier to the safety of allogeneic transplantation. Biol Blood Marrow Transplant 15: 162–168.

Renoncourt, Y., P. Carroll, P. Filippi, V. Arce, and S. Alonso. 1998. Neurons derived *in vitro* from ES cells express homeoproteins characteristic of motoneurons and interneurons. Mech Dev 79: 185–197.

Robertson, E., A. Bradley, M. Kuehn, and M. Evans. 1986. Germ-line transmission of genes introduced into cultured pluripotential cells by retroviral vector. Nature 323: 445–448.

Robertson, M.J., P. Gip, and D.V. Schaffer. 2008. Neural stem cell engineering: directed differentiation of adult and embryonic stem cells into neurons. Front Biosci 13: 21–50.

Rossi, D.J., C.H. Jamieson, and I.L. Weissman. 2008. Stems cells and the pathways to aging and cancer. Cell 132: 681–696.

Roy, N.S., C. Cleren, S.K. Singh, L. Yang, M.F. Beal, and S.A. Goldman. 2006. Functional engraftment of human ES cell-derived dopaminergic neurons enriched by coculture with telomerase-immortalized midbrain astrocytes. Nat Med 12: 1259–1268.

Sakurada, K., M. Ohshima-Sakurada, T.D. Palmer, and F.H. Gage. 1999. Nurr1, an orphan nuclear receptor, is a transcriptional activator of endogenous tyrosine hydroxylase in neural progenitor cells derived from the adult brain. Development 126: 4017–4026.

Santarelli, L., M. Saxe, C. Gross, A. Surget, F. Battaglia, S. Dulawa, N. Weisstaub, J. Lee, R. Duman, O. Arancio, C. Belzung, and R. Hen. 2003. Requirement of hippocampal neurogenesis for the behavioral effects of antidepressants. Science 301: 805–809.

Schuldiner, M., O. Yanuka, J. Itskovitz-Eldor, D.A. Melton, and N. Benvenisty. 2000. Effects of eight growth factors on the differentiation of cells derived from human embryonic stem cells. Proc Natl Acad Sci U S A 97: 11307–11312.

Schulz, T.C., S.A. Noggle, G.M. Palmarini, D.A. Weiler, I.G. Lyons, K.A. Pensa, A.C. Meedeniya, B.P. Davidson, N.A. Lambert, and B.G. Condie. 2004. Differentiation of human embryonic stem cells to dopaminergic neurons in serum-free suspension culture. Stem Cells 22: 1218–1238.

Studer, L., V. Tabar, and R.D. McKay. 1998. Transplantation of expanded mesencephalic precursors leads to recovery in parkinsonian rats. Nat Neurosci 1: 290–295.

Tabakow, P., W. Jarmundowicz, B. Czapiga, W. Fortuna, R. Miedzybrodzki, M. Czyz, J. Huber, D. Szarek, S. Okurowski, P. Szewczyk, A. Gorski, and G. Raisman. 2013. Transplantation of autologous olfactory ensheathing cells in complete human spinal cord injury. Cell Transplant 22: 1591–1612.

Takahashi, K., and S. Yamanaka. 2006. Induction of pluripotent stem cells from mouse embryonic and adult fibroblast cultures by defined factors. Cell 126: 663–676.

Takahashi, K., K. Tanabe, M. Ohnuki, M. Narita, T. Ichisaka, K. Tomoda, and S. Yamanaka. 2007. Induction of pluripotent stem cells from adult human fibroblasts by defined factors. Cell 131: 861–872.

Thomson, J.A., J. Itskovitz-Eldor, S.S. Shapiro, M.A. Waknitz, J.J. Swiergiel, V.S. Marshall, and J.M. Jones. 1998. Embryonic stem cell lines derived from human blastocysts. Science 282: 1145–1147.

U.S. Department of Health and Human Services report. 2013. http://aspe.hhs.gov/daltcp/napa/NatlPlan.pdf (last accessed on 02/06/2015).

van Gorp, S., M. Leerink, O. Kakinohana, O. Platoshyn, C. Santucci, J. Galik, E.A. Joosten, M. Hruska-Plochan, D. Goldberg, S. Marsala, K. Johe, J.D. Ciacci, and M. Marsala. 2013. Amelioration of motor/sensory dysfunction and spasticity in a rat model of acute lumbar spinal cord injury by human neural stem cell transplantation. Stem Cell Res Ther 4: 57.

Weiner, H.L. 2004. Multiple sclerosis is an inflammatory T-cell-mediated autoimmune disease. Arch Neurol 61: 1613–1615.

WHO Fact sheet No. 384. http://www.who.int/mediacentre/factsheets/fs384/en/ (last accessed on 02/06/2015).

World Alzheimer's report. 2014. http://www.alz.co.uk/research/WorldAlzheimerReport2014.pdf (last accessed on 02/06/2015).

Xu, L., J. Yan, D. Chen, A.M. Welsh, T. Hazel, K. Johe, G. Hatfield, and V.E. Koliatsos. 2006. Human neural stem cell grafts ameliorate motor neuron disease in SOD-1 transgenic rats. Transplantation 82: 865–875.

Yan, Y., D. Yang, E.D. Zarnowska, Z. Du, B. Werbel, C. Valliere, R.A. Pearce, J.A. Thomson, and S.C. Zhang. 2005. Directed differentiation of dopaminergic neuronal subtypes from human embryonic stem cells. Stem Cells 23: 781–790.

Yan, M., M. Sun, Y. Zhou, W. Wang, Z. He, D. Tang, S. Lu, X. Wang, S. Li, W. Wang, and H. Li. 2013. Conversion of human umbilical cord mesenchymal stem cells in Wharton's jelly to dopamine neurons mediated by the Lmx1a and neurturin *in vitro*: potential therapeutic application for Parkinson's disease in a rhesus monkey model. PLoS One 8: e64000.

Yan, Y., T. Ma, K. Gong, Q. Ao, X. Zhang, and Y. Gong. 2014. Adipose-derived mesenchymal stem cell transplantation promotes adult neurogenesis in the brains of Alzheimer's disease mice. Neural Regen Res 9: 798–805.

8

3D *in vitro* Models for Developmental Neurotoxicity (DNT)

Luc Stoppini,[1,a] *Jenny Sandström von Tobel,*[2,c] *Igor Charvet,*[1,b] *Marie-Gabrielle Zurich,*[2,d] *Lars Sundstrom*[3] *and Florianne Monnet-Tschudi*[2,e,*]

Introduction

Brain development is temporally and regionally determined by multiple processes (Purves and Williams, 2001). During development, stem and progenitor cells undergo progressive changes leading to fully differentiated neurons and glial cells, with characteristic molecular expression patterns, morphological and functional heterogeneity (Gritti and Bonfanti, 2007). From early neural commitment and neural

---

[1] Tissue Engineering Laboratory, HEPIA/HES-SO, Campus Biotech, 9, Chemin des mines, 1211, Geneva, Switzerland.
[a] E-mail: luc.stoppini@hesge.ch
[b] E-mail: igor.charvet@hesge.ch
[2] SCAHT and Department of Physiology, University of Lausanne, 7, rue du Bugnon, CH-1005 Lausanne, Switzerland.
[c] E-mail: Jenny.Sandstrom@unil.ch
[d] E-mail: mzurich@unil.ch
[e] E-mail: Florianne.Tschudi-Monnet@unil.ch
[3] Elizabeth Blackwell Institute for Health Research, Bristol University, Royal Fort House, Bristol BS 8 1UH.
E-mail: L.Sundstrom@bristol.ac.uk
* Corresponding author

tube formation, the main developmental steps are neurogenesis, cell migration, synaptogenesis, gliogenesis and myelination. In exposure to toxicants, any of these developmental steps can be considered to be specific windows of vulnerability.

The complex architecture of the adult central nervous system (CNS) is made up of three main neuroectodermal cell types: neurons and two types of glial cells; astrocytes and oligodendrocytes. Cell-cell interactions—neuron-glia and glia-glia—are involved in initiating and orchestrating critical developmental steps. For example, neurogenesis, migration, survival and differentiation, as well as axonal growth, synapse formation and function depend on neuron-astrocyte interactions (Stipursky *et al.*, 2012), whereas myelin formation is regulated by neuron-oligodendrocyte interactions (White and Kramer-Albers, 2014). The behaviour of oligodendrocyte precursor cells during development is influenced by cues secreted by astrocytes (Clemente *et al.*, 2013), for example, astrocyte produced platelet-derived growth factor-AA (PDGF-AA) coordinates the timing of oligodendrogliogenesis (Durand and Raff, 2000). Brain development is also depending on a fourth cell type; microglia. These immune cells of the brain originate from primitive macrophages produced by the yolk sack and enter the brain during the embryonic period (Gomez Perdiguero *et al.*, 2013). Even though their exact role during brain development has yet to be described, microglia seem to affect processes associated with neuronal proliferation and differentiation. Furthermore, in addition to playing a role in synaptic management, microglia are important for the pruning of dying neurons and in the clearance of debris (Harry, 2013). Cell differentiation during development relies not only on the intrinsic properties of cell progenitors but also on the extracellular environment, where the extracellular matrix (ECM) plays a critical role. For instance, microglial cells can modify the neural produced ECM to facilitate axonal pathfinding (Chamak *et al.*, 1994). Furthermore, astrocyte derived ECM can be used by oligodendrocyte precursor cells as substrate for migration (Clemente *et al.*, 2013). Released factors, such as neurotrophic factors of glial origin (Heacock *et al.*, 1986), neurotransmitters (Gaiarsa and Porcher, 2013; Holmes *et al.*, 1997), chemokines (Adler *et al.*, 2005), or cytokines with neurotrophic activity (Garden and Moller, 2006) can also mediate cell-cell communications regulating brain maturation.

Neurotoxicant-induced perturbations of these cell interactions will affect the establishment of brain structures and their relationships. The extension of these effects will depend on dose, duration, and developmental period of toxicant exposure. In addition, the consequences of neurotoxicant exposure can manifest both in a short-term perspective, i.e., immediately after exposure (e.g., during gestation, or at birth), or in a longer-term perspective, breaching a timespan from childhood until aging adults. The latter suggests that toxicant induced adverse effects could remain concealed over a long time period, and may only manifest when brain cognitive function is challenged, or with age-related decrements in neurological function (Barone *et al.*, 1995). This theory was put forward by Landrigan and coworkers in 2014, propounding that neurodegenerative disease could, in part, originate from a neurotoxicant exposure during brain development. Indeed, epidemiological studies and meta-analyses of animal experimental data reported long-term effects after exposure to the following toxicants: lead, methylmercury, polychlorinated biphenyls, arsenic, toluene, manganese, fluoride, chlorpyrifos, dichlorodiphenyltrichloroethane, tetrachloroethylene and the polybrominated diphenyl ethers (for review, (Grandjean and Landrigan, 2014a; Grandjean and Landrigan, 2014b)). Among these long-term, or delayed effects, are neurobehavioral disturbances such as attention-deficit and hyperactivity, which were observed both in adult rats (Moreira *et al.*, 2001) and in children (Finkelstein *et al.*, 1998) following lead exposure during pregnancy and lactation. An underlying mechanism for such effects might be the impairment of synapse formation and plasticity (Neal *et al.*, 2010; Stansfield *et al.*, 2012). Another process known to have long-term consequences is neuroinflammation, and several classes of neurotoxicants are known to trigger a neuroinflammatory process driven by microglial- and astrocyte reactivities (Monnet-Tschudi *et al.*, 2007). Indeed, the triggering of brain inflammation during gestation through the peripheral injection of a viral mimic in pregnant mice has shown to lead to increased deposition of amyloid peptide and tau hyperphosphorylation—hallmarks of Alzheimer's disease—in the aging offpsring (Krstic *et al.*, 2012). Furthermore, developmental exposure of rats and monkeys to lead acetate showed a delayed overexpression of the amyloid precursor protein and its amyloidogenic A-beta products (Zawia and Basha, 2005). Interestingly, *in vitro* studies in 3D rat brain cell cultures show that lead acetate exposure evokes an exacerbated microglial- and astrocyte reactivity in immature cultures, as compared to the more differentiated cultures (Zurich *et al.*, 2002). However, whether such early and intense neurotoxicant-induced neuroinflammation increases the risk of developing a neurodegenerative disease at a later time-point remains to be proven. Delayed adverse effects may also rely on other mechanisms of toxicity, such as epigenetic modifications of genes involved in the amyloid cascade, or oxidative stress (Zawia *et al.*, 2009). Oxidative stress is a common mechanism of toxicity to many chemically diverse toxicants (Crumpton *et al.*, 2000; Farina *et al.*, 2013). During development, toxicant-induced oxidative stress has been shown to disrupt mitogenic signalling, affecting both neurogenesis and gliogenesis (Li *et al.*, 2007). The high sensitivity of neural precursor cells to oxidative stress may be due to a limited content of anti-oxidants, similarly to what has already been described for oligodendrocyte progenitors (Hemdan and Almazan, 2007).

Brain in development is generally considered to be more susceptible to neurotoxic insults than the mature brain, due to the critical timing of the developmental events it undergoes, to the establishment of complex structures and to less effective protection mechanisms. Therefore, assessment of developmental neurotoxicity for the numerous new chemical substances released in the environment is of vital importance for public health.

Model Requirements for DNT

A good *in vitro* model for DNT must replicate the critical developmental steps such as cell proliferation, migration, process formation, as well as synapse formation and myelination. Therefore it is a prerequisite that cultures can be maintained for period long enough to reflect these critical maturation steps. Long-term culture maintenance is also required to address questions related to subchronic-/chronic neurotoxicity, and for the evaluation of delayed/long term effects. In addition to the timely aspect, cell-cell interactions are crucial for brain maturation. Therefore the presence of both neurons and glial cells is needed. Studies on differentiating human neural stem cells show augmented differentiation of both neurons and glia when grown in three dimensions as compared to monolayer cultures (Brannvall *et al.*, 2007; Choi *et al.*, 2014). The importance of a heterogenic cell culture model for DNT is further supported by the fact that both neurons and glial cells can be the primary targets of neurotoxic insults. Moreover, neuron-glia interactions modulate the uptake of toxicants (Monnet-Tschudi *et al.*, 2008; Zurich *et al.*, 1998) which in turn modify the toxic outcome (Zurich *et al.*, 2002; Zurich *et al.*, 2004). Neuron-glia crosstalk is furthermore essential for the triggering of a neuroinflammatory response, an adaptive mechanism that can have long-term consequences (Monnet-Tschudi *et al.*, 2010). Besides this, the relative proportion of neurons to glial cells, and glial to glial cells, must also be taken into account, as these proportions can modulate the release of bioactive molecules upon toxic exposure and thus modify the effect of a neurotoxic insult (Eskes *et al.*, 2002; Eskes *et al.*, 2003). Finally, the *in vitro* cellular structural organization could lead to the formation of so-called biological niches known to influence the behaviour of residing precursor cells. For instance, the mycotoxinochratoxin A differentially affects cell proliferation and homeostasis of neural precursors residing in immature-versus mature 3D brain cell cultures (Zurich and Honegger, 2011), illustrating the impact of the cellular microenvironment.

Three-dimensional (3D) Models Available or in Development for DNT

A recent workshop report described the state-of-the-art of 3D cultures of selected organs (liver, lung, skin and brain) and listed the existing 3D *in vitro* models for the brain: brain slices, layered co-culture, transwell culture, aggregating brain cell culture, engineered neural tissue (ENT) and secondary 3D cultures (Alepee *et al.*, 2014). In the present chapter, the authors have chosen to focus specifically on 3D models prepared from human stem- or neural precursor cells, and on aggregating rat brain cell cultures. The latter system has already been widely used for Neurotoxicology (NT) as well as DNT. It was involved in several European projects and was found to be predictive for detecting organ-specific toxicity (Prieto *et al.*, 2013; Zurich *et al.*, 2013). Noteworthy is, that any model that reaches a high maturation stage—hence presenting synapses, neuron-astrocyte interaction and compact myelin—could be used for both DNT and NT, depending on the maturation stage under which the exposure is undertaken.

3D human cultures in Matrigel

Matrigel is a support matrix containing solubilized basement membrane proteins (BD Matrigel; BD Biosciences). A first model using this synthetic extracellular matrix consists of mixing commercially available ReN Cell VM human neural (ReN) precursor cells with Matrigel to set up thin (about 100–300 μm) and thick-layer (about 4 mm) 3D models (Choi *et al.*, 2014). The differentiation capacity of ReN cells in 3D structures was shown by detectable expression of both neuronal markers microtubule-associated protein-2 (MAP2), Tubulin b3, neuron cell-adhesion molecule (NCAM) and synapsin 1; and glial markerglial fibrillary acidic protein (GFAP). Furthermore, cultures displayed extensive process formation after two to six weeks of differentiation (Choi *et al.*, 2014). These 3D culture conditions promote a higher degree of neuronal and glial differentiation than conventional 2D cultures (Choi *et al.*, 2014). A study by Li and co-workers (Li *et al.*, 2012) suggested that the porosity of the matrigel scaffold affects stem cell proliferation and maturation, where larger pore sizes seem to promote neural stem cell 3D differentiation. This model has been used to study amyloid and tau pathologies in relation to Alzheimer's disease and might be useful for DNT.

Another 3D culture system that has not yet been used for this purpose, but which could potentially serve as a DNT model are the so-called cerebral organoids. These are derived from human embryonic- or induced pluripotent stem cells, and were shown to even develop various discrete, although interdependent, brain regions, partially reproducing an organotypic organization (Lancaster and Knoblich, 2014; Lancaster *et al.*, 2013). The preparation of such organoides starts with the formation of embryoid bodies of either human embryonic stem cells (hESC) or human induced pluripotent stem cells (hiPSC). The embryoid bodies are transferred to droplets of Matrigel and maintained stationary for four days before being transferred into a spinning bioreactor (Lancaster *et al.*, 2013). This method accelerates the development of brain tissues, requiring only eight to ten days for the appearance of neural identity, and 20–30 days for the formation of defined brain regions. Cerebral organoids can survive up to ten months in culture (Lancaster *et al.*, 2013). This is currently the only culture system able to mimic sub-specification of brain regions, namely cortical regions, thus recapitulating many aspects of human cortical development, such as neurogenesis, migration and radial glia formation. Furthermore, this model also expresses specific hippocampal markers, making it the most complex *in vitro* model at present date.

3D human cultures on microporous membranes

Neural air-liquid interface types of cultures, called also engineered tissues (ENTs), were originally developed for organotypic culture of brain slices (Stoppini *et al.*, 1991). In this technique, cultures are grown directly at an air-liquid interface on a semi-permeable membrane, with the culture medium on the contralateral side of the membrane. This technique relies on capillarity forces to generate a small film of culture medium that covers the tissue, favouring extensive gas exchanges between tissue and air. The improved gas exchange allows the maintenance of relatively thick 3D cultures without hypoxic cell death. An additional advantage over other methods of growing 3D tissue cultures is that it is a static method. This is of interest for electrophysiological measurements or imaging since tissues do not need to be immobilized before such investigations can be carried out (Sundstrom *et al.*, 2012; van Vliet *et al.*, 2007).

This technique was used successfully with mouse ESC (Dubois-Dauphin *et al.*, 2010). With ESC of human origin, the obtained tissue was heterogeneous. However, the use of human ESC-derived neural precursor cells (NPC)—from the H1 and H9 cell lines (WiCell Research Institute, Madison Wisconsin, USA), and the HS-401 cell line (Karolinska Institute, Sweden)—allowed the generation of an organized 3D neural culture (Preynat-Seauve *et al.*, 2009). The principle of this modified approach is to add drops of neural cells in suspension (100K to 150K cells/μl) onto pre-cut patches of hydrophilic polytetrafluoroethylene (PTFE) membranes, which are placed on top of a culture plate insert, bearing from three to five 3D neural cultures (Figure 8.1A). Cultures are maintained at air-liquid interface within 6-well plates (Figure 8.1B and 8.1C) and are viable for several weeks up to several months (Preynat-Seauve *et al.*, 2009). After two months the 3D cultures form spherical caps, of two to three mm in diameter, and maximal height of 100 to 120 μm. Cells seem to organize within the 3D-structure, with a layer of GFAP positive cells (astrocytes) close to the PTFE porous membrane, and NeuN positive neuronal nuclei positioned above this layer (Figure 8.1E). A dense neurite network was found throughout the entire tissue, as shown by immunolabelling of neurofilaments (NF) (Figure 8.1D). After one month the neuronal marker tubulin β3 was expressed, whereas GFAP-labelled astrocytes were only detected after two months in culture (Figure 8.1E). So was the expression of the myelin-associated enzyme 2',3'-cyclic nucleotide 3'-phosphohydrolase (CNPase), suggesting the presence of differentiating oligodendrocytes. However, no myelin could be detected by the Luxol Blue and Bielchowski techniques.

Electrophysiological recordings using multi-electrode array systems (MEA) were used to assess functionality of the ENTs. Patches of PTFE membranes bearing the 3D neural tissues were placed upside down onto a MEA (Biocell-Interface cartridges). Spontaneous activity was monitored by eight independent electrodes. After three to four weeks in culture spike activities could be measured, and after six to eight weeks in culture, bursts were detected (Figure 8.1F). Bipolar stimulations, from 1000–4000 mV, were applied to the 3D neural tissues in order to generate evoked field potentials (EFP) (Figure 8.1G). The amplitude of the second EFP is slightly decreased when two pulses are paired in rapid succession leading to an inhibition, known as paired pulse inhibition (PPI) (Figure 8.1H). This stimulation technique is generally used to test for the presence of inhibitory networks.

This 3D culture system might be used for developmental neurotoxicity testing, mainly to evaluate the interferences of toxicants with the establishment and the functionality of neuronal networks.

A recent study combined the technique of the generation of neurosphere cultures in suspension and the ENTs. In a first step, NPC neurospheres were generated according to the free-floating technique (see below) and were in a second step transferred in the air-interface liquid culture device for further differentiation (Tieng *et al.*, 2014). Interestingly this study shows that three-dimensional structures were more successful than 2D cultures in rapidly generating a higher number of dopaminergic neurons (> 60% at day 28 of differentiation) in comparison to the 2D (20% of cells were tyrosine hydroxylase (TH) positive in day 25). The level of maturation of the TH-positive neurons was higher in 3D cultures, where many varicose neurites (morphologic sign of mature dopaminergic neurons *in vivo*) could be found. However, the differentiation potential of these neurons was variable, and depending on from which hiPSC-line the NPC were derived.

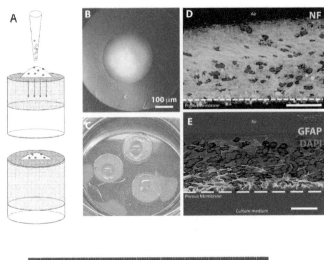

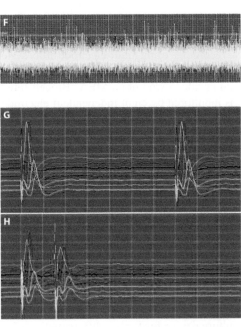

Figure 8.1. Air/liquid neural cultures derived from hESC. (A) Technique to generate 3D neural tissues onto porous membranes. A drop of 3–5 µl of NPC containing 100–150K cells is deposited onto the membrane. Cells will progressively compact due to capillary forces. (B) Macrophotography of 3D neural tissue grown onto porous membrane at air/liquid interface after 2 months in culture. (C) 3D neural tissues grown onto porous membranes within culture inserts. (D) Immunofluorescence staining of transversal sections of 3D neural tissue where neurofilaments (NF) can be visualized in green showing the abundant presence of neurites within the entire section (Bar: 50 µm). (E) Astrocytes labelled by GFAP could be found mostly located close to the synthetic PTFE membrane of the culture inserts, represented by the white dotted line, while neurons, immunostained by Neu-N, are distributed throughout the tissue. Nuclei were visualized by DAPI. (Bar: 50 µm). (F) Spontaneous activity recordings show individual spikes as well as the presence of bursts. (G) Evoked synaptic activity recorded by 8 selected channels in response to stimulation. (H) Paired-pulse inhibition could be observed when the second stimulation is induced from 70 ms to 15 ms.

3D free-floating cultures derived from rat brain cells, human embryonic stem cells (hESC), and human induced pluripotent stem cells (hiPSC)

Three-dimensional free-floating cultures can be prepared from rat or mouse foetal brains (Honegger *et al.*, 2011), from neural precursor cells derived from hESC (Hoelting *et al.*, 2013) or from iPSC (Hogberg *et al.*, 2013; Pamies *et al.*, 2014). These 3D structures are formed spontaneously by cell adhesion either under constant gyratory agitation or in low adherence plates. In these 3D systems, cells are not provided with a culture substrate, but produce their own extracellular matrix, giving rise to brain parenchyma with histotypic, but not organotypic organization, meaning that they contain neurons and glial cells organized in a tissue-like manner without reproducing recognizable discrete brain structures.

3D free-floating cultures derived from foetal rat brain cells

The 3D rat brain cell model (aggregating brain cell cultures) first developed by Moscona and coworkers (Moscona, 1961) has been widely characterized and used for neurotoxicological and DNT studies by Honegger and co-workers (Honegger *et al.*, 2011; Honegger and Zurich, 2011). It is prepared from embryonic day 16 (ED 16) rat foetal brains, which are mechanically dissociated and subcultured in a defined medium maintained in constant gyratory agitation. Each flask contains about 1.2 brains and can be split into 4 to 6 replicates for experiments. Most neurons are post-mitotic at culture initiation, whereas glial precursor cells proliferate during the first two weeks, whereafter they undergo extensive differentiation. After only 20 days in culture numerous synapses as well as compact myelin can be observed (Honegger *et al.*, 2011). Compact myelin, characterized by the alternations of dense and less dense (intra-period) lines observed by electron microscopy, is reported exclusively in these 3D rat brain cell cultures (Zurich *et al.*, 2003). Furthermore, these cultures contain the major brain cell types (Figure 8.2A–F), including several neuronal subtypes (Honegger *et al.*, 2011) and microglial cells. Microglial cells are derived from erythromyeloid precursors originating from the yolk sack (Kierdorf *et al.*, 2013), and at ED16 the foetal rat brain is already partially populated by microglia precursor cells. Within the *in vitro* 3D structure, these cells acquire their typical quiescent morphology (Figure 8.2F). In addition, highly differentiated cultures ($\geq$ DIV 20) also retain a small resident population exhibiting the characteristics of neural stem cells (NSCs) and related precursors (Zurich and Honegger, 2011).

In these 3D cultures, neurons and glial cells develop in a sequential manner mimicking the *in vivo* situation (Figure 8.2A–F). As an illustration the axonal marker NF-H is expressed progressively during time in culture (Figure 8.2A); myelin basic protein (MBP) immunoreactivity (Figure 8.2D) is—as is the case *in vivo*—detected before myelin oligodendrocyte glycoprotein (MOG) immunoreactivity, attesting for timely oligodendrocyte maturation (Figure 8.2E). An isoform of the Na/K ATPase pump was detected in these 3D aggregated cultures, whereas it was not expressed in monolayer cultures (Corthésy-Theulaz *et al.*, 1990), suggesting that the presence of

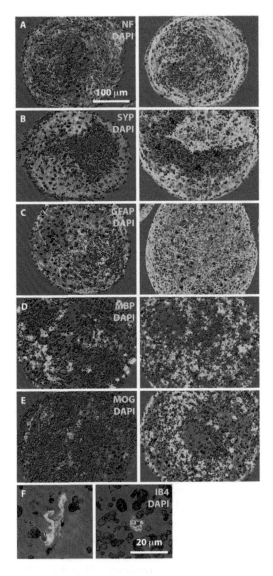

Figure 8.2. 3D free-floating cultures derived from rat brain cells. Neurons, astrocytes, oligodendrocytes and microglial cells were visualized by immunofluorescence for cell type-specific markers in an early maturation stage (day *in vitro* 15 (DIV 15)) and in a highly differentiated stage (DIV 35). Maturation of axons was detected by an increase in labelling for Neurofilaments of high molecular weight, the unphosphorylated and the phosphorylated form (NF) (mixed) (A); increase in immunolabelling for synaptophysin (SYP) was used as a marker of synaptogenesis (B). Maturation of astrocytes was assessed by an increase in immunolabelling for GFAP (C), and maturation of oligodendrocytes by immunostaining for myelin basic protein (MBP) (D) and myelin oligodendrocyte glycoprotein (MOG) (E). At DIV 15, MOG was less expressed than MBP, but their expression increased greatly with time in culture, attesting for the presence of myelinating oligodendrocytes (D and E). Presence of microglial cells was visualized by labelling with isolectin B4 of Griffoniasimplicifolia (IB4) coupled to FITC (F). Higher magnification shows the presence of very fine processes (left panel), whereas activated microglia have a rounded morphology (right panel). On all pictures, nuclei were visualized by Hoechts (blue).

all types of brain cells organized in a 3D architecture is an advantage for the cell-cell intercommunication necessary for timely expression of specific proteins, and more generally for brain maturation.

These cultures are of great interest for DNT since the effects of toxicants can be tested when proliferation of glial precursor cells, axonal elongation and early synaptogenesis and myelination are occurring. Due to the high degree of maturation reached by the cells, they can also be used for Neurotoxicology testing if treatment is applied when neuronal processes and axons are fully matured, synapses and compacted myelin are present. Comparison of toxic effects between immature-and highly differentiated cultures showed that immature cultures were generally more sensitive than differentiated cultures to certain heavy metals (thallium, cadmium, mercury chloride and methylmercury, trimethyl tin, triethyl tin, dibuthyl tin), to the organophosphorous compounds chlorpyrifos and parathion and their oxon forms, and to the food-contaminating mycotoxin, ochratoxin A. The reverse scenario was observed for bismuth (Eskes *et al.*, 1999; Monnet-Tschudi *et al.*, 1997; Monnet-Tschudi *et al.*, 1993; Monnet-Tschudi *et al.*, 1996; Monnet-Tschudi *et al.*, 1995; Monnet-Tschudi *et al.*, 2000). As this 3D culture system is devoid of a blood brain barrier (BBB), the maturation-dependent susceptibility is not due to an easier entrance of the toxicant, but rather to toxicodynamic properties of the toxic compound and how that resonates with the intrinsic properties of the different cell types at their specific state of maturation. (manuscript in preparation).

These 3D cultures also allow the analyses of delayed effects of toxicants, measured after a so-called recovery period, i.e., a period devoid of treatment (Sandstrom von Tobel *et al.*, 2014; Zurich *et al.*, 2002). This latter protocol is of significance for DNT, since adverse effects and/or functional deficits can appear long after a toxic exposure during brain development (Finkelstein *et al.*, 1998; Landrigan *et al.*, 2005; Moreira *et al.*, 2001). An illustration that such delayed effects can be measured in this *in vitro* model is a study on the recovery after paraquat dichloride exposure (Sandstrom von Tobel *et al.*, 2014). Three-dimensional rat brain cell cultures, sub-chronically exposed during an early maturation stage to non-cytotoxic concentrations of paraquat dichloride, could recover from the paraquat-induced adverse effect on neurons. Despite this neuronal recovery, markers of neurodegenerative M1-microglial phenotype were detected only at the end of the 20-day recovery period (Sandstrom von Tobel *et al.*, 2014), indicating that this exposure window has long-term consequences on neuroinflammatory processes.

3D free-floating cultures derived from human embryonic stem cells

Hoelting and coworkers (Hoelting *et al.*, 2013) were the first to report the development of a 3D culture system from hESC (WA09 line). Human ESCs were differentiated into neural precursor cells (NPCs) in adherent conditions on Matrigel-coated plates. Then NPCs were digested by dispase to form small clumps, which were transferred onto low-adhesion plates to form neurospheres. After two to four weeks markers specific for the different stages of neuronal differentiation were detected by RT-PCR: Tubulin β3, for immature neurons, double cortin for migrating neurons, TH for dopaminergic neurons, and syntaxin 1A and synaptophysin, for synaptogenesis (SYP). The mRNA expression of A2B5 was the only marker suggesting the presence of early precursor of glial cells. In addition, gene expression profiling (patterning markers such as Paired box 6 (PAX6), Homeodomain transcription factor EMX2 and LIM homeobox gene LHX2) suggested a predominantly forebrain-like differentiation. This model was used to test for the effects of the neurodevelopmental toxicant methylmercury chloride. Cultures were exposed for 48 h and 18 days at concentrations ranging from 10^{-9} M to 10^{-4} M. EC_{50} for cell viability was obtained at concentrations of $5.4\ 10^{-6}$ M and $1.0\ 10^{-6}$ M, respectively. After a 18-day exposure to a non-cytotoxic concentration of mercury (50 nM) the mRNA expression of neuronal markers (Tubulin β3 and Vesicular glutamate transporter 2 (VGLUT2)) decreased, demonstrating the impact of mercury on neuronal differentiation. This model was also used to test polyethylene nanoparticles, which were found to be incorporated and to exert toxic effects following 48-hour and 18-day exposure. Together these results show the suitability of this 3D model to detect toxicant-induced perturbations of neuronal development.

A similar 3D model is under development in the laboratory of Luc Stoppini (HEPIA, manuscript in preparation), using another hESC line (BAG-HES-GEW-0006). Here a similar differentiation protocol is used, but the 3D structures are formed in 96-well plates and thereafter pooled and maintained in culture flasks under constant gyratory agitation for further differentiation. A progressive differentiation was observed as the mRNA expression of neuronal, astrocyte and oligodendrocyte markers increased successively over time in culture. Neuronal markers *NFH* and synaptophysin were detected from two weeks of differentiation and onward, astrocyte marker *GFAP* from three weeks of differentiation and onward, and *MBP* from four weeks of differentiation and on. At the protein level, GFAP and MBP immunoreactivity was scarce after one month of differentiation, but after three months in culture both proteins were more widely detected (Figure 8.3C–D). Furthermore, electrical activity could be recorded by multi-electrodes arrays on 4-weeks old 3D cultures, indicating that these cultures form functional neuronal circuits.

The preliminary development of these 3D models from hESCs shows that the differentiation of NPCs is slower when derived from human than from murine cells. In addition, it seems that the differentiation potential varies from one cell line to the other (Sterthaus *et al.*, 2014). Despite the promising maturation of neurons and glial cells (Figure 8.3A–D) in these 3D structures, the absence of microglial cells makes them inadequate for the study of toxicant-induced neuroinflammatory responses. However, preliminary experiments showed feasibility of integrating, at the initiation of the 3D structure formation, human primary monocytes into the NPCs population. These cells developed processes, suggesting a possible differentiation into microglial cells (Figure 8.3E).

3D free-floating cultures derived from human induced pluripotent stem cells

Three protocols for the preparation of 3D free-floating cultures from human iPS have been proposed by Hogberg and co-workers (Hogberg *et al.*, 2013): The first protocol described allows for differentiation directly in the 3D-state, letting rosette aggregates

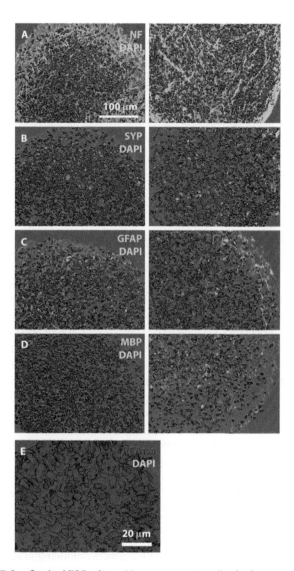

Figure 8.3. 3D free-floating hESC cultures. Neurons, astrocytes, oligodendrocytes and microglial-like cells were visualized by immunofluorescence for cell type-specific markers after one and three months in culture. Neurofilaments (NF), as marker of axonal processes (A); Synaptophysin (SYP), as marker of synaptogenesis (B); Glial Fibrillary Acidic Protein (GFAP), as marker of astrocytes (C), and Myelin Basic Protein (MBP), as marker of oligodendrocytes (D). Monocytes incorporated at the initiation of 3D culturing were maintained during one month and subsequently labelled with Ricinuscommunis agglutinin I (RCA_{120}). The presence of processes suggest a possible microglial differentiation (E). Nuclei were visualized by Hoechst (blue) (A–E).

composed of neural precursor cell (NPC) differentiate in a specific differentiation media. The second protocol uses NPCs grown adherently as a starting point, following dissociation NPCs are grown under constant gyratory agitation in 6-well plates for the 3D-culture formation. The third protocol describes a neuronal pre-differentiation of NPCs in monolayer, subsequent dissociation of the adherent cells and—again for 3D culture formation—gyratory agitation subculturing in 6-well. The latter protocol showed increased expression of neuronal markers (Tubβ3; NF-H; TH; and synapsin-1), suggesting on-going neuronal maturation in these 3D structures. Using the second protocol, Pamies and co-workers (Pamies *et al.*, 2014) obtained after four weeks of differentiation 3D structures reaching a diameter of 300 μm, and exhibiting immunoreactivity for neuronal markers NF-H and MAP2, and after eight weeks of differentiation detection of astrocyte marker GFAP.

Endpoints/Biomarkers for DNT Measurable in 3D Models

In 3D cultures, as in other types of cultures, general cytotoxicity can be assessed by a measurable decrease in protein content, by a decrease in intracellular lactate dehydrogenase (LDH) activity (Koh and Choi, 1987), and by a decrease in ATP content (Schildknecht *et al.*, 2009), or by the resazurin reduction assay (Zimmer *et al.*, 2011). Toxicant induced apoptosis could also be a measure of general adverse effect, where an increase in cleaved caspase 3 content (Jafari *et al.*, 2013), or an increase in cycloheximide-sensitive free nucleosomes would indicate an increase in apoptotic events. Measuring changes in frequency of terminal deoxynucleotidyl transferased UTP nick end labelling (TUNEL), a marker of DNA double-strand breaks, also gives a quantitative measure of the necrotic- and/or apoptotic cell population (Honegger *et al.*, 1998). However, as programmed cell death is a physiological event in the developing central nervous system (Naruse and Keino, 1995). The perturbation of this program may therefore also lead to specific developmental modifications. Indeed alterations of this process associate with several abnormalities (McFarland *et al.*, 2006; Sulik *et al.*, 1988).

In the same line, numerous other parameters can be measured to evaluate the subtle effects on developmental processes due to exposure to sub-cytotoxic concentrations. To evaluate potential interferences with cell proliferation, the rate of DNA synthesis can be determined by metabolic labelling with ^{14}C thymidine, or with bromodeoxyuridine (Miller and Nowakowski, 1988). The expression of the proliferation marker Ki67 is also widely used. To study potential adverse effects on migratory behaviour, it is possible to use so-called secondary 3D cultures, achieved by plating neurospheres onto a planar adhesive substrate, which allows outgrowth of radial glia and migrating neurons (Fritsche *et al.*, 2011).

Adverse effects on neuronal differentiation can be assessed by the analysis of a large battery of markers with spatio-temporal controlled expression during process formation, axonal elongation, synaptogenesis and neurotransmitter specification. Markers for the two first developmental events are often linked to the cytoskeleton, and include, but are not limited to, tubulin β3 (Tuj1), Microtubule-Associated Protein-2 (MAP-2), neurofilaments of light-, medium- and heavy (NF-L, M and H) molecular weight. For synaptogenesis and neuronal specification, markers comprise the different neurotransmitters and related proteins, such as enzymes involved in their synthesis and degradation, receptors or transporters, as well as specific synaptic proteins, for instance synaptophysin, synapsin, TH, or Vesicular glutamate transporter proteins

(V-GLUTs). A decrease of such markers would suggest toxicant interference with neuronal differentiation and/or function, as recently illustrated in 3D cultures of rat brain cells subchronically exposed to the herbicide paraquat (Sandstrom von Tobel *et al.*, 2014). Paraquat exposure correlated with decreased expression of NF-L, M and H, synaptophysin, as well as the dopaminergic neuron marker tyrosine hydroxylase (TH), suggesting a perturbation of neuronal differentiation, as previously reported *in vivo* (Eriksson, 1997; Miranda-Contreras *et al.*, 2005). The ultimate way to evaluate toxicant-induced adverse effects on neuronal circuitry is to measure electrical activity. Such functional analyses can be performed in 3D cultures using microelectrode arrays for extracellular recording and have been used successfully to detect acute neurotoxic effects (van Vliet *et al.*, 2007).

Astrocyte and oligodendrocyte differentiation markers can be used to monitor neurotoxic effects on glial maturation and on myelination. Mature astrocytes express specific proteins such as vimentin, glutamine synthetase, glial fibrillary acidic protein (GFAP), and S100 calcium-binding protein B (S100B). A decrease in the expression of these proteins would indicate a disruption of the astrocyte differentiation process. However, when expression of these proteins is upregulated it is an indication of astrocyte reactivity. Such an up-regulation has been reported upon neurotoxicant exposure and hence demonstrate neurotoxicant-induced astrogliosis (Eng *et al.*, 2000; Sandstrom von Tobel *et al.*, 2014; Zurich *et al.*, 2005). The last brain cell type to differentiate are the oligodendrocytes, and their markers appear temporally from oligodendrocyte precursors expressing transcription factor SOX10, platelet-derived growth factor receptor–α (PDGFRα), NG2 proteoglycan and oligodendrocyte marker O4, to myelinating oligodendrocytes, expressing Myelin Basic Protein (MBP) and Myelin Oligodendrocyte Glycoprotein (MOG), the latter involved in the compaction of the myelin sheath and hence expressed at the latest (Emery, 2010; Fancy *et al.*, 2011; Rowitch, 2004). Some toxins adversely affect oligodendrocytes, for instance the mycotoxin Fumonisin B1 specifically targets myelinating oligodendrocytes, but has no effects once myelin is compacted (Monnet-Tschudi *et al.*, 1999).

A further way of assessing DNT is to analyse toxicant-induced oxidative stress or neuroinflammation. Reactive oxygen species (ROS) formation has been measured in 3D cultures (Monnet-Tschudi *et al.*, 1997), however lipid peroxidation and increase in anti-oxidant defence seemed to be more sensitive markers of oxidative stress (Sandstrom von Tobel *et al.*, 2014). Neuroinflammation can be evaluated by an increase in the number and clustering of isolectin B4-labelled microglial cells, as well as by the expression of markers of the M1/neurodegenerative and M2/neuroprotective microglial phenotypes (Kigerl *et al.*, 2009; Perego *et al.*, 2011). Another assessment of the inflammatory state is to measure the expression and release of pro- and anti-inflammatory cytokines and chemokines. Notable is that neurodegenerative neuroinflammation, characterised by the presence of microglial M1-phenotype, can potentially develop long after cessation of toxicant treatment (Sandstrom von Tobel *et al.*, 2014). Another important parameter of the neuroinflammatory process is the reactivity of astrocytes, which can—as already mentioned above—be evidenced by an increased expression of GFAP, as well as by morphological changes such as shortening and thickening of astrocytes processes (Eng *et al.*, 2000).

Limitations of 3D Models

The ultimate goal of developing 3D brain cell models of human origin is to enhance human safety prediction and to, in as large extent as possible, replace animal testing, where data extrapolation to humans is often not discriminative enough. Therefore the enhanced safety prediction capacity of a human *in vitro* model would be of, not only great economical interest, but also of interest for predicting mechanistic effects that might differ between species (Baumann *et al.*, 2014; Pamies *et al.*, 2014; Tsuji and Crofton, 2012). However, none of the currently available 3D *in vitro* models are officially accepted for neurotoxicity risk assessment of drug- or environmental compounds.

In spite of its ability to detect organ toxicity, the 3D model might be less suitable to elucidate questions related to the primary target of a neurotoxicant, or to molecular mechanistic understanding of the neurotoxic insult, since the measured adversity is always rendered from a choir of signals of different cell-types residing in the structure.

Another concern may be the diffusion of neurotoxic substances of high molecular weight or of nanoparticles, as well as the oxygenation deficit in the centre of relatively large 3D structures (> 500 μm); therefore, the control of the size of the 3D structures is of importance.

An important limitation when considering risk assessment, however not specifically related to 3D cultures, is the absence of physiological barriers, in particular the blood-brain-barrier (BBB). However, during brain development the BBB is not fully formed, and other physiological barriers play a role in limiting the exchange of molecules between CSF and brain parenchyma—like for instance the cells a lining the ventricles (Whish *et al.*, 2015). The periphery of rat brain 3D-cultures bares resemblance to the organization found along the ventricular surface, with glial cells densely lining the outer surface of the sphere (unpublished observation). However, they do not contain the cerebral microvessel endothelium forming the BBB. In an attempt to vascularize these 3D structures, isolated brain capillaries were added to the cultures. The added microvessels did not populate the entire 3D structure, but tended to accumulate primarily in the centre, where oxygenation was the lowest (Juillerat-Jeanneret *et al.*, 2003). However, the incorporation of microvessels could not mimic a functional BBB.

To date, an integrated strategy is proposed (Blaauboer, 2003) comprising in first instance the use of *in vitro* models of BBB (Culot *et al.*, 2008) to test for the capability of a compound to cross or to affect BBB, before measuring its neurotoxic potential using preferentially 3D culture systems.

Potential Use of 3D Models for Risk Assessment

DNT is currently regulated by OECD guidelines (OECD, 2007) using conventional *in vivo* methods, which are expensive and time consuming (Bal-Price *et al.*, 2015a). Regulatory toxicology in the twenty-first century faces numerous challenges, such as the need for assessing an ever-increasing number of chemicals, while reducing animal use, costs, and time required for chemical testing. To address these challenges, there is an urgent need for developing and using *in vitro*, *in silico* and alternative species testing methods that allow faster risk evaluation and regulatory decisions (Bal-Price *et al.*, 2015a).

How toxic a compound is for a certain organism depends on the exposure route, the amount, the way in which the compound is taken up, how it distributes, its biotransformation, how it is eliminated from the organism (kinetics) and the intrinsic properties of the compound towards the organism (dynamics). These three elements: exposure, kinetics and dynamics form the basis of hazard and risk evaluations (Blaauboer, 2003). Toxicokinetics (TK) integrates information about the absorption, distribution in the body, metabolism and excretion (ADME) of a toxic substance. Under the new toxicity testing paradigm where *in vitro* testing is becoming more and more common, high quality *in vitro* TK data are crucial and indispensable for the *in vitro-in vivo* extrapolation (Coecke *et al.*, 2013).

The lack of kinetics information has limited the use of *in vitro* neurotoxicity data. However, recently 3D rat brain cell cultures have been successfully used to assess the biokinetic profile of several drugs. It was shown that chlorpromazine and amiodarone accumulated in the cells over a 12-day repeated exposure (Broeders *et al.*, 2013; Pomponio *et al.*, 2015), whereas diazepam and cyclosporine A did not (Bellwon *et al.*, 2015; Broeders *et al.*, 2014). Furthermore, the major oxidative metabolite of amiodarone was detected intracellularly, indicating the presence of drug metabolising enzymes (Pomponio *et al.*, 2015). Xenobiotic metabolism in the rat 3D brain cell cultures was also observed after exposure to tetrahydrocannabinol (THC). The production of the main THC metabolite was significantly higher in 3D cultures composed of neurons and glial cells, as compared to 3D cultures enriched for either neurons or glial cells, suggesting that both neuron-glia proportion as well as their interactions are involved in the regulation of drug metabolism (Monnet-Tschudi *et al.*, 2008).

A new concept that has been proposed to assist the regulators in decision-making is that of the adverse outcome pathway (AOP) (Ankley *et al.*, 2010). The AOP starts with a so-called *Molecular Initiating Event* (MIE), in which the compound of interest interacts with one or several biological target(s). This is followed by a series of *Key Events* (KEs), which are cellular, anatomical and/or functional changes in biological processes, which ultimately culminates in an *Adverse Outcome* (AO) manifested at organ, organism and/or population level (Bal-Price *et al.*, 2015b). The AOP concept has been developed as a tool for knowledge-based safety assessment, which relies on understanding mechanisms of toxicity, rather than simply observing its adverse outcomes.

Three-dimensional culture systems are well suited for the analyses of KEs due to their histotypic organization. Several KEs, such as oxidative stress, demyelination and neuroinflammation, have been described after toxicant exposure of 3D rat brain cell cultures (Bal-Price *et al.*, 2015b; Sandstrom von Tobel *et al.*, 2014; von Tobel *et al.*, 2014). Importantly, in this type of cultures Key Events Relationships (KERs) can also be evaluated for mechanistic understanding of the toxicity pathway. Despite a possible enhanced prediction capacity, the extrapolation of human *in vitro* data to the actual human exposure scenario remains a major challenge. Physiologically based toxicokinetic (PBTK) modelling is currently regarded as the most adequate approach to simulate the fate of compounds in the human body. The important recent development of toxicodynamics modelling will allow together with PBTK the estimation of a compound's critical amount/concentration on the critical site of action, which ideally would be the basis for hazard and risk assessments (Blaauboer, 2003; Blaauboer, 2010; Schilter *et al.*, 2014).

Perspectives: Multi-Organ on a Chip

In order to develop alternative testing methods that will allow the industry to timely address the increasing demands for reliable data to base risk assessments on, the idea of developing so called "multi-organs" has emerged (Kim *et al.*, 2014; Polini *et al.*, 2014). Recent advances in stem cell research, tissue engineering, biomaterial sciences, systems biology and micro-system technologies now make it possible to envisage the implementation of miniaturised devices, that host several micro-organs (brain, liver, kidney and others). These micro-organs would be generated from target differentiated human cells, and the different compartments would be connected through a micro-perfusion system (Pamies *et al.*, 2014; Zhang *et al.*, 2009; Zweigerdt *et al.*, 2014), allowing "organ intercommunication". Such an approach could allow carrying out toxicity tests in a model that more closely mimics the *in vivo* reactions in an organism. The proof-of-principle that human differentiated cell types can be hosted in separate compartments of a micro-chip that are interconnected through micro-fluidic channels has already been provided (Jang and Suh, 2010; Materne *et al.*, 2015). However, these conditions are established for use under small-scale research laboratory conditions and would not be applicable for large-scale industrial toxicological screenings. For such purposes, the system would need to provide conditions that allow the tissues to maintain a stable phenotype on the multi-organ device over time in culture.

Gregory T. Baxter and Robert Freedman (Baxter, 2009) have developed a cell-based microfluidic assay platform that meets these requirements. The system comprises an arrangement of separate but fluidic interconnected "organ" or "tissue" compartments. Each compartment would contain an *in vitro* "organ", composing of either primary cells or differentiated cells originating from hESC or iPSC. Microfluidic channels connect the different compartments and the flow of culture media would mimic the blood circulation system *in vivo*. A similar culture system has already been used by the group of Schuler with digestive tract and liver cells to simulate the first-pass metabolism of a reference drug, acetaminophen (Esch *et al.*, 2014; McAuliffe *et al.*, 2008). In their study Schuler and colleagues used immortalized human cell lines, which can present some of the characteristics of the *in situ* cells, however, they often lack several important functionalities as they are frequently derived from tumour cells. Deriving the tissues from hESCs or iPSCs could in part circumvent this problem. Another issue that needs to be addressed is the difference in cell morphology, physiology and responsiveness to drugs between cells grown in 2D versus in a 3D structure. For instance, comparing the toxicity responses of monolayer and 3D cultured cells in microscale perfusion devices clearly demonstrate that the 3D cultures more closely resemble cell responses observed *in vivo*. Yang and co-workers published a review on toxicity assays and assay endpoints useful for high-throughput cytotoxicity analysis in microfluidic devices (Yang *et al.*, 2008). The authors concluded that 3D cell cultures better mimic the *in vivo* response to the toxicant.

The final objective would be that the organ culture systems mimics the systemic response of a human body in reaction to compound exposure, while still maintaining the ability to analyse individual "organ" responses (Bhatia and Ingber, 2014). Therefore the tissue compartments must be interconnected in a sequential manner, mimicking the afferent and efferent blood flow in the body, but still respecting the restrictions for proper phenotype maintenance of the different tissues. Such an arrangement could allow sequential metabolism of a compound and observations of subsequent effects of metabolites on surrounding organs (Kimura *et al.*, 2014; Toh *et al.*, 2009). In conclusion, the next generation of *in vitro* tools for more predictive toxicity screening will be multi-organ human cell culture systems, which can more precisely emulate human response to drugs and that can better predict toxicity (Alepee *et al.*, 2014; Fabre *et al.*, 2014; Kim and Shin, 2014).

Acknowledgments

The authors thank HEPIA and SCAHT for their financial support for ongoing development of 3D structures from hESC and Dr. Nicolas Vuilleumier and Dr. Sabrina Pagano, University Hospital of Geneva, Switzerland, for providing human monocytes for incorporation into the hESC 3D structures.

References

Adler, M.W., E.B. Geller, X. Chen, and T.J. Rogers. 2005. Viewing chemokines as a third major system of communication in the brain. The AAPS Journal 7: E865–870.

Alepee, N., A. Bahinski, M. Daneshian, B. De Wever, E. Fritsche, A. Goldberg, J. Hansmann, T. Hartung, J. Haycock, H. Hogberg, L. Hoelting, J.M. Kelm, S. Kadereit, E. McVey, R. Landsiedel, M. Leist, M. Lubberstedt, F. Noor, C. Pellevoisin, D. Petersohn, U. Pfannenbecker, K. Reisinger, T. Ramirez, B. Rothen-Rutishauser, M. Schafer-Korting, K. Zeilinger, and M.G. Zurich. 2014. State-of-the-art of 3D cultures (organs-on-a-chip) in safety testing and pathophysiology. Altex 31: 441–477.

Ankley, G.T., R.S. Bennett, R.J. Erickson, D.J. Hoff, M.W. Hornung, R.D. Johnson, D.R. Mount, J.W. Nichols, C.L. Russom, P.K. Schmieder, J.A. Serrrano, J.E. Tietge, and D.L. Villeneuve. 2010. Adverse outcome pathways: a conceptual framework to support ecotoxicology research and risk assessment. Environmental toxicology and chemistry/SETAC 29: 730–741.

Bal-Price, A., K.M. Crofton, M. Leist, S. Allen, M. Arand, T. Buetler, N. Delrue, R.E. FitzGerald, T. Hartung, T. Heinonen, H. Hogberg, S.H. Bennekou, W. Lichtensteiger, D. Oggier, M. Paparella, M. Axelstad, A. Piersma, E. Rached, B. Schilter, G. Schmuck, L. Stoppini, E. Tongiorgi, M. Tiramani, F. Monnet-Tschudi, M.F. Wilks, T. Ylikomi, and E. Fritsche. 2015a. International STakeholder NETwork (ISTNET): creating a developmental neurotoxicity (DNT) testing road map for regulatory purposes. Arch Toxicol 89: 269–287.

Bal-Price, A., K.M. Crofton, M. Sachana, T.J. Shafer, M. Behl, A. Forsby, A. Hargreaves, B. Landesmann, P.J. Lein, J. Louisse, F. Monnet-Tschudi, A. Paini, A. Rolaki, A. Schrattenholz, C. Sunol, C. van Thriel, M. Whelan, and E. Fritsche. 2015b. Putative adverse outcome pathways relevant to neurotoxicity. Critical Reviews in Toxicology 45: 83–91.

Barone, S., Jr., M.E. Stanton, and W.R. Mundy. 1995. Neurotoxic effects of neonatal triethyltin (TET) exposure are exacerbated with aging. Neurobiol Aging 16: 723–735.

Baumann, J., M. Barenys, K. Gassmann, and E. Fritsche. 2014. Comparative human and rat "neurosphere assay" for developmental neurotoxicity testing. Curr Protoc Toxicol 59: 1–24.

Baxter, G.T. 2009. Hurel—an *in vivo*-surrogate assay platform for cell-based studies. Alternatives to laboratory animals: ATLA 37 Suppl 1: 11–18.

Bellwon, P., M. Culot, A. Wilmes, T. Schmidt, M.G. Zurich, L. Schultz, O. Schmal, A. Gramowski-Voss, D.G. Weiss, P. Jennings, A. Bal-Price, E. Testai, and W. Dekant. 2015. Cyclosporine

A kinetics in brain cell cultures and its potential of crossing the blood-brain barrier. Toxicol *In Vitro* 30: 166–175.

Bhatia, S.N., and D.E. Ingber. 2014. Microfluidic organs-on-chips. Nature Biotechnology 32: 760–772.

Blaauboer, B.J. 2003. The integration of data on physico-chemical properties, *in vitro*-derived toxicity data and physiologically based kinetic and dynamic as modelling a tool in hazard and risk assessment. A commentary. Toxicol Lett 138: 161–171.

Blaauboer, B.J. 2010. Biokinetic modeling and *in vitro-in vivo* extrapolations. J Toxicol Environ Health B Crit Rev 13: 242–252.

Brannvall, K., K. Bergman, U. Wallenquist, S. Svahn, T. Bowden, J. Hilborn, and K. Forsberg-Nilsson. 2007. Enhanced neuronal differentiation in a three-dimensional collagen-hyaluronan matrix. J Neurosci Res 85: 2138–2146.

Broeders, J.J., B.J. Blaauboer, and J.L. Hermens. 2013. *In vitro* biokinetics of chlorpromazine and the influence of different dose metrics on effect concentrations for cytotoxicity in Balb/c 3T3, Caco-2 and HepaRG cell cultures. Toxicol *In Vitro* 27: 1057–1064.

Broeders, J.J., J.L. Hermens, B.J. Blaauboer, and M.G. Zurich. 2014. The *in vitro* biokinetics of chlorpromazine and diazepam in aggregating rat brain cell cultures after repeated exposure. Toxicol *In Vitro* 30: 185–191.

Chamak, B., V. Morandi, and M. Mallat. 1994. Brain macrophages stimulate neurite growth and regeneration by secreting thrombospondin. J Neurosci Res 38: 221–233.

Choi, S.H., Y.H. Kim, M. Hebisch, C. Sliwinski, S. Lee, C. D'Avanzo, H. Chen, B. Hooli, C. Asselin, J. Muffat, J.B. Klee, C. Zhang, B.J. Wainger, M. Peitz, D.M. Kovacs, C.J. Woolf, S.L. Wagner, R.E. Tanzi, and D.Y. Kim. 2014. A three-dimensional human neural cell culture model of Alzheimer's disease. Nature 515: 274–278.

Clemente, D., M.C. Ortega, C. Melero-Jerez, and F. de Castro. 2013. The effect of glia-glia interactions on oligodendrocyte precursor cell biology during development and in demyelinating diseases. Front Cell Neurosci 7: 268.

Coecke, S., O. Pelkonen, S.B. Leite, U. Bernauer, J.G. Bessems, F.Y. Bois, U. Gundert-Remy, G. Loizou, E. Testai, and J.M. Zaldivar. 2013. Toxicokinetics as a key to the integrated toxicity risk assessment based primarily on non-animal approaches. Toxicol *In Vitro* 27: 1570–1577.

Corthésy-Theulaz, I., A.M. Mérillaz, P. Honegger, and B.C. Rossier. 1990. Na+-K+-ATPase gene expression during *in vitro* development of rat fetal forebrain. Am J Physiol 258: C1062–C1069.

Crumpton, T.L., F.J. Seidler, and T.A. Slotkin. 2000. Is oxidative stress involved in the developmental neurotoxicity of chlorpyrifos? Dev Brain Res 121: 189–195.

Culot, M., S. Lundquist, D. Vanuxeem, S. Nion, C. Landry, Y. Delplace, M.P. Dehouck, V. Berezowski, L. Fenart, and R. Cecchelli. 2008. An *in vitro* blood-brain barrier model for high throughput (HTS) toxicological screening. Toxicol *In Vitro* 22: 799–811.

Dubois-Dauphin, M.L., N. Toni, S.D. Julien, I. Charvet, L.E. Sundstrom, and L. Stoppini. 2010. The long-term survival of *in vitro* engineered nervous tissue derived from the specific neural differentiation of mouse embryonic stem cells. Biomaterials 31: 7032–7042.

Durand, B., and M. Raff. 2000. A cell-intrinsic timer that operates during oligodendrocyte development. BioEssays : News and Reviews in Molecular, Cellular and Developmental Biology 22: 64–71.

Emery, B. 2010. Regulation of oligodendrocyte differentiation and myelination. Science 330: 779–782.

Eng, L.F., R.S. Ghirnikar, and Y.L. Lee. 2000. Glial Fibrillary Acidic Protein: GFAP-Thirty-One Years (1969–2000). Neurochem Res 25: 1439–1451.

Eriksson, P. 1997. Developmental neurotoxicity of environmental agents in the neonate. NeuroToxicology 18: 719–726.

Esch, M.B., G.J. Mahler, T. Stokol, and M.L. Shuler. 2014. Body-on-a-chip simulation with gastrointestinal tract and liver tissues suggests that ingested nanoparticles have the potential to cause liver injury. Lab on a chip 14: 3081–3092.

Eskes, C., P. Honegger, T. Jones-Lepp, K. Varner, J.M. Matthieu, and F. Monnet-Tschudi. 1999. Neurotoxicity of dibutyltin in aggregating brain cell cultures. Toxicol *In Vitro* 13: 555–560.

Eskes, C., P. Honegger, L. Juillerat-Jeanneret, and F. Monnet-Tschudi. 2002. Microglial reaction induced by noncytotoxic methylmercury treatment leads to neuroprotection via interactions with astrocytes and IL-6 release. Glia 37: 43–52.

Eskes, C., L. Juillerat-Jeanneret, G. Leuba, P. Honegger, and F. Monnet-Tschudi. 2003. Involvement of microglia-neuron interactions in the tumor necrosis factor-alpha release, microglial activation, and neurodegeneration induced by trimethyltin. J Neurosci Res 71: 583–590.

Fabre, K.M., C. Livingston, and D.A. Tagle. 2014. Organs-on-chips (microphysiological systems): tools to expedite efficacy and toxicity testing in human tissue. Exp Biol Med (Maywood) 239: 1073–1077.

Fancy, S.P., J.R. Chan, S.E. Baranzini, R.J. Franklin, and D.H. Rowitch. 2011. Myelin regeneration: a recapitulation of development? Annu Rev Neurosci 34: 21–43.

Farina, M., D.S. Avila, J.B. da Rocha, and M. Aschner. 2013. Metals, oxidative stress and neurodegeneration: a focus on iron, manganese and mercury. Neurochem Int 62: 575–594.

Finkelstein, Y., M.E. Markowitz, and J.F. Rosen. 1998. Low-level lead-induced neurotoxicity in children: an update on central nervous system effects. Brain Res Rev 27: 168–176.

Fritsche, E., K. Gassmann, and T. Schreiber. 2011. Neurospheres as a model for developmental neurotoxicity testing. Methods Mol Biol 758: 99–114.

Gaiarsa, J.L., and C. Porcher. 2013. Emerging neurotrophic role of GABAB receptors in neuronal circuit development. Frontiers in Cellular Neuroscience 7: 206.

Garden, G.A., and T. Moller. 2006. Microglia biology in health and disease. J Neuroimmune Pharmacol 1: 127–137.

Gomez Perdiguero, E., C. Schulz, and F. Geissmann. 2013. Development and homeostasis of "resident" myeloid cells: the case of the microglia. Glia 61: 112–120.

Grandjean, P., and P.J. Landrigan. 2014a. Neurobehavioural effects of developmental toxicity. Lancet Neurol 13: 330–338.

Grandjean, P., and P.J. Landrigan. 2014b. Neurodevelopmental toxicity: still more questions than answers--authors' response. Lancet Neurol 13: 648–649.

Gritti, A., and L. Bonfanti. 2007. Neuronal-glial interactions in central nervous system neurogenesis: the neural stem cell perspective. Neuron Glia Biology 3: 309–323.

Harry, G.J. 2013. Microglia during development and aging. Pharmacology & Therapeutics 139: 313–326.

Heacock, A.M., A.R. Schonfeld, and R. Katzman. 1986. Hippocampal neurotrophic factor: characterization and response to denervation. Brain Res 363: 299–306.

Hemdan, S., and G. Almazan. 2007. Deficient peroxide detoxification underlies the susceptibility of oligodendrocyte progenitors to dopamine toxicity. Neuropharmacology 52: 1385–1395.

Hoelting, L., B. Scheinhardt, O. Bondarenko, S. Schildknecht, M. Kapitza, V. Tanavde, B. Tan, Q.Y. Lee, S. Mecking, M. Leist, and S. Kadereit. 2013. A 3-dimensional human embryonic stem cell (hESC)-derived model to detect developmental neurotoxicity of nanoparticles. Arch Toxicol 87: 721–733.

Hogberg, H.T., J. Bressler, K.M. Christian, G. Harris, G. Makri, C. O'Driscoll, D. Pamies, L. Smirnova, Z. Wen, and T. Hartung. 2013. Toward a 3D model of human brain development for studying gene/environment interactions. Stem cell research & therapy 4 Suppl 1: S4.

Holmes, C., S.A. Jones, T.C. Budd, and S.A. Greenfield. 1997. Non-cholinergic, trophic action of recombinant acetylcholinesterase on mid-brain dopaminergic neurons. J Neurosci Res 49: 207–218.

Honegger, P., B. Pardo, and F. Monnet-Tschudi. 1998. Muscimol-induced death of GABAergic neurons in rat brain aggregating cell cultures. Brain Res Dev Brain Res 105: 219–225.

Honegger, P., and M. Zurich. 2011. Preparation and use of serum-free aggregating brain cell cultures for routine neurotoxicity screening. pp. 105–128. *In*: M. Aschner, C. Sunol, and A. Bal-Price (eds.). Cell Culture Techniques. Humana Press, New York.

Honegger, P., A. Defaux, F. Monnet-Tschudi, and M.G. Zurich. 2011. Preparation, maintenance, and use of serum-free aggregating brain cell cultures. Methods Mol Biol 758: 81–97.

Jafari, P., O. Braissant, P. Zavadakova, H. Henry, L. Bonafe, and D. Ballhausen. 2013. Brain damage in methylmalonic aciduria: 2-methylcitrate induces cerebral ammonium accumulation and apoptosis in 3D organotypic brain cell cultures. Orphanet Journal of Rare Diseases 8: 4.

Jang, K.J., and K.Y. Suh. 2010. A multi-layer microfluidic device for efficient culture and analysis of renal tubular cells. Lab on a Chip 10: 36–42.

Juillerat-Jeanneret, L., F. Monnet-Tschudi, M.G. Zurich, S. Lohm, A.M. Duijvestijn, and P. Honegger. 2003. Regulation of peptidase activity in a three-dimensional aggregate model of brain tumor vasculature. Cell and Tissue Research 311: 53–59.

Kierdorf, K., D. Erny, T. Goldmann, V. Sander, C. Schulz, E.G. Perdiguero, P. Wieghofer, A. Heinrich, P. Riemke, C. Holscher, D.N. Muller, B. Luckow, T. Brocker, K. Debowski, G. Fritz, G. Opdenakker, A. Diefenbach, K. Biber, M. Heikenwalder, F. Geissmann, F. Rosenbauer, and M. Prinz. 2013. Microglia emerge from erythromyeloid precursors via Pu.1- and Irf8-dependent pathways. Nat Neurosci 16: 273–280.

Kigerl, K.A., J.C. Gensel, D.P. Ankeny, J.K. Alexander, D.J. Donnelly, and P.G. Popovich. 2009. Identification of two distinct macrophage subsets with divergent effects causing either neurotoxicity or regeneration in the injured mouse spinal cord. J Neurosci 29: 13435–13444.

Kim, D., X. Wu, A.T. Young, and C.L. Haynes. 2014. Microfluidics-based *in vivo* mimetic systems for the study of cellular biology. Accounts of Chemical Research 47: 1165–1173.

Kim, J., and M. Shin. 2014. An integrative model of multi-organ drug-induced toxicity prediction using gene-expression data. BMC bioinformatics 15 Suppl 16: S2.

Kimura, H., T. Ikeda, H. Nakayama, Y. Sakai, and T. Fujii. 2014. An on-chip small intestine-liver model for pharmacokinetic studies. J Lab Autom 20: 265–273.

Koh, J.Y., and D.W. Choi. 1987. Quantitative determination of glutamate mediated cortical neuronal injury in cell culture by lactate dehydrogenase efflux assay. Journal of Neuroscience Methods 20: 83–90.

Krstic, D., A. Madhusudan, J. Doehner, P. Vogel, T. Notter, C. Imhof, A. Manalastas, M. Hilfiker, S. Pfister, C. Schwerdel, C. Riether, U. Meyer, and I. Knuesel. 2012. Systemic immune challenges trigger and drive Alzheimer-like neuropathology in mice. J Neuroinflammation 9: 151.

Lancaster, M.A., M. Renner, C.A. Martin, D. Wenzel, L.S. Bicknell, M.E. Hurles, T. Homfray, J.M. Penninger, A.P. Jackson, and J.A. Knoblich. 2013. Cerebral organoids model human brain development and microcephaly. Nature 501: 373–379.

Lancaster, M.A., and J.A. Knoblich. 2014. Generation of cerebral organoids from human pluripotent stem cells. Nature Protocols 9: 2329–2340.

Landrigan, P.J., B. Sonawane, R.N. Butler, L. Trasande, R. Callan, and D. Droller. 2005. Early environmental origins of neurodegenerative disease in later life. Environ Health Perspect1 13: 1230–1233.

Li, H., A. Wijekoon, and N.D. Leipzig. 2012. 3D differentiation of neural stem cells in macroporous photopolymerizable hydrogel scaffolds. PLoS One 7: e48824.

Li, Z., T. Dong, C. Proschel, and M. Noble. 2007. Chemically diverse toxicants converge on Fyn and c-Cbl to disrupt precursor cell function. PLoS Biology 5: e35.

Materne, E.M., A.P. Ramme, A.P. Terrasso, M. Serra, P.M. Alves, C. Brito, D.A. Sakharov, A.G. Tonevitsky, R. Lauster, and U. Marx. 2015. A multi-organ chip co-culture of neurospheres and liver equivalents for long-term substance testing. J Biotechnol 205: 36–46.

McAuliffe, G.J., J.Y. Chang, R.P. Glahn, and M.L. Shuler. 2008. Development of a gastrointestinal tract microscale cell culture analog to predict drug transport. Molecular & Cellular Biomechanics: MCB 5: 119–132.

McFarland, K.N., S.R. Wilkes, S.E. Koss, K.S. Ravichandran, and J.W. Mandell. 2006. Neural-specific inactivation of ShcA results in increased embryonic neural progenitor apoptosis and microencephaly. J Neurosci 26: 7885–7897.

Miller, M.W., and R.S. Nowakowski. 1988. Use of bromodeoxyuridine-immunohistochemistry to examine the proliferation, migration and time of origin of cells in the central nervous system. Brain Res 457: 44–52.

Miranda-Contreras, L., R. Davila-Ovalles, P. Benitez-Diaz, Z. Pena-Contreras, and E. Palacios-Pru. 2005. Effects of prenatal paraquat and mancozeb exposure on amino acid synaptic transmission in developing mouse cerebellar cortex. Brain Res Dev Brain Res 160: 19–27.

Monnet-Tschudi, F., M.G. Zurich, and P. Honegger. 1993. Evaluation of the toxicity of different metal compounds in the developing brain using aggregating cell cultures as a model. Toxic *In Vitro* 7: 335–339.

Monnet-Tschudi, F., M.G. Zurich, B.M. Riederer, and P. Honegger. 1995. Effects of trimethyltin (TMT) on glial and neuronal cells in aggregate cultures: dependence on the developmental stage. Neurotoxicology 16: 97–104.

Monnet-Tschudi, F., M.G. Zurich, and P. Honegger. 1996. Comparison of the developmental effects of two mercury compounds on glial cells and neurons in aggregate cultures of rat telencephalon. Brain Res 741: 52–59.

Monnet-Tschudi, F., O. Sorg, P. Honegger, M.G. Zurich, A.C. Huggett, and B. Schilter. 1997. Effects of the naturally occurring food mycotoxin ochratoxin A on brain cells in culture. Neurotoxicology 18: 831–839.

Monnet-Tschudi, F., M.-G. Zurich, O. Sorg, J.-M. Matthieu, P. Honegger, and B. Schilter. 1999. The naturally occurring food mycotoxin fumonisin B1 impairs myelin formation in aggregating brain cell culture. NeuroToxicology 20: 41–48.

Monnet-Tschudi, F., M.G. Zurich, B. Schilter, L.G. Costa, and P. Honegger. 2000. Maturation-dependent effects of chlorpyrifos and parathion and their oxygen analogs on acetylcholinesterase and neuronal and glial markers in aggregating brain cell cultures. Toxicol Appl Pharmacol 165: 175–183.

Monnet-Tschudi, F., M.G. Zurich, and P. Honegger. 2007. Neurotoxicant-induced inflammatory response in three-dimensional brain cell cultures. Hum Exp Toxicol 26: 339–346.

Monnet-Tschudi, F., A. Hazekamp, N. Perret, M.G. Zurich, P. Mangin, C. Giroud, and P. Honegger. 2008. Delta-9-tetrahydrocannabinol accumulation, metabolism and cell-type-specific adverse effects in aggregating brain cell cultures. Toxicol Appl Pharmacol 228: 8–16.

Monnet-Tschudi, F., A. Defaux, and M.-G. Zurich. 2010. Probable involvement of heavy metal-induced neuroinflammation in neurodegeneration. pp. 1–13. *In*: S. Huang (ed.). Heavy Metals and Neurodegeneration. Research Signpost, Kerala, India.

Moreira, E.G., I. Vassilieff, and V.S. Vassilieff. 2001. Developmental lead exposure: behavioral alterations in the short and long term. Neurotoxicol Teratol 23: 489–495.

Moscona, A.A. 1961. Rotation-mediated histogenetic aggregation of dissociated cells: A quantifiable approachh to cell interactions *in vitro*. Exp Cell Res 22: 455–475.

Naruse, I., and H. Keino. 1995. Apoptosis in the developing CNS. Prog Neurobiol 47: 135–155.

Neal, A.P., K.H. Stansfield, P.F. Worley, R.E. Thompson, and T.R. Guilarte. 2010. Lead exposure during synaptogenesis alters vesicular proteins and impairs vesicular release: potential role of NMDA receptor-dependent BDNF signaling. Toxicol Sci 116: 249–263.

Pamies, D., T. Hartung, and H.T. Hogberg. 2014. Biological and medical applications of a brain-on-a-chip. Exp Biol Med (Maywood) 239: 1096–1107.

Perego, C., S. Fumagalli, and M.G. De Simoni. 2011. Temporal pattern of expression and colocalization of microglia/macrophage phenotype markers following brain ischemic injury in mice. J Neuroinflammation 8: 174.

Polini, A., L. Prodanov, N.S. Bhise, V. Manoharan, M.R. Dokmeci, and A. Khademhosseini. 2014. Organs-on-a-chip: a new tool for drug discovery. Expert Opinion on Drug Discovery 9: 335–352.

Pomponio, G., M.G. Zurich, L. Schultz, D.G. Weiss, L. Romanelli, A. Gramowski-Voss, E. Di Consiglio, and E. Testai. 2015. Amiodarone biokinetics, the formation of its major oxidative metabolite and neurotoxicity after acute and repeated exposure of brain cell cultures. Toxicol *In Vitro* 30:192–202.

Preynat-Seauve, O., D.M. Suter, D. Tirefort, L. Turchi, T. Virolle, H. Chneiweiss, M. Foti, J.A. Lobrinus, L. Stoppini, A. Feki, M. Dubois-Dauphin, and K.H. Krause. 2009. Development of human nervous tissue upon differentiation of embryonic stem cells in three-dimensional culture. Stem Cells 27: 509–520.

Prieto, P., A. Kinsner-Ovaskainen, S. Stanzel, B. Albella, P. Artursson, N. Campillo, R. Cecchelli, L. Cerrato, L. Diaz, E. Di Consiglio, A. Guerra, L. Gombau, G. Herrera, P. Honegger, C. Landry, J.E. O'Connor, J.A. Paez, G. Quintas, R. Svensson, L. Turco, M.G. Zurich, M.J. Zurbano, and A. Kopp-Schneider. 2013. The value of selected *in vitro* and *in silico* methods to predict acute oral toxicity in a regulatory context: results from the European Project ACuteTox. Toxicol *In Vitro* 27: 1357–1376.

Purves, D., and S.M. Williams. 2001. Neuroscience. Sinauer Associates, Sunderland, Mass. xviii, 681, 616, 683, 625 pp.

Rowitch, D.H. 2004. Glial specification in the vertebrate neural tube. Nat Rev Neurosci 5: 409–419.

Sandstrom von Tobel, J., D. Zoia, J. Althaus, P. Antinori, J. Mermoud, H.S. Pak, A. Scherl, and F. Monnet-Tschudi. 2014. Immediate and delayed effects of subchronic Paraquat exposure during an early differentiation stage in 3D-rat brain cell cultures. Toxicol Lett 230: 188–197.

Schildknecht, S., D. Poltl, D.M. Nagel, F. Matt, D. Scholz, J. Lotharius, N. Schmieg, A. Salvo-Vargas, and M. Leist. 2009. Requirement of a dopaminergic neuronal phenotype for toxicity of low concentrations of 1-methyl-4-phenylpyridinium to human cells. Toxicol Appl Pharmacol 241: 23–35.

Schilter, B., R. Benigni, A. Boobis, A. Chiodini, A. Cockburn, M.T. Cronin, E. Lo Piparo, S. Modi, A. Thiel, and A. Worth. 2014. Establishing the level of safety concern for chemicals in food without the need for toxicity testing. Regul Toxicol Pharmacol 68: 275–296.

Stansfield, K.H., J.R. Pilsner, Q. Lu, R.O. Wright, and T.R. Guilarte. 2012. Dysregulation of BDNF-TrkB signaling in developing hippocampal neurons by Pb(2+): implications for an environmental basis of neurodevelopmental disorders. Toxicol Sci 127: 277–295.

Sterthaus, O., A.C. Feutz, H. Zhang, F. Pletscher, E. Bruder, P. Miny, G. Lezzi, M. De Geyter, and C. De Geyter. 2014. Gene expression profiles of similarly derived human embryonic stem cell lines correlate with their distinct propensity to exit stemness and their different differentiation behavior in culture. Cellular Reprogramming 16: 185–195.

Stipursky, J., T.C. Spohr, V.O. Sousa, and F.C. Gomes. 2012. Neuron-astroglial interactions in cell-fate commitment and maturation in the central nervous system. Neurochem Res 37: 2402–2418.

Stoppini, L., P.A. Buchs, and D. Muller. 1991. A simple method for organotypic cultures of nervous tissue. Journal of Neuroscience Methods 37: 173–182.

Sulik, K.K., C.S. Cook, and W.S. Webster. 1988. Teratogens and craniofacial malformations: relationships to cell death. Development 103 Suppl: 213–231.

Sundstrom, L., T. Biggs, A. Laskowski, and L. Stoppini. 2012. OrganDots—an organotypic 3D tissue culture platform for drug development. Expert Opinion on Drug Discovery 7: 525–534.

Tieng, V., L. Stoppini, S. Villy, M. Fathi, M. Dubois-Dauphin, and K.H. Krause. 2014. Engineering of midbrain organoids containing long-lived dopaminergic neurons. Stem Cells and Development 23: 1535–1547.

Toh, Y.C., T.C. Lim, D. Tai, G. Xiao, D. van Noort, and H. Yu. 2009. A microfluidic 3D hepatocyte chip for drug toxicity testing. Lab on a Chip 9: 2026–2035.

Tsuji, R., and K.M. Crofton. 2012. Developmental neurotoxicity guideline study: issues with methodology, evaluation and regulation. Congenital Anomalies 52: 122–128.

van Vliet, E., L. Stoppini, M. Balestrino, C. Eskes, C. Griesinger, T. Sobanski, M. Whelan, T. Hartung, and S. Coecke. 2007. Electrophysiological recording of re-aggregating brain cell cultures on multi-electrode arrays to detect acute neurotoxic effects. Neurotoxicology 28: 1136–1146.

von Tobel, J.S., P. Antinori, M.G. Zurich, R. Rosset, M. Aschner, F. Gluck, A. Scherl, and F. Monnet-Tschudi. 2014. Repeated exposure to Ochratoxin A generates a neuroinflammatory response, characterized by neurodegenerative M1 microglial phenotype. Neurotoxicology 44C: 61–70.

Whish, S., K.M. Dziegielewska, K. Mollgard, N.M. Noor, S.A. Liddelow, M.D. Habgood, S.J. Richardson, and N.R. Saunders. 2015. The inner CSF-brain barrier: developmentally controlled access to the brain via intercellular junctions. Frontiers in Neuroscience 9: 16.

White, R., and E.M. Kramer-Albers. 2014. Axon-glia interaction and membrane traffic in myelin formation. Front Cell Neurosci 7: 284.

Yang, S.T., X. Zhang, and Y. Wen. 2008. Microbioreactors for high-throughput cytotoxicity assays. Current Opinion in Drug Discovery & Development 11: 111–127.

Zawia, N.H., and M.R. Basha. 2005. Environmental risk factors and the developmental basis for Alzheimer's disease. Rev Neurosci 16: 325–337.

Zawia, N.H., D.K. Lahiri, and F. Cardozo-Pelaez. 2009. Epigenetics, oxidative stress, and Alzheimer disease. Free Radic Biol Med 46: 1241–1249.

Zhang, C., Z. Zhao, N.A. Abdul Rahim, D. van Noort, and H. Yu. 2009. Towards a human-on-chip: culturing multiple cell types on a chip with compartmentalized microenvironments. Lab on a Chip 9: 3185–3192.

Zimmer, B., P.B. Kuegler, B. Baudis, A. Genewsky, V. Tanavde, W. Koh, B. Tan, T. Waldmann, S. Kadereit, and M. Leist. 2011. Coordinated waves of gene expression during neuronal differentiation of embryonic stem cells as basis for novel approaches to developmental neurotoxicity testing. Cell Death and Differentiation 18: 383–395.

Zurich, M.G., F. Monnet-Tschudi, M. Berode, and P. Honegger. 1998. Lead acetate toxicity *in vitro*: Dependence on the cell composition of the cultures. Toxicol *In Vitro* 12: 191–196.

Zurich, M.-G., C. Eskes, P. Honegger, M. Bérode, and F. Monnet-Tschudi. 2002. Maturation-dependent neurotoxicity of lead aceate *in vitro*: Implication of glial reactions. J Neurosc Res 70: 108–116.

Zurich, M.-G., F. Monnet-Tschudi, L.C. Costa, B. Schilter, and P. Honegger. 2003. Aggregating brain cell cultures for neurotoxicological studies. pp. 243–266. *In*: E. Tiffany-Castiglioni (ed.). *In vitro* Neurotoxicology: Principles and Challenges. Humana Press Inc., Totowa, NJ.

Zurich, M.G., P. Honegger, B. Schilter, L.G. Costa, and F. Monnet-Tschudi. 2004. Involvement of glial cells in the neurotoxicity of parathion and chlorpyrifos. Toxicol Appl Pharmacol 201: 97–104.

Zurich, M.G., S. Lengacher, O. Braissant, F. Monnet-Tschudi, L. Pellerin, and P. Honegger. 2005. Unusual astrocyte reactivity caused by the food mycotoxin ochratoxin A in aggregating rat brain cell cultures. Neuroscience 134: 771–782.

Zurich, M.G., and P. Honegger. 2011. Ochratoxin A at nanomolar concentration perturbs the homeostasis of neural stem cells in highly differentiated but not in immature three-dimensional brain cell cultures. Toxicol Lett 205: 203–208.

Zurich, M.G., S. Stanzel, A. Kopp-Schneider, P. Prieto, and P. Honegger. 2013. Evaluation of aggregating brain cell cultures for the detection of acute organ-specific toxicity. Toxicol *In Vitro* 27: 1416–1424.

Zweigerdt, R., I. Gruh, and U. Martin. 2014. Your heart on a chip: iPSC-based modeling of Barth-syndrome-associated cardiomyopathy. Cell Stem Cell 15: 9–11.

9

Neural Cell Lines (Lineage)

Qiang Gu

Introduction

The pioneer work by Ross Granville Harrison, who successfully grew cells outside the body over a century ago, formed the basis of modern cell culture techniques that allow the study of isolated living cells in an artificially controlled environment. Over the past century, advancements in knowledge, theory, practice and methodologies have made cell cultures indispensable tools for basic life science, biomedical and clinical research. To date, many aspects of cell culture techniques such as plate coatings, growth media and incubation conditions have been standardized, and a wide variety of cells can be isolated, purified, grown and maintained *in vitro*. In addition to some obvious advantages of cell culture systems compared to *in vivo* situations such as being simple, homogeneous cell populations, easy-controllable cell exposures and easy-manipulatable gene expressions, an important aspect of using cell cultures that deserves a notion here is to set alternative approaches to reduce animal use in experimentation. Due to the unique character of differentiated neurons that they are no longer dividing and non-multipliable, a number of cell sources have been explored with the goal to establish immortalized neural cell lines. These include cells derived from spontaneous or artificially-induced brain tumors, fusion of neurons to tumor cells, retroviral infection of neural precursor cells, or neurons transfected with oncogenes. To date a vast variety of cells lines have been developed that continuously express specific neurotransmitters and that are utilized as neuronal cell lines. In addition, several glial cell lines derived from glioma or glioblastoma have also been established and

Division of Neurotoxicology, HFT-132, National Center for Toxicological Research (NCTR), U.S. Food and Drug Administration (FDA), 3900 NCTR Rd, Jefferson, AR 72079.
E-mail: qiang.gu@fda.hhs.gov

characterized. Collectively these cell culture systems have been widely used for studies of neural growth, differentiation, synaptogenesis, gene expression, plasticity, regulation of transmitter release, signal transduction, pharmacology and toxicology. Given the wide variety of available cell lines and limited space allowed in this chapter, it would be impossible to review all neural cell lines. Therefore, this chapter will cover the most widely studied and well-characterized cell lines in neuroscience research (Table 9.1), and their applications and important results of those studies. An important objective and envision of developing neuronal cell lines in those early days was to generate mass quantities of a particular type of cells or cells carrying certain specific genes for clinical transplantation with the purpose to repair or cure neurological disorders. However, with the promising features of induced pluripotent stem cells, much effort and hope have been shifted towards utilizing neural stem cells. Because neural stem cells have been described in a preceding chapter (Chapter 6) in this book, they will not be covered here to avoid redundancy.

Table 9.1. Cell lines covered in this chapter.

Cell Origin	Species	Cell Line Name	Year Established*
Pheochromocytoma	Rat	PC12	1975
Neuroblastoma	Human	SH-SY5Y	1978
	Mouse	N1E-115	1972
		Neuro-2a	1969
	Rat	B104	1974
Medulloblastoma	Human	TE671	1977
Chronic Focal Encephalitis	Human	HCN-2	1994
Glioma	Human	U87MG	1968
	Rat (Wistar)	C6	1968
	Rat (Fisher)	F98	1971
		RG2	1973
Neuroblastoma-Glioma Hybrid	Mouse/Rat	NG108-115	1971
SV-40 Viral Transfected Brain Tumor	Mouse	CATH.a	1993

* Year of creation or when the first study utilizing the cell line was published.

Cell Lines of Neuronal Origin

PC12 cell line (rat pheochromocytoma)

The PC12 cell line is by far the most widely used cell line in neuroscience research, and the most widely used model of neuronal function and differentiation (Fujita *et al.*, 1989). PC12 cells were derived from rat adrenal gland with pheochromocytoma and introduced by Arthur Tischler and Lloyd Greene in the 1970s (Greene and Tischler, 1976; Tischler and Greene, 1975). PC12 cells synthesize neurotransmitters such as dopamine, norepinephrine, and acetylcholine (ACh). PC12 cells express both nicotinic and muscarinic ACh receptors and are responsive to ACh (Dichter *et al.*, 1977;

Greene and Rein, 1977; Jumblatt and Tischler, 1982). A variety of ion channels have been detected in PC12 cells including voltage-dependent Na^+-, Ca^{2+}- and K^+-channels, and Ca^{2+}-dependent K^+ channels (Arner and Stallcup, 1981; O'Lague and Huttner, 1980), which are localized in plasma membrane and involved in the propagation of electrical activity. PC12 cells have been shown to build functional synapses with cultured cells of skeletal muscle origin (Schubert *et al.*, 1977). The processes of PC12 cells contain varicosities that contain vesicular neurotransmitters which can be released in a quantal fashion upon electrical or chemical stimulations. Based on these characteristics, the PC12 cell line is considered a neuroendocrine or neuronal-type cell line and has been utilized as a model for studying neurosecretion (D'Alessandro and Meldolesi, 2010; Westerink and Ewing, 2008), and more specifically the regulation of neurotransmitter synthesis, storage and release. PC12 cells also express nerve growth factor (NGF) receptors and respond reversibly to NGF, i.e., in the presence of NGF, PC12 cells stop dividing and differentiate to neuronal phenotype, such as induction of electrical excitability, increase in expression of Na^+ channels and muscarinic ACh receptors, and outgrowth of long processes containing varicosities, whereas withdrawal of NGF will reverse the phenotype. These phenomena make the PC12 cell line the premiere model for the study of neurite outgrowth and neuronal differentiation (D'Alessandro and Meldolesi, 2010; Levi *et al.*, 1988). Since NGF application in a culture dish can be artificially well controlled, PC12 cells have been utilized for delineating molecular mechanisms underlying NGF action (Levi *et al.*, 1988). For example, NGF has been shown to increase the level of tubulin, microtubule associated proteins MAP-1, MAP-2 and tau, vimentin, and other neurofilament proteins (Black *et al.*, 1986; Drubin *et al.*, 1985; Greene *et al.*, 1983; Lee and Page, 1984). It is conceivable that neurite outgrowth may directly correlate with the biosynthesis of cytoskeleton proteins. PC12 cells also have been employed in research of stress, hypoxia, and neurotoxicity (Roth *et al.*, 2002; Spicer and Millhorn, 2003). Not only has the PC12 cell line helped the discovery of various phenomena associated with cell growth, differentiation and death, molecular analysis of these phenomena at different stages further revealed critical elements of several distinct and independent signal transduction pathways responsible for the underlying cellular and molecular mechanisms (Kontou *et al.*, 2009; Sombers *et al.*, 2002; Vaudry *et al.*, 2002). Another significant utilization of the PC12 cell line is in the area of *in vitro* pharmacological studies, since PC12 cells are extremely versatile following pharmacological manipulations and easy to culture. Such studies provide a wealth of information on cell proliferation and differentiation, on drug effects on vesicles, and on the regulation of exocytosis (Momboisse *et al.*, 2010). Successful foreign gene insertion and stable expression in PC12 cells have been achieved (Li *et al.*, 1999), with the hope that they could be useful for *in vivo* testing of transplantation for potential disease treatment (Yoshida *et al.*, 1999). In a recent study, PC12 cells were employed for developing and validating a new *in vitro* cytotoxicity assay (Gu *et al.*, 2014), which utilizes Fluoro-Jade C (FJ-C), a fluorescent label previously used for the assessment of neurodegeneration in fixed brain tissue sections for the labeling of degenerating cells in cell cultures following toxic insults (Figure 9.1). This new FJ-C based *in vitro* approach is simple and fast, has a high sensitivity and large dynamic range, and has the potential to be utilized in high-throughput toxicity screening of chemical compounds *in vitro*.

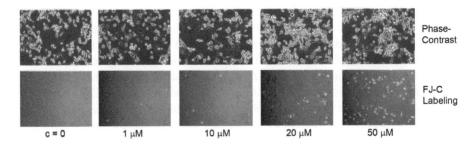

Phase-Contrast

FJ-C Labeling

c = 0   1 µM   10 µM   20 µM   50 µM

Figure 9.1. FJ-C labeling in PC12 cell cultures following 24 hr incubation with various concentrations of cadmium, a well-known toxicant, showing dose-dependent cytotoxicity as indicated by the increase number of FJ-C labeled cells. Reprint from Gu *et al.*, 2014.

SH-SY5Y cell line (human neuroblastoma)

The SH-SY5Y cell line was introduced in 1978 (Biedler *et al.*, 1978). It was a subclone of the SK-N-SH cell line that was derived from a bone marrow biopsy of a 4-year-old girl with neuroblastoma. SH-SY5Y cells are electrically excitable and possess a variety of voltage- and ligand-gated ion channels, for example tetrodotoxin-sensitive Na^+ channels, voltage-sensitive K^+ channels, and L-type and N-type voltage-sensitive Ca^{2+} channels (Forsythe *et al.*, 1992; Kennedy and Henderson, 1992; Morton *et al.*, 1992; Reeve *et al.*, 1992; Reeve *et al.*, 1994). SH-SY5Y cells also express a variety of transmitter receptors such as muscarinic ACh M_1, M_2 and M_3 receptors (Adem *et al.*, 1987; Lambert *et al.*, 1989a), nicotinic ACh receptors (Gould *et al.*, 1992), mu-opioid receptors (Seward *et al.*, 1991), and neuropeptide-Y receptors (McDonald *et al.*, 1994). Whole cell patch-clamp and intracellular Ca^{2+} imaging techniques have been applied to SH-SY5Y cell cultures and revealed that SH-SY5Y cells possess purinergic receptors that respond to extracellular ATP, ADT, UTP and UDP. Activation of purinergic $P2X_7$ receptors in SH-SY5Y cells result in a rapid elevation of intracellular Ca^{2+} through the opening of $P2X_7$ receptor-coupled voltage-dependent Ca^{2+} channels (Larsson *et al.*, 2002). SH-SY5Y cell line produces the neurotransmitter dopamine and also expresses dopamine transporters and receptors. Therefore, the SH-SY5Y cell line is considered primarily a dopaminergic-type neuron and is often used as a cellular model for Parkinson's disease (Xie *et al.*, 2010). SH-SY5Y cells can differentiate into several different phenotypes depending on differentiation agents such as retinoic acid (RA), brain derived neurotrophic factor (BDNF), and NGF (Edsjo *et al.*, 2003; Encinas *et al.*, 2000). When treated by RA and BNDF, SH-SY5Y increased expression of general cholinergic markers choline acetyltransferase and vesicular ACh transporter (Edsjo *et al.*, 2003). The SH-SY5Y cell line has been widely used for cellular studies of neuronal function, differentiation, regulation of transmitter release, degeneration, cytotoxicity and anti-tumor drug efficacy (Fasano *et al.*, 2008; Vaughan *et al.*, 1995). Similar to PC12 cells, SH-SY5Y cells were recently employed for developing and validating an FJ-C based *in vitro* assay for cytotoxicity (Figure 9.2).

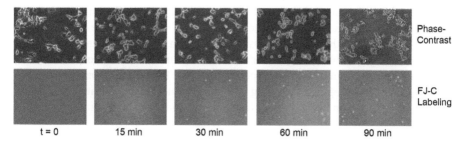

Figure 9.2. FJ-C labeling in SH-SY5Y cell cultures following 20 μM cadmium treatment for different durations showing time-dependent cytotoxicity as indicated by the increase number of FJ-C labeled cells. Reprint from Gu *et al.*, 2014.

N1E-115 cell line (mouse neuroblastoma)

The N1E-115 cell line was established in the early 1970s by cloning spontaneous mouse neuroblastoma tumor (Amano *et al.*, 1972). N1E-115 cells exhibit high levels of activity of the enzyme tyrosine hydroxylase and express a variety of transmitter receptors such as muscarinic ACh M_1, M_2 and M_4, histaminergic H_1, serotonergic 5-HT$_3$, and nicotinic ACh receptors, among others (Lenz *et al.*, 1994; Oortgiesen *et al.*, 1997; Richelson, 1990). N1E-115 cells also express other types of membrane receptors such as that for adenosine (Murphy and Byczko, 1990), angiotensin (Gilbert *et al.*, 1984), bradykinin (Snider and Richelson, 1984), melatonin (Bordt *et al.*, 2001), neurotensin (Gilbert and Richelson, 1984), purinergic P2X$_7$ (Schrier *et al.*, 2002), somatostatin (McKinney and Barrett, 1989), and tumor necrosis factor (TNF) (Sipe *et al.*, 1996). At the functional level, it has been shown that the N1E-115 cell line rapidly responds to melatonin stimulation and forms neurites within 24 hr (Bordt *et al.*, 2001), while exposure to extracellular TNF, ATP or adenosine can induce apoptosis (Schrier *et al.*, 2002; Schrier *et al.*, 2001; Sipe *et al.*, 1996). N1E-115 cells can be differentiated under conditions of low serum in the presence of dimethyl sulfoxide (DMSO) to produce morphologically mature neuronal-like cholinergic cells (Amano *et al.*, 1972). N1E-115 cells have been used to study transmitter receptor regulation, second messenger synthesis, electrophysiological changes, and cell death mechanisms. For example, most electrophysiological and pharmacological properties of 5-HT$_3$ receptors were initially characterized in N1E-115 cells (Lambert *et al.*, 1989b). In another study, muscarinic ACh M_1 receptors were detected intracellularly and in the plasma membrane of N1E-115 cells, and M_1 receptors in both locations appeared functionally active, suggesting additional signaling function of M_1 receptor intracellularly (Uwada *et al.*, 2011). Growth hormone and growth hormone receptors are abundantly expressed in N1E-115 cells (Grimbly *et al.*, 2009). Exposure of N1E-115 cells to mouse growth hormone induced neurite sprouting and increased axon growth, thereby providing a good model for the study of the effects of growth hormone on neurite outgrowth (Grimbly *et al.*, 2009). N1E-115 cells also have been employed for examining toxicant effects on viability, structure, and axonal transport of neurons (Brat and Brimijoin, 1992). In this model, N1E-115 cells were induced to extend neurites by elimination of serum from the medium, and extended neurites 2–5 μm in diameter and up to 400 µm in length on uncoated glass in serum-free medium. By measuring the fraction of cells bearing neurites and the mean length of the neurites in the presence and absence of a toxicant, the effect of a test compound on cell growth can be evaluated (Brat and Brimijoin, 1992). Organellar motility in neurites can be another indicator of healthiness concerning axonal transport. The N1E-115 cell line appears to be well suited for time-lapsed optical recording of organellar mobility in neurites, and yields valuable information on the toxicity of tested compounds (Brat and Brimijoin, 1992).

Neuro-2a cell line (mouse neuroblastoma)

Another mouse neuroblastoma cell line Neuro-2a (also called N2a) was established from a spontaneous tumor of a strain A albino mouse (Klebe and Ruddle, 1969). Neuro-2a cells produce large quantities of microtubular proteins including tubulin, actin and MAP-2, which are believed to play a role in a contractile system which is responsible for axoplasmic flow in nerve cells. Therefore, the Neuro-2a cell line has been useful for studies of the synthesis, assembly, and turnover of neurofilament proteins (Olmsted *et al.*, 1970; Wang *et al.*, 1996; Zimmermann *et al.*, 1987). The Neuro-2a cell line has made a significant contribution in helping to delineate signaling in regulation of neurite outgrowth and neuronal differentiation, especially mediated by the $Ga_{i/o}$ pathway (He *et al.*, 2006; Ma'ayan *et al.*, 2009). Neuro-2a cells have also been used to study neurotoxicity (LePage *et al.*, 2005), Alzheimer's disease (Provost, 2010), prion-related diseases (Kaneko *et al.*, 1997), and asymmetric division of mammalian cell lines (Ogrodnik *et al.*, 2014).

B104 cell line (rat neuroblastoma)

The B104 cell line was cloned from a rat neuroblastoma that was induced transplacentally with nitrosomethylurea injected 15 d after conception and harvested when the animal reached adulthood (Schubert *et al.*, 1974). Morphologically B104 cells are very long, thin, and usually bipolar (Schubert *et al.*, 1974). The B104 cell line exhibits many of the properties characteristic of differentiated neurons, such as generation of action potentials, synthesis of neurotransmitters such as ACh and GABA, and the presence of neurotransmitter receptors, neuron-specific 14-3-2 protein, and microtubule-associated proteins MAP-2 and MAP-3 (Huber *et al.*, 1985; Izant and McIntosh, 1980; Schubert *et al.*, 1975; Schubert *et al.*, 1974). The B104 cell line also expresses insulin-like growth factor (IGF) receptor I and II (Sturm *et al.*, 1989), and IGF-I is more potent than insulin at stimulating B104 cell replication in serum-free medium (Orlowski *et al.*, 1989). Furthermore, B104 cells, like the N1E-115 cell line, respond to removal of serum by rapidly extending neurites, a phenomenon that has been correlated with the preceding neuronal properties (Schubert *et al.*, 1974). When B104 cells were grown in the presence of 1 mM dibutyryl cAMP, they underwent morphological alterations that mimic neuronal differentiation (Schubert *et al.*, 1975). In cell culture, basic fibroblast growth factor and platelet-derived growth factor AA and BB were potent mitogens for B104 cells, because they promoted an increase in cell number even when the cells were grown in serum-free medium (Luo and Miller,

1997), while transforming growth factor-β1 inhibited the proliferation of B104 cells (Luo and Miller, 1999). All these properties make the B104 cell line suitable for studying regulation mechanisms of neuronal proliferation, growth and differentiation. The B104 cell line has also been studied for neurotoxicity and apoptosis (Heese *et al.*, 2000; Park *et al.*, 2001).

TE671 cell line (human medulloblastoma)

TE671 is a human cell line derived from a cerebellar tumor of a 6-year-old Caucasian girl (McAllister *et al.*, 1977), and appears to be composed of six different morphological types of cells (types I–VI) in varying percentages in culture (Zeltzer *et al.*, 1984). TE671 cells possess high glutamic acid decarboxylase activity, nicotinic ACh receptors, dopaminergic D_1 and D_3 receptors, and glutamatergic receptors, voltage-dependent Na^+- and K^+-channels, small conductance Ca^{2+}-activated K^+-channel SK3, and are capable of generating Na^+-dependent action potentials (Carignani *et al.*, 2002; Hamel *et al.*, 1981; Levavi-Sivan *et al.*, 1998; Stepulak *et al.*, 2009; Syapin *et al.*, 1982; Toral *et al.*, 1995). TE671 cells express neuron-specific enolase, gonadotropin-releasing hormone, and epidermal growth factor receptors, and are negative for the glial marker glial fibrillary acid protein (GFAP) (Hall *et al.*, 1990; Zeltzer *et al.*, 1984). Since medulloblastoma is one of the most common brain tumors of childhood, the TE671 cell line has become one of the *in vitro* experimental models to study neoplastic cells of brain tumors and therapeutic applications such as chemo-, radiation- and photodynamic therapy (Houchens *et al.*, 1983; Lipshutz *et al.*, 1994; Merlin *et al.*, 1991). At the cellular and molecular level, the TE671 cell line has been mostly studied for the expression, function, pharmacology and regulation of nicotinic ACh receptors, and because of the possession of functional nicotinic ACh receptors, has been used as a model for *in vitro* testing of potential therapeutic compounds against myasthenia gravis (Voltz *et al.*, 1991). The TE671 cell line has also been employed in pharmacology studies on ATP-sensitive K^+ channels (Miller *et al.*, 1999).

HCN-2 cell line (human encephalitis)

The HCN (human cortical neuron)-2 cell line was derived from cortical tissue removed from a 7-year-old girl undergoing a hemispherectomy for intractable seizures associated with chronic focal encephalitis (Ronnett *et al.*, 1994). HCN-2 cells can be induced to differentiate when cultured with NGF, dibutyryl cAMP, 1-isobutyl-3-methylxanthine, or phorbol ester (Ronnett *et al.*, 1994), and differentiated HCN-2 cells have a neuron-like morphology. The cells stain positively for a number of neuronal markers including neurofilament proteins and neuron-specific enolase. HCN-2 cells are also positive for tubulin, vimentin, somatostatin, glutamate, GABA, cholecystokinin-8, methionine enkephalin and vasoactive intestinal peptide, while they are negative for glial markers such as GFAP, S-100, and myelin basis protein (Ronnett *et al.*, 1994). The HCN-2 cell line has been proposed as a model for amyloid β neurotoxicity as exposure of differentiated HCN-2 cells to amyloid β can cause cell death (Zhang *et al.*, 1994). Using the HCN-2 cell line as a model, effects of oxidative stress on major cytoskeleton filaments, microfilaments, microtubule, and vimentin, which are thought to play an important role in neuronal growth and neurite outgrowth, have also been examined (Allani *et al.*, 2004). Because the HCN-2 cell line was created much later than other neuronal cell lines of neuroblastoma or medulloblastoma origins, it has yet to gain similar popularity and utilization.

Cell Lines of Glial Origin

U87MG cell line (human glioma)

The U87MG cell line was derived from a stage-IV glioma of a 44-year-old Caucasian man (Ponten and Macintyre, 1968). U87MG cells have epithelial morphology and express membrane receptors for several neurotransmitters including various glutamatergic subtype receptors NMDA- (NR2A, NR2B, NR2C, NR3A and NR3B), AMPA- (GluR2, GluR3, GluR4, GluR6 and GluR7), kainite- (KA1 and KA2), and metabotropic (mGluR1, mGluR2, mGluR3, mGluR5, mGluR6, mGluR7 and mGluR8) subtype receptors (Arcella *et al.*, 2005; Stepulak *et al.*, 2009; Yoshida *et al.*, 2006), as well as adrenergic β_1- and β_2-receptors (Sardi *et al.*, 2013) and muscarinic M_2 receptor (Ferretti *et al.*, 2013). U87MG cells also express plasma membrane receptors for basic fibroblast growth factor (Murphy and Knee, 1995), interleukin-1 and -6 (Goswami *et al.*, 1998; Gottschall *et al.*, 1991), growth hormone (Castro *et al.*, 2000), semaphorin (Rieger *et al.*, 2003), platelet-derived growth factor (Gross *et al.*, 2006), and cholecystokinin-B (Oikonomou *et al.*, 2008), as well as voltage-gated K^+-channels (Ru *et al.*, 2015). U87MG cells possess the glial marker GFAP (Ito *et al.*, 1989), synthesize and release some important macro-molecules such as granulocyte colony-stimulating factor (Tweardy *et al.*, 1987), basic fibroblast growth factor (Sato *et al.*, 1989), glial cell line-derived neurotrophic factor (Verity *et al.*, 1999), and TNF receptor-associated factor 6 (Peng *et al.*, 2013). U87MG cells also produce tenascin (Ventimiglia *et al.*, 1992), type VI collagen (Han *et al.*, 1994), interleukin-6 (Goswami *et al.*, 1998), somatostatin (Hirota *et al.*, 1998), class 3 semaphorins (Rieger *et al.*, 2003), platelet-derived growth factor (Gross *et al.*, 2006), and cholecystokinin (Oikonomou *et al.*, 2008). The entire genome of the U87MG cell line has been sequenced (Clark *et al.*, 2010). Activation of NMDA receptors increased proliferation of the U87MG cell line (Ramaswamy *et al.*, 2014), whereas TNF-α inhibits U87MG cell growth (Chen *et al.*, 1993). Differentiation of U87MG cells could be induced by treatment with dibutyryl cAMP (Hoffman *et al.*, 1994) or retinoid (Das *et al.*, 2009). Transplantation studies suggest that transforming growth factor-$\beta2$ promotes the invasion of U87MG cells (Wick *et al.*, 2001). The U87MG cell line also has platelet-activating activity (Bastida *et al.*, 1987). Different types of matrix metalloproteinases were detected in U87MG cells (Maruiwa *et al.*, 1993; Matsuzawa *et al.*, 1996), and modulation of matrix metalloproteinase activity has been shown to play an important role in U87MG cell proliferation, differentiation, and invasion (Ramaswamy *et al.*, 2014; Wick *et al.*, 2001). For this reason, matrix metalloproteinase has become a major target for anti-glioma therapy. The U87MG cell line has been used extensively as a model for studying glioma, including assessing drug toxicity and anti-tumor efficiency, cell cycle and apoptosis, glioma invasion, and vascularization. To date there are well over 1000 published papers utilizing this cell line for glioma research, both *in vitro* and xenografts. Experimental therapies have been conducted in animals using radiation-, chemo-, photodynamic, gene-, and immune-therapy, as well as caloric restriction, hypoxia, and hyperthermia.

C6 cell line (rat glioma)

The glial cell strain C6 was cloned from an outbred Wistar-Furth rat glial tumor induced by repeated injection of N-methylnitrosourea (Benda *et al.*, 1968). The C6 cell line expresses S-100 protein, GFAP, vimentin, and somatotrophin, and produces glyceryl phosphate dehydrogenase in response to glucocorticoids. The cell line has been extensively used as a brain tumor model in experimental neuro-oncology (Barth and Kaur, 2009; Grobben *et al.*, 2002). However, a serious side effect of the C6 cell line is the potential of evoking immune-response in host animals, thereby limiting its applications, especially in brain tumor survival studies. Despite this limitation the C6 cell line has provided a wealth of information on biological properties of brain tumors, such as effects of growth factors, extracellular matrix components, proteases, adhesion molecules, and secretion of tumor-derived factors (Grobben *et al.*, 2002). Molecular characterizations which compared changes in gene expression patterns between the C6 glioma and rat stem cell-derived astrocytes revealed that the changes in gene expressions observed in the C6 cell line were the most similar to those reported in human brain tumors (Sibenaller *et al.*, 2005). C6 cells have cancer stem cell-like characteristics, including self-renewal, the potential for multi-lineage differentiation *in vitro* and tumor formation *in vivo* (Shen *et al.*, 2008). The C6 cell line has also been used extensively for studies of glioblastoma growth, invasion, migration, blood-brain barrier disruption, capillary permeability, and neovascularization, and for evaluating new therapeutic modalities such as chemo-, radiation-, photodynamic, anti-angiogenic, oncolytic viral and gene-therapies, and treatments with proteasome inhibitors and various toxins (Barth and Kaur, 2009).

F98 and RG2 cell lines (rat glioma)

The F98 and RG2 cell lines were produced in the same laboratory in 1971 (Ko *et al.*, 1980). A pregnant Fischer rat on day 20 of gestation was inoculated with a single 50 mg/kg dose of N-ethyl-N-nitrosourea. The tumors were harvested and cloned *in vitro*. One of them was designated F98, and another RG2 (rat glioma 2) or RG-2 (the same clone was also called D74-RG2 or D74). The biological characteristics of both tumors closely resemble those of human glioblastoma, and have been used extensively for both *in vitro* and *in vivo* studies of a rat brain tumor. While both cell lines express vimentin, one difference between the two cell lines is that the F98 cell line demonstrated only a fraction of GFAP positive cells, whereas the RG2 cell line exhibited virtually no GFAP immunoreactivity in culture (Mathieu *et al.*, 2007; Reifenberger *et al.*, 1989). Another difference between the two cell lines is that F98 cells are weakly immunogenic while RG2 cells are non-immunogenic in syngeneic rats. Thus, both cell lines are very attractive models to investigate the mechanisms underlying glioma resistance to immunotherapy (Barth and Kaur, 2009). F98 and RG2 cell lines have been used for a variety of *in vivo* transplantation studies including vascular permeability, regional blood flow, tumor metabolism, tumor growth, and blood brain barrier disruption, anti-angiogenic, chemo-, radiation-, and gene-therapy (Barth and Kaur, 2009).

Other Neural Cell Lines

NG108-15 cell line (mouse/rat neuroblastoma-glioblastoma hybrid)

The NG108-15 cell line, originally named 108CC15, was developed by fusing mouse N18TG2 neuroblastoma cells with rat C6-BU-1 glioma cells in the presence of inactivated Sendai virus (Hamprecht, 1977). These cells display a considerable range of neuronal features and possess many of the functions of differentiated neurons, including extension of long processes, clear and dense core vesicles, excitable membranes, formation of functional synapses, express neurotransmitter synthesizing enzymes such as choline acetyltransferase and dopamine-β-hydroxylase, uptake systems for catecholamines, depolarization-induced Ca^{2+}-dependent release of ACh, and expression of transmitter receptors (Hamprecht *et al.*, 1985). NG108-15 cells have been widely used as a model system to examine transmembrane signaling processes in the nervous system. These cells are particularly suitable for such studies as they express a considerable range of receptors which can be demonstrated to couple to a variety of effector systems. The transmembrane signaling systems which have been examined in greatest detail in NG108-15 cells are stimulation and inhibition of adenylate cyclase (Klee *et al.*, 1985), activation of phosphoinositidase C, and regulation of voltage-sensitive Ca^{2+} channels (Tsunoo *et al.*, 1986). NG108-15 cells contain opioid receptors and adrenergic α_2 receptors (Klee and Nirenberg, 1974). Opiates initially inhibit adenylate cyclase activity in these cells, but after a period of adaptation the levels of cAMP return to normal (Sharma *et al.*, 1975). The cells are desensitized to the acute effects of opiates (tolerance) and if they are withdrawn, or an opioid antagonist is added, the cells respond with an overshoot of adenylate cyclase activity (Lee *et al.*, 1988). The adenylate cyclase overshoot response after chronic exposure to opiates has been proposed as a biochemical model for opiate dependence (Sharma *et al.*, 1975). Adaptation mechanisms similar to those for opiates have been described for α_2 receptor agonists in NG108-15 cells (Sabol and Nirenberg, 1979). NG108-15 cells also express bradykinin receptors and an external application of bradykinin produces a biphasic change in membrane potential (a hyperpolarization followed by a depolarization) accompanied by an increase in ACh release (Reiser and Hamprecht, 1982). This cell model allowed examination of physiological and biochemical consequences in the receptor-mediated regulation of neurotransmitter release at a single cell level (Higashida and Ogura, 1991). NG108-15 cells express several guanine nucleotide binding proteins (G-proteins) and a variety of heterotrimeric, plasma membrane-associated G-proteins coupled receptors, including G_s, G_i2, G_i3, and at least two variants of G_0, which allow the conversion of extracellular signals into intracellular responses (Milligan *et al.*, 1990). Overall, the NG108-15 cell line has been mostly used for studying expression and function of cross-membrane and intracellular signaling pathways.

CATH.a cell line (transgenic mouse)

The CATH.a cell line was established from cultures of a tumor that arose in the brain of a transgenic mouse carrying the SV-40 T antigen oncogene (Suri *et al.*, 1993). Although the CATH.a cell line continues to express T antigen, the cells exhibit many properties of locus coeruleus neurons. The cells express the catecholaminergic biosynthetic enzymes tyrosine hydroxylase and dopamine β-hydroxylase, neurofilaments, and synaptophysin (Suri *et al.*, 1993). The cell line possesses kappa- and delta-opioid receptors and responds to opioids by suppression of voltage-activated K^+-current (Baraban *et al.*, 1995; Bouvier *et al.*, 1998). Interestingly, dopamine produces a dose- and time-dependent increase in cell death in the CATH.a cell line, and the cell death is not mediated through dopaminergic receptors since selective receptor agonists have no effect on CATH.a cell viability (Masserano *et al.*, 1996). Brain-derived neurotrophic factor and glia cell line-derived neurotrophic factor reduced dopamine-induced cell death, while NGF, basic fibroblast growth factor, neurotrophin-4/5 and insulin had no protective effect on dopamine-induced cell death (Gong *et al.*, 1999). Since agents that activate the cAMP pathway lead to a decrease in expression of the cAMP response element-binding protein (CREB) mRNA in CATH.a cells, the cell line has made significant contributions in studying regulation mechanisms of CREB expression (Coven *et al.*, 1998; Widnell *et al.*, 1996; Widnell *et al.*, 1994).

Concluding Remarks

Available neural cell lines have provided relatively simple and well-controlled systems for elucidating basic biological processes and mechanisms governing neural growth, differentiation, signal transduction, and cell death. In addition, it can be concluded that all neural cell lines have been utilized in pharmacological and toxicological studies. Both *in vivo* transplantation and *in vitro* experiments further helped therapeutic development and evaluation for disease treatment. Despite the success and achievements of various cell lines, it remains essential to recognize their limitations, which include but are not limited to: (1) tumor origin and continued expression of oncogenes, which may have changed many cellular and molecular properties; (2) artificial media and growth conditions, which may have influenced signal transduction pathways and gene expression patterns; (3) mono-type of cells in the culture dish, which may lack certain features that require interactions with other types of cells; and (4) cells in a 2-dimensional mono-layer, which may behave differently than in a 3-dimensional environment. Looking forward it is anticipated that more genetically engineered cell lines will be developed that more closely represent neurons and glial cells, respectively, and the expression of particular genes can be artificially controlled. Co-cultures of multi-types of cells and 3-dimensional cell cultures will further mimic *in vivo* situations. Their applications could lead to a better understanding of molecular mechanisms underlying biological processes such as growth, differentiation, signal transduction and cell death, as well as more effective therapeutic treatments for brain tumors, neurological disorders and diseases beyond the nervous system.

Acknowledgements

The author's work cited in this chapter was supported by the U.S. Food and Drug Administration (FDA), National Center for Toxicological Research protocol numbers E746001 and E752401. This document has been reviewed in accordance with FDA policy and approved for publication. Approval does not signify that the contents necessarily reflect the position or opinions of the FDA nor does mention of trade names or commercial products constitute endorsement or recommendation for use. The findings and conclusions in this report are those of the authors and do not necessarily represent the views of the FDA.

References

Adem, A., M.E. Mattsson, A. Nordberg, and S. Pahlman. 1987. Muscarinic receptors in human SH-SY5Y neuroblastoma cell line: regulation by phorbol ester and retinoic acid-induced differentiation. Brain Res 430: 235–242.

Allani, P.K., T. Sum, S.G. Bhansali, S.K. Mukherjee, and M. Sonee. 2004. A comparative study of the effect of oxidative stress on the cytoskeleton in human cortical neurons. Toxicol Appl Pharmacol 196: 29–36.

Amano, T., E. Richelson, and M. Nirenberg. 1972. Neurotransmitter synthesis by neuroblastoma clones (neuroblast differentiation-cell culture-choline acetyltransferase-acetylcholinesterase-tyrosine hydroxylase-axons-dendrites). Proc Natl Acad Sci U S A 69: 258–263.

Arcella, A., G. Carpinelli, G. Battaglia, M. D'Onofrio, F. Santoro, R.T. Ngomba, V. Bruno, P. Casolini, F. Giangaspero, and F. Nicoletti. 2005. Pharmacological blockade of group II metabotropic glutamate receptors reduces the growth of glioma cells *in vivo*. Neuro Oncol 7: 236–245.

Arner, L.S., and W.B. Stallcup. 1981. Two types of potassium channels in the PC12 cell line. Brain Res 215: 419–425.

Baraban, S.C., E.W. Lothman, A. Lee, and P.G. Guyenet. 1995. Kappa opioid receptor-mediated suppression of voltage-activated potassium current in a catecholaminergic neuronal cell line. J Pharmacol Exp Ther 273: 927–933.

Barth, R.F., and B. Kaur. 2009. Rat brain tumor models in experimental neuro-oncology: the C6, 9L, T9, RG2, F98, BT4C, RT-2 and CNS-1 gliomas. J Neurooncol 94: 299–312.

Bastida, E., L. Almirall, G.A. Jamieson, and A. Ordinas. 1987. Cell surface sialylation of two human tumor cell lines and its correlation with their platelet-activating activity. Cancer Res 47: 1767–1770.

Benda, P., J. Lightbody, G. Sato, L. Levine, and W. Sweet. 1968. Differentiated rat glial cell strain in tissue culture. Science 161: 370–371.

Biedler, J.L., S. Roffler-Tarlov, M. Schachner, and L.S. Freedman. 1978. Multiple neurotransmitter synthesis by human neuroblastoma cell lines and clones. Cancer Res 38: 3751–3757.

Black, M.M., J.M. Aletta, and L.A. Greene. 1986. Regulation of microtubule composition and stability during nerve growth factor-promoted neurite outgrowth. J Cell Biol 103: 545–557.

Bordt, S.L., R.M. McKeon, P.K. Li, P.A. Witt-Enderby, and M.A. Melan. 2001. N1E-115 mouse neuroblastoma cells express MT1 melatonin receptors and produce neurites in response to melatonin. Biochim Biophys Acta 1499: 257–264.

Bouvier, C., D. Avram, V.J. Peterson, B. Hettinger, K. Soderstrom, T.F. Murray, and M. Leid. 1998. Catecholaminergic CATH.a cells express predominantly delta-opioid receptors. Eur J Pharmacol 348: 85–93.

Brat, D.J., and S. Brimijoin. 1992. A paradigm for examining toxicant effects on viability, structure, and axonal transport of neurons in culture. Mol Neurobiol 6: 125–135.

Carignani, C., R. Roncarati, R. Rimini, and G.C. Terstappen. 2002. Pharmacological and molecular characterisation of SK3 channels in the TE671 human medulloblastoma cell line. Brain Res 939: 11–18.

Castro, J.R., J.A. Costoya, R. Gallego, A. Prieto, V.M. Arce, and R. Senaris. 2000. Expression of growth hormone receptor in the human brain. Neurosci Lett 281: 147–150.

Chen, T.C., D.R. Hinton, M.L. Apuzzo, and F.M. Hofman. 1993. Differential effects of tumor necrosis factor-alpha on proliferation, cell surface antigen expression, and cytokine interactions in malignant gliomas. Neurosurgery 32: 85–94.

Clark, M.J., N. Homer, B.D. O'Connor, Z. Chen, A. Eskin, H. Lee, B. Merriman, and S.F. Nelson. 2010. U87MG decoded: the genomic sequence of a cytogenetically aberrant human cancer cell line. PLoS Genet 6: e1000832.

Coven, E., Y. Ni, K.L. Widnell, J. Chen, W.H. Walker, J.F. Habener, and E.J. Nestler. 1998. Cell type-specific regulation of CREB gene expression: mutational analysis of CREB promoter activity. J Neurochem 71: 1865–1874.

D'Alessandro, R., and J. Meldolesi. 2010. In PC12 cells, expression of neurosecretion and neurite outgrowth are governed by the transcription repressor REST/NRSF. Cell Mol Neurobiol 30: 1295–1302.

Das, A., N.L. Banik, and S.K. Ray. 2009. Molecular mechanisms of the combination of retinoid and interferon-gamma for inducing differentiation and increasing apoptosis in human glioblastoma T98G and U87MG cells. Neurochem Res 34: 87–101.

Dichter, M.A., A.S. Tischler, and L.A. Greene. 1977. Nerve growth factor-induced increase in electrical excitability and acetylcholine sensitivity of a rat pheochromocytoma cell line. Nature 268: 501–504.

Drubin, D.G., S.C. Feinstein, E.M. Shooter, and M.W. Kirschner. 1985. Nerve growth factor-induced neurite outgrowth in PC12 cells involves the coordinate induction of microtubule assembly and assembly-promoting factors. J Cell Biol 101: 1799–1807.

Edsjo, A., E. Lavenius, H. Nilsson, J.C. Hoehner, P. Simonsson, L.A. Culp, T. Martinsson, C. Larsson, and S. Pahlman. 2003. Expression of trkB in human neuroblastoma in relation to MYCN expression and retinoic acid treatment. Lab Invest 83: 813–823.

Encinas, M., M. Iglesias, Y. Liu, H. Wang, A. Muhaisen, V. Cena, C. Gallego, and J.X. Comella. 2000. Sequential treatment of SH-SY5Y cells with retinoic acid and brain-derived neurotrophic factor gives rise to fully differentiated, neurotrophic factor-dependent, human neuron-like cells. J Neurochem 75: 991–1003.

Fasano, M., T. Alberio, M. Colapinto, S. Mila, and L. Lopiano. 2008. Proteomics as a tool to investigate cell models for dopamine toxicity. Parkinsonism Relat Disord 14 Suppl 2: S135–138.

Ferretti, M., C. Fabbiano, M. Di Bari, C. Conte, E. Castigli, M. Sciaccaluga, D. Ponti, P. Ruggieri, A. Raco, R. Ricordy, A. Calogero, and A.M. Tata. 2013. M2 receptor activation inhibits cell cycle progression and survival in human glioblastoma cells. J Cell Mol Med 17: 552–566.

Forsythe, I.D., D.G. Lambert, S.R. Nahorski, and P. Lindsdell. 1992. Elevation of cytosolic calcium by cholinoceptor agonists in SH-SY5Y human neuroblastoma cells: estimation of the contribution of voltage-dependent currents. Br J Pharmacol 107: 207–214.

Fujita, K., P. Lazarovici, and G. Guroff. 1989. Regulation of the differentiation of PC12 pheochromocytoma cells. Environ Health Perspect 80: 127–142.

Gilbert, J.A., and E. Richelson. 1984. Neurotensin stimulates formation of cyclic GMP in murine neuroblastoma clone N1E-115. Eur J Pharmacol 99: 245–246.

Gilbert, J.A., M.A. Pfenning, and E. Richelson. 1984. The effect of angiotensins I, II, and III on formation of cyclic GMP in murine neuroblastoma clone N1E-115. Biochem Pharmacol 33: 2527–2530.

Gong, L., R.J. Wyatt, I. Baker, and J.M. Masserano. 1999. Brain-derived and glial cell line-derived neurotrophic factors protect a catecholaminergic cell line from dopamine-induced cell death. Neurosci Lett 263: 153–156.

Goswami, S., A. Gupta, and S.K. Sharma. 1998. Interleukin-6-mediated autocrine growth promotion in human glioblastoma multiforme cell line U87MG. J Neurochem 71: 1837–1845.

Gottschall, P.E., K. Koves, K. Mizuno, I. Tatsuno, and A. Arimura. 1991. Glucocorticoid upregulation of interleukin 1 receptor expression in a glioblastoma cell line. Am J Physiol 261: E362–368.

Gould, J., H.L. Reeve, P.F. Vaughan, and C. Peers. 1992. Nicotinic acetylcholine receptors in human neuroblastoma (SH-SY5Y) cells. Neurosci Lett 145: 201–204.

Greene, L.A., and A.S. Tischler. 1976. Establishment of a noradrenergic clonal line of rat adrenal pheochromocytoma cells which respond to nerve growth factor. Proc Natl Acad Sci U S A 73: 2424–2428.

Greene, L.A., and G. Rein. 1977. Release of (3H)norepinephrine from a clonal line of pheochromocytoma cells (PC12) by nicotinic cholinergic stimulation. Brain Res 138: 521–528.

Greene, L.A., R.K. Liem, and M.L. Shelanski. 1983. Regulation of a high molecular weight microtubule-associated protein in PC12 cells by nerve growth factor. J Cell Biol 96: 76–83.

Grimbly, C., B. Martin, E. Karpinski, and S. Harvey. 2009. Growth hormone production and action in N1E-115 neuroblastoma cells. J Mol Neurosci 39: 117–124.

Grobben, B., P.P. De Deyn, and H. Slegers. 2002. Rat C6 glioma as experimental model system for the study of glioblastoma growth and invasion. Cell Tissue Res 310: 257–270.

Gross, D., G. Bernhardt, and A. Buschauer. 2006. Platelet-derived growth factor receptor independent proliferation of human glioblastoma cells: selective tyrosine kinase inhibitors lack antiproliferative activity. J Cancer Res Clin Oncol 132: 589–599.

Gu, Q., S. Lantz-McPeak, H. Rosas-Hernandez, E. Cuevas, S.F. Ali, M.G. Paule, and S. Sarkar. 2014. *In vitro* detection of cytotoxicity using FluoroJade-C. Toxicol *In Vitro* 28: 469–472.

Hall, W.A., M.J. Merrill, S. Walbridge, and R.J. Youle. 1990. Epidermal growth factor receptors on ependymomas and other brain tumors. J Neurosurg 72: 641–646.

Hamel, E., I.E. Goetz, and E. Roberts. 1981. Glutamic acid decarboxylase and gamma-aminobutyric acid in Huntington's disease fibroblasts and other cultured cells, determined by a [3H]muscimol radioreceptor assay. J Neurochem 37: 1032–1038.

Hamprecht, B. 1977. Structural, electrophysiological, biochemical, and pharmacological properties of neuroblastoma-glioma cell hybrids in cell culture. Int Rev Cytol 49: 99–170.

Hamprecht, B., T. Glaser, G. Reiser, E. Bayer, and F. Propst. 1985. Culture and characteristics of hormone-responsive neuroblastoma X glioma hybrid cells. Methods Enzymol 109: 316–341.

Han, J., J.C. Daniel, N. Lieska, and G.D. Pappas. 1994. Immunofluorescence and biochemical studies of the type VI collagen expression by human glioblastoma cells *in vitro*. Neurol Res 16: 370–375.

He, J.C., S.R. Neves, J.D. Jordan, and R. Iyengar. 2006. Role of the Go/i signaling network in the regulation of neurite outgrowth. Can J Physiol Pharmacol 84: 687–694.

Heese, K., Y. Nagai, and T. Sawada. 2000. Induction of rat L-phosphoserine phosphatase by amyloid-beta (1-42) is inhibited by interleukin-11. Neurosci Lett 288: 37–40.

Higashida, H., and A. Ogura. 1991. Inositol trisphosphate/calcium-dependent acetylcholine release evoked by bradykinin in NG108-15 rodent hybrid cells. Ann N Y Acad Sci 635: 153–166.

Hirota, N., K. Matsumoto, M. Iida, H. Sakagami, and M. Takeda. 1998. Expression of somatostatin messenger RNA and receptor in cultured brain tumor cells. Anticancer Res 18: 3295–3297.

Hoffman, L.M., S.E. Brooks, M.R. Stein, and L. Schneck. 1994. Cyclic AMP causes differentiation and decreases the expression of neutral glycosphingolipids in cell cultures derived from a malignant glioma. Biochim Biophys Acta 1222: 37–44.

Houchens, D.P., A.A. Ovejera, S.M. Riblet, and D.E. Slagel. 1983. Human brain tumor xenografts in nude mice as a chemotherapy model. Eur J Cancer Clin Oncol 19: 799–805.

Huber, G., D. Alaimo-Beuret, and A. Matus. 1985. MAP3: characterization of a novel microtubule-associated protein. J Cell Biol 100: 496–507.

Ito, M., T. Nagashima, and T. Hoshino. 1989. Quantitation and distribution analysis of glial fibrillary acidic protein in human glioma cells in culture. J Neuropathol Exp Neurol 48: 560–567.

Izant, J.G., and J.R. McIntosh. 1980. Microtubule-associated proteins: a monoclonal antibody to MAP2 binds to differentiated neurons. Proc Natl Acad Sci U S A 77: 4741–4745.

Jumblatt, J.E., and A.S. Tischler. 1982. Regulation of muscarinic ligand binding sites by nerve growth factor in PC12 phaeochromocytoma cells. Nature 297: 152–154.

Kaneko, K., L. Zulianello, M. Scott, C.M. Cooper, A.C. Wallace, T.L. James, F.E. Cohen, and S.B. Prusiner. 1997. Evidence for protein X binding to a discontinuous epitope on the cellular prion protein during scrapie prion propagation. Proc Natl Acad Sci U S A 94: 10069–10074.

Kennedy, C., and G. Henderson. 1992. Chronic exposure to morphine does not induce dependence at the level of the calcium channel current in human SH-SY5Y cells. Neuroscience 49: 937–944.

Klebe, R.J., and F.H. Ruddle. 1969. Neuroblastoma: Cell culture analysis of a differentiating stem cell system. J Cell Biol 43: 69A.

Klee, W.A., and M. Nirenberg. 1974. A neuroblastoma times glioma hybrid cell line with morphine receptors. Proc Natl Acad Sci U S A 71: 3474–3477.

Klee, W.A., G. Milligan, W.F. Simonds, and B. Tocque. 1985. Opiate receptors in neuroblastoma x glioma hybrid cell lines: a system for investigating N,/N, interactions. Mol Asp Cell Regul 4: 117–129.

Ko, L., A. Koestner, and W. Wechsler. 1980. Morphological characterization of nitrosourea-induced glioma cell lines and clones. Acta Neuropathol 51: 23–31.

Kontou, M., W. Weidemann, K. Bork, and R. Horstkorte. 2009. Beyond glycosylation: sialic acid precursors act as signaling molecules and are involved in cellular control of differentiation of PC12 cells. Biol Chem 390: 575–579.

Lambert, D.G., A.S. Ghataorre, and S.R. Nahorski. 1989a. Muscarinic receptor binding characteristics of a human neuroblastoma SK-N-SH and its clones SH-SY5Y and SH-EP1. Eur J Pharmacol 165: 71–77.

Lambert, J.J., J.A. Peters, T.G. Hales, and J. Dempster. 1989b. The properties of 5-HT3 receptors in clonal cell lines studied by patch-clamp techniques. Br J Pharmacol 97: 27–40.

Larsson, K.P., A.J. Hansen, and S. Dissing. 2002. The human SH-SY5Y neuroblastoma cell-line expresses a functional P2X7 purinoceptor that modulates voltage-dependent Ca2+ channel function. J Neurochem 83: 285–298.

Lee, S., C.R. Rosenberg, and J.M. Musacchio. 1988. Cross-dependence to opioid and alpha 2-adrenergic receptor agonists in NG108-15 cells. FASEB J 2: 52–55.

Lee, V.M., and C. Page. 1984. The dynamics of nerve growth factor-induced neurofilament and vimentin filament expression and organization in PC12 cells. J Neurosci 4: 1705–1714.

Lenz, W., C. Petrusch, K.H. Jakobs, and C.J. van Koppen. 1994. Agonist-induced down-regulation of the m4 muscarinic acetylcholine receptor occurs without changes in receptor mRNA steady-state levels. Naunyn Schmiedebergs Arch Pharmacol 350: 507–513.

LePage, K.T., R.W. Dickey, W.H. Gerwick, E.L. Jester, and T.F. Murray. 2005. On the use of neuro-2a neuroblastoma cells versus intact neurons in primary culture for neurotoxicity studies. Crit Rev Neurobiol 17: 27–50.

Levavi-Sivan, B., B.H. Park, S. Fuchs, and C.S. Fishburn. 1998. Human D3 dopamine receptor in the medulloblastoma TE671 cell line: cross-talk between D1 and D3 receptors. FEBS Lett 439: 138–142.

Levi, A., S. Biocca, A. Cattaneo, and P. Calissano. 1988. The mode of action of nerve growth factor in PC12 cells. Mol Neurobiol 2: 201–226.

Li, S.H., A.L. Cheng, H. Li, and X.J. Li. 1999. Cellular defects and altered gene expression in PC12 cells stably expressing mutant huntingtin. J Neurosci 19: 5159–5172.

Lipshutz, G.S., D.J. Castro, R.E. Saxton, R.P. Haugland, and J. Soudant. 1994. Evaluation of four new carbocyanine dyes for photodynamic therapy with lasers. Laryngoscope 104: 996–1002.

Luo, J., and M.W. Miller. 1997. Basic fibroblast growth factor- and platelet-derived growth factor-mediated cell proliferation in B104 neuroblastoma cells: effect of ethanol on cell cycle kinetics. Brain Res 770: 139–150.

Luo, J., and M.W. Miller. 1999. Transforming growth factor beta1-regulated cell proliferation and expression of neural cell adhesion molecule in B104 neuroblastoma cells: differential effects of ethanol. J Neurochem 72: 2286–2293.

Ma'ayan, A., S.L. Jenkins, A. Barash, and R. Iyengar. 2009. Neuro2A differentiation by Galphai/o pathway. Sci Signal 2:cm1.

Maruiwa, H., Y. Sasaguri, M. Shigemori, M. Hirohata, and M. Morimatsu. 1993. A role of matrix metalloproteinases produced by glioma-cells. Int J Oncol 3: 1083–1088.

Masserano, J.M., L. Gong, H. Kulaga, I. Baker, and R.J. Wyatt. 1996. Dopamine induces apoptotic cell death of a catecholaminergic cell line derived from the central nervous system. Mol Pharmacol 50: 1309–1315.

Mathieu, D., R. Lecomte, A.M. Tsanaclis, A. Larouche, and D. Fortin. 2007. Standardization and detailed characterization of the syngeneic Fischer/F98 glioma model. Can J Neurol Sci 34: 296–306.

Matsuzawa, K., K. Fukuyama, P.B. Dirks, S. Hubbard, M. Murakami, L.E. Becker, and J.T. Rutka. 1996. Expression of stromelysin 1 in human astrocytoma cell lines. J Neurooncol 30: 181–188.

McAllister, R.M., H. Isaacs, R. Rongey, M. Peer, W. Au, S.W. Soukup, and M.B. Gardner. 1977. Establishment of a human medulloblastoma cell line. Int J Cancer 20: 206–212.

McDonald, R.L., P.F. Vaughan, and C. Peers. 1994. Muscarinic (M1) receptor-mediated inhibition of K(+)-evoked [3H]-noradrenaline release from human neuroblastoma (SH-SY5Y) cells via inhibition of L- and N-type Ca2+ channels. Br J Pharmacol 113: 621–627.

McKinney, M., and R.W. Barrett. 1989. Biochemical evidence for somatostatin receptors in murine neuroblastoma clone N1E-115. Eur J Pharmacol 162: 397–405.

Merlin, J.L., P. Chastagner, C. Marchal, B. Weber, and P. Bey. 1991. *In vitro* combination of high dose busulfan with radiotherapy on medulloblastoma cells: additive effect without potentiation. Anticancer Drugs 2: 465–468.

Miller, T.R., R.D. Taber, E.J. Molinari, K.L. Whiteaker, L.M. Monteggia, V.E. Scott, J.D. Brioni, J.P. Sullivan, and M. Gopalakrishnan. 1999. Pharmacological and molecular characterization of ATP-sensitive K+ channels in the TE671 human medulloblastoma cell line. Eur J Pharmacol 370: 179–185.

Milligan, G., F.R. McKenzie, S.J. McClue, F.M. Mitchell, and I. Mullaney. 1990. Guanine nucleotide binding proteins in neuroblastoma x glioma hybrid, NG108-15, cells. Regulation of expression and function. Int J Biochem 22: 701–707.

Momboisse, F., S. Ory, M. Ceridono, V. Calco, N. Vitale, M.F. Bader, and S. Gasman. 2010. The Rho guanine nucleotide exchange factors Intersectin 1L and beta-Pix control calcium-regulated exocytosis in neuroendocrine PC12 cells. Cell Mol Neurobiol 30: 1327–1333.

Morton, A.J., C. Hammond, W.T. Mason, and G. Henderson. 1992. Characterisation of the L- and N-type calcium channels in differentiated SH-SY5Y neuroblastoma cells: calcium imaging and single channel recording. Brain Res Mol Brain Res 13: 53–61.

Murphy, M.G., and Z. Byczko. 1990. Effects of membrane polyunsaturated fatty acids on adenosine receptor function in intact N1E-115 neuroblastoma cells. Biochem Cell Biol 68: 392–395.

Murphy, P.R., and R.S. Knee. 1995. Basic fibroblast growth factor binding and processing by human glioma cells. Mol Cell Endocrinol 114: 193–203.

O'Lague, P.H., and S.L. Huttner. 1980. Physiological and morphological studies of rat pheochromocytoma cells (PC12) chemically fused and grown in culture. Proc Natl Acad Sci U S A 77: 1701–1705.

Ogrodnik, M., H. Salmonowicz, R. Brown, J. Turkowska, W. Sredniawa, S. Pattabiraman, T. Amen, A.C. Abraham, N. Eichler, R. Lyakhovetsky, and D. Kaganovich. 2014. Dynamic JUNQ inclusion bodies are asymmetrically inherited in mammalian cell lines through the asymmetric partitioning of vimentin. Proc Natl Acad Sci U S A 111: 8049–8054.

Oikonomou, E., M. Buchfelder, and E.F. Adams. 2008. Cholecystokinin (CCK) and CCK receptor expression by human gliomas: Evidence for an autocrine/paracrine stimulatory loop. Neuropeptides 42: 255–265.

Olmsted, J.B., K. Carlson, R. Klebe, F. Ruddle, and J. Rosenbaum. 1970. Isolation of microtubule protein from cultured mouse neuroblastoma cells. Proc Natl Acad Sci U S A 65: 129–136.

Oortgiesen, M., R.G. van Kleef, and H.P. Vijverberg. 1997. Dual, non-competitive interaction of lead with neuronal nicotinic acetylcholine receptors in N1E-115 neuroblastoma cells. Brain Res 747: 1–8.

Orlowski, C.C., S.D. Chernausek, and R. Akeson. 1989. Actions of insulin-like growth factor-I on the B104 neuronal cell line: effects on cell replication, receptor characteristics, and influence of secreted binding protein on ligand binding. J Cell Physiol 139: 469–476.

Park, S.A., H.J. Park, B.I. Lee, Y.H. Ahn, S.U. Kim, and K.S. Choi. 2001. Bcl-2 blocks cisplatin-induced apoptosis by suppression of ERK-mediated p53 accumulation in B104 cells. Brain Res Mol Brain Res 93: 18–26.

Peng, Z., Y. Shuangzhu, J. Yongjie, Z. Xinjun, and L. Ying. 2013. TNF receptor-associated factor 6 regulates proliferation, apoptosis, and invasion of glioma cells. Mol Cell Biochem 377: 87–96.

Ponten, J., and E.H. Macintyre. 1968. Long term culture of normal and neoplastic human glia. Acta Pathol Microbiol Scand 74: 465–486.

Provost, P. 2010. Interpretation and applicability of microRNA data to the context of Alzheimer's and age-related diseases. Aging (Albany NY) 2: 166–169.

Ramaswamy, P., N. Aditi Devi, K. Hurmath Fathima, and N. Dalavaikodihalli Nanjaiah. 2014. Activation of NMDA receptor of glutamate influences MMP-2 activity and proliferation of glioma cells. Neurol Sci 35: 823–829.

Reeve, H.L., P.F. Vaughan, and C. Peers. 1992. Glibenclamide inhibits a voltage-gated K+ current in the human neuroblastoma cell line SH-SY5Y. Neurosci Lett 135: 37–40.

Reeve, H.L., P.F. Vaughan, and C. Peers. 1994. Calcium channel currents in undifferentiated human neuroblastoma (SH-SY5Y) cells: actions and possible interactions of dihydropyridines and omega-conotoxin. Eur J Neurosci 6: 943–952.

Reifenberger, G., T. Bilzer, R.J. Seitz, and W. Wechsler. 1989. Expression of vimentin and glial fibrillary acidic protein in ethylnitrosourea-induced rat gliomas and glioma cell lines. Acta Neuropathol 78: 270–282.

Reiser, G., and B. Hamprecht. 1982. Bradykinin induces hyperpolarizations in rat glioma cells and in neuroblastoma X glioma hybrid cells. Brain Res 239: 191–199.

Richelson, E. 1990. The use of cultured cells in the study of mood-normalizing drugs. Pharmacol Toxicol 66 Suppl 3: 69–75.

Rieger, J., W. Wick, and M. Weller. 2003. Human malignant glioma cells express semaphorins and their receptors, neuropilins and plexins. Glia 42: 379–389.

Ronnett, G.V., L.D. Hester, J.S. Nye, and S.H. Snyder. 1994. Human cerebral cortical cell lines from patients with unilateral megalencephaly and Rasmussen's encephalitis. Neuroscience 63: 1081–1099.

Roth, J.A., C. Horbinski, D. Higgins, P. Lein, and M.D. Garrick. 2002. Mechanisms of manganese-induced rat pheochromocytoma (PC12) cell death and cell differentiation. Neurotoxicology 23: 147–157.

Ru, Q., X. Tian, M.S. Pi, L. Chen, K. Yue, Q. Xiong, B.M. Ma, and C.Y. Li. 2015. Voltagegated K+ channel blocker quinidine inhibits proliferation and induces apoptosis by regulating expression of microRNAs in human glioma U87MG cells. Int J Oncol 46: 833–840.

Sabol, S.L., and M. Nirenberg. 1979. Regulation of adenylate cyclase of neuroblastoma x glioma hybrid cells by alpha-adrenergic receptors. I. Inhibition of adenylate cyclase mediated by alpha receptors. J Biol Chem 254: 1913–1920.

Sardi, I., L. Giunti, C. Bresci, A.M. Buccoliero, D. Degl'innocenti, S. Cardellicchio, G. Baroni, F. Castiglione, M.D. Ros, P. Fiorini, S. Giglio, L. Genitori, M. Arico, and L. Filippi. 2013. Expression of beta-adrenergic receptors in pediatric malignant brain tumors. Oncol Lett 5: 221–225.

Sato, Y., P.R. Murphy, R. Sato, and H.G. Friesen. 1989. Fibroblast growth factor release by bovine endothelial cells and human astrocytoma cells in culture is density dependent. Mol Endocrinol 3: 744–748.

Schrier, S.M., E.W. van Tilburg, H. van der Meulen, A.P. Ijzerman, G.J. Mulder, and J.F. Nagelkerke. 2001. Extracellular adenosine-induced apoptosis in mouse neuroblastoma cells: studies on involvement of adenosine receptors and adenosine uptake. Biochem Pharmacol 61: 417–425.

Schrier, S.M., B.I. Florea, G.J. Mulder, J.F. Nagelkerke, and I.J. AP. 2002. Apoptosis induced by extracellular ATP in the mouse neuroblastoma cell line N1E-115: studies on involvement of P2 receptors and adenosine. Biochem Pharmacol 63: 1119–1126.

Schubert, D., S. Heinemann, W. Carlisle, H. Tarikas, B. Kimes, J. Patrick, J.H. Steinbach, W. Culp, and B.L. Brandt. 1974. Clonal cell lines from the rat central nervous system. Nature 249: 224–227.

Schubert, D., W. Carlisle, and C. Look. 1975. Putative neurotransmitters in clonal cell lines. Nature 254: 341–343.

Schubert, D., S. Heinemann, and Y. Kidokoro. 1977. Cholinergic metabolism and synapse formation by a rat nerve cell line. Proc Natl Acad Sci U S A 74: 2579–2583.

Seward, E., C. Hammond, and G. Henderson. 1991. Mu-opioid-receptor-mediated inhibition of the N-type calcium-channel current. Proc Biol Sci 244: 129–135.

Sharma, S.K., W.A. Klee, and M. Nirenberg. 1975. Dual regulation of adenylate cyclase accounts for narcotic dependence and tolerance. Proc Natl Acad Sci U S A 72: 3092–3096.

Shen, G., F. Shen, Z. Shi, W. Liu, W. Hu, X. Zheng, L. Wen, and X. Yang. 2008. Identification of cancer stem-like cells in the C6 glioma cell line and the limitation of current identification methods. *In Vitro* Cell Dev Biol Anim 44: 280–289.

Sibenaller, Z.A., A.B. Etame, M.M. Ali, M. Barua, T.A. Braun, T.L. Casavant, and T.C. Ryken. 2005. Genetic characterization of commonly used glioma cell lines in the rat animal model system. Neurosurg Focus 19: E1.

Sipe, K.J., D. Srisawasdi, R. Dantzer, K.W. Kelley, and J.A. Weyhenmeyer. 1996. An endogenous 55 kDa TNF receptor mediates cell death in a neural cell line. Brain Res Mol Brain Res 38: 222–232.

Snider, R.M., and E. Richelson. 1984. Bradykinin receptor-mediated cyclic GMP formation in a nerve cell population (murine neuroblastoma clone N1E-115). J Neurochem 43: 1749–1754.

Sombers, L.A., T.L. Colliver, and A.G. Ewing. 2002. Differentiated PC12 cells: a better model system for the study of the VMAT's effects on neuronal communication. Ann N Y Acad Sci 971: 86–88.

Spicer, Z., and D.E. Millhorn. 2003. Oxygen sensing in neuroendocrine cells and other cell types: pheochromocytoma (PC12) cells as an experimental model. Endocr Pathol 14: 277–291.

Stepulak, A., H. Luksch, C. Gebhardt, O. Uckermann, J. Marzahn, M. Sifringer, W. Rzeski, C. Staufner, K.S. Brocke, L. Turski, and C. Ikonomidou. 2009. Expression of glutamate receptor subunits in human cancers. Histochem Cell Biol 132: 435–445.

Sturm, M.A., C.A. Conover, H. Pham, and R.G. Rosenfeld. 1989. Insulin-like growth factor receptors and binding protein in rat neuroblastoma cells. Endocrinology 124: 388–396.

Suri, C., B.P. Fung, A.S. Tischler, and D.M. Chikaraishi. 1993. Catecholaminergic cell lines from the brain and adrenal glands of tyrosine hydroxylase-SV40 T antigen transgenic mice. J Neurosci 13: 1280–1291.

Syapin, P.J., P.M. Salvaterra, and J.K. Engelhardt. 1982. Neuronal-like features of TE671 cells: presence of a functional nicotinic cholinergic receptor. Brain Res 231: 365–377.

Tischler, A.S., and L.A. Greene. 1975. Nerve growth factor-induced process formation by cultured rat pheochromocytoma cells. Nature 258: 341–342.

Toral, J., W. Hu, D. Critchett, A.J. Solomon, J.E. Barrett, P.T. Sokol, and M.R. Ziai. 1995. 5-HT3 receptor-independent inhibition of the depolarization-induced 86Rb efflux from human neuroblastoma cells, TE671, by ondansetron. J Pharm Pharmacol 47: 618–622.

Tsunoo, A., M. Yoshii, and T. Narahashi. 1986. Block of calcium channels by enkephalin and somatostatin in neuroblastoma-glioma hybrid NG108-15 cells. Proc Natl Acad Sci U S A 83: 9832–9836.

Tweardy, D.J., L.A. Cannizzaro, A.P. Palumbo, S. Shane, K. Huebner, P. Vantuinen, D.H. Ledbetter, J.B. Finan, P.C. Nowell, and G. Rovera. 1987. Molecular cloning and characterization of a cDNA for human granulocyte colony-stimulating factor (G-CSF) from a glioblastoma multiforme cell line and localization of the G-CSF gene to chromosome band 17q21. Oncogene Res 1: 209–220.

Uwada, J., A.S. Anisuzzaman, A. Nishimune, H. Yoshiki, and I. Muramatsu. 2011. Intracellular distribution of functional M(1) -muscarinic acetylcholine receptors in N1E-115 neuroblastoma cells. J Neurochem 118: 958–967.

Vaudry, D., P.J. Stork, P. Lazarovici, and L.E. Eiden. 2002. Signaling pathways for PC12 cell differentiation: making the right connections. Science 296: 1648–1649.

Vaughan, P.F., C. Peers, and J.H. Walker. 1995. The use of the human neuroblastoma SH-SY5Y to study the effect of second messengers on noradrenaline release. Gen Pharmacol 26: 1191–1201.

Ventimiglia, J.B., C.J. Wikstrand, L.E. Ostrowski, M.A. Bourdon, V.A. Lightner, and D.D. Bigner. 1992. Tenascin expression in human glioma cell lines and normal tissues. J Neuroimmunol 36: 41–55.

Verity, A.N., T.L. Wyatt, W. Lee, B. Hajos, P.A. Baecker, R.M. Eglen, and R.M. Johnson. 1999. Differential regulation of glial cell line-derived neurotrophic factor (GDNF) expression in human neuroblastoma and glioblastoma cell lines. J Neurosci Res 55: 187–197.

Voltz, R., R. Hohlfeld, A. Fateh-Moghadam, T.N. Witt, M. Wick, C. Reimers, B. Siegele, and H. Wekerle. 1991. Myasthenia gravis: measurement of anti-AChR autoantibodies using cell line TE671. Neurology 41: 1836–1838.

Wang, L.J., R. Colella, G. Yorke, and F.J. Roisen. 1996. The ganglioside GM1 enhances microtubule networks and changes the morphology of Neuro-2a cells *in vitro* by altering the distribution of MAP2. Exp Neurol 139: 1–11.

Westerink, R.H., and A.G. Ewing. 2008. The PC12 cell as model for neurosecretion. Acta Physiol (Oxf) 192: 273–285.

Wick, W., M. Platten, and M. Weller. 2001. Glioma cell invasion: regulation of metalloproteinase activity by TGF-beta. J Neurooncol 53: 177–185.

Widnell, K.L., D.S. Russell, and E.J. Nestler. 1994. Regulation of expression of cAMP response element-binding protein in the locus coeruleus *in vivo* and in a locus coeruleus-like cell line *in vitro*. Proc Natl Acad Sci U S A 91: 10947–10951.

Widnell, K.L., J.S. Chen, P.A. Iredale, W.H. Walker, R.S. Duman, J.F. Habener, and E.J. Nestler. 1996. Transcriptional regulation of CREB (cyclic AMP response element-binding protein) expression in CATH.a cells. J Neurochem 66: 1770–1773.

Xie, H.R., L.S. Hu, and G.Y. Li. 2010. SH-SY5Y human neuroblastoma cell line: *in vitro* cell model of dopaminergic neurons in Parkinson's disease. Chin Med J (Engl) 123: 1086–1092.

Yoshida, H., I. Date, T. Shingo, K. Fujiwara, Y. Miyoshi, T. Furuta, and T. Ohmoto. 1999. Evaluation of reaction of primate brain to grafted PC12 cells. Cell Transplant 8: 427–430.

Yoshida, Y., K. Tsuzuki, S. Ishiuchi, and S. Ozawa. 2006. Serum-dependence of AMPA receptor-mediated proliferation in glioma cells. Pathol Int 56: 262–271.

Zeltzer, P.M., S.L. Schneider, and D.D. Von Hoff. 1984. Morphologic, cytochemical and neurochemical characterization of the human medulloblastoma cell line TE671. J Neurooncol 2: 35–45.

Zhang, Z., G.J. Drzewiecki, J.T. Hom, P.C. May, and P.A. Hyslop. 1994. Human cortical neuronal (HCN) cell lines: a model for amyloid beta neurotoxicity. Neurosci Lett 177: 162–164.

Zimmermann, H.P., U. Plagens, and P. Traub. 1987. Influence of triethyl lead on neurofilaments *in vivo* and *in vitro*. Neurotoxicology 8: 569–577.

10

Advanced Cell Techniques to Study Developmental Neurobiology and Toxicology

Thomas Hartung,[1,a,2,*] *Helena T. Hogberg,*[1,b] *Marcel Leist,*[2,e] *David Pamies*[1,c] and *Lena Smirnova*[1,d]

Introduction

The developing nervous system is difficult to model *in vitro*. The complexity of the nervous system, the multitude of cells and the duration of this process represent tremendous challenges. However, in order to understand the various aspects of these processes and possible perturbations in disease or due to toxic or traumatic insults, it is necessary to make such cell models available. With the advent of stem cells, this field has given a boost to the development of organo-typic models, which promise to represent neural development and the possible targets of disruption.

---

[1] Center for Alternatives to Animal Testing (CAAT), Johns Hopkins Bloomberg School of Public Health,615 N. Wolfe Street, Room, 7032, Baltimore, Maryland, 21205.

[a] E-mail: thartun1@jhu.edu

[b] E-mail: hhogber2@jhu.edu

[c] E-mail: dpamies1@jhu.edu

[d] E-mail: lsmirno1@jhu.edu

[2] University of Konstanz, CAAT-Europe, Postbox 657, D-78467 Konstanz, Germany.

[e] E-mail: marcel.leist@uni-konstanz.de

* Corresponding author

Using patient-derived stem cells, the relevant genetic backgrounds can be studied and even combined with stressors, which might aggravate the manifestation of developmental disorders.

Key Cellular Events of Neural Development

The development of the central nervous system (CNS) is one of the most complex processes in the body and involves several critical key cellular events. The generation of neurons and cell migration starts in humans at six weeks of gestation and the development of the brain continues into the postnatal years. The weight of the brain keeps increasing even after puberty until approximately 20 years of age (Rodier, 1994). In general, a group of neuronal precursor cells are formed in the embryonic stage and migrate to a new location, where they begin to differentiate and become specialized. These processes are paralleled by glial events, such as myelination (Rodier, 1994). Interestingly, the fundamental principles underlying these key cellular events of brain development are remarkably conserved across species ranging from nematodes to humans and humans, which supports the use of alternative models (*in vitro* and non-mammalian animals). These events will be briefly introduced further.

Proliferation

As the CNS consists of numerous different cell types in a large quantity, proliferation of cells is one of the major events in the developing brain. It occurs over an elongated period with specific time windows for different cell types and brain structures, e.g., long motor neurons are developed before sensory neurons (Rodier, 1980). Moreover, the brainstem and diencephalon tend to form early, while the more complex layered structures like the cerebral cortex, hippocampus and cerebellum are developed over a longer time. This complex and precise development makes the brain more vulnerable to disturbance than all other developing organs.

Initially, neural progenitor cells divide symmetrically giving rise to two identical cells. Later on asymmetric division takes place where one cell stays in stem-cell fate and the other differentiates into a neuron. Over time, the population of neurons increases while progenitor cells decrease (Caviness and Takahashi, 1995). Neurogenesis takes place during development in the ventricular zone of the neural tube, while the mature brain has limited capacity for regeneration, which is restricted to two regions, the subventricular zone of the lateral ventricle and the dentate gyrus of the hippocampus (Rodier, 1994).

Cell migration

Following the proliferation stage, cells start to migrate to their final location in the brain. The most common migration is that of the neuronal progenitor cells, which migrate from the inner to the outer layer of the brain, which is why the deeper layers are formed first. However, the migration events occur at different time windows for different cell types, taking place over an extended period of time and continue for several months after birth. During migration the cell's cytoskeletal components get rearranged in response to extracellular molecules, e.g., neurotransmitters (Spitzer, 2006) followed by intracellular signaling pathways. The cells expand and contract their extension in order to move forward (Ayala *et al.*, 2007). Astroglial cells have an important role in the migration event as they guide the immature neurons to the right position and impact the formation of the complex network of axons and dendrites (Hatten and Liem, 1981; Wang *et al.*, 1994). If a group of neurons does not make the transition to the right position on time, they have difficulties in forming a network and cell-cell interactions, which can lead to impaired function and serious brain damage.

Neuronal differentiation

During migration, neuronal precursor cells start to differentiate into neurons with the expression of several specific genes that lead to the generation of, e.g., neurite outgrowths, expression of neurotransmitters, receptors and ultimately electrical activity (Eagleson *et al.*, 1997; Geschwind and Galaburda, 1985; McConnell, 1990). The neuronal network is formed to initiate cell-cell interactions (synapses) among neural cells or with other organs, e.g., muscles and glands. Initially, neurites (axons and dendrites) will increase in length and branching. One of the elongations grows faster and will obtain axonal features, while the rest will slow down and develop into dendrites (Dotti *et al.*, 1987; Dotti *et al.*, 1988). This event is influenced by different molecular signals such as neurotransmitters, hormones, calcium and electrical activity (Spitzer, 2006) that stimulate various signaling pathways (Goldberg, 2004).

For neurons to become electrically active, synaptogenesis (neuronal connections) needs to take place. This event occurs up to adolescence in humans (Mrzljak *et al.*, 1990; Uylings and van Eden, 1990) and over the first three weeks postnatal in rats (van Eden *et al.*, 1990). However, the neurons retain the capacity to form synapses through life and generally have more than a thousand connections per cell. Synapse formation involves biochemical and morphological changes in both the presynaptic and postsynaptic terminals and seems to impact neurotransmitter productions and the expression of receptors of the cells (Spitzer, 2006). As the neuronal differentiation is a lengthy and complex process, it is especially vulnerable to external interference (Audesirk G. and Audesirk T., 1998).

Glial differentiation

Glial cells are an important subpopulation of the CNS and include radial glia, microglia, astrocytes and oligodendrocytes. Radial glia cells have stem cell-like capacity and can differentiate into both neurons and astrocytes (Choi and Lapham, 1978; Tamamaki *et al.*, 2001). They provide support for the radial migration of neuronal precursor cells during the development of the brain.

Microglia are immune cells with phagocytic capacity and are responsible for removing dead cells and cellular debris from the brain (Cuadros and Navascues, 1998). However, microglial cells can also become reactive to a neuronal injury that triggers the release of proinflammatory factors such as cytokines and free radicals that often enhance the neuronal damage (Bal-Price and Brown, 2001). However, this is less common in the developing brain, while in the mature brain it is recognized as an indicator of a pathological response. Contrary to the other cells in the CNS, microglia cells are derived from the mesoderm germ layer that gives rise to connective tissue and blood cells. In most areas of the brain, radial glial cells and microglia development take place simultaneously with neuronal differentiation.

Inversely, astrocytes and oligodendrocytes mainly develop after the neuronal differentiation (Rice and Barone, 2000). The main role of astrocytes in the mature brain is to maintain the balance of the ionic and trophic environment. However, during development of the CNS, astrocytes also provide guidance for axons and synapse formation, and influence neuronal proliferation, migration and differentiation (Aschner *et al.*, 1999). Astrocytic differentiation is mainly influenced by growth factors, neurotransmitters and cytokines and involves several intracellular pathways (Bhat, 1995; Post and Brown, 1996).

Oligodendrocytes are responsible for the myelination of the axons in the CNS that provides electrical insulation and makes transmissions along the axon more rapid (Baumann and Pham-Dinh, 2001). One cell usually myelinates axons from multiple neurons, which is why the location of the oligodendrocyte is especially important. The process involves complex metabolic and biological events including adhesion, synthesis, accumulation of myelin sheets and axonal ion channel rearrangements. The myelination mainly takes place during the last trimester in humans and in the second week postnatal in rats (Rice and Barone, 2000), but is continuous even in the mature brain (Hunter *et al.*, 1997; Wiggins, 1982). Failure in the axonal development or myelin formation may lead to severe neuronal functional damage (Wiggins, 1986). In addition, oligodendrocytes are especially sensitive to disturbance as they work close to their metabolic capacity.

Apoptosis

During development of the CNS, neurons are produced in excess and therefore need to be regulated in order to establish the accurate final number of cells (Oppenheim, 1991; Raff *et al.*, 1993). This is accomplished by programmed cell death (apoptosis) mediated by gene induction, signal molecules and caspase activation (Gorman *et al.*, 1998). The event is initiated either by the release of cytochrome c from the mitochondria into the cytosol, by binding to the plasma-membrane death receptors or with an increase in intracellular calcium concentrations (Madden and Cotter, 2008; Shield, 1998; Wajant, 2002). This is followed by cleavage of a caspase leading to its activation. The cleaved caspase continues to cleave and activate additional caspases that finally stimulate key structural and nuclear proteins that cause the disassembly and death of the cell. Several caspases and their regulators have been identified and have been shown to be crucial for mammalian development (Cecconi *et al.*, 1998; Madden and Cotter, 2008), e.g., bcl-x, bax, caspase-9 and caspase-3 (Roth and D'Sa, 2001). Apoptosis takes place both pre- and postnatal in the CNS. It is believed that the type and amount of cell-cell connections is one factor that determines the fate of the cell, whether it gets to live or to die (Shield, 1998). Interference with this event has been suggested as a possible cause of autism (Corbett *et al.*, 2007; Sacco *et al.*, 2007).

Advanced Cellular Models and Techniques

The need for in vitro models to study developmental neurotoxicity

Only about 150 of over 100,000 substances on the market today have been subjected to the internationally agreed guideline studies for developmental neurotoxicity (DNT). This relatively small number of compounds contrasts strongly with the potential risk. Only a dozen compounds have been identified as definitive DNToxicants in humans, which include methyl mercury, lead, arsenic, PCBs, toluene, and ethanol (Grandjean and Landrigan, 2006), the list has been recently expanded to include manganese, fluoride, chlorpyrifos, dichlorodiphenyltrichloroethane, tetrachloroethylene, and the polybrominateddiphenyl ethers (Grandjean and Landrigan, 2014) (Figure 10.1). The developing brain is much more sensitive to chemical perturbation than the adult one, due to the still immature blood/brain-barrier, increased absorption versus low body weight, and diminished ability to detoxify exogenous chemicals (Adinolfi, 1985; NRC, 2000; Tilson, 2000). Moreover, CNS development is a complex and highly coordinated chain of events (see above), which occur within strictly controlled time frames and, therefore, each event creates a different window of vulnerability to xenobiotic exposure (Rice and Barone, 2000; Rodier, 1994). There is little potential to repair the disturbed neurodevelopment and it often leads to permanent consequences. Thus, there is concern about the lack of DNT data.

The main reasons for the lack of DNT data are the time consuming and costly guidelines for DNT (OECD, 2007; USEPA, 1998), which are based entirely on *in vivo* experiments. In rare cases, the animal-based DNT study is performed, but the data

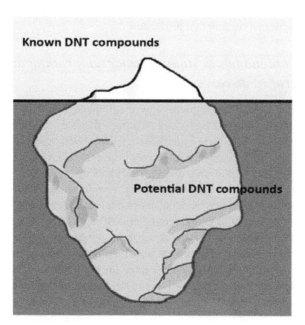

Figure 10.1. Only the top of the iceberg of potential developmental neurotoxic (DNT) compounds has actually been tested in current *in vivo* guidelines leading to very few compounds with known DNT effects.

can then be difficult to interpret, extrapolate to human health and, therefore, rarely contribute to regulation and risk assessment.

Thus, the identification of DNT effects of chemicals using a battery of human relevant high-throughput tests of modern toxicology is a priority (Bal-Price *et al.*, 2015a). To improve and accelerate DNT testing, experts in the field from industry, academia, and government have been discussing the current status and problems of DNT assessment, and have suggested promising alternative approaches to be included in an integrated testing strategy (Bal-Price *et al.*, 2012; Bal-Price *et al.*, 2010; Coecke *et al.*, 2007; Crofton *et al.*, 2011; Lein *et al.*, 2007).

The concept of an integrated testing strategy (ITS) (Hartung *et al.*, 2013; Leist *et al.*, 2014; Rovida *et al.*, 2015) utilizing cell-based *in vitro* approaches (Krewski *et al.*, 2010) is emerging rapidly in the field of toxicology. Bioinformatics plays a key role here by interpreting the high-content technologies and making sense of the output by modeling. With interdisciplinary collaborations, toxicology can take advantage of such expert knowledge. The challenge and the opportunity lie in the transition from mode of action (MoA) models to pathway modeling (Bouhifd *et al.*, 2015; Hartung and McBride, 2011; Kleensang *et al.*, 2014), i.e., increasing the resolution of analysis, and then going back to building ITS on this understanding of adverse outcome pathways (Bal-Price *et al.*, 2015b; Krug *et al.*, 2014; Smirnova *et al.*, 2014) and, ultimately, a systems integration of this mechanistic knowledge (Hartung *et al.*, 2012).

Moreover, environmental contaminations do not present themselves in isolation but as mixtures with possible synergistic effects. Traditional animal testing approaches are not suitable to test many combinations of doses and exposure times. New pathway-based tests, in contrast, could allow the identification of critical combinations and would provide better environmental protection.

Human relevant models to study neurodevelopmental disorders and neurodegenerative diseases

The aforementioned issues and shortcomings of animal models for DNT also apply to the studies of mechanisms and pathology of human diseases. Animal models do not reflect the human *in vivo* situation precisely enough. Many diseases are artificially induced, since they do not occur spontaneously in animals (Merkle and Eggan, 2013). For example, to model Parkinson's disease (PD) specific symptoms and pathology, animals are injected with the copmpounds methyl-phenyl-tetrahydropyridine (MPTP), 6-hydroxydopamine (6-OHDA), paraquat or rotenone (Blesa *et al.*, 2012). Interspecies differences create uncertainties in the extrapolation of results based on animal studies to the *in vivo* situation in humans. This may contribute to the small number of therapeutic compounds that have succeeded in clinical trials (Mullard, 2012; Scannell *et al.*, 2012).

Many neurodevelopmental disorders have an unknown etiology and a broad spectrum of symptoms that makes it difficult to model them in animals. Today one out of six children is diagnosed with a developmental disorder (Boyle *et al.*, 1994; Decoufle *et al.*, 2001; Schettler, 2001) that often involves the central nervous system (CNS). About 10–15% of all born children show disorders of neurobehavioral development (Grandjean and Landrigan, 2014), including learning disabilities, neurodevelopmental delays, autism spectrum disorder (ASD), and attention deficit and hyperactivity disorder (ADHD). ASD affects 1 in 68 in the US (CDC, http://www.cdc.gov/NCBDDD/ autism/data.html); In the United States, ADHD affects 14% of children (Landrigan *et al.*, 2012) and learning disabilities affect up to 10% of the children attending public schools (Schmid and Rotenberg, 2005). There is scientific evidence suggesting that such an increase is not only due to better assessment, reporting and reclassification of these disorders (Sullivan, 2005). ASD rates increased by 290% over the decade from 1997 to 2008 (Schmid and Rotenberg, 2005) (http://www.cdc.gov/NCBDDD/autism/ data.html). Both genetic and environmental factors appear to play a role. The US National Academy of Sciences (NAS) estimated that 3% of neurobehavioral disorders are caused directly by toxic environmental exposures and that another 25% are caused by interactions between environmental factors and inherited susceptibilities (NRC, 2000). It was estimated that about 4% of prescription drugs have been withdrawn from the market because of observed adverse neurological effects (Fung *et al.*, 2001), adding to such concerns. Evidence for the contribution of drugs derives from studies specifically linking autism to exposures in early pregnancy to thalidomide, misoprostol, and valproic acid (Landrigan, 2010). Research into gene/environment interactions is needed to model and predict developmental neurotoxicity of chemicals. Modern human relevant cell systems and technologies, including high-throughput testing of potential DNT compounds will facilitate and speed up risk assessment and identification of possible environmental/gene interactions leading to neurodevelopmental disorders and prevent a further increase of such diseases in the future.

Human induced pluripotent stem cells as cellular systems to study neurobiology and toxicology

There are several neural *in vitro* models already developed or under development today (Smirnova *et al.*, 2014). These include both simple two-dimensional cultures of cell lines to more complex primary cells, stem cells, 3D-cell cultures, and non-mammalian organisms (Figure 10.2). Many *in vitro* neurodegeneration models are based on rodent primary cell cultures or have been performed in human (SH-SY5Y) or rodent (PC12) transformed cell lines that have cancer origin, and therefore, are not stable in their chromosomal composition, do not reflect the true state of neural cells *in vivo* (Breier *et al.*, 2010; Zurich and Monnet-Tschudi, 2009) and are not optimal to study the molecular mechanism of disease and/or mode of action of a potential neurotoxicant. It is widely assumed that DNT is ultimately the consequence of the disturbance of relatively basic biological processes (Bal-Price *et al.*, 2012; Hogberg *et al.*, 2009; Kadereit *et al.*, 2012; Kuegler *et al.*, 2010), such as differentiation, proliferation, migration, and neurite growth. Therefore, several *in vitro* approaches have been established to focus on such biological activities (Frimat *et al.*, 2010; Harrill *et al.*, 2011; Hoelting *et al.*, 2013; Leist *et al.*, 2013; Radio and Mundy, 2008; Zimmer *et al.*, 2011).

The breakthrough discovery of induced pluripotent stem cells (iPSCs) has created new opportunities to study human neurodevelopment, diseases and toxicant effects directly in specific cell types, affected by the disease or toxicant (Takahashi *et al.*, 2007; Yu *et al.*, 2007). Fibroblasts from healthy and diseased individuals can

NS alternative in vitro models

| Inmortalized cell lines | Stem cells | non-mammalian organisms | 3D-cell cultures | Primary cells |

Figure 10.2. Alternative neural *in vitro* and non-mammalian models. We acknowledge Charlotte Mattson (PI OlovAndersson), at Department of Cell and Molecular Biology, Karolinska Institute, Sweden for photo of the zebrafish.

be reprogrammed into iPSCs, and subsequently be differentiated into all neural cell types. Recently somatic cells have even been directly converted into proliferating neural precursor cells (iNPCs) or induced neuronal cells (iN) bypassing the iPSC stage (Han *et al.*, 2012; Thier *et al.*, 2012; Vierbuchen *et al.*, 2010).

The human (h)iPSC-based approach overcomes interspecies differences, inaccessibility of primary neural tissue, and the shortcomings of cancer cell lines. This makes the use of hiPSC a valuable tool to study disease mechanisms, progression and pathology in different genetic backgrounds. Furthermore, increasing evidence demonstrates that iPSC is a promising platform for drug discovery and toxicological risk assessment (Bellin *et al.*, 2013; Grskovic *et al.*, 2011; Marchetto and Gage, 2012; Trounson *et al.*, 2012).

There are several iPSC-based disease models (Pamies *et al.*, 2014). The most promising and successful are monogenetic diseases, where phenotype can be linked to one gene. An example of such a model is iPSC derived from patients with spinal muscular atrophy (SMA), where disease phenotype (loss of motor neurons) can be recapitulated *in vitro* (Chang *et al.*, 2011; Ebert *et al.*, 2009). Other examples of successful iPSC-derived neurological disease models include Rett syndrome (Ananiev *et al.*, 2011; Dajani *et al.*, 2013; Farra *et al.*, 2012), familial dysautonomia (Lee *et al.*, 2009), spinocerebellar ataxia (Koch *et al.*, 2011), Fragile X Syndrome (Sheridan *et al.*, 2011) and Parkinson's disease (Nguyen *et al.*, 2011; Reinhardt *et al.*, 2013; Sanchez-Danes *et al.*, 2012).

Although diseases-derived iPSC lines contribute tremendously to the understanding of molecular mechanisms of disease development, it becomes challenging to model more complex neurological disorders caused by mutations in several genes and/or external factors (e.g., exposure to environmental toxicants or drugs). ASD is an example of such a disorder. The use of iPSCs will overcome many limitations of other sources, for example, ethical issues, limited accessibility, and restricted genetic backgrounds. In addition, iPSCs seem to be more stable, with higher neuronal differentiation efficiency than, for example, somatic stem cells (Caldwell *et al.*, 2001; Colombo *et al.*, 2006). It is important to note, however, that it is challenging to achieve full differentiation of any cell type from stem cells. In addition, the problems of generating iPSCs are that the percentage of reprogrammed cells is often low and reprogramming and differentiation protocols still require further optimization (still very time-consuming and often exhibit low reproducibility and efficiency of differentiation). Stem cells have been shown to be more genetically stable than other cell lines, but we have increasingly learned about their limitations in that respect, as well (Lund *et al.*, 2012; Mitalipova *et al.*, 2005; Steinemann *et al.*, 2013). The limitations experienced first are costs of culture and slow growth—many protocols require months, and labor, media, and supplement costs add up. The risk of infection increases unavoidably. We still do not obtain pure cultures and often require cell sorting, which, however, implies detachment of cells with the respective disruption of culture conditions and physiology.

Gene/environment interactions during neural development

Since many neurodevelopmental disorders, including ASD, do not have a clear or known genetic basis, they appear at least partially to be due to gene/environment interactions. Undoubtedly, there is a substantial genetic component to ASD etiology: gene mutations, copy number variants, and other genetic anomalies (Landrigan, 2010). ASD is a family of diseases with common phenotypes linked to a series of genetic anomalies, each of which is responsible for no more than 2–3% of cases while the total fraction of ASD attributable to genetic inheritance has been estimated to about 30–40% (Landrigan *et al.*, 2012). At the same time, findings from neuropathology, brain gene expression, twin and sibling concordance/recurrence risk analyses, as well as evidence from studies of now-rare teratogens, all suggest that environmental influences during the prenatal period also have a substantial impact on ASD risk. For this reason, there is a need for cell models, which allow the combination of genetic background and environmental exposures.

The effects of (epi)genetic backgrounds cannot be tested in standard animal tests, which use inbred rodents ("identical twins"). On the contrary, using iPSC from patients with developmental disorders, such as ASD, makes it possible to study the molecular mechanisms of the disease, and test substance sensitivity of different genetic backgrounds. Thus, using *in vitro* models based on iPSC may prove the hypothesis that gene/environmental interactions indeed contribute to development of certain disorders. Another advantage of iPSC as a model of drug sensitivity is that their differentiation *in vitro* is similar to the stages of brain development *in utero* (Nat and Dechant, 2011). Each stage of neurodevelopment is unique and displays different sensitivities to different xenobiotics. An *in vitro* model to test chemicals for toxicity during development must thus be able to screen possible toxicants at several of these different stages.

Organo-typic cell cultures

Cell cultures are prone to artifacts: Cells are pushed to survive in an artificial environment giving a selection of cells that is most adapted to the culture condition. Nevertheless, for a long time, scientists have been trying to reproduce biological functions in a dish in order to discern molecular mechanisms. However, due to the complexity of some tissues and organs, traditional *in vitro* models have several limitations. The recent development of organo-typic cell cultures has improved the *in vitro* models. These models including microphysiological systems are often three-dimensional (3D) cultures and co-cultures of different cell types, which enable the interaction of at least two or more cell types to better mimic tissue and organ complexity. These models have the potential to help us understand biological mechanisms as well as improve predictions of organism responses to medical treatment, disease agents and chemical exposure (Andersen *et al.*, 2014; Hartung, 2014) and also to provide more relevant results (Alepee *et al.*, 2014; Marx *et al.*, 2012). Traditional 2D cultures poorly represent the *in vivo* conditions, with reduced cell-to-cell interactions, reduced cell density and completely different structural properties. Organo-typic cultures on the other hand provide notable benefits: better cell differentiation, long-term stability, higher cell-to-cell interaction and a more *in vivo* like environment (Alepee *et al.*, 2014). However, these new models require special equipment, are more costly and time-consuming. In addition, the size of the 3D spheroid culture may decrease oxygen and nutrient supply and can lead to necrosis in the core (Lancaster *et al.*, 2013). The choice of cell source for the specific organo-typic culture presents a challenge. Tumor cell lines used as 3D cultures do not restore their genetic make-up; they have multiple point mutations, chromosomal aberrations, etc., which cannot be turned back. On the other side, primary human cells present an availability problem. The use of stem cells may be a solution to this problem (Hartung, 2013).

Organo-typic models of the nervous system

In vitro models are preferred for studying different molecular and cellular processes of neurogenesis, neurodegeneration and neurotoxicity, which is key to the elucidation of the biological mechanisms of brain development and function. CNS includes different cell types (different types of neurons, astrocytes, oligodendrocytes, microglia) with high spacial and temporal organization and architecture that makes it challenging to model *in vitro*. The use of more complex cell models such as organo-typic cultures may improve this state, as these models have the potential to better reflect the physiology and function of the brain and nervous system.

The presence of different cell types and the specific structure of the brain are crucial to mimic the biological mechanisms involved in CNS development and function. Brain structural organization affects the connectivity between cells, network formation and communication; it secondarily affects the electrical activity, metabolism and/or specific differentiation processes. An example of a 3D multicellular structure is the blood-brain barrier (BBB). Different 3D models have been established in this area (Vandenhaute *et al.*, 2012). Biological barriers (such as BBB and placenta) are important for studies in DNT and neurodevelopmental diseases (Giordano *et al.*, 2009). But there are other important aspects beside structural organization and cell composition. Specific cell structures are formed during brain development by precise cell compositions and gradients of different molecules that provide a highly defined environment that supports cell differentiation and function. Organo-typic cultures combined with other new techniques (bioprinting, organ-on-chip technologies and microfluidics) have the opportunity to advance the modeling of the CNS. Finally, cell maturation is a fundamental aspect to consider where 3D models often provide the possibility to study long-term processes. Enhanced survival in 3D is an important prerequisite for experiments addressing neurodegenerative diseases and chronic toxicity.

Two basic approaches are most commonly used to generate tissue-like cell cultures, in layers and spontaneous aggregation. Cells cultured in multilayer, have different neuronal and/or glial cells on top of each other (Schildknecht *et al.*, 2012; Viviani *et al.*, 1998). This can be performed in the normal dish, perforated membranes (Al Ahmad *et al.*, 2011), scaffolds (Lu *et al.*, 2012) or even at an air-liquid inter-phase (Dubois-Dauphin *et al.*, 2010; Preynat-Seauve *et al.*, 2009). 3D spheroid cultures can also be generated by spontaneous aggregation of dissociated cells (Hogberg *et al.*, 2013; Honegger *et al.*, 2011; Honegger *et al.*, 1979; Honegger and Richelson, 1976; Lancaster *et al.*, 2013) or by gravity using the hanging drop technique (Leist *et al.*, 2012).

Spheroid cultures without external material have many advantages. Organo-typical 3D aggregates can be formed from undifferentiated cells allowing them to proliferate and differentiate all together (Chiasson *et al.*, 1999; Kuhn and Svendsen, 1999). Moreover, neurospheres have the ability to produce their own extracellular matrix molecules (such as fibronectins and laminins), β-integrins and growth factors (Campos *et al.*, 2004; Lobo *et al.*, 2003). The cellular milieu of the neurosphere has been suggested to exhibit an *in vitro* microenvironment similar to the *in vivo* compartment (Conti and Cattaneo, 2010). Spheroids may also be plated onto a planar substrate. This technique has been used for electrophysiological studies and to follow differentiation during radial migration (Moors *et al.*, 2007; Moors *et al.*, 2009; van Vliet *et al.*, 2007). It is noteworthy that despite differentiation onto a surface, this culture technique presents some advantages over traditional 2D cells and monolayer cultures. Cells are differentiated in 3D with the consequent organization of organoid cell systems (Ahlenius and Kokaia, 2010; Bal-Price *et al.*, 2012; Liu and Sun, 2011; Schreiber *et al.*, 2010). Spheroid organo-typic cultures have been generated from different cell sources, e.g., induced pluripotent stem cells (Hogberg *et al.*, 2013) (Figure 10.3), neural stem cells from fetal brains, embryo carcinoma cell lines (Serra *et al.*, 2009; Serra *et al.*, 2007) (Serra *et al.*, 2007; Serra *et al.*, 2009) and primary rodent cells (van Vliet *et al.*, 2008).

Novel 3D *in vitro* models are continuously emerging in parallel areas (regenerative medicine, neuronal diseases, neurotoxicology, and developmental biology). The

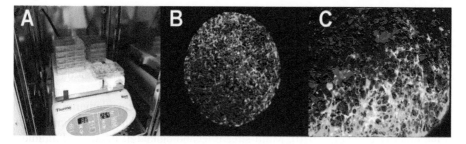

Figure 10.3. 3D organo-typic brain model derived from human induced pluripotent stem cells using gyratory shaking that lead to spontaneous aggregation. (A) Shows 3D cultures under constant gyratory movement. (B and C) Shows four weeks differentiated iPSC-derived neural spheroids stained positive for the neuronal marker Neurofilament 200 (NF200) (green) and dopaminergic marker Tyrosine hydroxylase (red).

combination of organo-typic culture with other technologies such as organ-on-chip, microfluidics, bioprinting, etc. is a powerful tool to study neuronal development and disease.

There are still some challenges in the application of 3D models, which have to be overcome. Some of them are specific for 3D models while others apply equally to 2D models. For instance, complex culture media do not necessarily recapitulate the cerebrospinal fluid (CSF) in brain. Also, the *in vitro* models usually do not have any barrier function that limits the access of toxicants the same way the blood-brain or the placental barrier do. A specific problem of 3D models may be limited compound diffusion to the cells, e.g., for larger models and high molecular mass compounds (e.g., nanoparticles), and limited accessibility of the cells far from the surface also needs to be considered.

Microfluidic systems

As mentioned above, in order to mimic the *in vivo* cell environment it is critical to reproduce biological functions. Traditional *in vitro* cultures poorly represent *in vivo* nutrients and oxygen perfusion. Medium change produces a drastic change of the cell environment when wastes, together with produced soluble factors are removed and new nutrients are replenished; in addition, thermal changes produce an important impact on the cells (Hartung, 2014). Moreover, long-term cell cultures present a higher risk for contamination, caused by the repeated manual intervention (Wu *et al.*, 2010). Microfluidics were established by combining micro-electro-mechanical systems (MEMS) and fluidic channels and have been used in numerous applications such as cell culture microenvironment control, collection of quantitative data, high-throughput analysis, constant supply of soluble factors and application of physiological stress (Kim *et al.*, 2014). This technology is able to handle small amounts of fluids in a variation of micrometer scales (Whitesides, 2006). Perfusion can make cultures more homeostatic by continuous access to nutrients, oxygen, and cytokines, and the removal of metabolic waste. Laminar flow within microchannels can be used to control all these soluble factors of the *in vitro* cell environment. Different examples of microfluidics can be found in the literature with the aim to study diverse processes such as migration (Cheng *et al.*, 2007), polarization (Shamloo *et al.*, 2008) and axonal injury (Dolle *et al.*, 2014). Moreover, microfluidics are used to generate different organ-on-chip approaches such as heart (Grosberg *et al.*, 2011), lung (Huh *et al.*, 2012) and intestine (Ramadan *et al.*, 2013) among others. In many cases, laminar flowing fluids are used to pattern cells and their microenvironment (Pedron *et al.*, 2015). In addition, multiple microenvironments can be applied simultaneously incorporating the cell cultures on an interface between two or more streams (Takayama *et al.*, 2001). For example, in the lung-on-chip developed by the Wyss Institute, endothelial cells are plated between two fluids, artificial blood in one microchannel and oxygen perfusion in the other (Huh *et al.*, 2010). Multi-stream technologies can be applied to study gradient and asymmetrically oriented signals (Xu *et al.*, 2013).

Bioprinting

These techniques are based on the generation of spatially-controlled cell patterns using biological 3D printing technologies and are believed to solve some problems related to synthetic tissue/organ formation, creation of extracellular matrixes, cell isolation, etc. (Yoo, 2015). Recently, the advances in tissue engineering technologies were able to generate tissues and parts of organs (Akhyari *et al.*, 2008; Hamilton *et al.*, 2014). Bioprinters differ from regular 3D printers due to some modifications that keep cells functional and viable during the formation of the construct. The standard technique is to seed cells onto solid and biodegradable scaffolds, and print different cell types in a specific conformation. However, there are several variations of bioprinting technologies (see below).

The structure, cell composition and biological function of different organs and tissues are extremely diverse. The production of artificial human organs is challenging due to the current limited knowledge about the composition of functional cells along with the appropriate epicellular environment (Yoo, 2015). 3D bioprinting often uses organ/tissue templates from different sources such as magnetic resonance imaging (MRI) or computerized tomography (CT) (Zhang and Zhang, 2015). Computer-aided design and manufacturing (CAD/CAM) technologies are commonly applied to the 3D bioprinting technologies in order to manufacture a structure with an accurate anatomical shape (Seol *et al.*, 2012). In recent years, the increment of open-source software algorithms and shared control hardware platforms has made these technologies more affordable (Yoo, 2015).

Bioprinting presents an important advantage in tissue/organ *in vitro* design, however, there are many challenges that need to be addressed besides the further development of these technologies such as, the optimization of bioscaffolds, increase of biomaterial availability, increase of similarities between biological tissues and bioprinted products, development of technologies to incorporate nutrients and oxygen supply and enhancement of the extracellular matrix incorporation (Billiet *et al.*, 2012; Seol *et al.*, 2012).

Currently there are three main bioprinting technologies available: Microextrusion 3D Bioprinting, Inkjet 3D Bioprinting and Laser-Assisted 3D Bioprinting.

- Microextrusion 3D Bioprinting is the most common due to its lower price. It consists of a temperature-controlled biomaterial dispensing system that generates continuous beads of material through 3 axes (Zhang and Zhang, 2015).
- Inkjet 3D Bioprinting (or drop-on-demand printer) is based on the incorporation of biomaterial droplets of variable size into the scaffold. Inkjet-based 3D bioprinting methods can produce high resolution structures due to the possibility of altering the drop size and density and due to the capacity to introduce concentration gradients of cells, materials, or growth factors throughout the droplet (Phillippi *et al.*, 2008).
- Laser-Assisted 3D Bioprinting allows the precise control of cell location and microenvironment. This technology is based on the Laser-Induced Forward Transfer (LIFT) principle, where the Laser-Assisted bioprinter is able to introduce

the biomaterial into a multilayer system made of a transparent glass slide onto which a thin layer of liquid (bio-ink) is applied (Devillard *et al.*, 2014).

Additional technologies have been developed in recent years. Extrusion of high-viscosity, for example, a technique that generates layer-by-layer structures of embedded cells on a novel hydrogel precursor material (via gelation) has recently been developed (Pati *et al.*, 2014). In order to increase resolution, the spherical cell-hydrogel aggregates technology has been developed, which requires high-precision technologies (Marga *et al.*, 2012).

Only a limited number of studies have successfully printed cells from the central nervous system. Rat primary embryonic hippocampal and cortical neurons were printed into soy agar and collagen hydrogels using a modified thermal Hewlett Packard (HP) 550C printer with 90% cell viability (Xu *et al.*, 2005). Moreover, the NT2 neuronal precursor cell line was printed together with prefabricated fibrin gels (Xu *et al.*, 2006). Recently, Inkjet 3D Bioprinting was applied to cells of the adult rat central nervous system (CNS), retinal ganglion cells (RGC) and glial cells; RGC/glial survival and RGC neurite outgrowth did not show any alterations after the printing process as compared to normal cultures (Lorber *et al.*, 2014). Nevertheless, to the best of our knowledge there is no bioprinting technology available so far that allows the generation of *in vitro* tissue or organ-like structures.

Organ-on-chip technologies

In recent years, bioengineers and biologists have developed the organ-on-chip approach. The increment activity in these cell culture technologies has been driven by the increased awareness of the limitations of traditional approaches and the need for more complex systems. Organ-on-chip by definition is the incorporation of organo-typical cultures into a chip in order to better simulate and measure biological functions of the organ-tissue. Novel platforms combining microfluidic and microfabrication technologies have the potential to simulate the function of an organ-on-chip. Microengineering has been used to mimic biological complexity by producing tissue-like structures (Khademhosseini *et al.*, 2006) and is able to precisely control different aspects of the cell environment. In the living organism, cells are presented in a controlled microenvironment with complex regulation of cells, intracellular factors and extracellular matrix molecules. Monolayer *in vitro* systems cannot reproduce the 3D environment sufficiently. Microscale approaches allow better control of the cell environment, by increased cell-cell, cell-matrix and cell-soluble factor interactions (Semino, 2003). Inclusion of microfluidics in the chip will further enhance perfusion of the functional organs (Selimovic *et al.*, 2013).

Different agencies are currently promoting programs to fund these types of technologies. One example is the collaborative R&D funding initiative by the US agencies NIH, DARPA and FDA, and another similar program sponsored by DTRA, where together $200 million have been made available over a period of 5 years (Huh *et al.*, 2010). These programs aim to model the structure and function of human organs (Hartung and Zurlo, 2012). The organ-on-chip approaches have not only focused on generating a more physiological environment for tissue/organ-like models, but also on mimicking the impact of physical factors such as stretch, peristaltic and pressure. Initially, these tools are being developed to help medical countermeasures for chemical and biological warfare and terrorism. Their broader use in drug discovery and toxicology is foreseeable. Numerous new applications appear daily (e.g., development of disease models). The interest shown in organs-on-chip has also been reflected in the biotechnology market. Different companies have developed and commercialized these approaches. Companies such a Nortis (US) have announced the fabrication of such technologies for 2015 and in Europe TissUse and MIMETAS are already selling "organs-on-chips" microfluidic platforms combined with organo-typic cultures. Other companies have emerged with focus on the development of organo-typic culture tools (e.g., InSphero).

Different organ-on-chip models have been published over the last years. One of the most famous examples is the lung-on-chip, developed by Wyss Institute, with a 3D model that reproduces breathing (Huh *et al.*, 2010). The device consists of human lung cells compartmentalized on PDMS microchannels to mimic alveolar-capillary barriers. The chip reproduces mechanical stretching on the capillary by applying vacuum to the side chambers (Huh *et al.*, 2010). Another example is the heart-on-chip that consists of cardiac myocytes isolated from ventricles in sub-millimeter-sized thin film cantilevers of soft elastomers on a channel fluidic microdevice (Agarwal *et al.*, 2013).

Brain-on-chip

Several organo-typic brain cultures have been developed (Lancaster *et al.*, 2013; Pamies *et al.*, 2014), however, not many have yet been incorporated into a chip. Recently, Dollé and collaborators (Dolle *et al.*, 2014) presented a brain-on-chip technology to study axonal responses to diffuse axonal injury. Hippocampal slices were isolated from brains of Sprague-Dawley rat pups and were placed on a flexible substrate (PDMS). Axon extensions were guided down the PDMS microchannels where they eventually exited and connected to the adjacent hippocampal slice. Injuries were induced by pressurizing the device and causing the PDMS to strain in a uniaxial direction. Images were taken before and after applying the uniaxial strain injury and axons were evaluated (Dolle *et al.*, 2014). Another brain-on-chip example is the microfluidic chip based on three-dimensional (3D) neurospheroids generated by Park and collaborators (Park *et al.*, 2015). The neurospheroids were generated from prenatal rat cortical neurons. This study showed that neurospheroids cultured with flow obtain more robust and complex neural networks than those cultured under static condition. This suggests that the perfusion of continuous nutrient, oxygen, and cytokine transport and the removal of metabolic wastes increase cell viability and maturation. However, only viability was evaluated in this study and not the other biological brain functions (Park *et al.*, 2015). In addition, some examples of organ-on-chip technologies can be found in the blood brain barrier research, with the use of different microfluidic technologies (Booth and Kim, 2012; Cucullo *et al.*, 2013; Griep *et al.*, 2013; Prabhakarpandian *et al.*, 2013; Yeon *et al.*, 2012).

Advanced Technologies to Measure Molecular Mechanisms of Biological Changes

Imaging techniques

Many different *in vitro* models can be used to study the CNS: monolayer cell culture, co-cultures, histotypic or organo-typic culture, stem cells, primary cultures, small organisms or even more complex like organ-on-chip technologies. Depending on the complexity of the model and the technology available for the parameter of interest there are different challenges in the measurements. As *in vitro* science is moving towards more representative complex systems, microscopy has faced some important limitations. For the quantitative analysis of 3D biological structures an excellent signal-to-noise ratio is needed. Moreover, other factors such as optical sectioning capability, large field of view, good spatial resolution and fast stack recording are required (Dickinson, 2006). Point-scanning techniques or wide-field microscopes are invaluable for studying complex models of the nervous system (Grienberger and Konnerth, 2012; Grutzendler *et al.*, 2002; Lichtman and Denk, 2011). Confocal microscopy represents the best commercial technique available to measure thick biological samples, however, the excitation light has to illuminate the entire biological sample, producing photo-bleaching and phototoxic effects in all planes (Pampaloni *et al.*, 2013). In addition, the reflection of the light makes it difficult to obtain good resolution and the limited penetration of the light makes it difficult to reach the deepest point of the tissue. As an alternative, two-photon microscopy presents a twofold higher penetration depth than confocal microscopy (Gilbert *et al.*, 2000), however, it has lower resolution (Stelzer *et al.*, 1994). Tomographic techniques, such as optical coherence tomography (OCT) and optical projection tomography (OPT), have been used in order to image large 3D samples (organs, tissues, 3D culture, etc.) and allow the imaging of objects from different angles and then combine them together. However, they still possess low spatial resolution (Huang *et al.*, 1991; Sharpe *et al.*, 2002).

Plane illumination microscopy (PIM) consists of side-on illumination of the sampler with a thin laminar sheet of light. Combining PIM with the new generation of scientific cameras (such as metal oxide silicon (sCMOS) cameras) that are able to obtain 200 image planes per second with a large number of pixels (Planchon *et al.*, 2011) will improve the speed, resolution, reduce the energy and avoid photo-bleaching and phototoxic effects (Keller and Ahrens, 2015). Light-sheet microscopy has been used for live-imaging of different dynamic neuronal processes such as tracking of neural progenitors in the developing nervous system (Amat *et al.*, 2014), whole-brain functional imaging at cellular resolution (Ahrens *et al.*, 2013; Panier *et al.*, 2013), and imaging sizeable volumes of the mammalian brain (Holekamp *et al.*, 2008).

Another novel technology is High-Content Imaging (HCI). HCI is the combination of automated microscopy with image analysis approaches to quantify multiple phenotypic and/or functional parameters in biological systems. Both phenotypic parameters such as size, cell morphology, axon length, nuclei, mitochondria or functional parameters such as signal transduction, gene expression and metabolism can be measured by using different fluorescent dyes, antibodies and gene reporters (van Vliet *et al.*, 2014). The major advantage of HCI is the quantity of data, which can be obtained from a single experiment. The application of multiple HCI read-outs from *in vitro* models can provide insight into the spatial distribution and dynamics of responses over time (Massoud and Gambhir, 2003) and allow identification of signaling pathways (Wink *et al.*, 2014). Different conditions (e.g., growth factors, toxic exposure, differentiation protocols and drug identification) can be studied in the same experiment to obtain a valuable quantity of data (Barbaric *et al.*, 2010; Buchser *et al.*, 2004; Henn *et al.*, 2009).

Automated imaging data is analyzed using imaging analysis algorithms that can be set up to discriminate a specific target element of the picture (e.g., nucleus, cell shape, axon elongation, branch points) or discriminate a group of parameters to define a specific situation (distinguish between cell population, detect double staining, etc.). Algorithms can be trained to define the exact size, shape and volume of biological structures (such as neurites and axons microtubules) or track cell movements over time, axon elongation and proliferation (van Vliet *et al.*, 2014).

Still, microscopy faces some challenges such as, limited fluorescent probes, biosensors and reporters on the market; incompatibility between different probes and reporters; limitation of imaging systems for complex biological systems (3D culture, non-adherent cultures, organ-on-chip technologies, etc.) and limitations of the software. Moreover, automated analyses are not always able to identify specific cell phenotypes or biological processes (Gerhardt *et al.*, 2001; Hansson *et al.*, 2000). However, new technologies are tackling these problems and have advanced in the last couple of years.

Multi-omics

Advanced technologies allow us to use multi-omics techniques including (toxico) genomics, epigenomics, proteomics, transcriptomics, metabolomics, and miRNomics (miRNA profiling) as an integrated approach for a global assessment of biological changes. These methods have the potential to study systems biology in terms of, e.g., development, disease, pharmacology and toxicology. Quantitative measurements with multi-omics technologies can bridge the gap between molecular initiating events and relevant adverse outcomes that will be a significant step towards a better understanding of the mechanisms underlying neurological disease development as well as DNT. Furthermore, a new testing strategy can be developed and established by demonstrating the potential of human organo-typic cell models in combination with multi-omic approaches to assess biological changes.

Neural differentiation during development or after perturbation by toxicant or drug exposure is commonly analyzed by gene expression assays. Based on these data, alterations associated with a specific developmental process or disease can be identified. This approach can also be useful to classify chemicals with regard to their toxicogenomic response, and to compare *in vivo* and *in vitro* data as well as data across different species. Currently, guidelines are being developed for experimental design and bioinformatic analysis to improve the quality of transcriptomic and toxicogenomic studies.

LC-MS-based Metabolomics have been commonly used for biomarker discovery in diagnosis, prognosis, and therapeutic response applications. For example, metabolic changes in biofluids from patients with ASD have been observed (Ratajczak, 2011). Metabolomics approaches are also increasingly being applied in toxicology (Bouhifd *et al.*, 2013; Ramirez *et al.*, 2013). Metabolomics has several advantages: it measures the phenotype in a disease or the final outcome after a toxic insult in a cascade of events, such as alterations of genes, transcripts, proteins and finally metabolites. Owing to the biological complexity, changes in one of them may or may not lead to changes in the others. Determining the final alteration (metabolites) enhances the possibility of understanding the actual disease or toxicity and to associate the effects with adverse outcomes or phenotypic changes. Cell-based metabolomics has the advantage that both extracellular and intracellular metabolites can be measured. Metabolites detected in extracellular samples (cell culture media), can be associated with *in vivo* identified biomarkers from biofluids, while the intracellular samples (cell lysate) can provide information about the toxic mechanism and key pathways on the cellular level.

In the last decade, the understanding of post-transcriptional regulation of gene expression has emerged, thanks to the discovery of miRNAs, small (~22 nt) non-coding regulatory RNA molecules, which bind to specific binding sites in 3'UTR of target mRNAs and repress their translation. The important role of miRNA networks in maintenance of crucial cellular processes and cascades, especially during development (Chua *et al.*, 2007; Rana, 2007), and that 50% of all miRNA are expressed in the brain, make them powerful regulators of brain development. Increasing number of publications elucidating the role of miRNAs in cellular response to environmental stress, including xenobiotics (reviewed in Smirnova *et al.*, 2012), make miRNA profiling an attractive tool to study cellular responses to toxicant exposure or in disease. miRNA microarrays or RNA sequencing are used to generate miRNA profiles that are then validated by real-time PCR.

Other epigenetic mechanisms, including DNA methylation, histone modification, and nucleosome remodeling regulate lineage-specific gene expressions by influencing chromatin structure and modulating DNA-protein interactions (Cantone and Fisher, 2013). Recent studies have revealed that dynamic regulation of epigenetic programming occurs during early development; methylation patterns are then mitotically inherited in somatic cells, thus preserving the cell identity of differentiated cells (Cantone and Fisher, 2013). Emerging data suggest that non-traditional roles for CpG and non-CpG (i.e., CpT, CpA, or CpC) methylation may be particularly important during brain development (Guo *et al.*, 2014; Kozlenkov *et al.*, 2014). In addition, imprinted genes heavily influence neurodevelopment and disruption of their brain-specific expression patterns can lead to a variety of behavioral phenotypes (reviewed in Plasschaert and Bartolomei, 2014). To identify candidate genomic regions with perturbations in DNA methylation, genome-wide DNA methylation profiles using Whole Genome Bisulfite Sequencing can be applied (WGBS). The high-content data from WGBS is then validated by targeted bisulfite clonal Sanger sequencing for imprinted regions and bisulfite pyrosequencing for other candidate regions.

Microelectrode arrays (MEAs) to study neuronal functionality

Neuronal functionality is most commonly assessed by measuring synaptic activity using traditional techniques such as Ca^{2+} live imaging or electrophysiology measurements by patch-clamping. However, these methods are not suitable for more complex 3D cell systems such as organo-typic and brain-on-chip models. Recording of extracellular field potential using microelectrode arrays (MEAs) instead is a useful tool in these models as it measures the collective activity of many cells.

The first recordings using MEAs were performed on neurons from the spinal cord of mice in the 1980's (Gross *et al.*, 1982) and since then the technique has evolved into better and easier recording systems. MEAs have mainly been used to characterize functionality and pharmacological responses of *in vitro* systems from primary cells, such as hippocampal slices and dissociated cultures from the spinal cord and cortex. However, recently MEAs have been used for more complex models such as 3D brain cell cultures and human stem cells (Makinen *et al.*, 2013; van Vliet *et al.*, 2007; Yla-Outinen *et al.*, 2010; Yla-Outinen *et al.*, 2014).

A significant advantage of MEA recordings is the possibility of measuring electrical activity without disturbing the cellular integrity. The same culture can therefore be measured repeatedly over a long period of time (Esposti *et al.*, 2009). Furthermore, neurons cultured *in vitro* reveal spontaneous electrical activity with the traditional spiking and bursting pattern as observed *in vivo*. In addition, they respond in a similar way to a wide variety of neurotransmitters and pharmacological agonists and antagonists (Gramowski *et al.*, 2006; Keefer *et al.*, 2001; Martinoia *et al.*, 2005; van Vliet *et al.*, 2007).

Recently, electrical activity measurements using MEAs have shown to be a promising tool for both neurotoxicity (Johnstone *et al.*, 2010; Mack *et al.*, 2014; Valdivia *et al.*, 2014) and DNT assessment (Hogberg *et al.*, 2011; Robinette *et al.*, 2011). MEAs also have the potential to be valuable in studies of neuronal disorders, especially with the opportunity to develop disease models using iPSCs.

Better cell systems also need better quality assurance

Good Cell Culture Practice (GCCP) (Coecke *et al.*, 2005) and publication guidance for *in vitro* studies (Leist *et al.*, 2010) are highly important for the implementation of complex cell systems like 3D cultures, microfluidic systems, and organ-on-chip. There is currently little standardization, harmonization and comparison of these cell models. However, such quality assurance will be necessary to make these new approaches valuable tools for basic research, drug discovery and safety assessments.

Cell cultures are prone to artifacts (Hartung, 2007)—far too many artificially chosen and difficult-to-control conditions influence our experiments. Therefore, quality assurance is an essential aspect of successful and reliable research projects. While Good Laboratory Practice (GLP)—at least originally—addressed only regulatory *in vivo* studies, and ISO guidance is not really specific for life science tools, neither addresses the key issue, i.e., the relevance of a test. This is the truly unique contribution of validation, which is far too infrequently applied in other settings.

Conclusions

Developmental neurobiology and neurotoxicity are areas in desperate need of predictive human cell-based test models. The complexity of the CNS requires advanced culture techniques to approximate this in the laboratory set-up. The advent of stem cell technologies, especially iPSC, and organo-typic cultures, has made new models available, which might help close this gap.

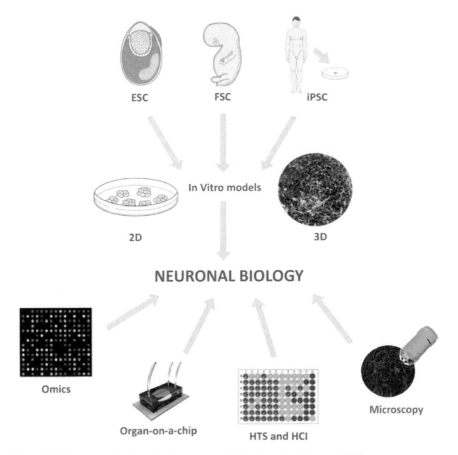

Figure 10.4. Summary of advanced cell systems and techniques to study neural biology.

References

Adinolfi, M. 1985. The development of the human blood-CSF-brain barrier. Dev Med Child Neurol 27: 532–537.

Agarwal, A., J.A. Goss, A. Cho, M.L. McCain, and K.K. Parker. 2013. Microfluidic heart on a chip for higher throughput pharmacological studies. Lab Chip 13: 3599–3608.

Ahlenius, H., and Z. Kokaia. 2010. Isolation and generation of neurosphere cultures from embryonic and adult mouse brain. Methods Mol Biol 633: 241–252.

Ahrens, M.B., M.B. Orger, D.N. Robson, J.M. Li, and P.J. Keller. 2013. Whole-brain functional imaging at cellular resolution using light-sheet microscopy. Nat Methods 10: 413–420.

Akhyari, P., H. Kamiya, A. Haverich, M. Karck, and A. Lichtenberg. 2008. Myocardial tissue engineering: the extracellular matrix. Eur J Cardiothorac Surg : Official Journal of the European Association for Cardio-thoracic Surgery 34: 229–241.

Al Ahmad, A., C.B. Taboada, M. Gassmann, and O.O. Ogunshola. 2011. Astrocytes and pericytes differentially modulate blood-brain barrier characteristics during development and hypoxic insult. J Cereb Blood Flow Metab : Official Journal of the International Society of Cerebral Blood Flow and Metabolism 31: 693–705.

Alepee, N., A. Bahinski, M. Daneshian, B. De Wever, E. Fritsche, A. Goldberg, J. Hansmann, T. Hartung, J. Haycock, H. Hogberg, L. Hoelting, J.M. Kelm, S. Kadereit, E. McVey, R. Landsiedel, M. Leist, M. Lubberstedt, F. Noor, C. Pellevoisin, D. Petersohn, U. Pfannenbecker, K. Reisinger, T. Ramirez, B. Rothen-Rutishauser, M. Schafer-Korting, K. Zeilinger, and M.G. Zurich. 2014. State-of-the-art of 3D cultures (organs-on-a-chip) in safety testing and pathophysiology. ALTEX 31: 441–477.

Amat, F., W. Lemon, D.P. Mossing, K. McDole, Y. Wan, K. Branson, E.W. Myers, and P.J. Keller. 2014. Fast, accurate reconstruction of cell lineages from large-scale fluorescence microscopy data. Nat Methods 11: 951–958.

Ananiev, G., E.C. Williams, H. Li, and Q. Chang. 2011. Isogenic pairs of wild type and mutant induced pluripotent stem cell (iPSC) lines from Rett syndrome patients as *in vitro* disease model. PloS One 6: e25255.

Andersen, M.E., K. Betts, Y. Dragan, S. Fitzpatrick, J.L. Goodman, T. Hartung, J. Himmelfarb, D.E. Ingber, A. Jacobs, R. Kavlock, K. Kolaja, J.L. Stevens, D. Tagle, D. Lansing Taylor, and D. Throckmorton. 2014. Developing microphysiological systems for use as regulatory tools—challenges and opportunities. ALTEX 31: 364–367.

Aschner, M., J.W. Allen, H.K. Kimelberg, R.M. LoPachin, and W.J. Streit. 1999. Glial cells in neurotoxicity development. Annu Rev Pharmacol Toxicol 39: 151–173.

Audesirk, G., and T. Audesirk. 1998. Neurite development. pp. 61–85. *In*: W. Slikker JRr. and C. LW. (eds.). Handbook of Developmental Neurotoxicity. Academic Press, San Diego, CA.

Ayala, R., T. Shu, and L.H. Tsai. 2007. Trekking across the brain: the journey of neuronal migration. Cell 128: 29–43.

Bal-Price, A., and G.C. Brown. 2001. Inflammatory neurodegeneration mediated by nitric oxide from activated glia-inhibiting neuronal respiration, causing glutamate release and excitotoxicity. J Neurosci: The Official Journal of the Society for Neuroscience 21: 6480–6491.

Bal-Price, A., K.M. Crofton, M. Leist, S. Allen, M. Arand, T. Buetler, N. Delrue, R.E. FitzGerald, T. Hartung, T. Heinonen, H. Hogberg, S.H. Bennekou, W. Lichtensteiger, D. Oggier, M. Paparella, M. Axelstad, A. Piersma, E. Rached, B. Schilter, G. Schmuck, L. Stoppini, E. Tongiorgi, M. Tiramani, F. Monnet-Tschudi, M.F. Wilks, T. Ylikomi, and E. Fritsche. 2015a. International STakeholder NETwork (ISTNET): creating a developmental neurotoxicity (DNT) testing road map for regulatory purposes. Arch Toxicol 89: 269–287.

Bal-Price, A., K.M. Crofton, M. Sachana, T.J. Shafer, M. Behl, A. Forsby, A. Hargreaves, B. Landesmann, P.J. Lein, J. Louisse, F. Monnet-Tschudi, A. Paini, A. Rolaki, A. Schrattenholz, C. Sunol, C. van Thriel, M. Whelan, and E. Fritsche. 2015b. Putative adverse outcome pathways relevant to neurotoxicity. Crit Rev Toxicol 45: 83–91.

Bal-Price, A.K., H.T. Hogberg, L. Buzanska, and S. Coecke. 2010. Relevance of *in vitro* neurotoxicity testing for regulatory requirements: challenges to be considered. Neurotoxicol Teratol 32: 36–41.

Bal-Price, A.K., S. Coecke, L. Costa, K.M. Crofton, E. Fritsche, A. Goldberg, P. Grandjean, P.J. Lein, A. Li, R. Lucchini, W.R. Mundy, S. Padilla, A.M. Persico, A.E. Seiler, and J. Kreysa. 2012. Advancing the science of developmental neurotoxicity (DNT): testing for better safety evaluation. ALTEX 29: 202–215.

Barbaric, I., P.J. Gokhale, and P.W. Andrews. 2010. High-content screening of small compounds on human embryonic stem cells. Biochemical Society Transactions 38: 1046–1050.

Baumann, N., and D. Pham-Dinh. 2001. Biology of oligodendrocyte and myelin in the mammalian central nervous system. Physiol Rev 81: 871–927.

Bellin, M., S. Casini, R.P. Davis, C. D'Aniello, J. Haas, D. Ward-van Oostwaard, L.G. Tertoolen, C.B. Jung, D.A. Elliott, A. Welling, K.L. Laugwitz, A. Moretti, and C.L. Mummery. 2013. Isogenic human pluripotent stem cell pairs reveal the role of a KCNH2 mutation in long-QT syndrome. The EMBO J 32: 3161–3175.

Bhat, N.R. 1995. Signal transduction mechanisms in glial cells. Dev Neurosci 17: 267–284.

Billiet, T., M. Vandenhaute, J. Schelfhout, S. Van Vlierberghe, and P. Dubruel. 2012. A review of trends and limitations in hydrogel-rapid prototyping for tissue engineering. Biomaterials 33: 6020–6041.

Blesa, J., S. Phani, V. Jackson-Lewis, and S. Przedborski. 2012. Classic and new animal models of Parkinson's disease. J Biomed Biotechnol 2012: 845618.

Booth, R., and H. Kim. 2012. Characterization of a microfluidic *in vitro* model of the blood-brain barrier (muBBB). Lab Chip 12: 1784–1792.

Bouhifd, M., T. Hartung, H.T. Hogberg, A. Kleensang, and L. Zhao. 2013. Review: toxicometabolomics. J Appl Toxicol : JAT 33: 1365–1383.

Bouhifd, M., M.E. Andersen, C. Baghdikian, K. Boekelheide, K.M. Crofton, A.J. Fornace, Jr., A. Kleensang, H. Li, C. Livi, A. Maertens, P.D. McMullen, M. Rosenberg, R. Thomas, M. Vantangoli, J.D. Yager, L. Zhao, and T. Hartung. 2015. The Human Toxome Project. ALTEX 32: 112–124.

Boyle, C.A., P. Decoufle, and M. Yeargin-Allsopp. 1994. Prevalence and health impact of developmental disabilities in US children. Pediatrics 93: 399–403.

Breier, J.M., K. Gassmann, R. Kayser, H. Stegeman, D. De Groot, E. Fritsche, and T.J. Shafer. 2010. Neural progenitor cells as models for high-throughput screens of developmental neurotoxicity: state of the science. Neurotoxicol Teratol 32: 4–15.

Buchser, W., M. Collins, T. Garyantes, R. Guha, S. Haney, V. Lemmon, Z. Li, and O.J. Trask. 2004. Assay development guidelines for image-based high content screening, high content analysis and high content imaging. *In*: G.S. Sittampalam, N.P. Coussens, H. Nelson, M. Arkin, D. Auld, C. Austin, B. Bejcek, M. Glicksman, J. Inglese, V. Lemmon, Z. Li, J. McGee, O. McManus, L. Minor, A. Napper, T. Riss, O.J. Trask, and J. Weidner (eds.). Assay Guidance Manual. Bethesda (MD).

Caldwell, M.A., X. He, N. Wilkie, S. Pollack, G. Marshall, K.A. Wafford, and C.N. Svendsen. 2001. Growth factors regulate the survival and fate of cells derived from human neurospheres. Nat Biotechnol 19: 475–479.

Campos, L.S., D.P. Leone, J.B. Relvas, C. Brakebusch, R. Fassler, U. Suter, and C. ffrench-Constant. 2004. Beta1 integrins activate a MAPK signalling pathway in neural stem cells that contributes to their maintenance. Development 131: 3433–3444.

Cantone, I., and A.G. Fisher. 2013. Epigenetic programming and reprogramming during development. Nat Struct Mol Biol 20: 282–289.

Caviness, V.S., Jr., and T. Takahashi. 1995. Proliferative events in the cerebral ventricular zone. Brain Dev 17: 159–163.

Cecconi, F., G. Alvarez-Bolado, B.I. Meyer, K.A. Roth, and P. Gruss. 1998. Apaf1 (CED-4 homolog) regulates programmed cell death in mammalian development. Cell 94: 727–737.

Chang, T., W. Zheng, W. Tsark, S. Bates, H. Huang, R.J. Lin, and J.K. Yee. 2011. Brief report: phenotypic rescue of induced pluripotent stem cell-derived motoneurons of a spinal muscular atrophy patient. Stem Cells 29: 2090–2093.

Cheng, S.Y., S. Heilman, M. Wasserman, S. Archer, M.L. Shuler, and M. Wu. 2007. A hydrogel-based microfluidic device for the studies of directed cell migration. Lab Chip 7: 763–769.

Chiasson, B.J., V. Tropepe, C.M. Morshead, and D. van der Kooy. 1999. Adult mammalian forebrain ependymal and subependymal cells demonstrate proliferative potential, but only subependymal cells have neural stem cell characteristics. J Neurosci : The Official Journal of the Society for Neuroscience 19: 4462–4471.

Choi, B.H., and L.W. Lapham. 1978. Radial glia in the human fetal cerebrum: a combined Golgi, immunofluorescent and electron microscopic study. Brain Res 148: 295–311.

Chua, J.H., A. Armugam, and K. Jeyaseelan. 2009. MicroRNAs: biogenesis, function and applications. Curr Opin Mol Ther 11: 189–199.

Coecke, S., M. Balls, G. Bowe, J. Davis, G. Gstraunthaler, T. Hartung, R. Hay, O.W. Merten, A. Price, L. Schechtman, G. Stacey, and W. Stokes. 2005. Guidance on good cell culture practice. a report of the second ECVAM task force on good cell culture practice. Alternatives to laboratory animals : Altern Lab Anim 33: 261–287.

Coecke, S., A.M. Goldberg, S. Allen, L. Buzanska, G. Calamandrei, K. Crofton, L. Hareng, T. Hartung, H. Knaut, P. Honegger, M. Jacobs, P. Lein, A. Li, W. Mundy, D. Owen, S. Schneider, E. Silbergeld, T. Reum, T. Trnovec, F. Monnet-Tschudi, and A. Bal-Price. 2007. Workgroup report: incorporating *in vitro* alternative methods for developmental neurotoxicity into international hazard and risk assessment strategies. Environ Health Perspect 115: 924–931.

Colombo, E., S.G. Giannelli, R. Galli, E. Tagliafico, C. Foroni, E. Tenedini, S. Ferrari, G. Corte, A. Vescovi, G. Cossu, and V. Broccoli. 2006. Embryonic stem-derived versus somatic neural stem cells: a comparative analysis of their developmental potential and molecular phenotype. Stem Cells 24: 825–834.

Conti, L., and E. Cattaneo. 2010. Neural stem cell systems: physiological players or *in vitro* entities? Nat Rev Neurosci 11: 176–187.

Corbett, B.A., A.B. Kantor, H. Schulman, W.L. Walker, L. Lit, P. Ashwood, D.M. Rocke, and F.R. Sharp. 2007. A proteomic study of serum from children with autism showing differential expression of apolipoproteins and complement proteins. Mol Psychiatry 12: 292–306.

Crofton, K.M., W.R. Mundy, P.J. Lein, A. Bal-Price, S. Coecke, A.E. Seiler, H. Knaut, L. Buzanska, and A. Goldberg. 2011. Developmental neurotoxicity testing: recommendations for developing alternative methods for the screening and prioritization of chemicals. ALTEX 28: 9–15.

Cuadros, M.A., and J. Navascues. 1998. The origin and differentiation of microglial cells during development. Prog Neurobiol 56: 173–189.

Cucullo, L., M. Hossain, W. Tierney, and D. Janigro. 2013. A new dynamic *in vitro* modular capillaries-venules modular system: cerebrovascular physiology in a box. BMC Neurosci 14: 18.

Dajani, R., S.E. Koo, G.J. Sullivan, and I.H. Park. 2013. Investigation of Rett syndrome using pluripotent stem cells. J Cell Biochem 114: 2446–2453.

Decoufle, P., C.A. Boyle, L.J. Paulozzi, and J.M. Lary. 2001. Increased risk for developmental disabilities in children who have major birth defects: a population-based study. Pediatrics 108: 728–734.

Devillard, R., E. Pages, M.M. Correa, V. Keriquel, M. Remy, J. Kalisky, M. Ali, B. Guillotin, and F. Guillemot. 2014. Cell patterning by laser-assisted bioprinting. Methods in Cell Biology 119: 159–174.

Dickinson, M.E. 2006. Multimodal imaging of mouse development: tools for the postgenomic era. Dev Dyn : An Official Publication of the American Association of Anatomists 235: 2386–2400.

Dolle, J.P., B. Morrison, 3rd, R.S. Schloss, and M.L. Yarmush. 2014. Brain-on-a-chip microsystem for investigating traumatic brain injury: Axon diameter and mitochondrial membrane changes play a significant role in axonal response to strain injuries. Technology 2: 106.

Dotti, C.G., G.A. Banker, and L.I. Binder. 1987. The expression and distribution of the microtubule-associated proteins tau and microtubule-associated protein 2 in hippocampal neurons in the rat *in situ* and in cell culture. Neuroscience 23: 121–130.

Dotti, C.G., C.A. Sullivan, and G.A. Banker. 1988. The establishment of polarity by hippocampal neurons in culture. J Neurosci : The Official Journal of the Society for Neuroscience 8: 1454–1468.

Dubois-Dauphin, M.L., N. Toni, S.D. Julien, I. Charvet, L.E. Sundstrom, and L. Stoppini. 2010. The long-term survival of *in vitro* engineered nervous tissue derived from the specific neural differentiation of mouse embryonic stem cells. Biomaterials 31: 7032–7042.

Eagleson, K.L., L. Lillien, A.V. Chan, and P. Levitt. 1997. Mechanisms specifying area fate in cortex include cell-cycle-dependent decisions and the capacity of progenitors to express phenotype memory. Development 124: 1623–1630.

Ebert, A.D., J. Yu, F.F. Rose, Jr., V.B. Mattis, C.L. Lorson, J.A. Thomson, and C.N. Svendsen. 2009. Induced pluripotent stem cells from a spinal muscular atrophy patient. Nature 457: 277–280.

Esposti, F., M.G. Signorini, S.M. Potter, and S. Cerutti. 2009. Statistical long-term correlations in dissociated cortical neuron recordings. IEEE Trans Neural Syst Rehabil Eng : a publication of the IEEE Engineering in Medicine and Biology Society 17: 364–369.

Farra, N., W.B. Zhang, P. Pasceri, J.H. Eubanks, M.W. Salter, and J. Ellis. 2012. Rett syndrome induced pluripotent stem cell-derived neurons reveal novel neurophysiological alterations. Mol Psychiatry 17: 1261–1271.

Frimat, J.P., J. Sisnaiske, S. Subbiah, H. Menne, P. Godoy, P. Lampen, M. Leist, J. Franzke, J.G. Hengstler, C. van Thriel, and J. West. 2010. The network formation assay: a spatially standardized neurite outgrowth analytical display for neurotoxicity screening. Lab Chip 10: 701–709.

Fung, M., A. Thornton, K. Mybeck, J. Wu, K. Hornbuckle, and E. Muniz. 2001. Evaluation of the characteristics of safety withdrawal of prescription drugs from worldwide pharmaceutical markets-1960 to 1999. Drug Inf J 35: 293–317.

Gerhardt, E., S. Kugler, M. Leist, C. Beier, L. Berliocchi, C. Volbracht, M. Weller, M. Bahr, P. Nicotera, and J.B. Schulz. 2001. Cascade of caspase activation in potassium-deprived cerebellar granule neurons: targets for treatment with peptide and protein inhibitors of apoptosis. Mol Cell Neurosci 17: 717–731.

Geschwind, N., and A.M. Galaburda. 1985. Cerebral lateralization. Biological mechanisms, associations, and pathology: III. A hypothesis and a program for research. Arch Neurol 42: 634–654.

Gilbert, R.J., M. Hoffman, A. Capitano, and P.T. So. 2000. Imaging of three-dimensional epithelial architecture and function in cultured CaCo2a monolayers with two-photon excitation microscopy. Microsc Res Tech 51: 204–210.

Giordano, G., T.J. Kavanagh, and L.G. Costa. 2009. Mouse cerebellar astrocytes protect cerebellar granule neurons against toxicity of the polybrominated diphenyl ether (PBDE) mixture DE-71. Neurotoxicol 30: 326–329.

Goldberg, J.L. 2004. Intrinsic neuronal regulation of axon and dendrite growth. Curr Opin Neurobiol 14: 551–557.

Gorman, A.M., S. Orrenius, and S. Ceccatelli. 1998. Apoptosis in neuronal cells: role of caspases. Neuroreport 9: R49–55.

Gramowski, A., K. Jugelt, S. Stuwe, R. Schulze, G.P. McGregor, A. Wartenberg-Demand, J. Loock, O. Schroder, and D.G. Weiss. 2006. Functional screening of traditional antidepressants with primary cortical neuronal networks grown on multielectrode neurochips. Eur J Neurosci 24: 455–465.

Grandjean, P., and P.J. Landrigan. 2006. Developmental neurotoxicity of industrial chemicals. Lancet 368: 2167–2178.

Grandjean, P., and P.J. Landrigan. 2014. Neurobehavioural effects of developmental toxicity. Lancet Neurol 13: 330–338.

Grienberger, C., and A. Konnerth. 2012. Imaging calcium in neurons. Neuron 73: 862–885.

Griep, L.M., F. Wolbers, B. de Wagenaar, P.M. ter Braak, B.B. Weksler, I.A. Romero, P.O. Couraud, I. Vermes, A.D. van der Meer, and A. van den Berg. 2013. BBB on chip: microfluidic platform to mechanically and biochemically modulate blood-brain barrier function. Biomed Microdevices 15: 145–150.

Grosberg, A., P.W. Alford, M.L. McCain, and K.K. Parker. 2011. Ensembles of engineered cardiac tissues for physiological and pharmacological study: heart on a chip. Lab Chip 11: 4165–4173.

Gross, G.W., A.N. Williams, and J.H. Lucas. 1982. Recording of spontaneous activity with photoetched microelectrode surfaces from mouse spinal neurons in culture. J Neurosci Meth 5: 13–22.

Grskovic, M., A. Javaherian, B. Strulovici, and G.Q. Daley. 2011. Induced pluripotent stem cells—opportunities for disease modelling and drug discovery. Nature Reviews. Drug Discovery 10: 915–929.

Grutzendler, J., N. Kasthuri, and W.B. Gan. 2002. Long-term dendritic spine stability in the adult cortex. Nature 420: 812–816.

Guo, J.U., Y. Su, J.H. Shin, J. Shin, H. Li, B. Xie, C. Zhong, S. Hu, T. Le, G. Fan, H. Zhu, Q. Chang, Y. Gao, G.L. Ming, and H. Song. 2014. Distribution, recognition and regulation of non-CpG methylation in the adult mammalian brain. Nat Neurosci 17: 215–222.

Hamilton, N., A.J. Bullock, S. Macneil, S.M. Janes, and M. Birchall. 2014. Tissue engineering airway mucosa: a systematic review. Laryngoscope 124: 961–968.

Han, D.W., N. Tapia, A. Hermann, K. Hemmer, S. Hoing, M.J. Arauzo-Bravo, H. Zaehres, G. Wu, S. Frank, S. Moritz, B. Greber, J.H. Yang, H.T. Lee, J.C. Schwamborn, A. Storch, and H.R. Scholer. 2012. Direct reprogramming of fibroblasts into neural stem cells by defined factors. Cell Stem Cell 10: 465–472.

Hansson, O., R.F. Castilho, G.S. Kaminski Schierle, J. Karlsson, P. Nicotera, M. Leist, and P. Brundin. 2000. Additive effects of caspase inhibitor and lazaroid on the survival of transplanted rat and human embryonic dopamine neurons. Exp Neurol 164: 102–111.

Harrill, J.A., B.L. Robinette, and W.R. Mundy. 2011. Use of high content image analysis to detect chemical-induced changes in synaptogenesis *in vitro*. Toxicol *In Vitro* : an international journal published in association with BIBRA 25: 368–387.

Hartung, T. 2007. Food for thought... on cell culture. ALTEX 24: 143–152.

Hartung, T., and M. McBride. 2011. Food for thought ... on mapping the human toxome. ALTEX 28: 83–93.

Hartung, T., and J. Zurlo. 2012. Alternative approaches for medical countermeasures to biological and chemical terrorism and warfare. ALTEX 29: 251–260.

Hartung, T., E. van Vliet, J. Jaworska, L. Bonilla, N. Skinner, and R. Thomas. 2012. Food for thought ... systems toxicology. ALTEX 29: 119–128.

Hartung, T. 2013. Look back in anger—what clinical studies tell us about preclinical work. ALTEX 30: 275–291.

Hartung, T., T. Luechtefeld, A. Maertens, and A. Kleensang. 2013. Integrated testing strategies for safety assessments. ALTEX 30: 3–18.

Hartung, T. 2014. 3D—a new dimension of *in vitro* research. Adv Drug Deliv Rev 69-70: vi.

Hatten, M.E., and R.K. Liem. 1981. Astroglial cells provide a template for the positioning of developing cerebellar neurons *in vitro*. J Cell Biol 90: 622–630.

Henn, A., S. Lund, M. Hedtjarn, A. Schrattenholz, P. Porzgen, and M. Leist. 2009. The suitability of BV2 cells as alternative model system for primary microglia cultures or for animal experiments examining brain inflammation. ALTEX 26: 83–94.

Hoelting, L., B. Scheinhardt, O. Bondarenko, S. Schildknecht, M. Kapitza, V. Tanavde, B. Tan, Q.Y. Lee, S. Mecking, M. Leist, and S. Kadereit. 2013. A 3-dimensional human embryonic stem cell (hESC)-derived model to detect developmental neurotoxicity of nanoparticles. Arch Toxicol 87: 721–733.

Hogberg, H.T., A. Kinsner-Ovaskainen, T. Hartung, S. Coecke, and A.K. Bal-Price. 2009. Gene expression as a sensitive endpoint to evaluate cell differentiation and maturation of the developing central nervous system in primary cultures of rat cerebellar granule cells (CGCs) exposed to pesticides. Toxicol Appl Pharmacol 235: 268–286.

Hogberg, H.T., T. Sobanski, A. Novellino, M. Whelan, D.G. Weiss, and A.K. Bal-Price. 2011. Application of micro-electrode arrays (MEAs) as an emerging technology for developmental neurotoxicity: evaluation of domoic acid-induced effects in primary cultures of rat cortical neurons. Neurotoxicol 32: 158–168.

Hogberg, H.T., J. Bressler, K.M. Christian, G. Harris, G. Makri, C. O'Driscoll, D. Pamies, L. Smirnova, Z. Wen, and T. Hartung. 2013. Toward a 3D model of human brain development for studying gene/environment interactions. Stem Cell Res Ther 4 (Suppl 1): S4: 1–7.

Holekamp, T.F., D. Turaga, and T.E. Holy. 2008. Fast three-dimensional fluorescence imaging of activity in neural populations by objective-coupled planar illumination microscopy. Neuron 57: 661–672.

Honegger, P., and E. Richelson. 1976. Biochemical differentiation of mechanically dissociated mammalian brain in aggregating cell culture. Brain Res 109: 335–354.

Honegger, P., D. Lenoir, and P. Favrod. 1979. Growth and differentiation of aggregating fetal brain cells in a serum-free defined medium. Nature 282: 305–308.

Honegger, P., A. Defaux, F. Monnet-Tschudi, and M.G. Zurich. 2011. Preparation, maintenance, and use of serum-free aggregating brain cell cultures. Meth Mol Biol 758: 81–97.

Huang, D., E.A. Swanson, C.P. Lin, J.S. Schuman, W.G. Stinson, W. Chang, M.R. Hee, T. Flotte, K. Gregory, C.A. Puliafito, and J.G. Fujimoto. 1991. Optical coherence tomography. Science 254: 1178–1181.

Huh, D., B.D. Matthews, A. Mammoto, M. Montoya-Zavala, H.Y. Hsin, and D.E. Ingber. 2010. Reconstituting organ-level lung functions on a chip. Science 328: 1662–1668.

Huh, D., D.C. Leslie, B.D. Matthews, J.P. Fraser, S. Jurek, G.A. Hamilton, K.S. Thorneloe, M.A. McAlexander, and D.E. Ingber. 2012. A human disease model of drug toxicity-induced pulmonary edema in a lung-on-a-chip microdevice. Sci Transl Med 4: 159ra147.

Hunter, S.F., J.A. Leavitt, and M. Rodriguez. 1997. Direct observation of myelination *in vivo* in the mature human central nervous system. A model for the behaviour of oligodendrocyte progenitors and their progeny. Brain : A J Neurol 120: 2071–2082.

Johnstone, A.F., G.W. Gross, D.G. Weiss, O.H. Schroeder, A. Gramowski, and T.J. Shafer. 2010. Microelectrode arrays: a physiologically based neurotoxicity testing platform for the 21st century. Neurotoxicol 31: 331–350.

Kadereit, S., B. Zimmer, C. van Thriel, J.G. Hengstler, and M. Leist. 2012. Compound selection for *in vitro* modeling of developmental neurotoxicity. Front Biosci 17: 2442–2460.

Keefer, E.W., A. Gramowski, and G.W. Gross. 2001. NMDA receptor-dependent periodic oscillations in cultured spinal cord networks. J Neurophysiol 86: 3030–3042.

Keller, P.J., and M.B. Ahrens. 2015. Visualizing whole-brain activity and development at the single-cell level using light-sheet microscopy. Neuron 85: 462–483.

Khademhosseini, A., R. Langer, J. Borenstein, and J.P. Vacanti. 2006. Microscale technologies for tissue engineering and biology. Proc Natl Acad Sci U S A 103: 2480–2487.

Kim, S.H., G.H. Lee, J.Y. Park, and S.H. Lee. 2014. Microplatforms for gradient field generation of various properties and biological applications. J Lab Autom.

Kleensang, A., A. Maertens, M. Rosenberg, S. Fitzpatrick, J. Lamb, S. Auerbach, R. Brennan, K.M. Crofton, B. Gordon, A.J. Fornace, Jr., K. Gaido, D. Gerhold, R. Haw, A. Henney, A. Ma'ayan, M. McBride, S. Monti, M.F. Ochs, A. Pandey, R. Sharan, R. Stierum, S. Tugendreich, C. Willett, C. Wittwehr, J. Xia, G.W. Patton, K. Arvidson, M. Bouhifd, H.T. Hogberg, T. Luechtefeld, L. Smirnova, L. Zhao, Y. Adeleye, M. Kanehisa, P. Carmichael, M.E. Andersen, and T. Hartung. 2014. t4 workshop report: Pathways of Toxicity. ALTEX 31: 53–61.

Koch, P., P. Breuer, M. Peitz, J. Jungverdorben, D. Kesavan, J. Poppe, J. Doerr, J. Ladewig, J. Mertens, T. Tuting, P. Hoffmann, T. Klockgether, B.O. Evert, U. Wullner, and O. Brustle. 2011. Excitation-induced ataxin-3 aggregation in neurons from patients with Machado-Joseph disease. Nature 480: 543–546.

Kozlenkov, A., P. Roussos, A. Timashpolsky, M. Barbu, S. Rudchenko, M. Bibikova, B. Klotzle, W. Byne, R. Lyddon, A.F. Di Narzo, Y.L. Hurd, E.V. Koonin, and S. Dracheva. 2014. Differences in DNA methylation between human neuronal and glial cells are concentrated in enhancers and non-CpG sites. Nucleic Acids Res 42: 109–127.

Krewski, D., D. Acosta, M. Andersen, H. Anderson, J.C. Bailar, K. Boekelheide, R. Brent, G. Charnley, V.G. Cheung, S. Green, K.T. Kelsey, N.I. Kerkvliet, A.A. Li, L. McCray, O. Meyer, R.D. Patterson, W. Pennie, R.A. Scala, G.M. Solomon, M. Stephens, J. Yager, L. Zeise, and S.C.T.T. Assess. 2010. Toxicity testing in the 21st century: a vision and a strategy. J Toxicol Env Heath B 13: 51–138.

Krug, A.K., S. Gutbier, L. Zhao, D. Poltl, C. Kullmann, V. Ivanova, S. Forster, S. Jagtap, J. Meiser, G. Leparc, S. Schildknecht, M. Adam, K. Hiller, H. Farhan, T. Brunner, T. Hartung, A. Sachinidis, and M. Leist. 2014. Transcriptional and metabolic adaptation of human neurons to the mitochondrial toxicant MPP(+). Cell Death Dis 5: e1222.

Kuegler, P.B., B. Zimmer, T. Waldmann, B. Baudis, S. Ilmjarv, J. Hescheler, P. Gaughwin, P. Brundin, W. Mundy, A.K. Bal-Price, A. Schrattenholz, K.H. Krause, C. van Thriel, M.S. Rao, S. Kadereit, and M. Leist. 2010. Markers of murine embryonic and neural stem cells, neurons and astrocytes: reference points for developmental neurotoxicity testing. ALTEX 27: 17–42.

Kuhn, H.G., and C.N. Svendsen. 1999. Origins, functions, and potential of adult neural stem cells. BioEssays: News and Reviews in Molecular, Cell Dev Biol 21: 625–630.

Lancaster, M.A., M. Renner, C.A. Martin, D. Wenzel, L.S. Bicknell, M.E. Hurles, T. Homfray, J.M. Penninger, A.P. Jackson, and J.A. Knoblich. 2013. Cerebral organoids model human brain development and microcephaly. Nature 501: 373–379.

Landrigan, P.J. 2010. What causes autism? Exploring the environmental contribution. Curr Opin Paediatr 22: 219–225.

Landrigan, P.J., L. Lambertini, and L.S. Birnbaum. 2012. A research strategy to discover the environmental causes of autism and neurodevelopmental disabilities. Environ Health Perspect 120: a258–260.

Lee, G., E.P. Papapetrou, H. Kim, S.M. Chambers, M.J. Tomishima, C.A. Fasano, Y.M. Ganat, J. Menon, F. Shimizu, A. Viale, V. Tabar, M. Sadelain, and L. Studer. 2009. Modelling pathogenesis and treatment of familial dysautonomia using patient-specific iPSCs. Nature 461: 402–406.

Lein, P., P. Locke, and A. Goldberg. 2007. Meeting report: alternatives for developmental neurotoxicity testing. Environ Health Perspect 115: 764–768.

Leist, M., L. Efremova, and C. Karreman. 2010. Food for thought ... considerations and guidelines for basic test method descriptions in toxicology. ALTEX 27: 309–317.

Leist, M., B.A. Lidbury, C. Yang, P.J. Hayden, J.M. Kelm, S. Ringeissen, A. Detroyer, J.R. Meunier, J.F. Rathman, G.R. Jackson, Jr., G. Stolper, and N. Hasiwa. 2012. Novel technologies and an overall strategy to allow hazard assessment and risk prediction of chemicals, cosmetics, and drugs with animal-free methods. ALTEX 29: 373–388.

Leist, M., A. Ringwald, R. Kolde, S. Bremer, C. van Thriel, K.H. Krause, J. Rahnenfuhrer, A. Sachinidis, J. Hescheler, and J.G. Hengstler. 2013. Test systems of developmental toxicity: state-of-the art and future perspectives. Arch Toxicol 87: 2037–2042.

Leist, M., N. Hasiwa, C. Rovida, M. Daneshian, D. Basketter, I. Kimber, H. Clewell, T. Gocht, A. Goldberg, F. Busquet, A.M. Rossi, M. Schwarz, M. Stephens, R. Taalman, T.B. Knudsen, J. McKim, G. Harris, D. Pamies, and T. Hartung. 2014. Consensus report on the future of animal-free systemic toxicity testing. ALTEX 31: 341–356.

Lichtman, J.W., and W. Denk. 2011. The big and the small: challenges of imaging the brain's circuits. Science 334: 618–623.

Liu, X.W., and J.W. Sun. 2011. Primary culture of neurospheres obtained from fetal mouse central nervous system and generation of inner ear hair cell immunophenotypes *in vitro*. J Laryngol Otol 125: 686–691.

Lobo, M.V., F.J. Alonso, C. Redondo, M.A. Lopez-Toledano, E. Caso, A.S. Herranz, C.L. Paino, D. Reimers, and E. Bazan. 2003. Cellular characterization of epidermal growth factor-expanded free-floating neurospheres. J Histochem Cytochem : Official Journal of the Histochemistry Society 51: 89–103.

Lorber, B., W.K. Hsiao, I.M. Hutchings, and K.R. Martin. 2014. Adult rat retinal ganglion cells and glia can be printed by piezoelectric inkjet printing. Biofabrication 6: 015001.

Lu, H.F., S.X. Lim, M.F. Leong, K. Narayanan, R.P. Toh, S. Gao, and A.C. Wan. 2012. Efficient neuronal differentiation and maturation of human pluripotent stem cells encapsulated in 3D microfibrous scaffolds. Biomaterials 33: 9179–9187.

Lund, R.J., E. Narva, and R. Lahesmaa. 2012. Genetic and epigenetic stability of human pluripotent stem cells. Nature reviews. Genetics 13: 732–744.

Mack, C.M., B.J. Lin, J.D. Turner, A.F. Johnstone, L.D. Burgoon, and T.J. Shafer. 2014. Burst and principal components analyses of MEA data for 16 chemicals describe at least three effects classes. Neurotoxicol 40: 75–85.

Madden, S.D., and T.G. Cotter. 2008. Cell death in brain development and degeneration: control of caspase expression may be key! Mol Neurobiol 37: 1–6.

Makinen, M., T. Joki, L. Yla-Outinen, H. Skottman, S. Narkilahti, and R. Aanismaa. 2013. Fluorescent probes as a tool for cell population tracking in spontaneously active neural networks derived from human pluripotent stem cells. J Neurosci Meth 215: 88–96.

Marchetto, M.C., and F.H. Gage. 2012. Modeling brain disease in a dish: really? Cell Stem Cell 10: 642–645.

Marga, F., K. Jakab, C. Khatiwala, B. Shepherd, S. Dorfman, B. Hubbard, S. Colbert, and F. Gabor. 2012. Toward engineering functional organ modules by additive manufacturing. Biofabrication 4: 022001.

Martinoia, S., L. Bonzano, M. Chiappalone, M. Tedesco, M. Marcoli, and G. Maura. 2005. *In vitro* cortical neuronal networks as a new high-sensitive system for biosensing applications. Biosens Bioelectron 20: 2071–2078.

Marx, U., H. Walles, S. Hoffmann, G. Lindner, R. Horland, F. Sonntag, U. Klotzbach, D. Sakharov, A. Tonevitsky, and R. Lauster. 2012. 'Human-on-a-chip' developments: a translational cutting-edge alternative to systemic safety assessment and efficiency evaluation of substances in laboratory animals and man? Altern Lab Anim : ATLA 40: 235–257.

Massoud, T.F., and S.S. Gambhir. 2003. Molecular imaging in living subjects: seeing fundamental biological processes in a new light. Genes Dev 17: 545–580.

McConnell, S.K. 1990. The specification of neuronal identity in the mammalian cerebral cortex. Experientia 46: 922–929.

Merkle, F.T., and K. Eggan. 2013. Modeling human disease with pluripotent stem cells: from genome association to function. Cell Stem Cell 12: 656–668.

Mitalipova, M.M., R.R. Rao, D.M. Hoyer, J.A. Johnson, L.F. Meisner, K.L. Jones, S. Dalton, and S.L. Stice. 2005. Preserving the genetic integrity of human embryonic stem cells. Nature Biotechnology 23: 19–20.

Moors, M., J.E. Cline, J. Abel, and E. Fritsche. 2007. ERK-dependent and -independent pathways trigger human neural progenitor cell migration. Toxicol Appl Pharmacol 221: 57–67.

Moors, M., T.D. Rockel, J. Abel, J.E. Cline, K. Gassmann, T. Schreiber, J. Schuwald, N. Weinmann, and E. Fritsche. 2009. Human neurospheres as three-dimensional cellular systems for developmental neurotoxicity testing. Environmental Health Perspectives 117: 1131–1138.

Mrzljak, L., H.B. Uylings, C.G. Van Eden, and M. Judas. 1990. Neuronal development in human prefrontal cortex in prenatal and postnatal stages. Progr Brain Res 85: 185–222.

Mullard, A. 2012. Oncology trials gear up for high-throughput sequencing. Nat Rev Drug Discov 11: 339–340.

Nat, R., and G. Dechant. 2011. Milestones of directed differentiation of mouse and human embryonic stem cells into telencephalic neurons based on neural development *in vivo*. Stem Cells Dev 20: 947–958.

Nguyen, H.N., B. Byers, B. Cord, A. Shcheglovitov, J. Byrne, P. Gujar, K. Kee, B. Schule, R.E. Dolmetsch, W. Langston, T.D. Palmer, and R.R. Pera. 2011. LRRK2 mutant iPSC-derived DA neurons demonstrate increased susceptibility to oxidative stress. Cell Stem Cell 8: 267–280.

NRC. 2000. National Research Council: Scientific frontiers in developmental toxicology and risk assessment, Washington DC.

OECD. 2007. Test No. 426: Developmental Neurotoxicity Study. OECD Publishing, http://dx.doi.org/10.1787/9789264067394-en.

Oppenheim, R.W. 1991. Cell death during development of the nervous system. Annu Rev Neurosci 14: 453–501.

Pamies, D., T. Hartung, and H.T. Hogberg. 2014. Biological and medical applications of a brain-on-a-chip. Exp Biol Med 9: pii: 1535370214537738.

Pampaloni, F., N. Ansari, and E.H. Stelzer. 2013. High-resolution deep imaging of live cellular spheroids with light-sheet-based fluorescence microscopy. Cell Tissue Res 352: 161–177.

Panier, T., S.A. Romano, R. Olive, T. Pietri, R. Sumbre, R. Candelier, and G. Debregeas. 2013. Fast functional imaging of multiple brain regions in intact zebrafish larvae using selective plane illumination microscopy. Front Neural Circuits 7: 65.

Park, J., B.K. Lee, G.S. Jeong, J.K. Hyun, C.J. Lee, and S.H. Lee. 2015. Three-dimensional brain-on-a-chip with an interstitial level of flow and its application as an *in vitro* model of Alzheimer's disease. Lab Chip 15: 141–150.

Pati, F., J. Jang, D.H. Ha, S. Won Kim, J.W. Rhie, J.H. Shim, D.H. Kim, and D.W. Cho. 2014. Printing three-dimensional tissue analogues with decellularized extracellular matrix bioink. Nat Commun 5: 3935.

Pedron, S., E. Becka, and B.A. Harley. 2015. Spatially gradated hydrogel platform as a 3D engineered tumor microenvironment. Adv Mater 27: 1567–1572.

Phillippi, J.A., E. Miller, L. Weiss, J. Huard, A. Waggoner, and P. Campbell. 2008. Microenvironments engineered by inkjet bioprinting spatially direct adult stem cells toward muscle- and bone-like subpopulations. Stem Cells 26: 127–134.

Planchon, T.A., L. Gao, D.E. Milkie, M.W. Davidson, J.A. Galbraith, C.G. Galbraith, and E. Betzig. 2011. Rapid three-dimensional isotropic imaging of living cells using Bessel beam plane illumination. Nat Meth 8: 417–423.

Plasschaert, R.N., and M.S. Bartolomei. 2014. Genomic imprinting in development, growth, behavior and stem cells. Development 141: 1805–1813.

Post, G.R., and J.H. Brown. 1996. G protein-coupled receptors and signaling pathways regulating growth responses. FASEB J : Official Publication of the Federation of American Societies for Experimental Biology 10: 741–749.

Prabhakarpandian, B., M.C. Shen, J.B. Nichols, I.R. Mills, M. Sidoryk-Wegrzynowicz, M. Aschner, and K. Pant. 2013. SyM-BBB: a microfluidic Blood Brain Barrier model. Lab on a Chip 13: 1093–1101.

Preynat-Seauve, O., D.M. Suter, D. Tirefort, L. Turchi, T. Virolle, H. Chneiweiss, M. Foti, J.A. Lobrinus, L. Stoppini, A. Feki, M. Dubois-Dauphin, and K.H. Krause. 2009. Development of human nervous tissue upon differentiation of embryonic stem cells in three-dimensional culture. Stem Cells 27: 509–520.

Radio, N.M., and W.R. Mundy. 2008. Developmental neurotoxicity testing *in vitro*: models for assessing chemical effects on neurite outgrowth. Neurotoxicol 29: 361–376.

Raff, M.C., B.A. Barres, J.F. Burne, H.S. Coles, Y. Ishizaki, and M.D. Jacobson. 1993. Programmed cell death and the control of cell survival: lessons from the nervous system. Science 262: 695–700.

Ramadan, Q., H. Jafarpoorchekab, C. Huang, P. Silacci, S. Carrara, G. Koklu, J. Ghaye, J. Ramsden, C. Ruffert, G. Vergeres, and M.A. Gijs. 2013. NutriChip: nutrition analysis meets microfluidics. Lab Chip 13: 196–203.

Ramirez, T., M. Daneshian, H. Kamp, F.Y. Bois, M.R. Clench, M. Coen, B. Donley, S.M. Fischer, D.R. Ekman, E. Fabian, C. Guillou, J. Heuer, H.T. Hogberg, H. Jungnickel, H.C. Keun, G. Krennrich, E. Krupp, A. Luch, F. Noor, E. Peter, B. Riefke, M. Seymour, N. Skinner, L. Smirnova, E. Verheij, S. Wagner, T. Hartung, B. van Ravenzwaay, and M. Leist. 2013. Metabolomics in toxicology and preclinical research. ALTEX 30: 209–225.

Rana, T.M. 2007. Illuminating the silence: understanding the structure and function of small RNAs. Nature reviews. Mol Cell Biol 8: 23–36.

Ratajczak, H.V. 2011. Theoretical aspects of autism: biomarkers—a review. J Immunotoxicol 8: 80–94.

Reinhardt, P., B. Schmid, L.F. Burbulla, D.C. Schondorf, L. Wagner, M. Glatza, S. Hoing, G. Hargus, S.A. Heck, A. Dhingra, G. Wu, S. Muller, K. Brockmann, T. Kluba, M. Maisel, R. Kruger, D. Berg, Y. Tsytsyura, C.S. Thiel, O.E. Psathaki, J. Klingauf, T. Kuhlmann, M. Klewin, H. Muller, T. Gasser, H.R. Scholer, and J. Sterneckert. 2013. Genetic correction of a LRRK2 mutation in human iPSCs links parkinsonian neurodegeneration to ERK-dependent changes in gene expression. Cell Stem Cell 12: 354–367.

Rice, D., and S. Barone, Jr. 2000. Critical periods of vulnerability for the developing nervous system: evidence from humans and animal models. Environ Health Perspect 108 Suppl 3: 511–533.

Robinette, B.L., J.A. Harrill, W.R. Mundy, and T.J. Shafer. 2011. *In vitro* assessment of developmental neurotoxicity: use of microelectrode arrays to measure functional changes in neuronal network ontogeny. Front Neuroeng 4: 1. doi: 10.3389/fneng.2011.00001.

Rodier, P.M. 1980. Chronology of neuron development: animal studies and their clinical implications. Dev Med Child Neurol 22: 525–545.

Rodier, P.M. 1994. Vulnerable periods and processes during central nervous system development. Environ Health Perspect 102 Suppl 2: 121–124.

Roth, K.A., and C. D'Sa. 2001. Apoptosis and brain development. Ment Retard Dev Disabil Res Rev 7: 261–266.

Rovida, C., N. Alepee, A.M. Api, D.A. Basketter, F.Y. Bois, F. Caloni, E. Corsini, M. Daneshian, C. Eskes, J. Ezendam, H. Fuchs, P. Hayden, C. Hegele-Hartung, S. Hoffmann, B. Hubesch, M.N. Jacobs,

J. Jaworska, A. Kleensang, N. Kleinstreuer, J. Lalko, R. Landsiedel, F. Lebreux, T. Luechtefeld, M. Locatelli, A. Mehling, A. Natsch, J.W. Pitchford, D. Prater, P. Prieto, A. Schepky, G. Schuurmann, L. Smirnova, C. Toole, E. van Vliet, D. Weisensee, and T. Hartung. 2015. Integrated Testing Strategies (ITS) for safety assessment. ALTEX 32: 25–40.

Sacco, R., R. Militerni, A. Frolli, C. Bravaccio, A. Gritti, M. Elia, P. Curatolo, B. Manzi, S. Trillo, C. Lenti, M. Saccani, C. Schneider, R. Melmed, K.L. Reichelt, T. Pascucci, S. Puglisi-Allegra, and A.M. Persico. 2007. Clinical, morphological, and biochemical correlates of head circumference in autism. Biol Psychiatry 62: 1038–1047.

Sanchez-Danes, A., Y. Richaud-Patin, I. Carballo-Carbajal, S. Jimenez-Delgado, C. Caig, S. Mora, C. Di Guglielmo, M. Ezquerra, B. Patel, A. Giralt, J.M. Canals, M. Memo, J. Alberch, J. Lopez-Barneo, M. Vila, A.M. Cuervo, E. Tolosa, A. Consiglio, and A. Raya. 2012. Disease-specific phenotypes in dopamine neurons from human iPS-based models of genetic and sporadic Parkinson's disease. EMBO Mol Med 4: 380–395.

Scannell, J.W., A. Blanckley, H. Boldon, and B. Warrington. 2012. Diagnosing the decline in pharmaceutical R&D efficiency. Nature Reviews. Nat Rev drug Discov 11: 191–200.

Schettler, T. 2001. Toxic threats to neurologic development of children. Environ Health Perspect 109 Suppl 6: 813–816.

Schildknecht, S., S. Kirner, A. Henn, K. Gasparic, R. Pape, L. Efremova, O. Maier, R. Fischer, and M. Leist. 2012. Characterization of mouse cell line IMA 2.1 as a potential model system to study astrocyte functions. ALTEX 29: 261–274.

Schmid, C., and J.S. Rotenberg. 2005. Neurodevelopmental toxicology. Neurol Clinics 23: 321–336.

Schreiber, T., K. Gassmann, C. Gotz, U. Hubenthal, M. Moors, G. Krause, H.F. Merk, N.H. Nguyen, T.S. Scanlan, J. Abel, C.R. Rose, and E. Fritsche. 2010. Polybrominated diphenyl ethers induce developmental neurotoxicity in a human *in vitro* model: evidence for endocrine disruption. Environ Heath Perspect 118: 572–578.

Selimovic, S., M.R. Dokmeci, and A. Khademhosseini. 2013. Organs-on-a-chip for drug discovery. Curr Opin Pharmacol 13: 829–833.

Semino, C.E. 2003. Can we build artificial stem cell compartments? J Biomed Biotech 2003: 164–169.

Seol, Y., T. Kang, and D. Cho. 2012. Solid freeform fabrication technology applied to tissue engineering with various biomaterials. Soft Matter 8: 1730–1735.

Serra, M., S.B. Leite, C. Brito, J. Costa, M.J. Carrondo, and P.M. Alves. 2007. Novel culture strategy for human stem cell proliferation and neuronal differentiation. J Neurosci Res 85: 3557–3566.

Serra, M., C. Brito, E.M. Costa, M.F. Sousa, and P.M. Alves. 2009. Integrating human stem cell expansion and neuronal differentiation in bioreactors. BMC Biotechnol 9: 82.

Shamloo, A., N. Ma, M.M. Poo, L.L. Sohn, and S.C. Heilshorn. 2008. Endothelial cell polarization and chemotaxis in a microfluidic device. Lab Chip 8: 1292–1299.

Sharpe, J., U. Ahlgren, P. Perry, B. Hill, A. Ross, J. Hecksher-Sorensen, R. Baldock, and D. Davidson. 2002. Optical projection tomography as a tool for 3D microscopy and gene expression studies. Science 296: 541–545.

Sheridan, S.D., K.M. Theriault, S.A. Reis, F. Zhou, J.M. Madison, L. Daheron, J.F. Loring, and S.J. Haggarty. 2011. Epigenetic characterization of the FMR1 gene and aberrant neurodevelopment in human induced pluripotent stem cell models of fragile X syndrome. PloS one 6: e26203.

Shield, M.A., and P. E. Mirkes. 1998. Apoptosis. pp. 159–188. *In*: Slikker, W.B., and L.W. Chang (eds.). Handbook of Developmental Neurotoxicity Academic Press, San Diego, CA.

Smirnova, L., A. Sittka, and A. Luch. 2012. On the role of low-dose effects and epigenetics in toxicology. EXS 101: 499–550.

Smirnova, L., H.T. Hogberg, M. Leist, and T. Hartung. 2014. Developmental neurotoxicity—challenges in the 21st century and *in vitro* opportunities. ALTEX 31: 129–156.

Spitzer, N.C. 2006. Electrical activity in early neuronal development. Nature 444: 707–712.

Steinemann, D., G. Gohring, and B. Schlegelberger. 2013. Genetic instability of modified stem cells—a first step towards malignant transformation? Am J Stem Cells 2: 39–51.

Stelzer, Ernst H.K., Stefan Hell, Steffen Lindek, Reinhold Stricker, Rainer Pick, Clemens Storz, Georg Ritter, and N. Salmon. 1994. Nonlinear absorption extends confocal fluorescence microscopy into the ultra-violet regime and confines the illumination volume. Opt Commun 104: 223–228.

Sullivan, M. 2005. Autism increase not a result of reclassification. Clin Psychiat News 68.

Takahashi, K., K. Tanabe, M. Ohnuki, M. Narita, T. Ichisaka, K. Tomoda, and S. Yamanaka. 2007. Induction of pluripotent stem cells from adult human fibroblasts by defined factors. Cell 131: 861–872.

Takayama, S., E. Ostuni, P. LeDuc, K. Naruse, D.E. Ingber, and G.M. Whitesides. 2001. Subcellular positioning of small molecules. Nature 411: 1016.

Tamamaki, N., K. Nakamura, K. Okamoto, and T. Kaneko. 2001. Radial glia is a progenitor of neocortical neurons in the developing cerebral cortex. Neurosci Res 41: 51–60.

Thier, M., P. Worsdorfer, Y.B. Lakes, R. Gorris, S. Herms, T. Opitz, D. Seiferling, T. Quandel, P. Hoffmann, M.M. Nothen, O. Brustle, and F. Edenhofer. 2012. Direct conversion of fibroblasts into stably expandable neural stem cells. Cell Stem Cell 10: 473–479.

Tilson, H.A. 2000. Neurotoxicology risk assessment guidelines: developmental neurotoxicology. Neurotoxicol 21: 189–194.

Trounson, A., K.A. Shepard, and N.D. DeWitt. 2012. Human disease modeling with induced pluripotent stem cells. Curr Opin Genet Dev 22: 509–516.

USEPA. 1998. Health effects guidelines OPPTS 870.6300 developmental neurotoxicity study Office of Prevention Pesticides and Toxic Substances, http://www.regulations.gov/#!documentDetail;D=EPA-HQ-OPPT-2009-0156-0042.nepis.epa.gov/Exe/ZyPURL.cgi?Dockey=P100IRWO.TXT [last accessede 13 July 2016].

Uylings, H.B., and C.G. van Eden. 1990. Qualitative and quantitative comparison of the prefrontal cortex in rat and in primates, including humans. Progr Brain Res 85: 31–62.

Valdivia, P., M. Martin, W.R. LeFew, J. Ross, K.A. Houck, and T.J. Shafer. 2014. Multi-well microelectrode array recordings detect neuroactivity of ToxCast compounds. Neurotoxicol 44: 204–217.

van Eden, C.G., J.M. Kros, and H.B. Uylings. 1990. The development of the rat prefrontal cortex. Its size and development of connections with thalamus, spinal cord and other cortical areas. Progr Brain Res 85: 169–183.

van Vliet, E., L. Stoppini, M. Balestrino, C. Eskes, C. Griesinger, T. Sobanski, M. Whelan, T. Hartung, and S. Coecke. 2007. Electrophysiological recording of re-aggregating brain cell cultures on multi-electrode arrays to detect acute neurotoxic effects. Neurotoxicol 28: 1136–1146.

van Vliet, E., S. Morath, C. Eskes, J. Linge, J. Rappsilber, P. Honegger, T. Hartung, and S. Coecke. 2008. A novel *in vitro* metabolomics approach for neurotoxicity testing, proof of principle for methyl mercury chloride and caffeine. Neurotoxicol 29: 1–12.

van Vliet, E., M. Daneshian, M. Beilmann, A. Davies, E. Fava, R. Fleck, Y. Jule, M. Kansy, S. Kustermann, P. Macko, W.R. Mundy, A. Roth, I. Shah, M. Uteng, B. van de Water, T. Hartung, and M. Leist. 2014. Current approaches and future role of high content imaging in safety sciences and drug discovery. ALTEX 31: 479–493.

Vandenhaute, E., E. Sevin, D. Hallier-Vanuxeem, M.P. Dehouck, and R. Cecchelli. 2012. Case study: adapting *in vitro* blood-brain barrier models for use in early-stage drug discovery. Drug Discov Today 17: 285–290.

Vierbuchen, T., A. Ostermeier, Z.P. Pang, Y. Kokubu, T.C. Sudhof, and M. Wernig. 2010. Direct conversion of fibroblasts to functional neurons by defined factors. Nature 463: 1035–1041.

Viviani, B., E. Corsini, C.L. Galli, and M. Marinovich. 1998. Glia increase degeneration of hippocampal neurons through release of tumor necrosis factor-alpha. Toxicol Appl Pharmacol 150: 271–276.

Wajant, H. 2002. The Fas signaling pathway: more than a paradigm. Science 296: 1635–1636.

Wang, L.C., D.H. Baird, M.E. Hatten, and C.A. Mason. 1994. Astroglial differentiation is required for support of neurite outgrowth. The J Neurosci : The Official Journal of the Society for Neuroscience 14: 3195–3207.

Whitesides, G.M. 2006. The origins and the future of microfluidics. Nature 442: 368–373.

Wiggins, R.C. 1982. Myelin development and nutritional insufficiency. Brain Res 257: 151–175.

Wiggins, R.C. 1986. Myelination: a critical stage in development. Neurotoxicol 7: 103–120.

Wink, S., S. Hiemstra, S. Huppelschoten, E. Danen, G. Niemeijer, G. Hendriks, H. Vrieling, B. Herpers, and B. van de Water. 2014. Quantitative high content imaging of cellular adaptive stress response pathways in toxicity for chemical safety assessment. Chem Res Toxicol 27: 338–355.

Wu, M.H., S.B. Huang, and G.B. Lee. 2010. Microfluidic cell culture systems for drug research. Lab Chip 10: 939–956.

Xu, B.Y., S.W. Hu, G.S. Qian, J.J. Xu, and H.Y. Chen. 2013. A novel microfluidic platform with stable concentration gradient for on chip cell culture and screening assays. Lab Chip 13: 3714–3720.

Xu, T., J. Jin, C. Gregory, J.J. Hickman, and T. Boland. 2005. Inkjet printing of viable mammalian cells. Biomaterials 26: 93–99.

Xu, T., C.A. Gregory, P. Molnar, X. Cui, S. Jalota, S.B. Bhaduri, and T. Boland. 2006. Viability and electrophysiology of neural cell structures generated by the inkjet printing method. Biomaterials 27: 3580–3588.

Yeon, J.H., D. Na, K. Choi, S.W. Ryu, C. Choi, and J.K. Park. 2012. Reliable permeability assay system in a microfluidic device mimicking cerebral vasculatures. Biomed Microdevices 14: 1141–1148.

Yla-Outinen, L., J. Heikkila, H. Skottman, R. Suuronen, R. Aanismaa, and S. Narkilahti. 2010. Human cell-based micro electrode array platform for studying neurotoxicity. Front Neuroeng 3: pii: 111. doi: 10.3389/fneng.2010.00111.

Yla-Outinen, L., T. Joki, M. Varjola, H. Skottman, and S. Narkilahti. 2014. Three-dimensional growth matrix for human embryonic stem cell-derived neuronal cells. J Tissue Eng Regen Med 8: 186–194.

Yoo, S.S. 2015. 3D-printed biological organs: medical potential and patenting opportunity. Expert Opin Ther Pat 25: 507-511.

Yu, J., M.A. Vodyanik, K. Smuga-Otto, J. Antosiewicz-Bourget, J.L. Frane, S. Tian, J. Nie, G.A. Jonsdottir, V. Ruotti, R. Stewart, Slukvin, II, and J.A. Thomson. 2007. Induced pluripotent stem cell lines derived from human somatic cells. Science 318: 1917–1920.

Zhang, X., and Y. Zhang. 2015. Tissue engineering applications of three-dimensional bioprinting. Cell Biochem Biophys 72: 777–782.

Zimmer, B., P.B. Kuegler, B. Baudis, A. Genewsky, V. Tanavde, W. Koh, B. Tan, T. Waldmann, S. Kadereit, and M. Leist. 2011. Coordinated waves of gene expression during neuronal differentiation of embryonic stem cells as basis for novel approaches to developmental neurotoxicity testing. Cell Death Differ 18: 383–395.

Zurich, M.G., and F. Monnet-Tschudi. 2009. Contribution of *in vitro* neurotoxicology studies to the elucidation of neurodegenerative processes. Brain Res Bull 80: 211–216.

11

Anesthesia and Neural Plasticity

Laszlo Vutskits

Introduction

Every day worldwide, millions of patients receive general anesthesia for surgical and diagnostic procedures. The principal goals of general anesthesia are to provide loss of consciousness or sedation along with related anxiolysis and amnesia, analgesia and, if needed, muscle relaxation. These goals are usually achieved using a combination of drugs, and are aimed to protect the patient from harmful stimuli of many kinds during the perioperative period. Drugs which dose-dependently induce sedation or loss of consciousness are called general anesthetics (GA). While the exact mechanisms of actions through which GA induce loss of consciousness are to be yet incompletely elucidated, it is nevertheless well established that these drugs act on a wide variety of ligand-gated ion channels and, thereby, exert major influence on several neurotransmitter systems, including Υ-amino butyric acid (GABA) and glutamate (Rudolph and Antkowiak, 2004). These neurotransmitters play a major role in the maintenance of central nervous system (CNS) homeostasis and are of primary importance to determine the excitation/inhibition balance in neuronal networks (Akerman and Cline, 2007; Baroncelli *et al.*, 2011; Eichler and Meier, 2008; Nguyen *et al.*, 2001). Hence, GA-induced pharmacological interference with the physiological patterns of neurotransmission might rapidly induce dysregulation of brain homeostasis, which, in turn, could lead to loss of consciousness.

Department of Anesthesiology, Pharmacology and Intensive Care, University Hospital of Geneva, 4, rue Gabrielle-Perret-Gentil, 1211 Geneva 4, Switzerland.
E-mail: laszlo.vutskits@unige.ch

While exposure to general anesthesia has been initially considered to induce a rapidly reversible state of CNS function, increasing number of both clinical observations and laboratory studies conducted during the past few decades suggest that this might not be the case (Lin *et al.*, 2014). There is now ample experimental evidence, along with some human clinical data, suggesting that even a single and relatively short exposure to GA might generate lasting effects extending far beyond the perioperative period (Lei *et al.*, 2014; Lin *et al.*, 2014). Importantly, these effects have been suggested to represent both protective and toxic or even of therapeutic value (Vutskits, 2012; Vutskits, 2014). Indeed, the very same GA have been demonstrated to induce neurotoxicity when administered at critical periods of development, neuroprotection in the event of ischemic brain injury and even clinically relevant therapeutics in some psychiatric disorders such as depression. The ensemble of these observations thus strongly suggests that the impact of GA on the CNS depends on the pathophysiological context in which any given organism is exposed to these drugs. If so, GA should be considered as context-dependent modulators of neuronal homeostasis and thereby of neuronal plasticity. The aim of this chapter is to provide insights into this concept. First, to better understand the concept of how and why GA should be considered as context-dependent modulators of neural plasticity, some definitions on plasticity will be provided. This will be followed by presenting the potential impact of these drugs on the developing central nervous system. Mechanisms of plasticity in the context of anesthesia exposure in the adult brain will then be discussed. Finally recent insights into the therapeutic role of general anesthesia in major depressive disorders will be given. Wherever possible, fundamental molecular and cellular mechanisms linking neural plasticity to these phenomena will be discussed.

Neural Plasticity versus Neurotoxicity: Definitions

The word plasticity stems from the greek word "plasticos" (πλάστικός) that depicts the capability of an object or an organism of being shaped. Accordingly, neural plasticity can be defined as changes in neuronal systems due to changed environment or altered homeostatic conditions. These changes can involve both structural and molecular alterations, like neuronal loss or, in contrast, the generation of new neurons. Synaptic plasticity is an important aspect of neuronal plasticity and depicts the ability of a neuron or neuronal networks to change the strength of their connections in order to adapt to changed environment. This can be achieved via modulation of receptor number and/or composition of synapses as well as via changes in the morphology or even in the number of synaptic contacts. Synaptic plasticity has strong physiological relevance not only during CNS development but also in adults where it is generally considered to underlie fundamental mechanisms such as cognition and memory. Last but not least, synaptic plasticity does play a major role in recovery following lesions of the brain. Understanding the mechanisms guiding plasticity of neural networks is therefore of utmost importance.

It is very important to realize the fundamental differences between the concepts of "plasticity" and "toxicity". Indeed, toxicity derives from the Greek word "toxicon" (τοξικόν), which refers to the poison the solders put on their arrows (toxon - τοξόν) to kill the enemy. Thus, neurotoxicity rather reflects the degree to which a substance can cause physical harm or damage to the nervous system such as cell death. This is of course in striking contrast to plasticity used by biological systems to adapt their continuously changing environment.

Anesthesia and the Developing Brain: An Emerging Public Health Concern

The possibility that exposure to anesthesia during early life may induce long-term adverse effects on the central nervous system is an intensely discussed issue with potential public health implications (Rappaport *et al.*, 2015; Servick, 2014). There is experimental evidence in laboratory animals, from rodents to non-human primates, that exposure to GA can induce cell death, impaired neurogenesis and synaptic development as well as cognitive deficits in crude behavioral tests (Lin *et al.*, 2014). These effects depend on the developmental stage at which the animals exposed to GA and the length of exposure. While most currently used GA, with some possible exceptions, have been shown to induce cell death at distinct stages of brain development, we do not know whether there is a causal relationship between this phenomenon and altered neurocognitive function. Also, since the majority of these experiments have been performed in the absence of surgery, we do not know if GA in that context exacerbate injury or, in contrast, would protect from the surgical insult. The situation is by far more complicated in the human setting. There is a weak and mixed evidence for an association between anesthesia/surgery in early childhood and increased risk of poor neurodevelopmental outcome (Davidson *et al.*, 2015; Lei *et al.*, 2014). However, we have largely incomplete knowledge of which children are at greatest risk of poor developmental outcome in this context. Most importantly, we do not know whether the observed association is indeed linked to GA *per se* or other factors like surgery-induced inflammation and problems linked to anesthesia management might also contribute to these findings. At present, our knowledge is based on results obtained from retrospective human cohort studies with a large number of confounding factors. Given that adequate analgesia and reduction of perioperative stress response plays a major role in patient outcome following surgery (Anand and Hickey, 1992), it is definitely clear that anesthesia should not be withheld from young children at the current state of our knowledge. Neither there is evidence that any anesthetic regimen is better or worse in this context. Future research, both fundamental/translational and clinical is needed to provide answers to these issues.

Biological Rational for Anesthetics-induced Modulation of Neuronal Plasticity during Brain Development

General anesthetics have been demonstrated to influence neurotransmitter signaling through a wide variety of ligand-gated ionotropic transmembrane receptors (Rudolph and Antkowiak, 2004). Many of these receptors are expressed from early stages of development and were proposed to act as potent morphogens during the assembly of the CNS guiding both progenitor proliferation, cell migration and neuronal differentiation (Nguyen *et al.*, 2001; Owens and Kriegstein, 2002; Represa and Ben-Ari, 2005; Waters and Machaalani, 2004). A specific ontogenic progression in the expression of different receptor subunit genes characterizes the subunit assembly of the majority of ionotropic receptors suggesting that these protein complexes might respond differently to endogenous and exogenous ligands at distinct stages of neural development (Herlenius and Lagercrantz, 2004; Lujan *et al.*, 2005). For example, in humans, AMPA/kainate receptors are expressed in the neural tube as early as the 5th gestational week, while $GABA_A$ receptors appear around the 6th and NMDA receptors around the 10th week of gestation (Bardoul *et al.*, 1998; Ritter *et al.*, 2001). Anatomical mapping of ionotropic receptor subunits in the developing rodent CNS reveals a complex, transient and region-specific expression pattern of these proteins, and it is now well established that signaling characteristics of these receptors in the embryonic/early postnatal period differ markedly from those expressed in the adult brain (Hutcheon *et al.*, 2004; Laurie *et al.*, 1992; Poulter *et al.*, 1992; Yu *et al.*, 2006). This, in turn, implies that the effects of anesthetics on neural plasticity might depend upon the developmental stage at which neural systems are exposed to these drugs.

In the vertebrate CNS, formation of synaptic contacts takes place over a protracted period of development, extending from early embryonic life into adulthood. The most intense phase of synaptogenesis, also called the brain growth spurt, is highly species-dependent. In the rodent cerebral cortex, this period is restricted to a relatively narrow time window taking place between the second and fourth postnatal weeks (De Felipe *et al.*, 1997; Juraska, 1982). The brain region-specific temporal evolution pattern of synaptogenesis is by far the most described in the rhesus monkey, where an up to 17-fold increase in the number of synapses occurs within a few months during the perinatal period (Rakic *et al.*, 1994). In humans, based on analysis of postmortem tissue, a comparable extension in the number of synapses is observed between the third trimester of pregnancy and the first few years of postnatal life (Huttenlocher, 1979; Huttenlocher and Dabholkar, 1997; Huttenlocher and de Courten, 1987; Huttenlocher *et al.*, 1982; Petanjek *et al.*, 2011). In all these aforementioned species, the perinatal exponential synaptic growth leads to an initial overproduction of synapses and then to a selective pruning of synaptic contacts extending from puberty to adolescents and beyond (De Felipe *et al.*, 1997; Juraska, 1982; Petanjek *et al.*, 2011; Rakic *et al.*, 1994).

Early life experience plays a decisive role to prime life-long behavioral patterns. This has been convincingly demonstrated by the pioneering work of Spalding regarding filial imprinting in domestic chicken (Spalding, 1872), and later by Lorenz who demonstrated that incubator-hatched greylag geese would imprint on the first suitable stimulus they detect within a well-defined temporal window between 13–16 hours following hatching (Lorenz, 1958). These initial observations, providing convincing evidence on the persistent impact of brief stimuli when applied at critical developmental stages, were subsequently confirmed by a large amount of experimental and clinical data suggesting periods of extreme vulnerability in the majority of developing cognitive, motor and sensory systems (Hensch, 2004). Today, critical periods are defined as brain region-specific and temporally restricted time windows when brain circuits that subserve a given function are particularly receptive to acquiring certain kinds of information, or even need that instructive signal for their continued normal development (Hensch, 2004). Since this enhanced receptivity implies high modulability of developing neural circuitry, critical periods are characterized by significantly heightened levels of neural plasticity.

Experience translates into neuronal activity that, in turn, exerts a preponderant epigenetic influence on the development of neural circuitry (Zhang and Poo, 2001). Indeed, while type-specific morphology of CNS neurons as well as their initial connectivity patterns are genetically determined (Jan and Jan, 2003), neuronal activity triggers a variety of molecular pathways involved in axon pathfinding, dendritic differentiation and synaptogenesis aimed to refine developing synaptic networks (Zhang and Poo, 2001). Electrical activity has both direct and indirect effects on intracellular Ca2+ homeostasis and on the transduction efficacy of non-ionophoric membrane proteins (Cooper *et al.*, 1998; Ghosh and Greenberg, 1995). Importantly, synaptic currents-generated focal electric fields can guide electrophoretic or electroosmotic migration of various intracellular components, such as receptors and transporters, into the cell membrane or in within the cytoplasm (Poo and Young, 1990; Ranck, 1964). This, in turn, would lead to gradients in the cellular distribution patterns of these molecules, facilitating thereby specific molecular interactions and localizations at synaptic sites.

Pharmacological interference with physiological patterns of neuronal activity during critical periods of development can have significant long-term effects on neural circuitry assembly and function (Hensch, 2004). Through drug-induced targeting of GABAergic and glutamatergic neurotransmission, as achieved by general anesthetics, both hyperexcitation and silencing of neural networks have been shown to disrupt critical period plasticity, leading thereby to markedly altered information processing in the CNS (Hensch, 2005). A recent series of experiments shed light on the particular role of $GABA_A$ receptor-mediated neurotransmission on the control of timing regarding the onset and the duration of critical period plasticity (Hensch, 2005; Hensch *et al.*, 1998). Indeed, the onset of critical periods can be brought forward by enhancing GABAergic neurotransmission with benzodiazepines or by promoting the maturation of interneurons through excess of brain-derived neurotrophic factor (BDNF) expression. Conversely, delaying maturation of GABAergic systems delays the onset of plasticity. These findings are of fundamental importance and constitute the first steps towards understanding how critical period plasticity can be manipulated in order to enable the development of novel training paradigms and therapies for efficacious life-long learning or recovery following brain injuries.

Impact of General Anesthetics on Synaptic Plasticity during the Brain Growth Spurt

The aforementioned decisive role of neural activity in shaping brain development along with the fact that GA can exert major modulatory effects on physiological patterns neural activity raise the plausible hypothesis that these drugs could display a major impact on developing neuronal networks. A substantial amount of initial research investigated this possibility in the context of environmental exposure of pregnant operating room personnel to anesthesia gases such as halothane. Using experimental paradigms where rats were chronically exposed to low concentrations of halothane in utero and/or during the early postnatal period, the ensemble of these studies revealed that administration of this drug during CNS development permanently impaired dendritic arbor development, decreased synapse density and induced significant functional deficits in learning and behavior (Chang *et al.*, 1974; Quimby *et al.*, 1974; Quimby *et al.*, 1975; Uemura *et al.*, 1985).

The question of whether and how anesthetics impact on differentiating neural networks during the brain growth spurt has only been addressed during the last few years. One difficulty to study this question is that, as described above, developmental synaptogenesis is species-specifically actively ongoing over a period of weeks to years, an issue that has important implications regarding experimental design and extrapolation of animal data to human relevance. Indeed, since significant evolution in ionotropic receptor systems and transporters is observed during the brain growth spurt (Lujan *et al.*, 2005), functional characteristics of ionotropic receptor-mediated neurotransmissions can be fundamentally different between earlier and later stages of this period (Nusser, 2012). Of utmost importance, due to high intracellular concentrations of chloride compared to the extracellular space, $GABA_A$ receptor-mediated neurotransmission is depolarizing during the early phase of the brain growth spurt. The functional switch towards the hyperpolarizing actions of this neurotransmitter is linked to the developmental expression of the K+ - Cl- cotransporter KCC2, which extrudes actively intracellular Cl- from neurons (Kaila *et al.*, 2014; Rivera *et al.*, 1999). Significant increase in the expression pattern of this protein occurs between the 10th and 14th postnatal days in rodents and from around the last trimester of pregnancy in humans (Kaila *et al.*, 2014; Vanhatalo *et al.*, 2005). The ensemble of these observations thus strongly suggests a differential impact of GA on neurotransmitter systems and related neural network formation at distinct stages of the brain growth spurt. Hence, in laboratory investigations aimed to evaluate potential anesthesia-related adverse neurodevelopmental effects, focusing on one sole experimental time point most probably does not allow us to draw general conclusions on the effect of a given drug on synaptic growth.

In a seminal series of experiments, the group of Jevtovic-Todorovic exposed 7-day-old rat pups to a mixture of midazolam—nitrous oxide—isoflurane for 6 hours (Jevtovic-Todorovic *et al.*, 2003). In addition to observe significantly increased neuronal apoptosis one day following this treatment, they also demonstrated impairment in hippocampal neuronal networks by electrophysiological recordings using the long-term potentiation paradigm 3 weeks post anesthesia exposure (Jevtovic-Todorovic *et al.*, 2003). These latter results indicated that single anesthesia exposure during early stages of the brain growth spurt induces persistent alterations in synaptic function. In line with these data, the same group have subsequently demonstrated anesthesia induced neuropil scarcity, mitochondrial degeneration and a 30–40% decrease of synaptic profiles in the developing subiculum when evaluated two weeks after the anesthesia exposure (Lunardi *et al.*, 2010). In depth morphometric analysis of synaptic structure revealed a comparable loss of both excitatory and inhibitory synapses in the anesthesia exposed group. Also, the relative ratios between axo-spinous and axo-dendritic synaptic contacts between drug exposed and control littermates remained comparable, further suggesting a lasting and non-selective synaptotoxicity of anesthetics under these experimental conditions. In recent works, the authors also demonstrated a significant decrease of mitochondrial profiles in presynaptic terminals following the same anesthesia regimen, and this was correlated with highly disturbed synaptic transmission in the subiculum as well as in the CA1 region of the hippocampal formation. Since the important roles of mitochondria in fueling synaptogenesis and maintaining synaptic plasticity are well established, these results suggest that GA-induced interference with mitochondrial homeostasis might play a central role in anesthetics-induced impairment of synaptogenesis and thus of neural network function (Sanchez *et al.*, 2011). They also raise the possibility that pharmacological protection of mitochondria during general anesthesia protects developing neural circuitry (Boscolo *et al.*, 2013). In line with this hypothesis, recent laboratory work suggests that mitochondria in fact could be a potential therapeutic target to abolish or at least temper GA-induced morphological damage and related cognitive impairment in the developing brain (Ben-Shachar and Laifenfeld, 2004; Boscolo *et al.*, 2012).

Further insights into the molecular mechanisms involved in GA-induced synaptic loss during the early stages of the brain growth spurt from the group of Dr. Patel. These studies, conducted principally on 5–7-day-old primary neuronal cultures isolated from neonatal rodent brains, reveal a central role for GA-induced modulation of tissue plasminogen activator (tPA) release and a relatedly altered brain-derived neurotrophic factor (BDNF) signaling in anesthesia-induced synaptic loss and neuroapoptosis (Head *et al.*, 2009; Lemkuil *et al.*, 2011; Pearn *et al.*, 2012). Indeed, in this model, both isoflurane and propofol have been shown to reduce presynaptic release of tPA that is needed to convert plasminogen to plasmin which, in turn, is required to cleave constitutively released pro-BDNF to mature BDNF. While mature BDNF acts on its high affinity tyrosine kinase B receptor to promote cell survival and synaptic plasticity, pro-BDNF primarily binds to the low affinity neurotrophin receptor p75 (p75NTR). Activation of p75NTR, via the downstream effector RhoA kinase, induces actin depolarization and related synapse loss and cell death. In addition to providing a potential framework for future neuroprotective approaches, these results also bring important arguments in favor of the possibility that GA are powerful modulators of highly relevant signaling pathways implicated in the modulation of neural plasticity.

Another approach to study morpho-functional aspects of synaptic plasticity is to focus on dendritic spines, representing primary postsynaptic sites of excitatory synaptic contacts onto pyramidal neurons (Holtmaat and Svoboda, 2009). These structures are particularly well-suited for studying morpho-functional aspects of synaptogenesis in the developing as well as in the adult brain (Arellano *et al.*, 2007a). Given that less than 4% of dendritic spines lack synapses (Arellano *et al.*, 2007b), changes in dendritic spine densities reflect alterations in the amount of excitatory synaptic inputs onto these principal cells. Importantly, correlation between spine morphology and function is now well established. Among them, spine head volume has been shown to be directly proportional to the size of postsynaptic density, the presynaptic number of docked vesicles, and thus the releasable pool of neurotransmitters (Harris and Stevens, 1989; Schikorski and Stevens, 2001). Iontophoretic injection of Lucifer Yellow into pyramidal neurons is an extensively validated method to reliably reveal dendritic spine morphology (Benavides-Piccione *et al.*, 2002; Briner *et al.*, 2010; Briner *et al.*, 2011). Using this technique, and in line with existing ultrastructural data, our research group have recently shown a more than tenfold increase in dendritic spine densities of layer 5 pyramidal neurons during the first postnatal month in the rat medial prefrontal cortex (Briner *et al.*, 2011). To assess how GA affect this intense dendritic spine growth on these neurons, rat pups were exposed to propofol at distinct developmental time points throughout the brain growth spurt. These experiments revealed that the effect of propofol on dendritic spines depends on the developmental stage at which this drug is administered. In fact, when administered at postnatal day 5 or 10, even a single dose of propofol induced a significant decrease in the number of dendritic spines, while an important increase in spine density was observed when the drug was administered at later stages of the peak synaptogenic period, such as postnatal day 15 or 20 (Briner *et al.*, 2011). Of utmost importance, the acute effects of propofol exposure on dendritic spines during early postnatal life persisted into adulthood in these animals, suggesting thereby that even brief exposure to this drug could permanently alter circuitry assembly and function. Since propofol, similarly to the majority of currently used GA, is a powerful potentiator of $GABA_A$ receptor-mediated neurotransmission, one plausible hypothesis is that these differential effects depend on the developmental switch regarding the functional modalities of GABAergic neurotransmission. To gain insights into this issue, we have thus precociously expressed KCC2 in a subset of pyramidal neurons by means of in utero electroporation resulting in an increased efficacy of chloride extrusion and thereby of GABAergic inhibition, in these cells. This genetic manipulation completely prevented propofol-induced dendritic spine and synaptic loss during early stages of the brain growth spurt (Jevtovic-Todorovic *et al.*, 2013). These results mean that KCC2-dependent developmental increase in the efficacy of $GABA_A$R-mediated inhibition is a major determinant of the age-dependent actions of propofol, and probably of other anesthetics acting on the $GABA_A$R, on synapse assembly in the developing cerebral cortex.

The developmental stage-dependent effects of anesthetics on circuitry development might also be related to the important changes in $GABA_A$ receptor subunit composition during early postnatal life which, in turn, significantly modify GABAergic inhibitory tone modalities (Bosman *et al.*, 2002; Fritschy *et al.*, 1994; Laurie *et al.*, 1992; Yu *et al.*, 2006). In this context, a role for the α1 subunit of this receptor complex in excitatory synaptogenesis has recently been suggested (Heinen *et al.*, 2003). In the rodent brain, the expression of the α1 subunit is upregulated around PND 14, and genetic deletion of this subunit results in significantly impaired dendritic spine development (Heinen *et al.*, 2003). It is also conceivable that, independent of their actions on the $GABA_A$ receptor complex, anesthetics can also modify the secretion or signaling pathways of trophic factors that are directly involved in synaptogenesis. Amongst these molecules, brain-derived neurotrophic factor is an interesting candidate and, as mentioned above, a dual role for this neurotrophin has recently been suggested to explain the developmental stage-dependent effects of anesthetics on synaptogenesis (Head *et al.*, 2009). Finally, it is important to note that the early postnatal period is characterized by major changes in the expression patterns and functions of several molecular classes, including, amongst others, adhesion molecules, growth factor and neurotransmitter signaling pathways, all of which are known to be important determinants of synaptogenesis. Clearly, a large number of further studies are needed to investigate these important issues.

The fact that propofol rapidly induces significant increases in dendritic spine densities during the brain growth spurt gives rise to the possibility that GA could also exert an important impact on the assembly of neural circuitry at developmental stages when these drugs no longer induce neuroapoptosis anymore. To further extend this possibility, in a related series of works, we have shown that other anesthetics, such as volatile agents, midazolam or ketamine exert similar boosting effects on dendritic spine densities (Briner *et al.*, 2010; De Roo *et al.*, 2009). The phenomenon of GA-induced increase in dendritic spine densities is not restricted to the medial prefrontal cortex but could also be found in other cortical regions as well as in the hippocampus. Most importantly, using quantitative electron microscopy, we have shown that increases in the number of dendritic spines indeed reflect an increase in the number of excitatory synaptic profiles (Briner *et al.*, 2011). Altogether, these observations further strengthen the view that GA can act as powerful modulators of synaptogenesis during critical periods of brain development.

During the early postnatal period, dendritic spines display intense protrusive motility resulting in the possibility of spine formation/elimination in within the time scale of minutes. Post-hoc analysis of dendritic spine and synapse density or shape will thus not allow us to gain insight into how anesthetics modulate these highly dynamic processes. On the other hand, since *in vivo* tracing of dendritic spine plasticity is only feasible under general anesthesia, these experiments would be hindered by the necessity of exposing the animals to anesthetics. To circumvent this issue, we took advantage of organotypic hippocampal slice cultures where the three-dimensional organization of neural circuitry is preserved (Stoppini *et al.*, 1991) and the temporal dynamics of dendritic spines can be traced using two-photon confocal microscopy (De Roo *et al.*, 2008a; De Roo *et al.*, 2008b). Using this approach, we have shown that the GA-induced increase in spine density is mediated through an increased rate of protrusion formation, a better stabilization of newly formed spines that, ultimately, led to an increased formation of functional synapses (De Roo *et al.*, 2009).

Findings that both anesthetics that enhance $GABA_A$ receptor-mediated inhibition and those blocking NMDA receptor mediated excitation promoted spine growth and synaptogenesis suggests that GA-induced interference with physiological patterns of E/I balance is a major determinant of this drug-induced synaptic growth process. This possibility is in line with previous observations where prolonged pharmacological interference with excitatory or inhibitory receptors impaired spine densities and synapse properties (Adesnik *et al.*, 2008; Luthi *et al.*, 2001; Mateos *et al.*, 2007; Muller *et al.*, 1993; Ultanir *et al.*, 2007; Zha *et al.*, 2005). Compared to these previous reports, our observations demonstrate that spine growth triggered by changes in E/I balance can be extremely fast, and is achieved by blockade of both AMPA and NMDA receptors, indicating that it is very sensitive to NMDA-dependent mechanisms (De Roo *et al.*, 2009). The functional relevance of this anesthesia-triggered, E/I balance-mediated massive reorganization of synaptic circuitry during the brain growth spurt remains to be determined. It is tempting to speculate that the fast regulation of synapse number could adjust the level of excitatory activity required for ensuring plasticity-mediated mechanisms. This could be particularly important at times when synapse turnover is high and selection of appropriate contacts is a major issue.

Whether anesthetics-induced perturbations of synaptogenesis are translated to behavioral or cognitive alterations is an important issue. In this context, previous experimental work suggests a causal link between early anesthesia exposure and impaired cognitive testing in multiple behavioral paradigms at later periods (Fredriksson *et al.*, 2007; Jevtovic-Todorovic *et al.*, 2003; Stratmann *et al.*, 2009). How these changes are related to impaired neural circuitry development remains to be explored. A growing body of human pathological evidence indicates that many psychiatric and neurological disorders, ranging from mental retardation and autism to Alzheimer's disease and addiction, are accompanied by alterations in spine morphology and synapse number (Penzes *et al.*, 2011; van Spronsen and Hoogenraad, 2010). It is thus tempting to speculate that the anesthetics-induced alterations in synaptic density and structure might stand as a morphological basis underlying impaired neurocognitive performance associated with early anesthesia exposure in humans (Flick *et al.*, 2011; Sprung *et al.*, 2012; Wilder *et al.*, 2009).

Most experimental research in the context of developmental anesthesia neuroplasticity has been focused on the excitatory components of neural transmission and much less is known about how anesthetics influence the development of GABAergic systems. The functional units of these networks are GABA-releasing interneurons, representing 10%–25% of all neurons in the cerebral cortex (Le Magueresse and Monyer, 2013). These cells are primarily generated in embryonic life in the ganglionic eminences of the ventral telencephalon, from where they migrate toward the cerebral cortex to control and orchestrate the activity of pyramidal neurons. Compared to principal cells, GABAergic interneurons display a wide variety of morphological and physiological characteristics, rendering their classification rather challenging (Petilla Interneuron Nomenclature *et al.*, 2008). Additionally, interneurons display a high degree of functional plasticity by means of environmentally-induced changes in the expression patterns of several intracellular calcium binding proteins such as parvalbumin, calbindin or calretinin (Markram *et al.*, 2004). This form of GABAergic plasticity has been suggested to play an important role during critical periods of early postnatal life when immature neural circuitry is particularly sensitive to extrinsic cues (Hensch, 2004). To gain insights into this issue, in a recent series of experiments, we evaluated whether and how the benzodiazepine midazolam influences expression patterns of calcium binding proteins in developing GABAergic interneurons (Osterop *et al.*, 2015). We found that exposure at P5 led to a subsequent increase in the number of parvalbumin positive neurons in lower cortical layers, and midazolam administration at P15 increased the number of both parvalbumin and calretinin expressing neurons 5 days following exposure in the rat medial prefrontal cortex. The physiological significance of these findings remains to be determined.

General Anesthetics and Neural Plasticity in the Adult Brain

In the majority of mammalian species, including humans, generation of adult-born neurons takes places in two distinct brain regions: the anterior subventricular zone and the hippocampal dentate gyrus (Abrous *et al.*, 2005; Curtis *et al.*, 2007; Eriksson *et al.*, 1998). An increasing number of experimental studies support a role for adult neurogenesis in memory processing of the adult brain (Abrous *et al.*, 2005; Deng *et al.*, 2010; Koehl and Abrous, 2011). Newly generated neurons from the adult anterior subventricular zone migrate into the olfactory bulb where they are involved in olfactory memories (Arenkiel, 2010; Breton-Provencher *et al.*, 2009; Mandairon *et al.*, 2011), whereas adult-born dentate granule cells integrate into hippocampal circuitry and play a role in mediating complex forms of spatial and associative memories (Deng *et al.*, 2010; Koehl and Abrous, 2011). Significant alterations in the rate of adult neurogenesis have been reported in experimental models of depression, epilepsy, neurodegenerative disorders and stroke (Kokaia, 2011; Samuels and Hen, 2011; Winner *et al.*, 2011). Although direct evidence is currently lacking, animal data strongly suggest a causal link between altered neurogenesis and cognitive dysfunction (Villeda *et al.*, 2011).

Neural network activity plays a crucial role in hippocampal neurogenesis (Deng *et al.*, 2010). Initial differentiation of newly born dentate granule cells relies on tonic activation by ambient GABA. These cells then establish GABAergic synapses and thereby start to integrate into the hippocampal network during the second week after their birth while glutamatergic synaptic inputs are detectable from the third week (Deng *et al.*, 2010). The physiological balance between E/I signaling is of paramount importance for appropriate dentate granule cell development and function (Liu, 2004; Marin-Burgin *et al.*, 2012). Related to this context, an important series of recent works evaluated how anesthetics interfere with hippocampal neurogenesis (Stratmann *et al.*, 2010; Stratmann *et al.*, 2009). In these experiments, exposure of 7-day-old rats to isoflurane decreased neuronal progenitor proliferation and caused persistent deficit of hippocampal function as revealed by fear conditioning and spatial reference memory tasks. In contrast, isoflurane-induced interference with neuronal homeostasis in 2-month-old rats increased progenitor proliferation and neuronal differentiation and this was correlated with improved memory function. Interestingly, the same treatment paradigm did not seem to interfere with hippocampal neurogenesis and function in aged rats. An important issue in the context of adult neurogenesis is to determine whether and how anesthetics modify the integration of adult neurogenic zone-derived newly born neurons into the existing functional circuitry in these brain regions. To study this question, we labeled progenitors of hippocampal neurons using a retroviral vector expressing the green fluorescent protein in adult mice. We then exposed these animals to a 6-hour-long propofol anesthesia at eleven or seventeen days after the labeling procedure, corresponding to two functionally distinct stage of hippocampal progenitor differentiation (Krzisch *et al.*, 2013). Our results revealed that propofol impaired the survival and maturation of adult-born neurons in an age-dependent manner by inducing reduced dendritic maturation of 17-day-old but not that of 11-day-old neurons. These data are further strengthened with new observations demonstrating that it is most probably the age of the neuron rather than the post conceptual age that determines vulnerability to anesthesia exposure (Hofacer *et al.*, 2013).

The Influence of General Anesthesia on Experimental Designs in Young Animals

The fact that, at least during the brain growth spurt, anesthetics rapidly and significantly impair dendritic spine growth and synaptogenesis have major technical implications regarding experimental design in studies where young animals should undergo general anesthesia during the early postnatal period for surgery or, most importantly, for live imaging of dendritic spine plasticity. For example, recent studies using *in vivo* 2-photon confocal imaging have reported important pruning in spine number between the second and fourth week of life in mice (Holtmaat *et al.*, 2005). This is, however, at variance with *in vivo* studies based on classical staining methods demonstrating a continuous increase in spine densities during this very same period (De Felipe *et al.*, 1997; Micheva and Beaulieu, 1996; Petit *et al.*, 1988). One plausible explanation for this apparent discrepancy is that the in vivo imaging studies have been carried out in mice that had undergone a long anesthesia for the preparation and image acquisition during the 2-photon imaging approach. This manipulation probably boosted spine density in these young mice and, as shown by our studies (Briner *et al.*, 2010; Briner *et al.*, 2011; De Roo *et al.*, 2009), analysis in animals anesthetized around postnatal day 15 then suggest a subsequent pruning of spines over the next two weeks, while analysis of mice that did not undergo anesthesia conversely indicate a progressive increase in spine density. This example, among probably many others, underlines the importance of anesthesia as an important source of misinterpretation of experimental data during the early postnatal period.

Based on the developmental stage-dependent differential effects of anesthetics on synaptic growth, assessing multiple developmental time points in experimental studies is of utmost importance in view of human relevance. Initial seminal works revealing the possibility of anesthesia-related developmental neurotoxicity focused on 7-day-old rodent pups (Ikonomidou *et al.*, 2000; Ikonomidou *et al.*, 1999; Jevtovic-Todorovic *et al.*, 2003). Although it is very difficult to properly extrapolate developmental stages between species, recent studies attempting to draw temporal correlation between rodent and human brain development suggest that the developmental stage of the 7-day-old rodent brain corresponds to the maturational state of the human brain at the very beginning of the third trimester of pregnancy (Clancy *et al.*, 2001; Clancy *et al.*, 2007). In line with these observations, electron microscopic data from the rodent cerebral cortex demonstrates only a low number of synaptic contacts in 7-day-old pups while the most intense phase of synaptogenesis occurs between the 15th and 20th postnatal days (De Felipe *et al.*, 1997). Similar intensity of synaptic growth in the human cerebral cortex takes place during the first year postnatal life (Huttenlocher and Dabholkar, 1997). Together, these data suggest that administration of anesthetics to 7-day-old rodent pups might reflect drug exposure of mid-term fetuses or very premature neonates in humans while experiments conducted in 15- to 20-day-old animals would be more relevant to a period situated somewhere during the first few years on the human scale.

General Anesthetics as Therapeutics in Major Depressive Disorders

Clinical observations, accumulating over the past 30 years, suggest that GA could exert important therapeutic effects under some psychopathological conditions such as major depressive disorders. The idea for such a possibility appeared with the advent of electroconvulsive therapy (ECT), where seizures were most often induced under general anesthesia. Since general anesthesia, in itself, is a robust modulator of electrical activity in the brain, determining whether the therapeutic value of ECT is indeed linked to the passage of electrode-delivered electricity or to the impact of general anesthetics on neural networks is of utmost importance. Results from early observational studies aimed to address this problem were conflicting and open to important methodological criticism due to very low sample size and to the lack of adequate comparison groups (Brill *et al.*, 1959; McDonald *et al.*, 1966; Miller *et al.*, 1953). The first randomized trial, where 70 patients who fulfilled indications for ECT were randomly allocated either to a course of 8 simulated ECT (i.e., barbiturate anesthesia alone) or to a course of 8 real ECTs (under barbiturate anesthesia) nevertheless showed no significant long-term difference between the anesthesia only and the ECT group when evaluated by psychiatrists unaware of the treatment paradigm (Johnstone *et al.*, 1980).

It has been suggested that the brief period of electrical silence in the cerebral cortex following ECT-induced grand mal seizure could be a crucial biological determinant for the therapeutic effects of ECT (Ottosson, 1962; Volavka *et al.*, 1972). Based on this assumption, the effects of isoflurane-induced cortical burst suppression have been studied in treatment-refractory depressed patients. In a first series of observations, global psychopathological assessment demonstrated rapid improvement following this narcotherapy in 9 out of 11 patients (Langer *et al.*, 1985), and this has been confirmed by another independent study (Carl *et al.*, 1988). Following these observations, a prospective double-blind study compared high isoflurane concentration-based narcotherapy (associated with cortical electrical silence) with ECT (in the presence of standard isoflurane concentrations for anesthesia maintenance) in drug-refractory severely depressed women (10 patients per group) (Langer *et al.*, 1995). Patients from each group were subjected to treatment twice a week for a total of 6 sessions. Interestingly, rapid antidepressant effects of the first treatment session were only significant in the isoflurane narcotherapy group. At subsequent sessions, the effects were comparable between the two groups. During follow-up, patients in the isoflurane group improved further, while those in the ECT group tended to relapse.

Recent data indicate robust and rapid antidepressant effects of subanesthetic concentrations of ketamine. In a seminal randomized, placebo-controlled, double-blind crossover study, subjects with major depression were given intravenous infusion of ketamine at subanesthetic doses (0.5 mg/kg) or placebo on two test days a week apart (Zarate *et al.*, 2006). Significant improvement in the 21-item Hamilton Depression Rating Scale was found as early as 2 hours following ketamine but not placebo injection, and this amelioration was maintained for at least one week in more than one third of patients. These observations are supported by several subsequent works (Covvey *et al.*, 2012). Therefore, they raise powerful new therapeutic options in a pathology that is of major public health concern and where weeks or months are required for standard medications to be effective. New experimental observations have provided cellular and molecular insights into the mechanisms underlying these antidepressant effects of ketamine (Jernigan *et al.*, 2011; Li *et al.*, 2010). In these studies, ketamine rapidly activated the mammalian target of rapamycin (mTOR) pathway, leading to an increased number of dendritic spines and improved behavioral response in a rodent model of depression. Ketamine has also been shown to reduce the phosphorylation of eukaryotic elongation factor 2 (eEF2) and a related de-suppression of translation of BDNF, which, in turn, induces fast-acting behavioral antidepressant-like effects (Autry *et al.*, 2011). Altogether, these observations clearly suggest a potential for GA to modulate synaptic plasticity in the adult brain. They could also provide potential therapeutic value of these drugs in a variety of neuropsychiatric disorders with impaired neural plasticity such as schizophrenia (Ben-Shachar and Laifenfeld, 2004) or even Alzheimer's disease (Baloyannis, 2011). Investigating further on these concepts is clearly of fundamental importance.

Conclusions

Clinical observations along with experimental research over the past few decades provided us with a wealth of convincing data suggesting that GA, in addition to provide loss of consciousness to facilitate surgery, also exert a powerful, context-dependent modulatory impact on the CNS. During development, exposure to these drugs can induce life long changes in neural circuitry and, therefore, in function. In adults, GA have been shown to interfere with cellular and molecular events subserving the maintenance of plasticity. On one hand, this could explain the potentially, although as yet unproved, harmful effects of these drugs on cognitive function in the elderly. On the other hand, the possibility to modulate neural plasticity by GA could provide us with therapeutic tools to improve function. This has been convincingly demonstrated in major depressive disorders and raises the, at least theoretical, basis of neuroprotection in other CNS disorders and insults. Altogether, these observations tend to switch the view of anesthesia neurotoxicity toward the concept of context-dependent modulation of neural plasticity by GA, and, thereby, offer a new level of understanding on how these drugs affect the CNS.

References

Abrous, D.N., M. Koehl, and M. Le Moal. 2005. Adult neurogenesis: from precursors to network and physiology. Physiol Rev 85: 523–569.

Adesnik, H., G. Li, M.J. During, S.J. Pleasure, and R.A. Nicoll. 2008. NMDA receptors inhibit synapse unsilencing during brain development. Proc Natl Acad Sci U S A 105: 5597–5602.

Akerman, C.J., and H.T. Cline. 2007. Refining the roles of GABAergic signaling during neural circuit formation. Trends Neurosci 30: 382–389.

Anand, K.J., and P.R. Hickey. 1992. Halothane-morphine compared with high-dose sufentanil for anesthesia and postoperative analgesia in neonatal cardiac surgery. N Engl J Med 326: 1–9.

Arellano, J.I., R. Benavides-Piccione, J. Defelipe, and R. Yuste. 2007a. Ultrastructure of dendritic spines: correlation between synaptic and spine morphologies. Front Neurosci 1: 131–143.

Arellano, J.I., A. Espinosa, A. Fairen, R. Yuste, and J. DeFelipe. 2007b. Non-synaptic dendritic spines in neocortex. Neuroscience 145: 464–469.

Arenkiel, B.R. 2010. Adult neurogenesis supports short-term olfactory memory. J Neurophysiol 103: 2935–2937.

Autry, A.E., M. Adachi, E. Nosyreva, E.S. Na, M.F. Los, P.F. Cheng, E.T. Kavalali, and L.M. Monteggia. 2011. NMDA receptor blockade at rest triggers rapid behavioural antidepressant responses. Nature 475: 91–95.

Baloyannis, S.J. 2011. Mitochondria are related to synaptic pathology in Alzheimer's disease. Int J Alzheimers Dis 2011: 305395.

Bardoul, M., C. Levallois, and N. Konig. 1998. Functional AMPA/kainate receptors in human embryonic and foetal central nervous system. J Chem Neuroanat 14: 79–85.

Baroncelli, L., C. Braschi, M. Spolidoro, T. Begenisic, L. Maffei, and A. Sale. 2011. Brain plasticity and disease: a matter of inhibition. Neural Plast 2011: 286073.

Ben-Shachar, D., and D. Laifenfeld. 2004. Mitochondria, synaptic plasticity, and schizophrenia. Int Rev Neurobiol 59: 273–296.

Benavides-Piccione, R., I. Ballesteros-Yanez, J. DeFelipe, and R. Yuste. 2002. Cortical area and species differences in dendritic spine morphology. J Neurocytol 31: 337–346.

Boscolo, A., J.A. Starr, V. Sanchez, N. Lunardi, M.R. DiGruccio, C. Ori, A. Erisir, P. Trimmer, J. Bennett, and V. Jevtovic-Todorovic. 2012. The abolishment of anesthesia-induced cognitive impairment by timely protection of mitochondria in the developing rat brain: the importance of free oxygen radicals and mitochondrial integrity. Neurobiol Dis 45: 1031–1041.

Boscolo, A., D. Milanovic, J.A. Starr, V. Sanchez, A. Oklopcic, L. Moy, C.C. Ori, A. Erisir, and V. Jevtovic-Todorovic. 2013. Early exposure to general anesthesia disturbs mitochondrial fission and fusion in the developing rat brain. Anesthesiology 118: 1086–1097.

Bosman, L.W., T.W. Rosahl, and A.B. Brussaard. 2002. Neonatal development of the rat visual cortex: synaptic function of GABAA receptor alpha subunits. J Physiol 545: 169–181.

Breton-Provencher, V., M. Lemasson, M.R. Peralta, 3rd, and A. Saghatelyan. 2009. Interneurons produced in adulthood are required for the normal functioning of the olfactory bulb network and for the execution of selected olfactory behaviors. J Neurosci 29: 15245–15257.

Brill, N.Q., E. Crumpton, S. Eiduson, H.M. Grayson, L.I. Hellman, and R.A. Richards. 1959. Relative effectiveness of various components of electroconvulsive therapy; an experimental study. AMA Arch Neurol Psychiatry 81: 627–635.

Briner, A., M. De Roo, A. Dayer, D. Muller, W. Habre, and L. Vutskits. 2010. Volatile anesthetics rapidly increase dendritic spine density in the rat medial prefrontal cortex during synaptogenesis. Anesthesiology 112: 546–556.

Briner, A., I. Nikonenko, M. De Roo, A. Dayer, D. Muller, and L. Vutskits. 2011. Developmental Stage-dependent persistent impact of propofol anesthesia on dendritic spines in the rat medial prefrontal cortex. Anesthesiology 115: 282–293.

Carl, C., W. Engelhardt, G. Teichmann, and G. Fuchs. 1988. Open comparative study with treatment-refractory depressed patients: electroconvulsive therapy—anesthetic therapy with isoflurane (preliminary report). Pharmacopsychiatry 21: 432–433.

Chang, L.W., A.W. Dudley, Jr., Y.K. Lee, and J. Katz. 1974. Ultrastructural changes in the nervous system after chronic exposure to halothane. Exp Neurol 45: 209–219.

Clancy, B., R.B. Darlington, and B.L. Finlay. 2001. Translating developmental time across mammalian species. Neuroscience 105: 7–17.

Clancy, B., B.L. Finlay, R.B. Darlington, and K.J. Anand. 2007. Extrapolating brain development from experimental species to humans. Neurotoxicology 28: 931–937.

Cooper, D.M., M.J. Schell, P. Thorn, and R.F. Irvine. 1998. Regulation of adenylyl cyclase by membrane potential. J Biol Chem 273: 27703–27707.

Covvey, J.R., A.N. Crawford, and D.K. Lowe. 2012. Intravenous ketamine for treatment-resistant major depressive disorder. Ann Pharmacother 46: 117–123.

Curtis, M.A., M. Kam, U. Nannmark, M.F. Anderson, M.Z. Axell, C. Wikkelso, S. Holtas, W.M. van Roon-Mom, T. Bjork-Eriksson, C. Nordborg, J. Frisen, M. Dragunow, R.L. Faull, and P.S. Eriksson. 2007. Human neuroblasts migrate to the olfactory bulb via a lateral ventricular extension. Science 315: 1243–1249.

Davidson, A.J., K. Becke, J. de Graaff, G. Giribaldi, W. Habre, T. Hansen, R.W. Hunt, C. Ing, A. Loepke, M.E. McCann, G.D. Ormond, A. Pini Prato, I. Salvo, L. Sun, L. Vutskits, S. Walker, and N. Disma. 2015. Anesthesia and the developing brain: a way forward for clinical research. Paediatr Anaesth 25: 447–452.

De Felipe, J., P. Marco, A. Fairen, and E.G. Jones. 1997. Inhibitory synaptogenesis in mouse somatosensory cortex. Cereb Cortex 7: 619–634.

De Roo, M., P. Klauser, P. Mendez, L. Poglia, and D. Muller. 2008a. Activity-dependent PSD formation and stabilization of newly formed spines in hippocampal slice cultures. Cereb Cortex 18: 151–161.

De Roo, M., P. Klauser, and D. Muller. 2008b. LTP promotes a selective long-term stabilization and clustering of dendritic spines. PLoS Biol 6: e219.

De Roo, M., P. Klauser, A. Briner, I. Nikonenko, P. Mendez, A. Dayer, J.Z. Kiss, D. Muller, and L. Vutskits. 2009. Anesthetics rapidly promote synaptogenesis during a critical period of brain development. PLoS One 4: e7043.

Deng, W., J.B. Aimone, and F.H. Gage. 2010. New neurons and new memories: how does adult hippocampal neurogenesis affect learning and memory? Nat Rev Neurosci 11: 339–350.

Eichler, S.A., and J.C. Meier. 2008. E-I balance and human diseases—from molecules to networking. Front Mol Neurosci 1: 2.

Eriksson, P.S., E. Perfilieva, T. Bjork-Eriksson, A.M. Alborn, C. Nordborg, D.A. Peterson, and F.H. Gage. 1998. Neurogenesis in the adult human hippocampus. Nat Med 4: 1313–1317.

Flick, R.P., S.K. Katusic, R.C. Colligan, R.T. Wilder, R.G. Voigt, M.D. Olson, J. Sprung, A.L. Weaver, D.R. Schroeder, and D.O. Warner. 2011. Cognitive and behavioral outcomes after early exposure to anesthesia and surgery. Pediatrics 128: e1053–1061.

Fredriksson, A., E. Ponten, T. Gordh, and P. Eriksson. 2007. Neonatal exposure to a combination of N-methyl-D-aspartate and gamma-aminobutyric acid type A receptor anesthetic agents potentiates apoptotic neurodegeneration and persistent behavioral deficits. Anesthesiology 107: 427–436.

Fritschy, J.M., J. Paysan, A. Enna, and H. Mohler. 1994. Switch in the expression of rat GABAA-receptor subtypes during postnatal development: an immunohistochemical study. J Neurosci 14: 5302–5324.

Ghosh, A., and M.E. Greenberg. 1995. Calcium signaling in neurons: molecular mechanisms and cellular consequences. Science 268: 239–247.

Harris, K.M., and J.K. Stevens. 1989. Dendritic spines of CA 1 pyramidal cells in the rat hippocampus: serial electron microscopy with reference to their biophysical characteristics. J Neurosci 9: 2982–2997.

Head, B.P., H.H. Patel, I.R. Niesman, J.C. Drummond, D.M. Roth, and P.M. Patel. 2009. Inhibition of p75 neurotrophin receptor attenuates isoflurane-mediated neuronal apoptosis in the neonatal central nervous system. Anesthesiology 110: 813–825.

Heinen, K., R.E. Baker, S. Spijker, T. Rosahl, J. van Pelt, and A.B. Brussaard. 2003. Impaired dendritic spine maturation in GABAA receptor alpha1 subunit knock out mice. Neuroscience 122: 699–705.

Hensch, T.K., M. Fagiolini, N. Mataga, M.P. Stryker, S. Baekkeskov, and S.F. Kash. 1998. Local GABA circuit control of experience-dependent plasticity in developing visual cortex. Science 282: 1504–1508.

Hensch, T.K. 2004. Critical period regulation. Annu Rev Neurosci 27: 549–579.

Hensch, T.K. 2005. Critical period plasticity in local cortical circuits. Nat Rev Neurosci 6: 877–888.

Herlenius, E., and H. Lagercrantz. 2004. Development of neurotransmitter systems during critical periods. Exp Neurol 190 Suppl 1: S8–21.

Hofacer, R.D., M. Deng, C.G. Ward, B. Joseph, E.A. Hughes, C. Jiang, S.C. Danzer, and A.W. Loepke. 2013. Cell age-specific vulnerability of neurons to anesthetic toxicity. Ann Neurol 73: 695–704.

Holtmaat, A., and K. Svoboda. 2009. Experience-dependent structural synaptic plasticity in the mammalian brain. Nat Rev Neurosci 10: 647–658.

Holtmaat, A.J., J.T. Trachtenberg, L. Wilbrecht, G.M. Shepherd, X. Zhang, G.W. Knott, and K. Svoboda. 2005. Transient and persistent dendritic spines in the neocortex *in vivo*. Neuron 45: 279–291.

Hutcheon, B., J.M. Fritschy, and M.O. Poulter. 2004. Organization of GABA receptor alpha-subunit clustering in the developing rat neocortex and hippocampus. Eur J Neurosci 19: 2475–2487.

Huttenlocher, P.R. 1979. Synaptic density in human frontal cortex—developmental changes and effects of aging. Brain Res 163: 195–205.

Huttenlocher, P.R., C. De Courten, L.J. Garey, and H. Van der Loos. 1982. Synaptic development in human cerebral cortex. Int J Neurol 16-17: 144–154.

Huttenlocher, P.R., and C. de Courten. 1987. The development of synapses in striate cortex of man. Hum Neurobiol 6: 1–9.

Huttenlocher, P.R., and A.S. Dabholkar. 1997. Regional differences in synaptogenesis in human cerebral cortex. J Comp Neurol 387: 167–178.

Ikonomidou, C., F. Bosch, M. Miksa, P. Bittigau, J. Vockler, K. Dikranian, T.I. Tenkova, V. Stefovska, L. Turski, and J.W. Olney. 1999. Blockade of NMDA receptors and apoptotic neurodegeneration in the developing brain. Science 283: 70–74.

Ikonomidou, C., P. Bittigau, M.J. Ishimaru, D.F. Wozniak, C. Koch, K. Genz, M.T. Price, V. Stefovska, F. Horster, T. Tenkova, K. Dikranian, and J.W. Olney. 2000. Ethanol-induced apoptotic neurodegeneration and fetal alcohol syndrome. Science 287: 1056–1060.

Jan, Y.N., and L.Y. Jan. 2003. The control of dendrite development. Neuron 40: 229–242.

Jernigan, C.S., D.B. Goswami, M.C. Austin, A.H. Iyo, A. Chandran, C.A. Stockmeier, and B. Karolewicz. 2011. The mTOR signaling pathway in the prefrontal cortex is compromised in major depressive disorder. Prog Neuropsychopharmacol Biol Psychiatry 35: 1774–1779.

Jevtovic-Todorovic, V., R.E. Hartman, Y. Izumi, N.D. Benshoff, K. Dikranian, C.F. Zorumski, J.W. Olney, and D.F. Wozniak. 2003. Early exposure to common anesthetic agents causes widespread neurodegeneration in the developing rat brain and persistent learning deficits. J Neurosci 23: 876–882.

Jevtovic-Todorovic, V., A.R. Absalom, K. Blomgren, A. Brambrink, G. Crosby, D.J. Culley, G. Fiskum, R.G. Giffard, K.F. Herold, A.W. Loepke, D. Ma, B.A. Orser, E. Planel, W. Slikker, Jr., S.G. Soriano, G. Stratmann, L. Vutskits, Z. Xie, and H.C. Hemmings, Jr. 2013. Anaesthetic neurotoxicity and neuroplasticity: an expert group report and statement based on the BJA Salzburg Seminar. Br J Anaesth 111: 143–151.

Johnstone, E.C., J.F. Deakin, P. Lawler, C.D. Frith, M. Stevens, K. McPherson, and T.J. Crow. 1980. The Northwick Park electroconvulsive therapy trial. Lancet 2: 1317–1320.

Juraska, J.M. 1982. The development of pyramidal neurons after eye opening in the visual cortex of hooded rats: a quantitative study. J Comp Neurol 212: 208–213.

Kaila, K., T.J. Price, J.A. Payne, M. Puskarjov, and J. Voipio. 2014. Cation-chloride cotransporters in neuronal development, plasticity and disease. Nat Rev Neurosci 15: 637–654.

Koehl, M., and D.N. Abrous. 2011. A new chapter in the field of memory: adult hippocampal neurogenesis. Eur J Neurosci 33: 1101–1114.

Kokaia, M. 2011. Seizure-induced neurogenesis in the adult brain. Eur J Neurosci 33: 1133–1138.

Krzisch, M., S. Sultan, J. Sandell, K. Demeter, L. Vutskits, and N. Toni. 2013. Propofol anesthesia impairs the maturation and survival of adult-born hippocampal neurons. Anesthesiology 118: 602–610.

Langer, G., J. Neumark, G. Koinig, M. Graf, and G. Schonbeck. 1985. Rapid psychotherapeutic effects of anesthesia with isoflurane (ES narcotherapy) in treatment-refractory depressed patients. Neuropsychobiology 14: 118–120.

Langer, G., R. Karazman, J. Neumark, B. Saletu, G. Schonbeck, J. Grunberger, R. Dittrich, W. Petricek, P. Hoffmann, L. Linzmayer, P. Anderer, and K. Steinberger. 1995. Isoflurane narcotherapy in depressive patients refractory to conventional antidepressant drug treatment. A double-blind comparison with electroconvulsive treatment. Neuropsychobiology 31: 182–194.

Laurie, D.J., W. Wisden, and P.H. Seeburg. 1992. The distribution of thirteen GABAA receptor subunit mRNAs in the rat brain. III. Embryonic and postnatal development. J Neurosci 12: 4151–4172.

Le Magueresse, C., and H. Monyer. 2013. GABAergic interneurons shape the functional maturation of the cortex. Neuron 77: 388–405.

Lei, S.Y., M. Hache, and A.W. Loepke. 2014. Clinical research into anesthetic neurotoxicity: does anesthesia cause neurological abnormalities in humans? J Neurosurg Anesthesiol 26: 349–357.

Lemkuil, B.P., B.P. Head, M.L. Pearn, H.H. Patel, J.C. Drummond, and P.M. Patel. 2011. Isoflurane neurotoxicity is mediated by p75NTR-RhoA activation and actin depolymerization. Anesthesiology 114: 49–57.

Li, N., B. Lee, R.J. Liu, M. Banasr, J.M. Dwyer, M. Iwata, X.Y. Li, G. Aghajanian, and R.S. Duman. 2010. mTOR-dependent synapse formation underlies the rapid antidepressant effects of NMDA antagonists. Science 329: 959–964.

Lin, E.P., S.G. Soriano, and A.W. Loepke. 2014. Anesthetic neurotoxicity. Anesthesiol Clin 32: 133–155.

Liu, G. 2004. Local structural balance and functional interaction of excitatory and inhibitory synapses in hippocampal dendrites. Nat Neurosci 7: 373–379.

Lorenz, K.Z. 1958. The evolution of behavior. Sci Am 199: 67–74 passim.

Lujan, R., R. Shigemoto, and G. Lopez-Bendito. 2005. Glutamate and GABA receptor signalling in the developing brain. Neuroscience 130: 567–580.

Lunardi, N., C. Ori, A. Erisir, and V. Jevtovic-Todorovic. 2010. General anesthesia causes long-lasting disturbances in the ultrastructural properties of developing synapses in young rats. Neurotox Res 17: 179–188.

Luthi, A., L. Schwyzer, J.M. Mateos, B.H. Gahwiler, and R.A. McKinney. 2001. NMDA receptor activation limits the number of synaptic connections during hippocampal development. Nat Neurosci 4: 1102–1107.

Mandairon, N., S. Sultan, M. Nouvian, J. Sacquet, and A. Didier. 2011. Involvement of newborn neurons in olfactory associative learning? The operant or non-operant component of the task makes all the difference. J Neurosci 31: 12455–12460.

Marin-Burgin, A., L.A. Mongiat, M.B. Pardi, and A.F. Schinder. 2012. Unique processing during a period of high excitation/inhibition balance in adult-born neurons. Science 335: 1238–1242.

Markram, H., M. Toledo-Rodriguez, Y. Wang, A. Gupta, G. Silberberg, and C. Wu. 2004. Interneurons of the neocortical inhibitory system. Nat Rev Neurosci 5: 793–807.

Mateos, J.M., A. Luthi, N. Savic, B. Stierli, P. Streit, B.H. Gahwiler, and R.A. McKinney. 2007. Synaptic modifications at the CA3-CA1 synapse after chronic AMPA receptor blockade in rat hippocampal slices. J Physiol 581: 129–138.

McDonald, I.M., M. Perkins, G. Marjerrison, and M. Podilsky. 1966. A controlled comparison of amitriptyline and electroconvulsive therapy in the treatment of depression. Am J Psychiatry 122: 1427–1431.

Micheva, K.D., and C. Beaulieu. 1996. Quantitative aspects of synaptogenesis in the rat barrel field cortex with special reference to GABA circuitry. J Comp Neurol 373: 340–354.

Miller, D.H., J. Clancy, and E. Cumming. 1953. A comparison between unidirectional current nonconvulsive electrical stimulation given with Reiter's machine, standard alternating current electro-shock (Cerletti method), and pentothal in chronic schizophrenia. Am J Psychiatry 109: 617–620.

Muller, M., B.H. Gahwiler, L. Rietschin, and S.M. Thompson. 1993. Reversible loss of dendritic spines and altered excitability after chronic epilepsy in hippocampal slice cultures. Proc Natl Acad Sci U S A 90: 257–261.

Nguyen, L., J.M. Rigo, V. Rocher, S. Belachew, B. Malgrange, B. Rogister, P. Leprince, and G. Moonen. 2001. Neurotransmitters as early signals for central nervous system development. Cell Tissue Res 305: 187–202.

Nusser, Z. 2012. Differential subcellular distribution of ion channels and the diversity of neuronal function. Curr Opin Neurobiol 22: 366–371.

Osterop, S.F., M.A. Virtanen, J.R. Loepke, B. Joseph, A.W. Loepke, and L. Vutskits. 2015. Developmental stage-dependent impact of midazolam on calbindin, calretinin and parvalbumin expression in the immature rat medial prefrontal cortex during the brain growth spurt. Int J Dev Neurosci 45: 19–28.

Ottosson, J.O. 1962. Seizure characteristics and therapeutic efficiency in electroconvulsive therapy: an analysis of the antidepressive efficiency of grand mal and lidocaine-modified seizures. J Nerv Ment Dis 135: 239–251.

Owens, D.F., and A.R. Kriegstein. 2002. Developmental neurotransmitters? Neuron 36: 989–991.

Pearn, M.L., Y. Hu, I.R. Niesman, H.H. Patel, J.C. Drummond, D.M. Roth, K. Akassoglou, P.M. Patel, and B.P. Head. 2012. Propofol neurotoxicity is mediated by p75 neurotrophin receptor activation. Anesthesiology 116: 352–361.

Penzes, P., M.E. Cahill, K.A. Jones, J.E. VanLeeuwen, and K.M. Woolfrey. 2011. Dendritic spine pathology in neuropsychiatric disorders. Nat Neurosci 14: 285–293.

Petanjek, Z., M. Judas, G. Simic, M.R. Rasin, H.B. Uylings, P. Rakic, and I. Kostovic. 2011. Extraordinary neoteny of synaptic spines in the human prefrontal cortex. Proc Natl Acad Sci U S A 108: 13281–13286.

Petilla Interneuron Nomenclature, G., G.A. Ascoli, L. Alonso-Nanclares, S.A. Anderson, G. Barrionuevo, R. Benavides-Piccione, A. Burkhalter, G. Buzsaki, B. Cauli, J. Defelipe, A. Fairen, D. Feldmeyer, G. Fishell, Y. Fregnac, T.F. Freund, D. Gardner, E.P. Gardner, J.H. Goldberg, M. Helmstaedter, S. Hestrin, F. Karube, Z.F. Kisvarday, B. Lambolez, D.A. Lewis, O. Marin, H. Markram, A. Munoz, A. Packer, C.C. Petersen, K.S. Rockland, J. Rossier, B. Rudy, P. Somogyi, J.F. Staiger, G. Tamas, A.M. Thomson, M. Toledo-Rodriguez, Y. Wang, D.C. West, and R. Yuste. 2008. Petilla terminology: nomenclature of features of GABAergic interneurons of the cerebral cortex. Nat Rev Neurosci 9: 557–568.

Petit, T.L., J.C. LeBoutillier, A. Gregorio, and H. Libstug. 1988. The pattern of dendritic development in the cerebral cortex of the rat. Brain Res 469: 209–219.

Poo, M.M., and S.H. Young. 1990. Diffusional and electrokinetic redistribution at the synapse: a physicochemical basis of synaptic competition. J Neurobiol 21: 157–168.

Poulter, M.O., J.L. Barker, A.M. O'Carroll, S.J. Lolait, and L.C. Mahan. 1992. Differential and transient expression of GABAA receptor alpha-subunit mRNAs in the developing rat CNS. J Neurosci 12: 2888–2900.

Quimby, K.L., L.J. Aschkenase, R.E. Bowman, J. Katz, and L.W. Chang. 1974. Enduring learning deficits and cerebral synaptic malformation from exposure to 10 parts of halothane per million. Science 185: 625–627.

Quimby, K.L., J. Katz, and R.E. Bowman. 1975. Behavioral consequences in rats from chronic exposure to 10 PPM halothane during early development. Anesth Analg 54: 628–623.

Rakic, P., J.P. Bourgeois, and P.S. Goldman-Rakic. 1994. Synaptic development of the cerebral cortex: implications for learning, memory, and mental illness. Prog Brain Res 102: 227–243.

Ranck, J.B., Jr. 1964. Synaptic "learning" due to electroosmosis: a theory. Science 144: 187–189.

Rappaport, B.A., S. Suresh, S. Hertz, A.S. Evers, and B.A. Orser. 2015. Anesthetic neurotoxicity—clinical implications of animal models. N Engl J Med 372: 796–797.

Represa, A., and Y. Ben-Ari. 2005. Trophic actions of GABA on neuronal development. Trends Neurosci 28: 278–283.

Ritter, L.M., A.S. Unis, and J.H. Meador-Woodruff. 2001. Ontogeny of ionotropic glutamate receptor expression in human fetal brain. Brain Res Dev Brain Res 127: 123–133.

Rivera, C., J. Voipio, J.A. Payne, E. Ruusuvuori, H. Lahtinen, K. Lamsa, U. Pirvola, M. Saarma, and K. Kaila. 1999. The K+/Cl- co-transporter KCC2 renders GABA hyperpolarizing during neuronal maturation. Nature 397: 251–255.

Rudolph, U., and B. Antkowiak. 2004. Molecular and neuronal substrates for general anaesthetics. Nat Rev Neurosci 5: 709–720.

Samuels, B.A., and R. Hen. 2011. Neurogenesis and affective disorders. Eur J Neurosci 33: 1152–1159.

Sanchez, V., S.D. Feinstein, N. Lunardi, P.M. Joksovic, A. Boscolo, S.M. Todorovic, and V. Jevtovic-Todorovic. 2011. General anesthesia causes long-term impairment of mitochondrial morphogenesis and synaptic transmission in developing rat brain. Anesthesiology 115: 992–1002.

Schikorski, T., and C.F. Stevens. 2001. Morphological correlates of functionally defined synaptic vesicle populations. Nat Neurosci 4: 391–395.

Servick, K. 2014. Biomedical research. Researchers struggle to gauge risks of childhood anesthesia. Science 346: 1161–1162.

Spalding, D.A. 1872. On instinct. Nature 6: 485–486.

Sprung, J., R.P. Flick, S.K. Katusic, R.C. Colligan, W.J. Barbaresi, K. Bojanic, T.L. Welch, M.D. Olson, A.C. Hanson, D.R. Schroeder, R.T. Wilder, and D.O. Warner. 2012. Attention-deficit/hyperactivity disorder after early exposure to procedures requiring general anesthesia. Mayo Clin Proc 87: 120–129.

Stoppini, L., P.A. Buchs, and D. Muller. 1991. A simple method for organotypic cultures of nervous tissue. J Neurosci Methods 37: 173–182.

Stratmann, G., J.W. Sall, L.D. May, J.S. Bell, K.R. Magnusson, V. Rau, K.H. Visrodia, R.S. Alvi, B. Ku, M.T. Lee, and R. Dai. 2009. Isoflurane differentially affects neurogenesis and long-term neurocognitive function in 60-day-old and 7-day-old rats. Anesthesiology 110: 834–848.

Stratmann, G., J.W. Sall, J.S. Bell, R.S. Alvi, L. May, B. Ku, M. Dowlatshahi, R. Dai, P.E. Bickler, I. Russell, M.T. Lee, M.W. Hrubos, and C. Chiu. 2010. Isoflurane does not affect brain cell death, hippocampal neurogenesis, or long-term neurocognitive outcome in aged rats. Anesthesiology 112: 305–315.

Uemura, E., E.D. Levin, and R.E. Bowman. 1985. Effects of halothane on synaptogenesis and learning behavior in rats. Exp Neurol 89: 520–529.

Ultanir, S.K., J.E. Kim, B.J. Hall, T. Deerinck, M. Ellisman, and A. Ghosh. 2007. Regulation of spine morphology and spine density by NMDA receptor signaling *in vivo*. Proc Natl Acad Sci U S A 104: 19553–19558.

van Spronsen, M., and C.C. Hoogenraad. 2010. Synapse pathology in psychiatric and neurologic disease. Curr Neurol Neurosci Rep 10: 207–214.

Vanhatalo, S., J.M. Palva, S. Andersson, C. Rivera, J. Voipio, and K. Kaila. 2005. Slow endogenous activity transients and developmental expression of K+-Cl- cotransporter 2 in the immature human cortex. Eur J Neurosci 22: 2799–2804.

Villeda, S.A., J. Luo, K.I. Mosher, B. Zou, M. Britschgi, G. Bieri, T.M. Stan, N. Fainberg, Z. Ding, A. Eggel, K.M. Lucin, E. Czirr, J.S. Park, S. Couillard-Despres, L. Aigner, G. Li, E.R. Peskind, J.A. Kaye, J.F. Quinn, D.R. Galasko, X.S. Xie, T.A. Rando, and T. Wyss-Coray. 2011. The ageing systemic milieu negatively regulates neurogenesis and cognitive function. Nature 477: 90–94.

Volavka, J., S. Feldstein, R. Abrams, R. Dornbush, and M. Fink. 1972. EEG and clinical change after bilateral and unilateral electroconvulsive therapy. Electroencephalogr Clin Neurophysiol 32: 631–639.

Vutskits, L. 2012. General anesthesia: a gateway to modulate synapse formation and neural plasticity? Anesth Analg 115: 1174–1182.

Vutskits, L. 2014. General anesthetics in brain injury: friends or foes? Curr Pharm Des 20: 4203–4210.

Waters, K.A., and R. Machaalani. 2004. NMDA receptors in the developing brain and effects of noxious insults. Neurosignals 13: 162–174.

Wilder, R.T., R.P. Flick, J. Sprung, S.K. Katusic, W.J. Barbaresi, C. Mickelson, S.J. Gleich, D.R. Schroeder, A.L. Weaver, and D.O. Warner. 2009. Early exposure to anesthesia and learning disabilities in a population-based birth cohort. Anesthesiology 110: 796–804.

Winner, B., Z. Kohl, and F.H. Gage. 2011. Neurodegenerative disease and adult neurogenesis. Eur J Neurosci 33: 1139–1151.

Yu, Z.Y., W. Wang, J.M. Fritschy, O.W. Witte, and C. Redecker. 2006. Changes in neocortical and hippocampal GABAA receptor subunit distribution during brain maturation and aging. Brain Res 1099: 73–81.

Zarate, C.A., Jr., J.B. Singh, P.J. Carlson, N.E. Brutsche, R. Ameli, D.A. Luckenbaugh, D.S. Charney, and H.K. Manji. 2006. A randomized trial of an N-methyl-D-aspartate antagonist in treatment-resistant major depression. Arch Gen Psychiatry 63: 856–864.

Zha, X.M., S.H. Green, and M.E. Dailey. 2005. Regulation of hippocampal synapse remodeling by epileptiform activity. Mol Cell Neurosci 29: 494–506.

Zhang, L.I., and M.M. Poo. 2001. Electrical activity and development of neural circuits. Nat Neurosci 4 Suppl: 1207–1214.

12

Blood-Brain Barrier (BBB)

Balabhaskar Prabhakarpandian,[1] *Syed Ali*[2,]* and *Kapil Pant*[3]

Introduction

Delivery of neuroprotective or therapeutic agents to specific regions of the brain presents a major challenge, largely due to the presence of the Blood-Brain Barrier (BBB). Physiologically, the BBB consists of an intricate network of vascular endothelial cells (ECs) that isolate the central nervous system (CNS) from systemic blood circulation except the circumventricular organs. A combination of physical and biochemical barriers establishes the BBB endothelium as quite distinct from other endothelia (Abbruscato and Davis, 1999; Audus and Borchardt, 1986; Audus and Borchardt, 1987; Joo, 1992; Reese and Karnovsky, 1967). The BBB is formed by capillary endothelial cells, surrounded by basal lamina and astrocytic perivascular endfeet with astrocytes providing the cellular link to the neurons (Figure 12.1A). The astrocyte endfeet form an envelope around the blood vessels and are attached to the basement membrane tightly by their adhesion molecules. The basement membrane is composed of extracellular matrix molecules such as type IV collagen, laminins, fibronectin, heparan sulfates, and proteoglycans. An intact basement

[1] Biomedical Technology, CFD Research Corporation, 701 McMillian Way, Huntsville, AL 35806.
  E-mail: prabhakar.pandian@cfdrc.com
[2] Division of Neurotoxicology, HFT-132, National Center for Toxicological Research (NCTR), Food and Drug Administration (FDA), 3900 NCTR Rd, Jefferson, AR 72079.
  E-mail: syed.ali@fda.hhs.gov
[3] Biomedical Technology, CFD Research Corporation, 701 McMillian Way NW, Huntsville, AL 35806.
  E-mail: kapil.pant@cfdrc.com
* Corresponding author

membrane provides structural support to the cells and is also critical in delivering communicative signals between the intravascular components and the glial/neuronal cells in addition to nutritional support from the blood stream (Iadecola, 2004).

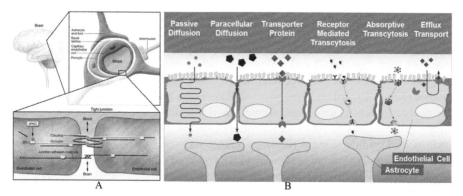

Figure 12.1. (A) Anatomy of the Blood-Brain Barrier (BBB). The vessel lumen is lined with endothelial cells (ECs), which are lined with pericytes and astrocytic end-feet to form tight junctions (TJs). (B) Transport pathways across the BBB. Formation of tight junctions leads to dramatic reduction in paracellular transport and the dominant transport mechanisms are passive diffusion for small molecules or mediated transport for larger molecules. Figure A republished with permission of the American Society of Clinical Investigation, from "Hypertensive encephalopathy and the blood-brain barrier: is δPKC a gatekeeper?" Wen-Hai Chou and Robert O. Messing, Journal of Clinical Investigation, Volume 118, Issue 1, 2008; permission conveyed through Copyright Clearance Center, Inc. Figure B adapted from Wikimedia Commons.

The microcapillary endothelium is characterized by the presence of tight junctions, a lack of fenestrations and minimal pinocytotic vesicles (Hawkins and Davis, 2005), which limit the transport pathways across the BBB (Figure 12.1B). Small (< 500 Da) lipid-soluble substances such as alcohol, narcotics and anticonvulsants, are believed to pass through the BBB with relative ease. However, for most other substances, tight junctions between the cerebral ECs form a diffusion barrier. Even in the event of successful crossing of the barrier, the drug efflux pump P-glycoprotein (P-gp), an important component of the BBB, pumps unrecognized substrates from the CNS thereby limiting exposure. The tight junction consists of transmembrane proteins (claudin, occludin, and junction adhesion molecules) and cytoplasmic accessory proteins. Claudins form dimers and bind to claudins on adjacent endothelial cells to establish the primary gate of the tight junction. The main functions of occludin appear to be to regulate the electrical resistance across the barrier and decrease paracellular permeability. Zonula occuldens proteins (ZO-1, ZO-2, and ZO-3) serve as recognition proteins for tight junction placement and connect transmembrane proteins to the actin cytoskeleton. Dissociation of ZO-1 has been shown to be associated with increased barrier permeability (Abbruscato and Davis, 1999; Kago *et al.*, 2006).

It has been estimated that the tight junctions of the BBB prevents the brain from taking up 100% of the large (> 1 kDa) molecule therapeutics comprising of genes and recombinant proteins as well as more than 98% of potential neurotherapeutics (Pardridge, 2005) comprising of small molecule proteins and peptides (500–1 kDa). The development of therapeutics that can cross the BBB is a formidable challenge, and an important one, considering that a large number of emerging therapeutics are based on peptides, recombinant proteins and genes, and are macromolecular in nature.

Recently, a study (Shityakov *et al.*, 2015) evaluated the ability of multi-wall carbon nanotubes (MWCNTs) functionalized with fluorescein isothiocyanate (FITC) as potential targeting delivery system to permeate the BBB and they found that these MWCNTs–FITC conjugate are able to penetrate microvascular cerebral endothelial monolayers without causing any toxicity to the BBB. In other studies, MWCNTs modified with angiopep-2 and single wall carbon nanotubes (SWCNT) enhanced immunotherapy using CpG oligodendronucleotides inhibited tumor growth in glioma models (Ren *et al.*, 2012; Zhao *et al.*, 2011). These and other studies suggest that designing a successful nanomaterial conjugated with drugs that cross the BBB without effecting its integrity could be the potential treatment for many neurological diseases such as Parkinson's Disease (PD), Alzheimer's Disease (AD), Multiple Sclerosis (MS), Amyotrophic Lateral Sclerosis (ALS), and even Traumatic Brain Injury (TBI).

In parallel to the efforts focused on drug delivery across the BBB, there has been a recent interest and surge of literature describing the effects of nanoparticles (NPs) on the BBB. Most of these studies are performed *in vitro* using various cell lines and devices/systems. Some of them have been described in the later part of this chapter. Physicochemical properties play an important role in the effects of these NPs on the BBB. Factors such as surface, size, charge, and coating can make a difference in the nanotoxicity of any NPs on the BBB. Several recent studies have reported that various metallic colloidal nanoparticles (such as silver, gold and copper) can produce significant changes in the BBB integrity (Trickler *et al.*, 2014; Trickler *et al.*, 2010; Trickler *et al.*, 2011; Trickler *et al.*, 2012). These reports suggest that smaller sizes of silver nanoparticles (Ag-NPs) are more toxic to the BBB integrity, compared to a larger size. Ag-NPs can produce significant morphological changes in rBMECs (Figure 12.2A) and these changes could be due to higher accumulation of Ag-NPs in rBMECs (Figure 12.2B). Further studies also showed size and dose dependent viability in these rBMECs (Figure 12.2C). Ag-NPs also produce size dependent changes in several neuroinflammatory markers such a TNF, PGE2, IL-1B (Trickler *et al.*, 2014; Trickler *et al.*, 2010). However, other NPs such as gold (Au-NPs) did not show any changes in any of neuroinflammatory markers, cell viability and membrane permeability (Trickler *et al.*, 2011). Copper nanoparticles (Cu-NPs), on the other hand, have a different and more pronounced effect on the BBB integrity (Trickler *et al.*, 2012).

Mechanism of action of nanomaterials toxicity is still not clear, however, one hypothesis puts forward that these NPs generate oxidative stress and reactive oxygen species (ROS). Several reports suggested that both *in vitro* and *in vivo*, Ag-NPs and Cu-NPs triggered ROS production (Rahman *et al.*, 2009; Sharma *et al.*, 2009a; Trickler *et al.*, 2005). Injections of Ag-NPs or Cu-NPs in rats produced BBB dysfunction, astrocytes swelling and neuronal degeneration along with leakage of radioiodine and Evan's blue dye, marker of increase permeability of the BBB (Sharma *et al.*, 2009b; Wang *et al.*, 2009).

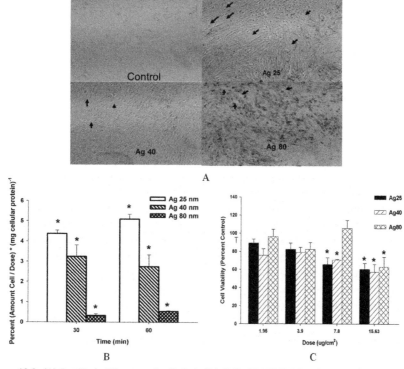

Figure 12.2. (A) Rat Brain Microvascular Endothelial Cells (rBMEC) Monolayer Integrity Following Exposure to Various Silver Nanoparticles Cell monolayers were treated with of various size silver nanoparticles for 24 hours. Control – media alone, Ag 25–25 nm silver nanoparticles, Ag 40–40 nm silver nanoparticles, Ag 80–80 nm silver nanoparticles. (B) Size Dependent Silver Nanoparticle Accumulation in rBMECs. (C) Silver Nanoparticle Size and Dose dependent Cell Viability (Trickler *et al.*, Silver nanoparticle induced blood-brain barrier inflammation and increased permeability in primary rat brain microvessel endothelial cells, Toxicol Sci, 2010, 118(1): 160–170, by permission of Oxford University Press).

Cells of the Blood-Brain Barrier

In recent years, the somewhat limiting notion that the Blood-Brain Barrier (BBB) is defined by the vascular cells alone has been replaced by the concept of the "neurovascular unit" (or NVU), which aims to recognize that the complex interplay between the various cells of the BBB such as endothelial cells, pericytes, glia, and the neuronal cells determines the structure and function of the brain.

Endothelial cells

The endothelial cells of the brain vasculature are significantly different from the endothelial cells of different organs. Brain vascular endothelial cells are extremely thin with the luminal and abluminal membranes separated by less than 300 nm. The endothelial cells lack fenestrations and are held together by tight junctions (TJs) resulting in high transendothelial electrical resistance (TEER). The presence of these TJs is responsible for development of the polarized membranes that regulate transport across the BBB using both active and efflux transporters. The inter-endothelial space is characterized by a junctional complex comprising of TJ and adherens junction (AJ) which primarily restrict the permeability of the barrier. The transmembrane molecules regulating the TJ include claudins, occludin and junctional adhesion molecule (Warner *et al.* 2014). Brain endothelial cells also synthetize special molecules such as the multidrug-resistance protein P-glycoprotein (P-gp). This transporter machinery is highly dependent on ATP generation, hence the reason for copious amount of mitochondria in these endothelial cells. Finally, the brain endothelial cells have minimal expression of adhesion molecules thereby regulating the immune cells trafficking across the BBB.

Claudins are 20- to 24-kDa proteins with four transmembrane domains. The extracellular loops of claudins interact via homophilic and heterophilic interactions between endothelial cells to form the primary component of the TJ with claudins-1 and -5 being responsible for formation of the TJ. Occludin is a 65-kDa phosphoprotein with no common sequence homology with claudins. They have four transmembrane domains with the COOH-terminal and NH_2 terminal oriented towards the cytoplasmic domain. The two extracellular loops of occludin and claudin originating from neighboring cells form the paracellular barrier of TJ.

Adherens junctions (AJs) are composed of a cadherin–catenin complex and its associated proteins responsible for the formation of adhesive contacts between endothelial cells. The cytoplasmic domains of cadherins bind to actin cytoskeleton via α-catenin which is bound to the sub-membranal plaque proteins β-catenin and ϒ-catenin. AJ components including cadherin, alpha-actinin, and vinculin (α-catenin analog) are known to interact with other cytoplasmic proteins, particularly ZO-1 and catenins, to determine the fate of the TJ formation.

Cytoplasmic proteins link membrane proteins to actin for the maintenance of structural and functional integrity of the endothelium. Cytoplasmic proteins primarily involved in TJ formation include zonula occludens proteins namely ZO-1 (220 kDa), ZO-2 (160 kDa), and ZO-3 (130 kDa). All of these proteins have sequence homology and belong to class of proteins called membrane-associated guanylate kinase-like protein (MAGUKs). They contain three PDZ domains (PDZ1, PDZ2, and PDZ3), one SH3 domain, and one guanyl kinase-like (GUK) domain. These domains function as protein binding molecules and thus play a role in organizing proteins at the plasma membrane. The PDZ1 domain of ZO-1, ZO-2, and ZO-3 has been reported to bind directly to COOH-terminal of claudins. Actin binds to COOH-terminal of ZO-1 and ZO-2, and this complex cross-links transmembrane elements and thus provides structural support to the endothelial cells.

Junction Adhesion Molecules (JAMs) are immunoglobulin superfamily members that form homotypic interactions at tight junctions in brain endothelial cells. JAMs have been shown to regulate leukocyte extravasation as well as paracellular permeability. The transmembrane components of the TJ at the BBB include junctional adhesion molecule (JAM-1), occludin, and the claudins. JAM-1 is a 40-kDa member of the IgG superfamily and is believed to mediate the early attachment of adjacent cell membranes via homophilic interactions. JAM-1 is composed of a single membrane-spanning chain with a large extracellular domain. Related proteins JAM-2 and JAM-3 are also present in endothelial tissues but their function is still not fully understood.

Astrocytes

Astrocytes (star-like cells) are the most numerous and diverse glial cells of the brain and envelop > 99% of the endothelium. Expression of glial fibrillary acidic protein (GFAP) is commonly used as a specific marker for the identification of astrocytes. They extend polarized cellular processes providing a link between the neuronal processes and the endothelium of the blood vessels. They also regulate the contraction and dilation of vascular smooth muscle cells (SMCs) to control blood flow in response to neuronal activity. Interaction of astrocytes with endothelial cells greatly enhances TJ. Astrocytes are also known to secrete a large number of substances including peptides, growth factors and chemokines which modulate the BBB function. These changes are driven by changes in the TJs formation specifically of the proteins such as ZO-1, occludin and claudin in addition to TGF-β1 and glial cell-derived neurotrophic factor (GDNF). Astrocytes are also known to modulate the expression and localization of transporters such as P-gp or multi-drug resistance protein (MRP) as well as BBB-specific enzymes.

Pericytes

Pericytes are flat, undifferentiated, contractile connective tissue cells that form a key component of the vascular walls. Brain vasculature has high content of these cells and the pericytes to endothelial cells ratio is known to correlate with the barrier properties. They also extend long cellular processes along the abluminal surface of the endothelium that can often span several endothelial cell bodies. Pericytes are also actively involved in maintenance of the integrity of the vessel, vasoregulation and restricted BBB permeability. However, the effect of pericytes on BBB modulation is not fully understood. Pericytes also secrete factors that modulate BBB such as TGF-β and GDNF that increase production of tight junction proteins comprising of claudins, occludin and cytoplasmic proteins. Therefore, similar to astrocytes, pericytes are capable of modulating TJs by their interaction with the endothelial cells. Pericytes play significant role in regulating other responses such as angiogenesis, wound healing, immune and cell trafficking.

Neurons

The brain's neurovascular unit comprises of the complex interactions between endothelial cells, pericytes, astrocytes and neurons. Neuronal projections are in contact with astrocytes and pericytes thereby allowing neuronal mediators to affect cerebral blood flow and vessel dynamics. However, the precise physiological influences of neuronal interactions and signaling onto the BBB still remain unclear.

In Vitro Blood-Brain Barrier (BBB) Models

Panula *et al.* first demonstrated that brain endothelial cells could be maintained in culture as a prototype *in vitro* BBB model (Panula *et al.*, 1978). However, over time, these primary cultures of Central Nervous System (CNS) endothelial cells lose many of the characteristics of the *in vivo* phenotype, suggesting that factors inherent to the *in vivo* environment are required to maintain an optimally functioning BBB. For example, it has been shown that brain endothelial cells become highly impermeable and exhibit extensive tight junction formation when cultured in the presence of astrocytes or astrocyte-conditioned medium (ACM) (Janzer and Raff, 1987; Rubin *et al.*, 1991). The formation of these tight junctions in *in vitro* BBB models is often characterized by the expression levels of one or more of the tight junction proteins. In addition to biochemical assays, some devices feature electrodes to measure trans-endothelial electrical resistance (TEER) (Biegel *et al.*, 1995; Matthes *et al.*, 2011) as an indicator of barrier permeability.

Transwell apparatus

Transwell apparatus based devices are the most commonly used model of the BBB (Figure 12.3). They feature a porous membrane that separates two chambers, in which endothelial cells and astrocytes are cultured (Czupalla *et al.*, 2014). These *in vitro* cellular models have had a beneficial impact on a diverse range of scientific fields ranging from classical pharmacodynamics, pharmacokinetic and toxicological research, to drug design and discovery (Mizuno *et al.*, 2003), as well as nanotechnology (Lockman *et al.*, 2003). However, these models do not reproduce the BBB microenvironment comprising of endothelial cells continuously exposed to

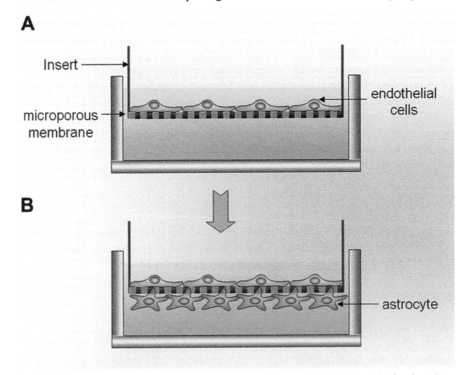

Figure 12.3. Transwell apparatus based BBB Model. Endothelial cells and astrocytes (and pericytes) are cultured on either side of a membrane in a static environment (Xu *et al.*, 2013)-Reproduced by permission of Frontiers in Pharmacology, http://dx.doi.org/10.3389/fphar.2013.00140.

shear generated by the flow of blood across their apical surfaces. These static assays (Biegel and Pachter, 1994; de Vries *et al.*, 1996; Fitsanakis *et al.*, 2005; Vu *et al.*, 2009) do not represent important morphological and physiological aspects of the microcirculatory environment in the brain and as a consequence provide little correlation or predictive ability in comparison with *in vivo* data.

Hollow fiber bioreactor

Seeking to overcome the limitations of the static incubation approaches, researchers have developed a hollow fiber apparatus (Cucullo *et al.*, 2005; Cucullo *et al.*, 2002; Santaguida *et al.*, 2006). In this device, brain endothelial cells are cultured in the lumen of hollow fibers (mimicking capillary vessels) and exposed to flow while astrocytes are seeded in the extra-luminal compartment to form the BBB (Figure 12.4). The results obtained from the device for TEER and biochemical assays are very encouraging. However, the dimensions of the device are not representative of the microvasculature. Secondly, high flow rates and large volumes (~ ml) are needed to maintain physiological shear in the system. Finally, real-time visualization is a significant challenge with hollow fiber based systems.

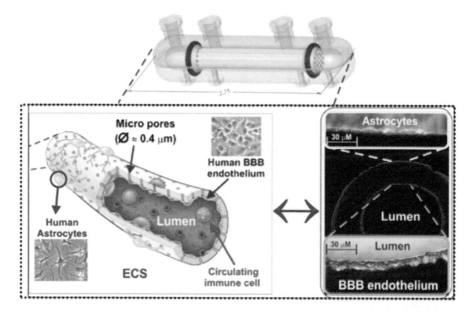

Figure 12.4. Hollow Fiber based BBB Model. Endothelial cells are cultured under flow inside the lumen of the hollow fiber and the astrocytes (and pericytes) are cultured on the outside wall of the hollow fiber. Reprinted from Palmiotti *et al.* (2014), with kind permission of Springer Science and Business Media.

Microfluidic (Transwell) device

A recent review (Ribeiro *et al.*, 2010) provided a chronological history of *in vitro* BBB models, including both static and dynamic models, but overlooked the rising field of microfluidics based cell culture models (Wolff *et al.*, 2015). Endothelial cells have been cultured in flow based devices for over a decade for studying vascular biology (van der Meer *et al.*, 2009). These flow based devices ranged from larger parallel plate flow chambers to microfluidic based *in vivo* mimetic flow chambers. Endothelial cells of different origin can be readily cultured and monitored real-time for functional assays (Prabhakarpandian *et al.*, 2011). Taking advantage of these microfabrication technologies, a microfluidic BBB model was recently demonstrated (Booth and Kim, 2012) using a co-culture of endothelial cells and astrocytes cultured on a membrane (Figure 12.5). This model featured a top-bottom architecture, which limits the ability to simultaneously visualize in real-time both the vascular and neuronal sides of the BBB.

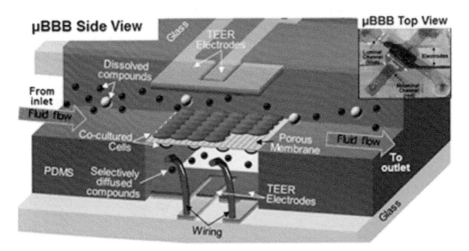

Figure 12.5. Microfluidic "Transwell" BBB Model. As with the traditional Transwell apparatus, endothelial cells and astrocytes (and pericytes) are cultured on either side a membrane. However, endothelial cells are subject to fluid flow to mimic the *in vivo* like phenotype (Booth and Kim, 2012)-Reproduced by permission of The Royal Society of Chemistry, http://dx.doi.org/10.1039/c2lc40094d.

Synthetic Microvascular Blood-Brain Barrier (SyM-BBB)

To overcome these limitations, a microfluidics based Synthetic Microvasculature model of the Blood-Brain Barrier (SyM-BBB) was developed (Prabhakarpandian *et al.*, 2013). The SyM-BBB model comprises of microchannels partitioned into two side-by-side chamber (Figure 12.6) by utilizing pillars or posts which mimic the use of membranes in conventional models. The posts are separated by small gaps (3–10 µm, similar to Transwell membranes), which ensures communication between the two chambers. This approach is more manufacture-friendly than other top-bottom microfluidic approaches to the creation of two chamber assays, which typically use a filter or micromachined insert between the top and the bottom chambers. An immortalized rat brain endothelial cell line; RBE4 (Couraud *et al.*, 2003; Roux *et al.*, 1994) was cultured in the apical chamber under fluidic shear conditions and in continuous contact with astrocyte-conditioned medium (ACM) in the basolateral chamber. RBE4 cells preserve the endothelial phenotype, show differentiation characteristics of brain endothelium in the presence of glial factors express P-glycoprotein (Begley *et al.*, 1996; Roux *et al.*, 1994) and have been used for wide range of *in vitro* assays to characterize endothelial function and transport (Aschner *et al.*, 2006; dos Santos *et al.*, 2010; Fitsanakis *et al.*, 2005; Fitsanakis *et al.*, 2006; Santos *et al.*, 2013; Yang and Aschner, 2003). The basolateral chamber can house neuronal components, such as astrocytes leading to a more accurate structural representation of the BBB. This model has been demonstrated (Deosarkar *et al.*, 2015) for co-culture of primary endothelial cells and astrocytes and validated against *in vivo* permeability data.

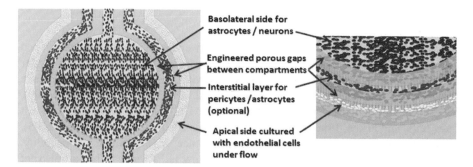

Figure 12.6. SyM-BBB Model. Endothelial cells, pericytes, and astrocytes are cultured in individual chambers separated by engineered porous gaps in a side-by-side architecture. Flow and other environmental parameters in a cell compartment may be controlled individually and independent of other chambers.

Other models

The discussion on *in vitro* models in this chapter is not intended to be exhaustive. Certainly, there are variations and other models that researchers have used to study BBB physiology and transport. For instance, hanging drop methodology, commonly used to develop 3D tumor spheroids or 3D organotypic cultures, is gaining traction in development of the BBB model. A key differentiation compared to the other models being used is self-assembly of the interacting cells (e.g., endothelial, astrocytes and pericytes) rather than the arranged architecture of the cellular layers. A key limitation is the ability to generate a vascular blood flow environment necessary to quantify permeability across the barrier in addition to quantitation of therapeutic distribution and measurement of TEER values.

Cells for *In Vitro* Blood-Brain Barrier (BBB) Models

Primary cells

Primary brain endothelial cells are the closest in resembling the *in vivo* phenotype (Bernas *et al.*, 2010). They also show characteristic brain endothelial markers, robust expression of tight and adherens junction proteins and functional properties (such as, active transporters, low permeability and high TEER formation). In addition, depending on the species used for isolation of the endothelial cells, they can be used for multiple experiments with ease. However, there are significant hurdles for use of primary cells and these include the time and technically skilled staff for isolation of pure population of brain endothelial cells, batch to batch variability especially with large mammalian samples (e.g., porcine, bovine and human). Although, several commercial entities are providing the brain endothelial cells, cost of these primary cells and the quick loss of *in vivo* phenotypic properties following multiple passages *in vitro* is a significant concern. To overcome these limitations, researchers often use cells lines that have been immortalized from primary brain endothelial cells from different species.

Cell lines

There have been a large number (> 25) of brain endothelial cell lines developed in the past three decades from mouse, rat, porcine, bovine and human for a better representation in an *in vitro* model. The cell lines with most published studies include RBE4 rat cell line (Roux *et al.*, 1994), b.End3 mouse cell line (Montesano *et al.*, 1990), SV-BEC bovine cell line (Durieu-Trautmann *et al.*, 1991), PBMEC/C1-2 porcine cell line (Teifel and Friedl, 1996) and hCMEC/D3 human cell line (Weksler *et al.*, 2005). However, these cell lines lack the *in vivo* relevance of paracellular permeability, exhibit discontinuous tight junction and have very low TEER values. They are also not amenable for culture under fluid flow conditions which are very relevant for obtaining physiological relevant and predictive information.

Stem cell derived cells

Stem cells have been recently identified as a natural source for the development of the BBB models. Specifically, induced pluripotent stem cells (iPSCs) can be readily differentiated into brain endothelial cells under well controlled conditions (Lippmann *et al.*, 2012). These endothelial cells demonstrate unique BBB endothelium properties such as uniform tight junction formation, active transporters with polarizable property and high TEER values ($\sim 1000s \, \Omega \times cm^2$). Use of iPS cells also enables development of pure culture of not only endothelial cells, but also astrocytes (Lippmann *et al.*, 2013) and pericytes (Lippmann *et al.*, 2013) which are difficult to obtain for a human population. Co-culture iPSC derived brain endothelial cells with astrocytes and pericytes highlighted TEER values (Lippmann *et al.*, 2014) in the range of $5,000 \, \Omega \times cm^2$ which brings it closer to the *in vivo* scenario.

Analytical Methods

The *in vitro* Blood-Brain Barrier (BBB) models may be interrogated using a variety of methods to evaluate and investigate the structure and function of the BBB. These methods include optical, electrical or biochemical measurements for quantifying the tight junction formation, permeation, and transport across the BBB.

Cell morphology

Cell morphology is the most obvious indicator of the cell culture in an *in vitro* model. Figure 12.7 shows a co-culture of endothelial cells (RBE4, an immortalized rat brain cell line) with astrocytes. Astrocytes were cultured on Matrigel™ and endothelial cells were cultured on fibronectin. Figure 12.7A shows confluent layer of endothelial cells cultured on the device with integrated electrodes for TEER measurements. Figure 12.7B shows endothelial cells stained with Hoechst, a nuclear stain. Figure 12.7C shows endothelial cell stained with CD31 marker. Figure 12.7D shows astrocytes stained for glial fibrillary acidic protein (GFAP) marker, nucleus and actin indicating fully functional environment. Figure 12.7E shows co-cultured images.

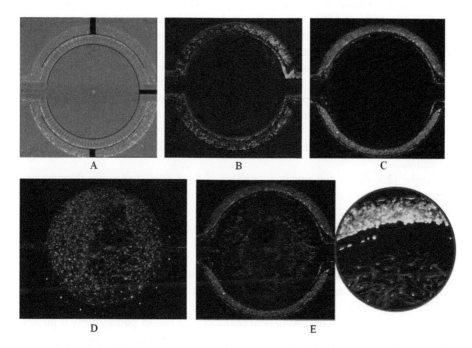

Figure 12.7. Endothelial cells and astrocytes culture in SyM-BBB (A) Phase contrast image showing confluent layer of endothelial cells on devices integrated with electrodes (B) Nucleus (Hoechst) stained endothelial cells (C) CD31 stained endothelial cells (D) Astrocytes stained with glial fibrillary acidic protein stained, Hoechst and actin (E) Image of co-culture of astrocytes and endothelial cells with magnified view of the cells and barrier.

Permeability assays

Permeation studies may be carried out using optical or radio methods which involved molecular agents (fluorescent or radio-labeled small-to-medium molecules), or electrical methods, or cells or particles, which can be used to model transport of particles across the BBB.

Fluorescent dextran permeation

Confluent layer of cells in SyM-BBB—with and without Astrocyte Conditioned Medium (ACM)—were perfused with fluorescent-dextran (3000–5000 Da, Sigma Aldridge) at a concentration of 0.5 mg/ml on the apical side at a flow rate of 0.1 µl/min. At regular time points, the entire device was scanned using an automated stage (LEP Ltd.) and imaged using a cooled charge-coupled device (CCD) camera (Roper Coolsnap HQ2) to quantify the dextran permeation in the basolateral side. The images were post-processed using NIKON Elements to yield the intensity ratio between the apical chamber and the basolateral chamber.

For the Transwell experiments, RBE4s were plated on a 3.0 µm polyester membrane insert (same size as the gaps in SyM-BBB). The cells were subsequently maintained in ACM and RBE4 media for 48 hours similar to SyM-BBB. At the end of 48 hours, 150 µl of fresh media with fluorescent dextran was added to the cells and 600 µl of media was added to the bottom chamber. At regular intervals starting at 1 minute, 100 µl of the volume was withdrawn from the bottom chamber to assess the fluorescent intensity. The withdrawn volume was replenished with fresh media to eliminate any volume dependent permeation. Experiments were repeated for RBE4 with no ACM and with no cells in the device. The intensity profiles of RBE4 (no ACM) and RBE4 (with ACM) were normalized with the intensity profiles of the cell free conditions to determine temporal evolution of fluorescent-dextran permeation.

Figure 12.8 shows the intensity profiles of SyM-BBB following fluorescent-dextran perfusion. The optically clear compartments in SyM-BBB allow real-time monitoring of the permeation of dextran, which is not possible with Transwell chambers. The top panel shows the intensity profiles at regular time points, indicating that by 60 min, the basolateral chamber of the cell-free SyM-BBB was completely perfused with fluorescent dextran. However, in the case of SyM-BBB with cells but no ACM and SyM-BBB with cells and ACM, there was a significant difference in the dextran permeation, even at 60 min. The presence of cells, even in the absence of ACM, significantly decreases the permeability of the barrier as reflected by lower fluorescent dextran intensity levels in the basolateral chamber. The barrier permeability is further reduced by the use of the ACM. The decrease in dextran permeation can be attributed to the presence of cell layer and the formation of tight junctions in the presence of ACM, which block the 3 µm pores thereby reducing the diffusion of the molecules from the apical chamber to the basolateral chamber. Lower permeation in ACM treated SyM-BBB than the non-ACM treated RBE4 cells suggest that the presence of ACM and flow have an additive effect, which cannot be attained under static conditions.

Neutrophil migration

Transendothelial migration of leukocytes on dysfunctional BBB may also be studied. Endothelial cells cultured in the SyM-BBB device were activated with TNF-alpha, a well-known inflammatory protein, to disrupt the tight junction formation. A potent chemoattractant, Formyl-Methionyl-Leucyl-Phenylalanine (fMLP) was injected in

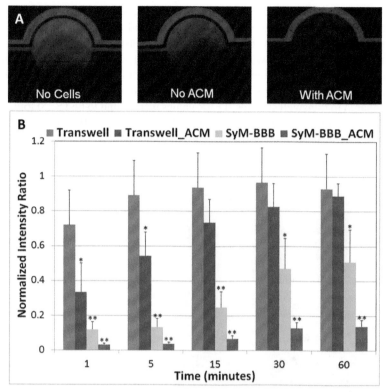

Figure 12.8. Comparison of fluorescent-dextran permeation. (A) Fluorescent microscopy images showing highly permeable BBB in the absence of cells. With cells, the permeability is reduced, but it is only in the presence of Astrocyte Conditioned Medium (ACM), that significant tighter barrier is formed. (b) Tighter junctions are formed in SyM-BBB compared to Transwell. The barrier integrity is also maintained over a significantly longer duration. Adapted from Prabhakarpandian *et al.*, 2013 by permission of The Royal Society of Chemistry, http://dx.doi.org/ 10.1039/C2LC41208J.

the basolateral side while leukocyte surrogates (HL-60 cells) were injected into the apical side. Migration of leukocytes towards the basolateral chamber was monitored in real-time.

Figure 12.9 shows time lapse images of leukocyte migration in SyM-BBB with increased migration with time. Note that a majority of the leukocytes get transported on the apical side without interacting with the activated endothelium located at the vessel wall. The leukocytes that do interact with the endothelium exhibit the entire leukocyte adhesion cascade—including rolling on the activated endothelium, firm adhesion and arrest, followed by transendothelial migration towards the chemoattractant in the basolateral chamber.

TEER measurements

One of the common methods to determine the effectiveness of the BBB is to determine the transendothelial electrical resistance (TEER) which is a quantitative measurement of the resistance across the endothelial cells. TEER values have been found to increase from < 500 Ω × cm² for immortalized endothelial cell lines to ~ 1000 Ω × cm² for primary endothelial cells and endothelial cells derived from stem cells. Co-culture of endothelial cells with ACM or astrocytes also increases the TEER values in static systems. The presence of flow in the system has led to a significant increase in the TEER values for both cell lines and primary cells compared to basal levels under static conditions. For instance, the DIV-BBB model (Cucullo *et al.*, 2008) showed that compared to transwell systems, TEER values were found to be 15–20 fold higher under fluidic shear conditions. A recent study (Lippman *et al.*, 2014) showed that use of stem cell derived cells under the presence of retinoic acid reached unprecedented high levels of TEER (~ 5000 Ω × cm²) *in vitro*.

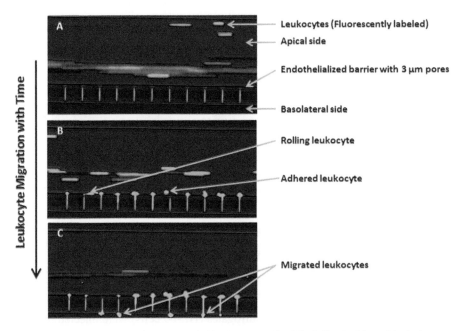

Figure 12.9. Time-lapse images of leukocyte migration across Blood-Brain Barrier. Most of the leukocytes get transported on the apical side without interacting with the activated endothelium. The leukocytes that do interact with the endothelium exhibit the entire leukocyte adhesion cascade—rolling, adhesion, and migration.

Proteomics

Protein analysis may be carried out using one of three methods—*in situ* immunostaining of the cells, extraction of the supernatant from the model for further analysis, or harvesting the cells from the model to subject them to off-line measurements.

Tight junction proteins

Figure 12.10A-B shows cells stained for tight junction proteins in the SyM-BBB device, whereas Figure 12.10C shows the Western Blot analysis of the harvested cells for selected tight junction and transporter proteins. The RBE4 cells were cultured in

SyM-BBB in the presence and absence of ACM as mentioned before. As a control, RBE4 were also cultured in Transwell chambers both in presence and absence of ACM. Cells were harvested in trypsin and washed with phosphate-buffered saline (PBS), ice cold RIPA (Radio Immuno Precipitation Assay) buffer (150 mM sodium chloride, 1.0% NP-40, 0.5% sodium deoxycholate, 0.1% SDS, 50 mMTris, pH 8.0) and an additional mixture of protease inhibitors were added to cells and sonicated. Protein concentration of cell homogenates was determined with bicinchoninic acid (BCA™, Pierce) protein assay. Samples were loaded onto 10% sodium dodecyl sulfate (SDS)-polyacrylamide gels, and the separated proteins were transferred to a polyvinylidene fluoride (PVDF) membrane. Subsequently, the membranes were incubated for 1 hour with blocking solution (5% fat free dry milk in PBS) followed by incubation overnight at 4°C with primary antibody for tight junction molecules zonula occludens-1 (ZO-1) and claudin-1 and efflux transporter P-glycoprotein (P-gp). Following a 3X wash, the membranes were further incubated with secondary antibody for 2 hours at room temperature. Final detection was performed with enhanced chemiluminescence methodology (Super Signal West Dura Extended Duration Substrate) with the intensity of the chemiluminescence signal normalized with β-actin expression.

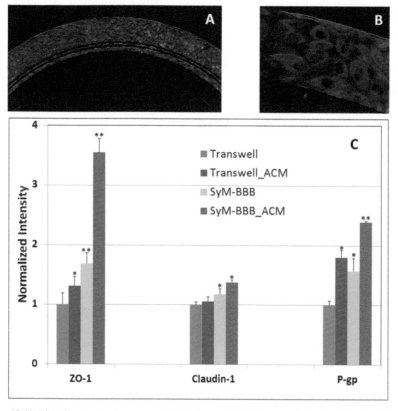

Figure 12.10. Tight junction protein expression (A-B) ZO-1 staining. (C) Western Blot analysis of tight junction proteins shows significant expression of ZO-1, Claudin-1 and P-gp. Adapted from Prabhakarpandian *et al.*, 2013 by permission of The Royal Society of Chemistry, http://dx.doi.org/ 10.1039/C2LC41208J.

Expression levels of the tight junction proteins (ZO-1 and claudin-1) and the efflux transporter system (P-gp) are shown in Figure 12.10. The upregulation was found to be maximal for ZO-1 (P < 0.01), but other molecules (Claudin-1 and P-gp) were also upregulated significantly (P < 0.05 and P < 0.01, respectively) under flow in SyM-BBB with ACM amplifying the signal significantly. This is likely due to the additive effect of flow with ACM, which will be prevalent under *in vivo* conditions. Since ZO-1 dissociation has been associated with increased barrier permeability (Kago *et al.*, 2006) and claudins (Abbruscato *et al.*, 2002) establish the primary gate of the tight junction, the minimal dextran permeability and the upregulation of these proteins show that tight junctions are formed in SyM-BBB.

Transporter proteins

Efflux activity of Rhodamine 123 in RBE4 cells cultured in the presence and absence of ACM in SyM-BBB was investigated (Figure 12.11). In order to demonstrate drug inhibition studies, we used Verapamil, an L-type calcium channel blocker of the phenylalkylamine class and a potent inhibitor of P-gp. Cells in SyM-BBB were equilibrated with 150 ng/ml of Rhodamine 123 (Sigma Aldridge) for 20 min at 37°C. After washing the cells with PBS containing 0.1% Bovine Serum Albumin (BSA), the cells were imaged for Rhodamine permeation. Subsequently, the dye was allowed to efflux for two hours at 37°C to calculate the efflux rates. RBE4 cells pre-incubated with verapamil (5 μM) for 30 min before the efflux served as a model for P-gp inhibition.

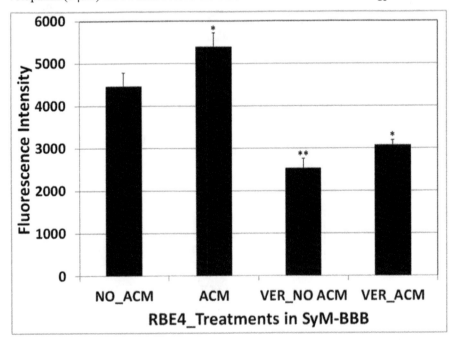

Figure 12.11. Rhodamine 123 efflux in SyM-BBB in the absence and presence of astrocyte conditioned medium (ACM) and verapamil (VER). Reproduced from Prabhakarpandian *et al.*, 2013 by permission of The Royal Society of Chemistry, http://dx.doi.org/10.1039/C2LC41208J.

The presence of ACM in SyM-BBB allowed significant (P < 0.05) uptake and increased efflux compared with the endothelial cells in the absence of ACM. In addition, although Verapamil reduces the efflux of Rhodamine 123, the efflux increases in the presence of ACM. This shows that P-gp is functionally active in SyM-BBB and that it efficiently regulates the efflux activity. Furthermore, these studies establish the modulation of the functional characteristics of SyM-BBB by known transport blockers.

Genomics

Tight junction genes

It is well know that shear stress induces tight junction formation in endothelial cells. There have been several studies focused on identifying these shear dependent changes from a genomic perspective. One of the earlier studies on the effect of flow on brain endothelial cells at a genomic level was conducted by Cucullo *et al.* (Cucullo *et al.*, 2011), whose findings highlighted significant shear dependent upregulation of tight junction and adherens genes (Figure 12.12). This upregulation in gene expression was corroborated by studying the permeability of a variety of molecules as well as fold changes in expression levels of corresponding tight junction and adherens protein.

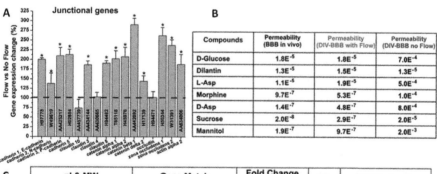

B

Compounds	Permeability (BBB in vivo)	Permeability (DIV-BBB with Flow)	Permeability (DIV-BBB no Flow)
D-Glucose	$1.8E^{-5}$	$1.8E^{-5}$	$7.0E^{-4}$
Dilantin	$1.3E^{-5}$	$1.5E^{-5}$	$1.3E^{-5}$
L-Asp	$1.1E^{-5}$	$1.9E^{-5}$	$5.0E^{-4}$
Morphine	$9.7E^{-7}$	$5.3E^{-7}$	$1.0E^{-4}$
D-Asp	$1.4E^{-7}$	$4.8E^{-7}$	$8.0E^{-4}$
Sucrose	$2.0E^{-6}$	$2.9E^{-7}$	$2.0E^{-5}$
Mannitol	$1.9E^{-7}$	$9.7E^{-7}$	$2.0E^{-3}$

C

pI & MW	Gene Match	Fold Change (Flow/No Flow)	±Se	Protein family
pI: 5.01, MW: 82565.47	Cadherin 5	2.13	0.172	Adherens Junctions
pI: 4.43, MW: 82030.01	Cadherin 2, N-cadherin	2.04	0.136	
pI: 4.83, MW: 78320.93	Cadherin 1, E-cadherin	2.00	0.23	
pI: 5.77, MW: 59143.79	Occludin	2.47	0.147	Tight Junctions
pI: 5.32, MW: 61991.18	Claudin 5	5.91	0.390	

Figure 12.12. Shear stress promotes tight junction (TJ) formation in endothelial cells (A) Comparison of the level of TJ and adherens junction RNA in endothelial cells grown under flow vs. static condition. (B) The tightness of the vascular endothelial bed was confirmed by permeability measurements. (C) Gene expression findings were also supported by comparative analysis of TJ and adherens junction proteins (Cucullo *et al.*, 2011). Reproduced with permissions from Biomed Central.

Toxicogenomic response

One of the primary motivations for developing a BBB model is to study diseases states and to perform toxicological investigations. As an example, this section highlights the type of studies that may be carried out. In this case, we studied the genomic level response of Human Umbilical Vein Endothelial Cells (HUVECs), a common endothelial cell model, to Doxorubicin, a well-known chemotherapeutic. Figure 12.13A shows the total number of significant genes that were identified at 30 min, 4 hr and 24 hr while Figure 12.13B shows the temporal changes up- and down-regulated genes at the various time points.

Seven transcripts (MT1H, TIE1, RPL26P37, MT1E, FADD, MT1JP and NSUN2) were found to be commonly up/down-regulated at all three time points. While all seven genes were upregulated at the early stage (30 min), most of them (five out of seven) were down-regulated by 24 hr. Among other shared transcripts, 34 are common to the first two time points (30 min and 4 hr) while 36 transcripts are common to the second and the third (4 hr and 24 hr) time points. The third time point has the largest number of down regulated genes. This is a near reversal of the early-stage response where most of genes are up-regulated and indicates that by 24 hours, the treated cells have down-regulated their internal signaling and metabolic demands.

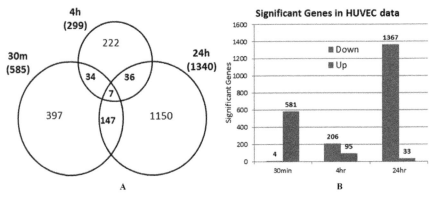

Figure 12.13. (A) Venn diagram of significantly-expressed genes at different time points for doxorubicin Treatment of endothelial (HUVEC) cells. (B) Number of up- and down-regulated genes at each time-point.

These genes were used to query the KEGG (Kyoto Encyclopedia of Genes and Genomes) database to identify the cellular pathways that are activated in response to the treatment by the drug. All differentially regulated pathways based on the expressed genes can be analyzed to identify pathways with a statistical significant difference ($P < 0.05$, shown in Table 12.1).

The canonical pathways that are activated at each of the time points were used to construct pathway models which are shown in Figure 12.14. CFDRC-developed bioinformatics tool, SBMLForge was used to merge multiple canonical pathways into a single large pathway model in an automated fashion. These figures show the most connected components of these pathways. As it can be seen, the complexity of the pathway significantly increases by 24 hours owing to the activation of a large number of pathways.

Table 12.1. List of statistically significant (P < 0.05) pathways that are activated in HUVECs treated with doxorubicin at different time points.

List of Pathways Activated at 30 minutes

Pathway ID	KEGG Pathway Name	P-value
hsa03010.xml	Ribosome	2.9e-14
hsa00190.xml	Oxidative Phosphorylation	8.5e-4
hsa05012.xml	Parkinson's disease	6.2e-3
hsa04120.xml	Ubiquitin mediated proteolysis	1.1e-2
hsa05016.xml	Huntington's Disease	1.6e-2
hsa05010.xml	Alzheimer's Disease	3.6e-2
hsa04114.xml	Oocyte Meiosis	3.7e-2
hsa03050.xml	Proteasome	4.3e-2
hsa03010.xml	Ribosome	2.9e-14

List of Pathways Activated at 4 hours

Pathway ID	KEGG Pathway Name	P-value
hsa04350.xml	TGF-beta signaling pathway	6.5e-4
hsa04110.xml	Cell cycle	1.2e-3
hsa04115.xml	P53 signaling pathway	2.8e-2
hsa00970.xml	Aminoacyl-tRNA biosynthesis	3.2e-2
hsa05220.xml	Chronic myeloid leukemia	3.9e-2
hsa04520.xml	Adherens junction	4.2e-2
hsa05200.xml	Pathways in cancer	5.2e-2

List of Pathways Activated at 24 hours

Pathway ID	KEGG Pathway Name	P-value
hsa05012.xml	Parkinson's disease	7.6e-21
hsa00190.xml	Oxidative Phosphorylation	7.3e-19
hsa03010.xml	Ribosome	7.e-19
hsa05016.xml	Huntington's Disease	2.8E-17
hsa05010.xml	Alzheimer's Disease	2.1E-13
hsa03050.xml	Proteasome	1.5E-10
hsa05130.xml	Pathogenic Escherichia coli infection	3.8E-7
hsa03040.xml	Spliceosome	7.2E-7
hsa04260.xml	Cardiac muscle contraction	2.3E-5
hsa00010.xml	Glycolysis / Gluconeogenesis	6.7E-3
hsa04810.xml	Regulation of actin cytoskeleton	1.1E-2
hsa04666.xml	Fc gamma R-mediated phagocytosis	1.5E-2
hsa04670.xml	Leukocyte transendothelial migration	1.6E-2
hsa04540.xml	Gap junction	1.8E-2
hsa04722.xml	Neurotrophin signaling pathway	2.6E-2
hsa03030.xml	DNA Replication	2.7e-2
hsa00240.xml	Pyrimidine metabolism	3.1e-2
hsa00020.xml	Citrate cycle (TCA cycle)	3.4e-2
hsa00030.xml	Pentose phosphate pathway	3.7e-2
hsa00062.xml	Fatty acid elongation in mitochondria	4.1e-2
hsa00270.xml	Cysteine and methonine metabolism	5.4e-2
hsa05110.xml	Vibrio cholerase infection	5.6e-2
hsa03020.xml	RNA Polymerase	6.1e-2
hsa04110.xml	Cell cycle	8.6e-2

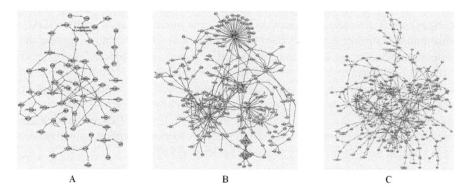

A   B   C

Figure 12.14. Pathway networks constructed from gene expression data of doxorubicin-treated HUVECs at: (A) 30 min (B) 4 hr (C) 24 hr. Genes with unlinked cellular pathways are not shown.

Future Perspectives

Understanding the dynamics of the Blood-Brain Barrier (BBB) is critical to the development of novel therapeutics that aim to target Central Nervous System (CNS) disorders. The BBB is also critical to further our understanding of the way environmental toxicants impact the brain. *In vitro* models provide a controlled environment, which can not only elucidate the mechanistic underpinnings of the barrier pathophysiology, but also accelerate the collective insight into the structure and function of the BBB.

There are several methodologies available or currently under development for reproducing a robust and accurate *in vitro* model of the BBB. These range from the classical Transwell apparatus to hollow fiber bioreactors to newly emerging microfluidic models. These models aim to capture the dynamics of the BBB—for instance, by incorporating flow, or the realism of the BBB—for instance, by incorporating 3D co-culture or tri-culture of cells. However, there is a lack of consensus on the best model and parameters to be used. In this context, we believe there are two steps that need to be taken to enable widespread use of the *in vitro* BBB models in scientific and translational research. First, the *in vitro* models need to be validated against *in vivo* data, both animal as well as human, on a number of different elements including morphology, permeability, proteomics, gene expression and drug delivery, before they can be reliably used in practice. Second, a combination of models may be the best solution moving forward, with the model to be used will depend on the scientific question being posed.

The models discussed in this chapter focus on the BBB, with emphasis on the ability to reproduce cell-cell interactions and the microenvironment. The xenobiotic interactions with the BBB manifest themselves via neurological responses in the brain or the entire CNS. So, future developments should target the development and integration of additional elements of the brain or the CNS, so as to facilitate an integrative response at the physiological level. Another key *in vivo* element that needs to be faithfully captured is that in the body, the drug or the toxin may interact with the BBB or the brain via multiple routes—direct as well as indirect (mediated by other organs). In order to capture the dynamics of *in vivo* behavior such as ADME (absorption, distribution, metabolism and elimination), it is critical to capture the presence of other organs in the body. Propelled by recent technological developments in the field of 'organ on a chip' or OOC and 'human on a chip' or HOC, it is indeed within reach to bioengineer *in vitro* platforms with sufficient complexity to capture these organ-organ interactions. For instance, the SyM-BBB architecture may be engineered to build other organoid models such as liver, heart, kidney, etc., which can then be coupled in the proper *in vivo* like architecture to yield a synthetic environment for evaluating drug (or toxicant) transport, toxicity and efficacy.

Acknowledgements

The authors have no conflict of interest. The contents of this manuscript do not necessarily reflect the views and policies of the U.S. Food & Drug Administration, and the mention of trade names or commercial products does not constitute endorsement or recommendation for use.

References

Abbruscato, T.J., and T.P. Davis. 1999. Protein expression of brain endothelial cell E-cadherin after hypoxia/aglycemia: influence of astrocyte contact. Brain Res 842: 277–286.

Abbruscato, T.J., S.P. Lopez, K.S. Mark, B.T. Hawkins, and T.P. Davis. 2002. Nicotine and cotinine modulate cerebral microvascular permeability and protein expression of ZO-1 through nicotinic acetylcholine receptors expressed on brain endothelial cells. J Pharm Sci 91: 2525–2538.

Aschner, M., V.A. Fitsanakis, A.P. dos Santos, L. Olivi, and J.P. Bressler. 2006. Blood-brain barrier and cell-cell interactions: methods for establishing *in vitro* models of the blood-brain barrier and transport measurements. Methods Mol Biol (Clifton, N.J.) 341: 1–15.

Audus, K.L., and R.T. Borchardt. 1986. Characteristics of the large neutral amino acid transport system of bovine brain microvessel endothelial cell monolayers. J Neurochem 47: 484–488.

Audus, K.L., and R.T. Borchardt. 1987. Bovine brain microvessel endothelial cell monolayers as a model system for the blood-brain barrier. Ann N Y Acad Sci 507: 9–18.

Begley, D.J., D. Lechardeur, Z.D. Chen, C. Rollinson, M. Bardoul, F. Roux, D. Scherman, and N.J. Abbott. 1996. Functional expression of P-glycoprotein in an immortalised cell line of rat brain endothelial cells, RBE4. J Neurochem 67: 988–995.

Bernas, M.J., F.L. Cardoso, S.K. Daley, M.E. Weinand, A.R. Campos, A.J. Ferreira, J.B. Hoying, M.H. Witte, D. Brites, Y. Persidsky, S.H. Ramirez, and M.A. Brito. 2010. Establishment of primary cultures of human brain microvascular endothelial cells to provide an *in vitro* cellular model of the blood-brain barrier. Nat Protoc 5: 1265–1272.

Biegel, D., and J.S. Pachter. 1994. Growth of brain microvessel endothelial cells on collagen gels: applications to the study of blood-brain barrier physiology and CNS inflammation. *In vitro* cellular & developmental biology. Animal 30a: 581–588.

Biegel, D., D.D. Spencer, and J.S. Pachter. 1995. Isolation and culture of human brain microvessel endothelial cells for the study of blood-brain barrier properties *in vitro*. Brain Res 692: 183–189.

Booth, R., and H. Kim. 2012. Characterization of a microfluidic *in vitro* model of the blood-brain barrier (muBBB). Lab Chip 12: 1784–1792.

Couraud, P.O., J. Greenwood, F. Roux, and P. Adamson. 2003. Development and characterization of immortalized cerebral endothelial cell lines. Methods Mol Med 89: 349–364.

Cucullo, L., M.S. McAllister, K. Kight, L. Krizanac-Bengez, M. Marroni, M.R. Mayberg, K.A. Stanness, and D. Janigro. 2002. A new dynamic *in vitro* model for the multidimensional study of astrocyte-endothelial cell interactions at the blood-brain barrier. Brain Res 951: 243–254.

Cucullo, L., B. Aumayr, E. Rapp, and D. Janigro. 2005. Drug delivery and *in vitro* models of the blood-brain barrier. Curr Opin Drug Discov Devel 8: 89–99.

Cucullo, L., M. Hossain, V. Puvenna, N. Marchi, and D. Janigro. 2011. The role of shear stress in Blood-Brain Barrier endothelial physiology. BMC Neuroscience 12: 40.

Czupalla, C.J., S. Liebner, and K. Devraj. 2014. *In vitro* models of the blood-brain barrier. Methods Mol Biol (Clifton, N.J.) 1135: 415–437.

de Vries, H.E., M.C. Blom-Roosemalen, A.G. de Boer, T.J. van Berkel, D.D. Breimer, and J. Kuiper. 1996. Effect of endotoxin on permeability of bovine cerebral endothelial cell layers *in vitro*. J Pharmacol Exp Ther 277: 1418–1423.

Deosarkar, S.P., B. Prabhakarpandian, B. Wang, J.B. Sheffield, B. Krynska, and M.F. Kiani. 2015. A Novel Dynamic Neonatal Blood-Brain Barrier on a Chip. PLoS One 10(11): e0142725.

dos Santos, A.P., D. Milatovic, C. Au, Z. Yin, M.C. Batoreu, and M. Aschner. 2010. Rat brain endothelial cells are a target of manganese toxicity. Brain Res 1326: 152–161.

Durieu-Trautmann, O., N. Foignant-Chaverot, J. Perdomo, P. Gounon, A.D. Strosberg, and P.O. Couraud. 1991. Immortalization of brain capillary endothelial cells with maintenance of structural characteristics of the blood-brain barrier endothelium. *In vitro* Cellular & Developmental Biology : J Tissue Cult Assoc 27a: 771–778.

Fitsanakis, V.A., G. Piccola, J.L. Aschner, and M. Aschner. 2005. Manganese transport by rat brain endothelial (RBE4) cell-based transwell model in the presence of astrocyte conditioned media. J Neurosci Res 81: 235–243.

Fitsanakis, V.A., G. Piccola, J.L. Aschner, and M. Aschner. 2006. Characteristics of manganese (Mn) transport in rat brain endothelial (RBE4) cells, an *in vitro* model of the blood-brain barrier. Neurotoxicology 27: 60–70.

Hawkins, B.T., and T.P. Davis. 2005. The blood-brain barrier/neurovascular unit in health and disease. Pharmacol Rev 57: 173–185.

Iadecola, C. 2004. Neurovascular regulation in the normal brain and in Alzheimer's disease. Nature Rev. Neurosci 5: 347–360.

Janzer, R.C., and M.C. Raff. 1987. Astrocytes induce blood-brain barrier properties in endothelial cells. Nature 325: 253–257.

Joo, F. 1992. The cerebral microvessels in culture, an update. J Neurochem 58: 1–17.

Kago, T., N. Takagi, I. Date, Y. Takenaga, K. Takagi, and S. Takeo. 2006. Cerebral ischemia enhances tyrosine phosphorylation of occludin in brain capillaries. Biochem Biophys Res Commun 339: 1197–1203.

Lippmann, E.S., S.M. Azarin, J.E. Kay, R.A. Nessler, H.K. Wilson, A. Al-Ahmad, S.P. Palecek, and E.V. Shusta. 2012. Derivation of blood-brain barrier endothelial cells from human pluripotent stem cells. Nat Biotechnol 30: 783–791.

Lippmann, E.S., A. Al-Ahmad, S.P. Palecek, and E.V. Shusta. 2013. Modeling the blood-brain barrier using stem cell sources. Fluids and Barriers of the CNS 10: 2.

Lippmann, E.S., A. Al-Ahmad, S.M. Azarin, S.P. Palecek, and E.V. Shusta. 2014. A retinoic acid-enhanced, multicellular human blood-brain barrier model derived from stem cell sources. Sci Rep 4: 4160.

Lockman, P.R., J. Koziara, K.E. Roder, J. Paulson, T.J. Abbruscato, R.J. Mumper, and D.D. Allen. 2003. *In vivo* and *in vitro* assessment of baseline blood-brain barrier parameters in the presence of novel nanoparticles. Pharm Res 20: 705–713.

Matthes, F., P. Wolte, A. Bockenhoff, S. Huwel, M. Schulz, P. Hyden, J. Fogh, V. Gieselmann, H.J. Galla, and U. Matzner. 2011. Transport of arylsulfatase A across the blood-brain barrier *in vitro*. J Biol Chem 286: 17487–17494.

Mizuno, N., T. Niwa, Y. Yotsumoto, and Y. Sugiyama. 2003. Impact of drug transporter studies on drug discovery and development. Pharmacol Rev 55: 425–461.

Montesano, R., M.S. Pepper, U. Mohle-Steinlein, W. Risau, E.F. Wagner, and L. Orci. 1990. Increased proteolytic activity is responsible for the aberrant morphogenetic behavior of endothelial cells expressing the middle T oncogene. Cell 62: 435–445.

Panula, P., F. Joo, and L. Rechardt. 1978. Evidence for the presence of viable endothelial cells in cultures derived from dissociated rat brain. Experientia 34: 95–97.

Pardridge, W.M. 2005. The blood-brain barrier: bottleneck in brain drug development. NeuroRx : NeuroTherapeutics 2: 3–14.

Prabhakarpandian, B., M.C. Shen, K. Pant, and M.F. Kiani. 2011. Microfluidic devices for modeling cell-cell and particle-cell interactions in the microvasculature. Microvasc Res 82: 210–220.

Prabhakarpandian, B., M.C. Shen, J.B. Nichols, I.R. Mills, M. Sidoryk-Wegrzynowicz, M. Aschner, and K. Pant. 2013. SyM-BBB: a microfluidic Blood Brain Barrier model. Lab Chip 13: 1093–1101.

Rahman, M.F., J. Wang, T.A. Patterson, U.T. Saini, B.L. Robinson, G.D. Newport, R.C. Murdock, J.J. Schlager, S.M. Hussain, and S.F. Ali. 2009. Expression of genes related to oxidative stress in the mouse brain after exposure to silver-25 nanoparticles. Toxicol Lett 187: 15–21.

Reese, T.S., and M.J. Karnovsky. 1967. Fine structural localization of a blood-brain barrier to exogenous peroxidase. J Cell Biol 34: 207–217.

Ren, J., S. Shen, D. Wang, Z. Xi, L. Guo, Z. Pang, Y. Qian, X. Sun, and X. Jiang. 2012. The targeted delivery of anticancer drugs to brain glioma by PEGylated oxidized multi-walled carbon nanotubes modified with angiopep-2. Biomater 33: 3324–3333.

Ribeiro, M.M., M.A. Castanho, and I. Serrano. 2010. *In vitro* blood-brain barrier models—latest advances and therapeutic applications in a chronological perspective. Mini Rev Med Chem 10: 262–270.

Roux, F., O. Durieu-Trautmann, N. Chaverot, M. Claire, P. Mailly, J.M. Bourre, A.D. Strosberg, and P.O. Couraud. 1994. Regulation of gamma-glutamyl transpeptidase and alkaline phosphatase activities in immortalized rat brain microvessel endothelial cells. J Cell Physiol 159: 101–113.

Rubin, L.L., D.E. Hall, S. Porter, K. Barbu, C. Cannon, H.C. Horner, M. Janatpour, C.W. Liaw, K. Manning, J. Morales, L.I. Tanner, K.J. Tomaselli, and F. Bard. 1991. A cell culture model of the blood-brain barrier. J Cell Biol 115: 1725–1735.

Santaguida, S., D. Janigro, M. Hossain, E. Oby, E. Rapp, and L. Cucullo. 2006. Side by side comparison between dynamic versus static models of blood-brain barrier *in vitro*: a permeability study. Brain Res 1109: 1–13.

Santos, D., M.C. Batoreu, M. Aschner, and A.P. Marreilha dos Santos. 2013. Comparison between 5-aminosalicylic acid (5-ASA) and para-aminosalicylic acid (4-PAS) as potential protectors against Mn-induced neurotoxicity. Biol Trace Elem Res 152: 113–116.

Sharma, H.S., S.F. Ali, S.M. Hussain, J.J. Schlager, and A. Sharma. 2009a. Influence of engineered nanoparticles from metals on the blood-brain barrier permeability, cerebral blood flow, brain edema and neurotoxicity. An experimental study in the rat and mice using biochemical and morphological approaches. J Nanosci and Nanotechnol 9: 5055–5072.

Sharma, H.S., S.F. Ali, Z.R. Tian, S.M. Hussain, J.J. Schlager, P.O. Sjoquist, A. Sharma, and D.F. Muresanu. 2009b. Chronic treatment with nanoparticles exacerbate hyperthermia induced blood-brain barrier breakdown, cognitive dysfunction and brain pathology in the rat. Neuroprotective effects of nanowired-antioxidant compound H-290/51. J Nanosci and Nanotechnol 9: 5073–5090.

Shityakov, S., E. Salvador, G. Pastorin, and C. Forster. 2015. Blood-brain barrier transport studies, aggregation, and molecular dynamics simulation of multiwalled carbon nanotube functionalized with fluorescein isothiocyanate. Int J Nanomedicine 10: 1703–1713.

Teifel, M., and P. Friedl. 1996. Establishment of the permanent microvascular endothelial cell line PBMEC/C1-2 from porcine brains. Exp Cell Res 228: 50–57.

Trickler, W.J., W.G. Mayhan, and D.W. Miller. 2005. Brain microvessel endothelial cell responses to tumor necrosis factor-alpha involve a nuclear factor kappa B (NF-kappaB) signal transduction pathway. Brain Res 1048: 24–31.

Trickler, W.J., S.M. Lantz, R.C. Murdock, A.M. Schrand, B.L. Robinson, G.D. Newport, J.J. Schlager, S.J. Oldenburg, M.G. Paule, W. Slikker, Jr., S.M. Hussain, and S.F. Ali. 2010. Silver nanoparticle induced blood-brain barrier inflammation and increased permeability in primary rat brain microvessel endothelial cells. Toxicological Sciences : Toxicol Sci 118: 160–170.

Trickler, W.J., S.M. Lantz, R.C. Murdock, A.M. Schrand, B.L. Robinson, G.D. Newport, J.J. Schlager, S.J. Oldenburg, M.G. Paule, W. Slikker, Jr., S.M. Hussain, and S.F. Ali. 2011. Brain microvessel endothelial cells responses to gold nanoparticles: *In vitro* pro-inflammatory mediators and permeability. Nanotoxicology 5: 479–492.

Trickler, W.J., S.M. Lantz, A.M. Schrand, B.L. Robinson, G.D. Newport, J.J. Schlager, M.G. Paule, W. Slikker, A.S. Biris, S.M. Hussain, and S.F. Ali. 2012. Effects of copper nanoparticles on rat cerebral microvessel endothelial cells. Nanomedicine (London, England) 7: 835–846.

Trickler, W.J., S.M. Lantz-McPeak, B.L. Robinson, M.G. Paule, W. Slikker, Jr., A.S. Biris, J.J. Schlager, S.M. Hussain, J. Kanungo, C. Gonzalez, and S.F. Ali. 2014. Porcine brain microvessel endothelial cells show pro-inflammatory response to the size and composition of metallic nanoparticles. Drug Metab Rev 46: 224–231.

van der Meer, A.D., A.A. Poot, M.H. Duits, J. Feijen, and I. Vermes. 2009. Microfluidic technology in vascular research. J Biomed Biotechnol 2009: 823148.

Vu, K., B. Weksler, I. Romero, P.O. Couraud, and A. Gelli. 2009. Immortalized human brain endothelial cell line HCMEC/D3 as a model of the blood-brain barrier facilitates *in vitro* studies of central nervous system infection by Cryptococcus neoformans. Eukaryot Cell 8: 1803–1807.

Wang, J., M.F. Rahman, H.M. Duhart, G.D. Newport, T.A. Patterson, R.C. Murdock, S.M. Hussain, J.J. Schlager, and S.F. Ali. 2009. Expression changes of dopaminergic system-related genes in PC12 cells induced by manganese, silver, or copper nanoparticles. Neurotoxicology 30: 926–933.

Warner, D.S., M.L. James, D.T. Laskowitz, and E.F. Wijdicks. 2014. Translational research in acute central nervous system injury: lessons learned and the future. JAMA Neurol 71: 1311–1318.

Weksler, B.B., E.A. Subileau, N. Perriere, P. Charneau, K. Holloway, M. Leveque, H. Tricoire-Leignel, A. Nicotra, S. Bourdoulous, P. Turowski, D.K. Male, F. Roux, J. Greenwood, I.A. Romero, and P.O. Couraud. 2005. Blood-brain barrier-specific properties of a human adult brain endothelial cell line. FASEB Journal 19: 1872–1874.

Wolff, A., M. Antfolk, B. Brodin, and M. Tenje. 2015. *In Vitro* Blood-brain barrier models-An overview of established models and new microfluidic approaches. J Pharm Sci 104(9): 2727–46.

Yang, J., and M. Aschner. 2003. Developmental aspects of blood-brain barrier (BBB) and rat brain endothelial (RBE4) cells as *in vitro* model for studies on chlorpyrifos transport. Neurotoxicology 24: 741–745.

Zhao, D., D. Alizadeh, L. Zhang, W. Liu, O. Farrukh, E. Manuel, D.J. Diamond, and B. Badie. 2011. Carbon nanotubes enhance CpG uptake and potentiate antiglioma immunity. Clin Cancer Res 17: 771–782.

13

Neurotrophins and Alzheimer's Disease

Jyotshna Kanungo

Introduction

Neurotrophins belong to a family of proteins that are essential for the nervous system development, function and survival of neurons. These secreted proteins are trophic/growth factors, which not only promote the differentiation, growth, maintenance and survival of neurons, but also play critical roles in synaptic plasticity (Chao and Lee, 2004; Chao *et al.*, 2006). Members of the neurotrophin family include the nerve growth factor (NGF), the brain-derived neurotrophic factor (BDNF), neurotrophin 3 (NT-3) and neurotrophin 4/5 (NT-4/5) (Allen *et al.*, 2013; Reichardt, 2006; Skaper, 2012).

The most studied neurotrophins are NGF, BDNF, and NT-3. Their specific receptors are tropomyosin related kinase (Trk) receptor- TrkA (NGF), TrkB (BDNF), and TrkC (NT-3). There is also the p75 receptor, which is common to all Trk-expressing neurons (Chao and Hempstead, 1995; Huang and Reichardt, 2001) (Figure 13.1). In adults, p75NTR expression is relatively low (Friedman, 2010). All the neurotrophins are originally translated as larger pro-forms and upon proteolytic cleavage, form the mature neurotrophins that exist in the cells as homodimers (Aloe *et al.*, 2012). The role of neurotrophins in the survival of developing neurons in the peripheral nervous system (PNS) has been well established (Crowley *et al.*, 1994; Smeyne *et al.*, 1994; Zweifel *et al.*, 2005) whereas their roles in developing neurons in the central nervous system (CNS) were recognized later (Baydyuk and Xu, 2014).

Division of Neurotoxicology, HFT-132, National Center for Toxicological Research (NCTR), U.S. Food and Drug Administration (FDA), 3900 NCTR Rd, Jefferson, AR 72079.
E-mail: jyotshnabala.kanungo@fda.hhs.gov

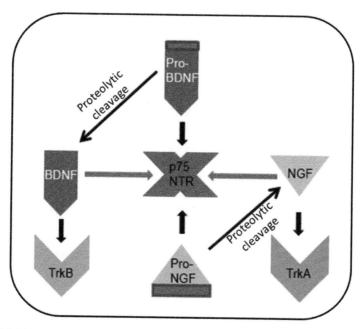

Figure 13.1. The neurotrophins, NGF and BDNF and their receptors. NGF and BDNF are produced by proteolytic cleavage of their respective proforms, ProNGF and ProBDNF. Although NGF and BDNF bind to TrkA and TrkB, respectively, they also can bind to the low affinity receptor, p75NTR. P75NTR also acts as a receptor for proNGF and proBDNF.

Alzheimer's disease (AD) is a CNS neurodegenerative disease. The pathological presentation of AD, the leading cause of senile dementia, involves regionalized neuronal death and an accumulation of neuronal and extracellular lesions termed neurofibrillary tangles and senile plaques, respectively (Smith and Perry, 1997). Several independent hypotheses have been proposed to link the pathological lesions and neuronal cytopathology with, among others, apolipoprotein E genotype (Corder *et al.*, 1993; Roses, 1995), hyperphosphorylation of cytoskeletal proteins (neurofilaments and Tau) (Trojanowski *et al.*, 1993), and amyloid-β (Aβ) metabolism (Selkoe, 1997). However, none of these theories alone is sufficient by itself to explain the diversity of biochemical and pathological abnormalities of AD.

It's expected that the prevalence of AD will increase dramatically over the next decade. AD accounts for between 60% and 70% of dementias; which translates to 1–2% of the US population. Reductions in these neurotrophic factors have been associated with a number of neurodegenerative disorders including AD (Allen *et al.*, 2013). Preclinical drug development directed towards enhancing neurotrophin signaling are considered to provide novel therapeutic approaches for neurodegenerative disorders (Steiner and Nath, 2014). Efforts toward developing successful therapies to treat AD include replenishing endogenous neurotrophic factors by supplying BDNF, NGF or functional mimetics. This chapter summarizes the studies involving neurotrophins, especially the most common ones, NGF and BDNF and their potential therapeutic values in treating AD.

NGF

NGF was the first neurotrophin to be identified after the discovery of its ability to promote neurite growth when applied to explanted dorsal root ganglia (Cowan, 2001). NGF is initially translated as a precursor polypeptide, proNGF, which is cleaved extracellularly by plasmin and matrix metalloproteases to produce the mature form of NGF (Lee *et al.*, 2001b). NGF functions in the neuronal development and survival of specific neurons in the CNS and PNS and was identified as a survival factor during the period of programmed cell death for small sensory neurons (Huang and Reichardt, 2001; Mendell, 1999). NGF transcripts are highly expressed in the spinal cord and dorsal root ganglia (DRG) (Sobue *et al.*, 1998). NGF is produced in the cortex and hippocampus and NGF-null mice exhibit significant sympathetic and sensory deficits. These mice only survive a few weeks after birth (Crowley *et al.*, 1994; Huang and Reichardt, 2001) and during their few weeks of survival, the mice possess an intact cholinergic basal forebrain (Crowley *et al.*, 1994). On the other hand, mice, heterozygous for NGF knockout, survive to adulthood, display memory deficits with a significantly reduced cholinergic innervation of the hippocampus (Chen *et al.*, 1997). These findings suggest that NGF is essential during maturity of the cholinergic–hippocampal system.

The "neurotrophic factor hypothesis" proposed that developing neurons will not survive unless they can compete for limited supplies of a target-derived neurotrophic factor, which was suggested then to be NGF (Yuen *et al.*, 1996). Consistent with this hypothesis, target tissues of sensory and sympathetic fibers were shown to produce NGF that was taken up by the fiber terminals, and then transported back to the neuron cell body, a process critical for the neuron's survival. NGF facilitates electrophysiological long-term potentiation (LTP) and its blockage inhibits LTP in rat cholinergic septohippocampal system. The retrograde transport of NGF from the hippocampal sites of synthesis to projecting cholinergic neurons nested in the basal forebrain has been shown to modulate the cholinergic system within the innervated targets and thus can indirectly control neuronal plasticity leading to learning and memory retention (Conner *et al.*, 2009).

NGF receptors

NGF exerts its biological effects by binding and activating the Trk receptor, TrkA (Reichardt, 2006). Just like NGF knockout mice, mice with homozygous knockout of TrkA although perinatally survive only a few weeks, have an intact cholinergic basal forebrain (Huang and Reichardt, 2001; Smeyne *et al.*, 1994) with significantly reduced cholinergic basal forebrain's hippocampal and cortical projections (Smeyne *et al.*, 1994). In the first week, these mice have relatively smaller postnatal cholinergic neurons and in a month, there is significant depletion in neuron numbers associated with abnormal cholinergic innervation of the hippocampus (Fagan *et al.*, 1996). The homodimeric NGF also binds to a low affinity p75 pan neurotrophin receptor (p75NTR) (Aloe *et al.*, 2012). p75NTR is a member of the tumor necrosis factor (TNF) receptor superfamily and does not contain a catalytic motif (Gruss and Dower, 1995).

NGF-activated TrkA promotes neuronal survival, differentiation and cell growth (Allen *et al.*, 2013). The p75NTR alone in the absence of TrkA and NGF, can promote neuronal apoptosis via activation of the TNF receptor associated factor (Trafs), NFκB (nuclear factor-κB) and ceramide (Blochl and Blochl, 2007) (Figure 13.2). Without the Trk receptors or the neurotrophins, p75NTR can trigger signaling pathways leading to apoptosis (Blochl and Blochl, 2007).

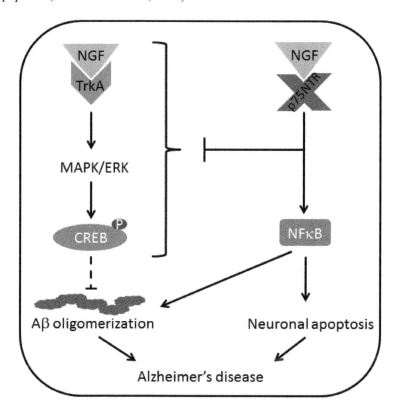

Figure 13.2. Major NGF signaling pathways that induce CREB phosphorylation via the MAPK/ERK activation, and NFκB activation. Binding of either NGF/TrkA or NGF/p75NTR can differentially modulate two different signaling pathways; the latter when activated can inhibit the former. Deregulation of both the pathways can lead to neuronal apoptosis and induce Aβ oligomerization, primers for developing pathological symptoms observed in AD.

NGF-mediated signaling pathways

NGF promotes cell survival and proliferation effects by binding to its receptor TrkA and inducing the TrkA-mediated signaling. The activation of cyclic AMP response element-binding protein (CREB) downstream of TrkA signaling inhibits the oligomerization and thus, formation of Aβ and therefore, Aβ-induced toxicity (Defelice and Ferreira, 2002; Gong *et al.*, 2004). On the other hand, the activation of NGF/p75NTR signaling inhibits NGF/TrkA signaling (Fortress *et al.*, 2011; Song *et al.*, 2010) (Figure 13.2). Exogenous NGF injection has been shown to activate TrkA signaling, which then activates the MAPK/ERK pathway in aging rats (Williams *et al.*, 2007). These results indicated a possible mechanism of NGF signaling associated with memory loss in the CNS, since CREB is an important substrate of ERK signaling and impaired CREB phosphorylation is believed to be a pathological component in AD (Scott Bitner, 2012).

In the absence of Trks and neurotrophins, p75NTR may elicit apoptosis via ceramide and NFκB activation through TNF receptor associated factors (Gruss and Dower, 1995). p75NTR can bind to both the mature and proforms of the neurotrophins, and along with other signaling proteins, can promote neuronal out-growth, proliferation and cell death (Chao *et al.*, 2006). TrkA/p75NTR signaling imbalance can contribute to amyloidosis and neurodegeneration since inhibition of TrkA signaling shifts NGF signaling toward p75NTR signaling (Capsoni *et al.*, 2010) (Figure 13.2). Activated p75NTR can mediate apoptosis and amyloidosis (Costantini *et al.*, 2005a; Costantini *et al.*, 2005b; Hashimoto *et al.*, 2004). Furthermore, p75NTR-mediated signaling pathway reportedly leads to c-jun N-terminal kinase (JNK) activation, increased expressions of the tumor suppressor, p53 and cleaved-caspase 3 (Diarra *et al.*, 2009; Kaplan and Miller, 2000; Kenchappa *et al.*, 2010; Suhl *et al.*, 2012). Memantine, an NMDA receptor antagonist activates NGF/TrkA signaling, possibly by increasing endogenous NGF levels while inhibiting the p75NTR signaling, JNK activation, and increase in p53 and cleaved-caspase 3 levels (Liu *et al.*, 2014).

BDNF

BDNF is synthesized from the precursor proBDNF (Teng *et al.*, 2010). Intracellular cleavage of proBDNF to mature BDNF can occur via furin or by other proprotein convertases and extracellular cleavage occurs by metalloproteases or plasmin (Greenberg *et al.*, 2009). BDNF is essential for normal neuronal development and multiple functions in the adult brain (Yoshii and Constantine-Paton, 2010). In addition to promoting survival and inducing synaptic plasticity, BDNF also regulates adult neurogenesis (Bath *et al.*, 2012). BDNF is the most widely distributed neurotrophin in the CNS (Tapia-Arancibia *et al.*, 2008) and plays a major role in regulating axonal and dendritic growth and guidance, long-term potentiation and participation in neurotransmitter release (Jeanneteau *et al.*, 2010).

The expression levels of BDNF vary during development. BDNF is expressed by a number of cells in the mouse hippocampus on embryonic day 15.5 (E15.5), and by E17 it is found in the piriform cortex, hippocampus, thalamus, hypothalamus and amygdala, but in only a few cells in the cortex and none in the striatum (Baydyuk *et al.*, 2013). In the adult brain, BDNF is found in many regions, including the cerebral cortex, basal forebrain, striatum, hippocampus, hypothalamus, brainstem, and cerebellum (Conner *et al.*, 1997). While both *Bdnf* mRNA and BDNF protein are present in most brain regions including the cortex, in the striatum BDNF protein levels are high but *Bdnf* mRNA is almost absent (Ma *et al.*, 2012). In mice, BDNF is expressed in the substantia nigra at P0 and E16.5 specifically in neurons positive for tyrosine hydroxylase (Baydyuk *et al.*, 2013). In adult rodents, BDNF mRNA is located in all of the cell layers of the hippocampus, but BDNF is primarily synthesized in granule cells and is anterogradely transported to its axons. Levels of BDNF protein in whole hippocampus are relatively low at birth and increase during early life (Harte-Hargrove *et al.*, 2013).

During the development of the cerebral cortex and hippocampus, BDNF triggers the differentiation of neural stem cells into mature neurons, and promotes the survival of nascent neuronal progenitors (Barnabe-Heider and Miller, 2003). At synapses, BDNF signaling enhances LTP (Saareleinen *et al.*, 2003) and may involve gene-expressions induced by the cAMP-response-element binding protein (CREB) (Ernfors *et al.*, 2003). In humans, a common single nucleotide polymorphism of BDNF (Val66Met BDNF) has been associated with deficits in social and cognitive functions (Baj *et al.*, 2013).

BDNF receptors

BDNF binds and activates the TrkB receptor (Reichardt, 2006). In mice, TrkB is present in high levels in the brain during development and its expression is variable among specific cell types. In adults, TrkB expression is observed in specific regions harboring neuronal subtypes. Both TrkB mRNA and protein are present in the developing and adult striatum (Baydyuk *et al.*, 2011; Yan *et al.*, 1997). BDNF and TrkB are highly expressed in the hippocampus, where BDNF exerts effects on the neuronal structure, function and plasticity of hippocampal neurons by activation of TrkB (Baydyuk and Xu, 2014; Yoshii and Constantine-Paton, 2010). *TrkB* mRNA and protein levels generally decrease with age (Silhol *et al.*, 2005). While mature BDNF and TrkB expression increase during development, proBDNF and p75NTR follow an opposite pattern (Yang *et al.*, 2009). Both proBDNF and BDNF bind to p75NTR (Deinhardt and Chao, 2014). Because of the relatively high levels of p75NTR and proBDNF found during early development, it has been suggested that proBDNF plays an important role in the developing brain to reduce neuronal number and refine dendrites and axons of emerging circuits (Teng *et al.*, 2010; Yang *et al.*, 2009).

Deletion of either the *TrkB* or *Bdnf* gene leads to cell atrophy, dendritic degeneration, neuronal loss and learning deficits as shown in the excitatory neurons of the dorsal forebrain (Gorski *et al.*, 2003). ProBDNF negatively regulates neuronal remodeling, synaptic transmission, and synaptic plasticity in hippocampus (Yang *et al.*, 2014a). In the human brain, there are three main isoforms of TrkB: a full-length and two truncated forms, which lack the tyrosine kinase domain. Full-length TrkB can initiate signaling pathways leading to survival, differentiation and synaptic plasticity, while its truncated forms can inhibit these effects by heterodimerization with it (Wong *et al.*, 2012). More recently, sortilin (a type-1 receptor expressed in neurons) has been shown to be a receptor for the pro-neurotrophins including proBDNF but not the mature neurotrophins like BDNF (Glerup *et al.*, 2014). ProBDNF binds to sortilin in the presence of p75NTR when released by cultured neurons and proBDNF induced-apoptosis requires the cell surface interaction of proBDNF with sortilin and p75NTR (Teng *et al.*, 2010).

BDNF-mediated signaling pathways

Upon binding to BDNF, activated full-length TrkB triggers multiple intracellular signaling cascades through protein-protein interactions (Smith, 2014) (Figure 13.3). The three major pathways, activated by TrkB are (1) the mitogen activated protein kinase/extracellular signal-regulated kinase (MAPK/ERK) pathway that activates regulators of protein translation; (2) the phosphatidylinositol 3-kinase (PI 3-kinase) pathway that activates AKT, which mediates anti-apoptotic effects; and the PLC-g pathway that leads to the production of diacylglycerol and an increase in intracellular calcium, and as a result the activation of CAM kinases and PKC (Smith, 2014). By activating these diverse signaling cascades in neurons, BDNF can regulate neuronal development and survival, initiation of neurite outgrowth and path-finding (Orefice *et al.*, 2013). Upon BDNF binding, autophosphorylation of TrkB occurs, leading to activation of signaling cascades such as the MAPK pathway (Yoshii and Constantine-Paton, 2010). BDNF-mediated transcriptional changes are thought to be initiated by the PI-3 kinase and/or Ras/MAPK/ERK pathways (Yoshii and Constantine-Paton, 2010). CREB is also a major target of BDNF signaling (Minichiello, 2009). These signaling pathways culminate in activation of several transcription factors thus leading to many effects of BDNF on hippocampal function. Truncated TrkB has been shown to have complex biological effects including neuronal development and calcium signaling (Minichiello, 2009).

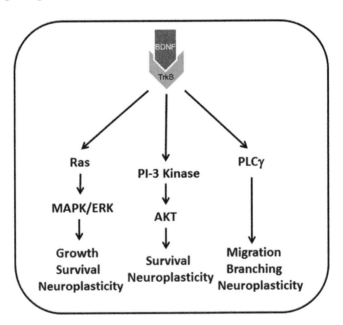

Figure 13.3. BDNF upon binding to its specific receptor, TrkB, activates major signaling pathways, Ras/MAPK/ERK, PI-3 kinase/AKT, and PLCγ. These pathways can lead to neuronal survival, axonal branching, neuronal migration and neuroplasticity.

BDNF binding to p75NTR activates intracellular signaling cascades including NFκB and JNK signaling pathways (Teng *et al.*, 2010). ProBDNF-activated p75NTR leads to the regulation of cell survival, pruning of neuronal processes and modulation of synaptic plasticity in hippocampal area CA1 (Teng *et al.*, 2010). BDNF activated TrkB has been shown to have effects opposite to those of proBDNF activated p75NTR

(Hempstead, 2006). While BDNF activated TrkB appears to facilitate LTP of the Schaffer collateral synapse in CA1 (Bramham and Messaoudi, 2005; Rex *et al.*, 2006), proBDNF bound p75NTR facilitates long-term depression (LTD) (Woo *et al.*, 2005). BDNF activated TrkB promotes spine growth and complexity (Tyler *et al.*, 2002) whereas proBDNF activated p75NTR initiates pruning and process retraction (Teng *et al.*, 2010). p75NTR can also interact with Trk receptors to alter the affinity of mature neurotrophins (Esposito *et al.*, 2001).

In a transgenic mouse model, exercise has been shown to normalize the dendritic outgrowth of neurons and increase the expression of hippocampal BDNF and normalize the hyperactivation of cyclin-dependent kinase 5 (Cdk5). Exercise has trophic activity on the neuronal lineage which is mediated by Cdk5 modulation of the BDNF pathway (Lee *et al.*, 2013). Hyperactivated Cdk5 has been implicated in neurodegenerative diseases including AD (Su and Tsai, 2011).

Alzheimer's Disease (AD)

AD is a CNS neurodegenerative disease, which was first described in 1906 by Dr. Alois Alzheimer. The pathological presentation of Alzheimer's disease, the leading cause of senile dementia, involves regionalized neuronal death and an accumulation of intraneuronal and extracellular lesions termed neurofibrillary tangles and senile plaques, respectively (Smith and Perry, 1997). A progressive neurodegenerative disease, AD is clinically characterized by cognitive impairments and pathologically characterized by deposits of Aβ peptides as senile plaques in the AD brain (Walsh and Selkoe, 2007). AD patients exhibit neuronal loss in the hippocampus with progressive cognitive deficits (Chao *et al.*, 2006; Hefti *et al.*, 1996; Mufson *et al.*, 2000; Tuszynski *et al.*, 2005). As mentioned earlier several independent hypotheses have been proposed to link the pathological lesions and neuronal cytopathology with, among others, apolipoprotein E genotype (Corder *et al.*, 1993; Roses *et al.*, 1995); hyperphosphorylation of cytoskeletal proteins (Trojanowski *et al.*, 1993), and Aβ metabolism (Selkoe, 1997); tauopathy (Lee *et al.*, 2001a); inflammation (Weiner and Selkoe, 2002) and oxidative stress (Perry *et al.*, 2002). However, as stated before none of these theories alone is sufficient by itself to explain the diversity of biochemical and pathological abnormalities of AD. Furthermore, attempts to mimic the disease by a perturbation of one of these elements using cell or animal models, including transgenic animals, do not result in similar pathological alterations. The most striking case is that when Aβ plaques are deposited in some transgenic rodent models over-expressing Aβ-protein precursor (Hsiao and Chiu, 1996), there is no neuronal loss—a seminal feature of Alzheimer's disease. Aβ peptide, usually 40 or 42 amino acids in length, is produced from the amyloid precursor protein (APP) by proteolytic cleavage. In most cases of AD, during the progression of dementia, the neurofibrillary tangles appear first in the transentorhinal cortex, then in the hippocampus and finally in various other parts of the neocortex (Braak and Braak, 1996). MRI (magnetic resonance imaging) studies based on longitudinal analysis of an asymptomatic population show alterations in the basal forebrain and the entorhinal cortex four years prior to the onset of symptoms (Hall *et al.*, 2008).

A majority of the cases of AD are 'sporadic' (www.alz.org). A number of risk factors that are either genetic or environmental or both, are likely the causative factors. However, rare familial cases of AD (FAD) occur with a much earlier onset albeit with similar symptoms and neuropathology observed in the sporadic cases (www.alz.org). Genes associated with FAD are APP and presenilin 1/2 (PSEN1 and PSEN2). PSEN1 and PSEN2 act as the catalytic parts of gamma secretase and specific mutations in these genes can cause FAD, by inducing increased production of total Aβ or an increase in the Aβ42: 40 ratio (Borchelt *et al.*, 1996). These disease causing mutations have been the basis of producing transgenic animals as AD models (Howlett, 2011).

The amyloid hypothesis states that all AD pathology and symptoms arise from the altered biology of Aβ, thus implying a direct correlation between the presence of Aβ and occurrence of neurofibrillary tangles (Hardy and Allsop, 1991). There is no direct evidence yet to challenge this hypothesis (Goate and Hardy, 2012). In AD, aberrant hyperphosphorylation of proteins, such as Tau and NF is prominent and is considered to be the consequence of the perturbation in the balance between the activities of kinases and phosphatases (Gong *et al.*, 2000). MAPK (Trojanowski *et al.*, 1993), glycogen synthase kinase-3 (GSK-3) (Picklo *et al.*, 2002), and Cdk5 (Su and Tsai, 2011) pathways are known to be involved in Tau and NF phosphorylation.

Data obtained from biopsy samples have shown that in early-stage AD patients, cholinergic function, acetylcholine synthesis and choline acetyltransferase (ChAT) activity were significantly reduced (Sims *et al.*, 1983). Moreover, > 90% of cholinergic neurons of the basal forebrain in rats and humans stain positive to a p75NTR antibody (Allen *et al.*, 1989). In AD brain, neurons positive for p75NTR were relatively spared despite a 50–90% loss of cortical ChAT activity (Allen *et al.*, 1989). This observation supported neurotrophins as possible therapeutic agents for support of cholinergic innervation of the cortex and hippocampus.

Neurotrophins and the pathogenesis of AD

A number of studies link cholinergic changes in AD to changes in the neurotrophins (Schindowski *et al.*, 2008). Although the deficit of BDNF or NGF is not the initial trigger for the AD disease process, it may lead to increased neuronal loss adding to the symptoms (Figures 13.2 and 13.3). BDNF is important for the survival of dopaminergic and motor neurons. Both NGF and BDNF are required for cholinergic neuron survival, whereas BDNF plays a critical role in the survival and functioning of serotonergic, hippocampal and cortical neurons (Allen *et al.*, 2013). Multiple studies link reduction in BDNF levels to a number of neurodegenerative conditions including AD (Hock *et al.*, 2000; Holsinger *et al.*, 2000; Peng *et al.*, 2005).

NGF deficiency in the brain induces apoptosis, death and dysfunction of neurons, and accelerates Aβ deposits and Aβ-induced toxicity (Colafrancesco and Villoslada, 2011; Counts and Mufson, 2005). In rodents and primates, NGF deficiency leads to shrunken cholinergic neurons and memory deficits. NGF's physiological role in synaptic plasticity, neocortical and hippocampal learning and memory processes has also been reported (Bonhoeffer, 1996; Conner *et al.*, 2009). Degeneration of NGF-dependent basal forebrain cholinergic neurons (BFCNs) characterizes AD pathology. However, no reduction in NGF in the cortex or hippocampus of the AD patients could be seen by ELISA (Allen *et al.*, 1991). On the contrary, there was an increase in NGF (Crutcher *et al.*, 1993), which indicates that the retrograde transport of NGF could be compromised in AD (Allen *et al.*, 2013). Later, proNGF but not NGF, was shown to be the dominant form in the human cortex (Fahnestock *et al.*, 2001). The level of proNGF is doubled in the frontal and occipital cortex and in hippocampus in late-stage AD and an increase of 40–50% has been reported in cases with mild cognitive impairment (Peng *et al.*, 2004). In addition, in human AD brains, altered balance in the expression of protease complexes (e.g., plasminogen/plasmin, tissue plasminogen activator, Neuroserpin, matrix metalloproteinase-9) resulting in a deficit of proNGF/NGF was observed (Cuello and Bruno, 2007). A transgenic mouse model (expressing an antibody to NGF) of cholinergic loss showed a reduction in ChBF neurons and exhibited memory deficits (Capsoni *et al.*, 2000). The antibody showed increased binding affinity to the mature NGF than to the proNGF resulting in an increase in the proNGF/NGF ratio, which may have contributed to the presentation of an AD-like pathology. In a transgenic AD mouse model expressing APP and PSEN1, a mutant form of NGF (P61S/R100E) delivered intranasally blunted the progress of neurodegeneration and behavioral deficits (Capsoni *et al.*, 2012) suggesting a link between NGF and AD.

In aged animals, an increase in APP processing occurs due to the higher expression of p75NTR as compared to TrkA (Costantini *et al.*, 2005b). Aβ pathway directly induces TrkA phosphorylation and indirectly promotes NGF secretion in cultured neurons (Bulbarelli *et al.*, 2009). There is also a potential link between p75NTR expression and changes occurring in AD (Costantini *et al.*, 2005a; Costantini *et al.*, 2005b), possibly due to direct binding between p75NTR and Aβ 1-42 peptide (Coulson *et al.*, 2009). p75NTR mediates APP promoter activity, leading to an increase of secreted APP (Ge and Lahiri, 2002). Cultured hippocampal neurons, following exposure to NGF for 48 h, undergo apoptosis upon NGF withdrawal and show increased APP processing and both intra- and extra-cellular Aβ accumulation (Matrone *et al.*, 2008b). These findings suggest that an alteration in TrkA function resulting from aberrant p75 processing with a concomitant increase in Aβ production, may be due to NGF signaling impairment that can lead to neuronal death (Figure 13.2).

Transgenic mice expressing recombinant antibodies to NGF show loss of basal forebrain cholinergic neurons, Tau hyperphosphorylation, accumulation of Aβ plaques, synaptic plasticity deficits, and an imbalance between proNGF and NGF signaling, hallmarks of a neurodegenerative phenotype observed in human sporadic forms of AD (Cattaneo and Calissano, 2012). Significantly reduced levels of NGF are also seen in the basal forebrain of older animals and AD patients that display cell shrinkage, reduction of nerve fiber density and down-regulation of transmitter-associated enzymes with a decrease of cholinergic transmission (Svendsen *et al.*, 1991; Venero *et al.*, 1994). Since NGF mRNA level is unchanged, the impairment of NGF transport along with the degeneration of cholinergic neurons may represent one of the earliest events during AD progression (Mufson *et al.*, 1995).

NGF regulates Tau synthesis either directly by up-regulation of gene transcription or indirectly by its ubiquitination and proteosomal degradation (Babu *et al.*, 2005; Sadot *et al.*, 1996). Tau phosphorylation is also regulated by NGF deprivation *in vivo* (Capsoni *et al.*, 2002). Tau controls the microtubule-dependent axonal transport in a concentration-dependent manner (Dixit *et al.*, 2008) thus affecting intracellular trafficking of neurotrophins (Mandelkow *et al.*, 2003). In aged rats, retrograde microtubule-dependent NGF signaling is reportedly diminished as compared to young rats (Niewiadomska *et al.*, 2011). These findings suggest that the failure of Tau-mediated axonal transport could curtail trophic support in aged or AD brains (Niewiadomska *et al.*, 2011). Retrograde labeling of basal forebrain has shown an altered distribution of phospho-Tau, GSK3-β, and TrkA protein in aged cholinergic neurons (Niewiadomska *et al.*, 2011). An Aβ-mediated down-regulation of BDNF retrograde trafficking in the neurons of a transgenic mouse (Tg2576) model of AD has been reported, whereas no obvious change in Tau phosphorylation was detected (Poon *et al.*, 2011). Moreover, the loss of the N-terminal 25 amino acids of Tau has been shown in the cellular and animal models of AD induced by disrupted NGF signaling (Corsetti *et al.*, 2008). In NGF-deprived hippocampal primary neurons, an early transient and site-specific GSK-3β-mediated Tau hyper-phosphorylation of two AD-related epitopes (Ser262 and Thr231) has been correlated with an activation of the endogenous amyloidogenic pathway (Matrone *et al.*, 2008a). A relationship between AD pathology and both the TrkA and p75NTR signaling is supported by a study in which a transgenic mouse (TgMNAC13) that expressed an anti-TrkA antibody displayed cholinergic function deficit and amyloid- but not tau-associated pathology (Capsoni *et al.*, 2010).

A potential link between BDNF and AD pathophysiology presented by increased production of Aβ peptide and Tau hyperphosphorylation has been suggested (Arancibia *et al.*, 2008; Carlino *et al.*, 2013). BDNF regulates the processing of APP by stimulating the non-amyloidogenic processing pathway (Rohe *et al.*, 2009). This may produce beneficial effects such as reduction in the production of Aβ peptides and the release of the secreted form of APP that has neurotrophic and neuroprotective effects (Nishitomi *et al.*, 2006; Thornton *et al.*, 2006). In primary neuronal cultures, pretreatment with BDNF protects against the cytotoxic effects of Aβ (Arancibia *et al.*, 2008). The administration of BDNF is also able to restore changes in neurons pretreated with Aβ (Aliaga *et al.*, 2010). Sublethal doses of Aβ down-regulate BDNF expression in cultured cortical neurons and impair BDNF intra-cellular trafficking (Poon *et al.*, 2011). A report showing the increased production of BDNF by astrocytes rescuing neuritic degeneration in differentiated human neuroblastoma cells suggests a compensatory event to maintain adequate neurotrophic support to neurons in the initial stages of AD pathology (Kimura *et al.*, 2014). Up-regulation of *BDNF* mRNA expression and protein synthesis is also found in microglia and astrocytes in the vicinities of amyloid plaques in a transgenic mouse model of AD (Burbach *et al.*, 2004). BDNF and phospho-CREB immunoreactivity is decreased in correlation to age-dependent decline of neurogenesis (Hattiangady *et al.*, 2005). Patients with AD show a reduced BDNF concentration in the CSF (cerebrospinal fluid) as compared to healthy older subjects without cognitive impairment (Laske *et al.*, 2007; Li *et al.*, 2009; Zhang *et al.*, 2008) and higher BDNF serum levels are associated with a slower rate of cognitive decline in AD patients (Laske *et al.*, 2011). It has been shown that BDNF and exercise enhance DNA repair in cerebral cortical and hippocampal neurons by a mechanism involving CREB suggesting that impaired synaptic activity and associated reductions in BDNF signaling may compromise BER (base excision repair) resulting in the accumulation of oxidative DNA damage (Yang *et al.*, 2014b).

Neurotrophin-based therapy for AD

Based on preclinical and clinical studies, the roles of NGF and BDNF in cortical and hippocampal neuron plasticity and behavior open up new avenues in the AD therapeutic strategy (Allen *et al.*, 2013). The major objective in this direction has been to prevent Aβ production or clear Aβ from the brain. Preclinical studies on NGF protein administration as an AD therapeutic were later translated to a Phase I clinical trial begun in 2001 (Tuszynski *et al.*, 2005) and it was determined safe to proceed to Phase II trials. In this context, the safe, long-term and noninvasive delivery of recombinant human NGF to the brain to provide cholinergic trophic support is considered to be an encouraging therapeutic tool to treat cognitive impairment in AD (Cattaneo and Calissano, 2012).

In vivo studies also show promise towards the use of NGF as a therapeutic agent to treat AD (Williams *et al.*, 2006). In an AD mouse model, intranasal administration of NGF and an oral administration of two AchE inhibitors, ganstigmine and donepezil have been shown to improve memory and cognition (De Rosa *et al.*, 2005; Origlia *et al.*, 2006). Small molecule p75NTR ligands that can block Aβ-induced activation of signaling cascades associated with AD pathology by targeting calpain/Cdk5, GSK3β, JNK signaling, Tau phosphorylation and blunt Aβ-dependent AKT and CREB inactivation would have the ability to reverse synaptic impairment and inhibit Aβ-induced neuronal loss (Longo *et al.*, 2007; Yang *et al.*, 2008). Recently, a number of small molecule NGF agonists with high affinity to TrkA were characterized and were found to be active in NGF-dependent assays (Scarpi *et al.*, 2012) indicating that they possess NGF-like potency.

In order to develop novel therapeutic agents for neurodegenerative diseases, investigations to delineate the underlying mechanisms of BDNF upregulation in neurons and to induce biosynthesis of both proBDNF and BDNF using natural compounds are ongoing (Numakawa, 2014). BDNF has the potential to facilitate recovery from neurodegenerative diseases (Lynch *et al.*, 2007; Zuccato and Cattaneo, 2009). A number of preclinical studies have suggested that BDNF may be potent against both age-related as well as AD-associated pathology. Delivery of BDNF into the supracallosal gyrus or the third ventricle showed substantial protection from Aβ when given concurrently (Arancibia *et al.*, 2008). *In vivo*, BDNF improved disease-related symptoms in rodents and primate models of advanced age and AD (Nagahara *et al.*, 2009a). In AD transgenic mice, BDNF introduced by gene delivery, although did not affect the clearance of Aβ plaques, reversed synapse loss and restored cell signaling, learning and memory (Nagahara *et al.*, 2009b). In aged Fischer rats, BDNF infusions significantly improved spatial learning and memory in water maze performance (Nagahara *et al.*, 2009b). In a primate model showing age-related neurodegeneration and cognitive decline with neuronal function deficits but no extensive cell death, injections of lentiviral vector expressing BDNF showed a significant improvement in performance. Elevated levels of BDNF were also found in the hippocampus, suggesting anterograde transport (Nagahara *et al.*, 2009b). In a rat model of aging, in which entorhinal-associated cognitive function declined with no neuron loss, BDNF administration prevented cell death and reduced memory deficits (Nagahara *et al.*, 2009b). These findings establish BDNF delivery as a means for treating AD.

However, due to a variety of signaling cascades that BDNF regulates, whether other non-targeted signaling pathways alter to a detriment due to excess BDNF in the brain remains a concern (Murray and Holmes, 2011). BDNF virally delivered by injection into subcortical white matter induced BDNF levels and reversed memory deficits and synaptic degeneration in a mouse model of AD (Iwasaki *et al.*, 2012).

A small molecule 7, 8-dihydroxyflavone, which was characterized as a TrkB agonist was shown to activate downstream pathways similar to TrkB-mediated signaling, and protect neurons from kainic acid-induced apoptosis (Jang *et al.*, 2010). Delivery of BDNF to the CNS has been successfully performed using stem cells. In an aged triple transgenic mouse model of AD expressing pathogenic forms of APP, PSEN, and Tau when treated with hippocampal neural stem cell transplantation, spatial learning and memory deficits were reversed, which involved BDNF-mediated enhancement of hippocampal synaptic density (Blurton-Jones *et al.*, 2009). Additionally, in a transgenic AD animal model, neural stem cell transplantation reportedly increased hippocampal BDNF levels resulting in higher synaptic density, and thus could restore hippocampus-dependent cognition (Blurton-Jones *et al.*, 2009).

Summary

Neurotrophins, such as BDNF and NGF, are important trophic factors that play critical roles in adulthood as well as during nervous system development. Subsets of neurons use either or both of these trophic factors for proper functioning. Studies show that a reduction in NGF or BDNF levels may be responsible for at least some of the symptoms of AD. Therefore, each of these factors has been used as a neuro-restorative therapy in animal models of AD or in clinical trials. Since the delivery of these proteins into the human brain has obvious difficulties, small molecules and mimetics, such as the p75NTR antagonists and Trk agonists, are being developed as cost-effective and accessible therapies. For NGF, BDNF and their mimetics, understanding their mechanism of action and the effective modes of delivery to the brain still remains a focus of many ongoing investigations. A critical approach during the AD disease process would involve intact neurotrophin receptors and their functional downstream pathways. In addition, timing of such therapy is another issue that needs consideration, because once a neuron dies, its replacement may not be feasible. Hence, for neurotrophin supplementation therapies to be effective, treatments need to begin before significant neuronal loss. Since AD is a very complex disease and its cause, as is known thus far, involves multiple pathways going awry at the same time or in succession, designing a neurotrophin-based therapy demands careful assessment. Nonetheless, the impact of neurotrophin-based therapies for AD is making great strides as continued insight keeps coming from ongoing studies.

Disclaimer

This document has been reviewed in accordance with United States Food and Drug Administration (FDA) policy and approved for publication. Approval does not signify that the contents necessarily reflect the position or opinions of the FDA, nor does mention of trade names or commercial products constitute endorsement or recommendation for use. The findings and conclusions in this report are those of the author and do not necessarily represent the views of the FDA.

References

Aliaga, E., M. Silhol, N. Bonneau, T. Maurice, S. Arancibia, and L. Tapia-Arancibia. 2010. Dual response of BDNF to sublethal concentrations of beta-amyloid peptides in cultured cortical neurons. Neurobiol Dis 37: 208–217.

Allen, S.J., D. Dawbarn, M.G. Spillantini, M. Goedert, G.K. Wilcock, T.H. Moss, and F.M. Semenenko. 1989. Distribution of beta-nerve growth factor receptors in the human basal forebrain. J Comp Neurol 289: 626–640.

Allen, S.J., S.H. MacGowan, J.J. Treanor, R. Feeney, G.K. Wilcock, and D. Dawbarn. 1991. Normal beta-NGF content in Alzheimer's disease cerebral cortex and hippocampus. Neurosci Lett 131: 135–139.

Allen, S.J., J.J. Watson, D.K. Shoemark, N.U. Barua, and N.K. Patel. 2013. GDNF, NGF and BDNF as therapeutic options for neurodegeneration. Pharmacol & Ther 138: 155–175.

Aloe, L., M.L. Rocco, P. Bianchi, and L. Manni. 2012. Nerve growth factor: from the early discoveries to the potential clinical use. J Transl Med 10: 239.

Arancibia, S., M. Silhol, F. Mouliere, J. Meffre, I. Hollinger, T. Maurice, and L. Tapia-Arancibia. 2008. Protective effect of BDNF against beta-amyloid induced neurotoxicity *in vitro* and *in vivo* in rats. Neurobiol Dis 31: 316–326.

Babu, J.R., T. Geetha, and M.W. Wooten. 2005. Sequestosome 1/p62 shuttles polyubiquitinated tau for proteasomal degradation. J Neurochem 94: 192–203.

Baj, G., D. Carlino, L. Gardossi, and E. Tongiorgi. 2013. Toward a unified biological hypothesis for the BDNF Val66Met-associated memory deficits in humans: a model of impaired dendritic mRNA trafficking. Front Neurosci 7: 188.

Bath, K.G., M.R. Akins, and F.S. Lee. 2012. BDNF control of adult SVZ neurogenesis. Dev Psychobiol 54: 578–589.

Baydyuk, M., T. Russell, G.Y. Liao, K. Zang, J.J. An, L.F. Reichardt, and B. Xu. 2011. TrkB receptor controls striatal formation by regulating the number of newborn striatal neurons. Proc Nat Acad of Sci U S A 108: 1669–1674.

Baydyuk, M., Y. Xie, L. Tessarollo, and B. Xu. 2013. Midbrain-derived neurotrophins support survival of immature striatal projection neurons. J Neurosci 33: 3363–3369.

Baydyuk, M., and B. Xu. 2014. BDNF signaling and survival of striatal neurons. Front Cell Neurosci 8: 254.

Blochl, A., and R. Blochl. 2007. A cell-biological model of p75NTR signaling. J Neurochemistry 102: 289–305.

Blurton-Jones, M., M. Kitazawa, H. Martinez-Coria, N.A. Castello, F.J. Muller, J.F. Loring, T.R. Yamasaki, W.W. Poon, K.N. Green, and F.M. LaFerla. 2009. Neural stem cells improve cognition via BDNF in a transgenic model of Alzheimer disease. Proc Nat Acad Sci U S A 106: 13594–13599.

Bonhoeffer, T. 1996. Neurotrophins and activity-dependent development of the neocortex. Curr Opinion Neurobiol 6: 119–126.

Borchelt, D.R., G. Thinakaran, C.B. Eckman, M.K. Lee, F. Davenport, T. Ratovitsky, C.M. Prada, G. Kim, S. Seekins, D. Yager, H.H. Slunt, R. Wang, M. Seeger, A.I. Levey, S.E. Gandy, N.G. Copeland, N.A. Jenkins, D.L. Price, S.G. Younkin, and S.S. Sisodia. 1996. Familial Alzheimer's disease-linked presenilin 1 variants elevate Abeta1-42/1-40 ratio *in vitro* and *in vivo*. Neuron 17: 1005–1013.

Braak, H., and E. Braak. 1996. Development of Alzheimer-related neurofibrillary changes in the neocortex inversely recapitulates cortical myelogenesis. Acta Neuropathol 92: 197–201.

Bramham, C.R., and E. Messaoudi. 2005. BDNF function in adult synaptic plasticity: the synaptic consolidation hypothesis. Progr Neurobiol 76: 99–125.

Bulbarelli, A., E. Lonati, E. Cazzaniga, F. Re, S. Sesana, D. Barisani, G. Sancini, T. Mutoh, and M. Masserini. 2009. TrkA pathway activation induced by amyloid-beta (Abeta). Mol Cell Neurosci 40: 365–373.

Burbach, G.J., R. Hellweg, C.A. Haas, D. Del Turco, D. Deicke, D. Abramowski, M. Jucker, M. Staufenbiel, and T. Deller. 2004. Induction of brain-derived neurotrophic factor in plaque-associated glial cells of aged APP23 transgenic mice. J Neurosci 24: 2421–2430.

Capsoni, S., G. Ugolini, A. Comparini, F. Ruberti, N. Berardi, and A. Cattaneo. 2000. Alzheimer-like neurodegeneration in aged antinerve growth factor transgenic mice. Proc Nat Acad Sci U S A 97: 6826–6831.

Capsoni, S., S. Giannotta, and A. Cattaneo. 2002. Beta-amyloid plaques in a model for sporadic Alzheimer's disease based on transgenic anti-nerve growth factor antibodies. Mol Cell Neurosci 21: 15–28.

Capsoni, S., C. Tiveron, G. Amato, D. Vignone, and A. Cattaneo. 2010. Peripheral neutralization of nerve growth factor induces immunosympathectomy and central neurodegeneration in transgenic mice. J Alz Dis 20: 527–546.

Capsoni, S., N.M. Carucci, and A. Cattaneo. 2012. Pathogen free conditions slow the onset of neurodegeneration in a mouse model of nerve growth factor deprivation. J Alz Dis 31: 1–6.

Carlino, D., M. De Vanna, and E. Tongiorgi. 2013. Is altered BDNF biosynthesis a general feature in patients with cognitive dysfunctions? Neuroscientist 19: 345–353.

Cattaneo, A., and P. Calissano. 2012. Nerve growth factor and Alzheimer's disease: new facts for an old hypothesis. Mol Neurobiol 46: 588–604.

Chao, M.V., and B.L. Hempstead. 1995. p75 and Trk: a two-receptor system. Trends Neurosci 18: 321–326.

Chao, M.V., and F.S. Lee. 2004. Neurotrophin survival signaling mechanisms. J Alz Dis : JAD 6: S7–11.

Chao, M.V., R. Rajagopal, and F.S. Lee. 2006. Neurotrophin signalling in health and disease. Clin Sci (Lond) 110: 167–173.

Chen, K.S., M.C. Nishimura, M.P. Armanini, C. Crowley, S.D. Spencer, and H.S. Phillips. 1997. Disruption of a single allele of the nerve growth factor gene results in atrophy of basal forebrain cholinergic neurons and memory deficits. J Neurosci 17: 7288–7296.

Colafrancesco, V., and P. Villoslada. 2011. Targeting NGF pathway for developing neuroprotective therapies for multiple sclerosis and other neurological diseases. Arch Ital Biol 149: 183–192.

Conner, J.M., J.C. Lauterborn, Q. Yan, C.M. Gall, and S. Varon. 1997. Distribution of brain-derived neurotrophic factor (BDNF) protein and mRNA in the normal adult rat CNS: evidence for anterograde axonal transport. J Neurosci 17: 2295–2313.

Conner, J.M., K.M. Franks, A.K. Titterness, K. Russell, D.A. Merrill, B.R. Christie, T.J. Sejnowski, and M.H. Tuszynski. 2009. NGF is essential for hippocampal plasticity and learning. J Neurosci 29: 10883–10889.

Corder, E.H., A.M. Saunders, W.J. Strittmatter, D.E. Schmechel, P.C. Gaskell, G.W. Small, A.D. Roses, J.L. Haines, and M.A. Pericak-Vance. 1993. Gene dose of apolipoprotein E type 4 allele and the risk of Alzheimer's disease in late onset families. Science 261: 921–923.

Corsetti, V., G. Amadoro, A. Gentile, S. Capsoni, M.T. Ciotti, M.T. Cencioni, A. Atlante, N. Canu, T.T. Rohn, A. Cattaneo, and P. Calissano. 2008. Identification of a caspase-derived N-terminal tau fragment in cellular and animal Alzheimer's disease models. Mol Cell Neurosci 38: 381–392.

Costantini, C., F. Rossi, E. Formaggio, R. Bernardoni, D. Cecconi, and V. Della-Bianca. 2005a. Characterization of the signaling pathway downstream p75 neurotrophin receptor involved in beta-amyloid peptide-dependent cell death. J Mol Neurosci 25: 141–156.

Costantini, C., R. Weindruch, G. Della Valle, and L. Puglielli. 2005b. A TrkA-to-p75NTR molecular switch activates amyloid beta-peptide generation during aging. Biochem J 391: 59–67.

Coulson, E.J., L.M. May, A.M. Sykes, and A.S. Hamlin. 2009. The role of the p75 neurotrophin receptor in cholinergic dysfunction in Alzheimer's disease. Neuroscientist 15: 317–323.

Counts, S.E., and E.J. Mufson. 2005. The role of nerve growth factor receptors in cholinergic basal forebrain degeneration in prodromal Alzheimer disease. J Neuropathol Exp Neurol 64: 263–272.

Cowan, W.M. 2001. Viktor Hamburger and Rita Levi-Montalcini: the path to the discovery of nerve growth factor. Annl Rev Neurosci 24: 551–600.

Crowley, C., S.D. Spencer, M.C. Nishimura, K.S. Chen, S. Pitts-Meek, M.P. Armanini, L.H. Ling, S.B. McMahon, D.L. Shelton, A.D. Levinson, and H.S. Phillips. 1994. Mice lacking nerve growth factor display perinatal loss of sensory and sympathetic neurons yet develop basal forebrain cholinergic neurons. Cell 76: 1001–1011.

Crutcher, K.A., S.A. Scott, S. Liang, W.V. Everson, and J. Weingartner. 1993. Detection of NGF-like activity in human brain tissue: increased levels in Alzheimer's disease. J Neurosci 13: 2540–2550.

Cuello, A.C., and M.A. Bruno. 2007. The failure in NGF maturation and its increased degradation as the probable cause for the vulnerability of cholinergic neurons in Alzheimer's disease. Neurochem Res 32: 1041–1045.

De Rosa, R., A.A. Garcia, C. Braschi, S. Capsoni, L. Maffei, N. Berardi, and A. Cattaneo. 2005. Intranasal administration of nerve growth factor (NGF) rescues recognition memory deficits in AD11 anti-NGF transgenic mice. Proc Nat Acad Sci U S A 102: 3811–3816.

Defelice, F.G., and S.T. Ferreira. 2002. Physiopathological modulators of amyloid aggregation and novel pharmacological approaches in Alzheimer's disease. An Acad Bras Cienc 74: 265–284.

Deinhardt, K., and M.V. Chao. 2014. Shaping neurons: Long and short range effects of mature and proBDNF signalling upon neuronal structure. Neuropharmacology 76 Pt C: 603–609.

Diarra, A., T. Geetha, P. Potter, and J.R. Babu. 2009. Signaling of the neurotrophin receptor p75 in relation to Alzheimer's disease. Biochem Biophys Res Commun 390: 352–356.

Dixit, R., J.L. Ross, Y.E. Goldman, and E.L. Holzbaur. 2008. Differential regulation of dynein and kinesin motor proteins by tau. Science 319: 1086–1089.

Esposito, D., P. Patel, R.M. Stephens, P. Perez, M.V. Chao, D.R. Kaplan, and B.L. Hempstead. 2001. The cytoplasmic and transmembrane domains of the p75 and Trk A receptors regulate high affinity binding to nerve growth factor. J Biol Chem 276: 32687–32695.

Fagan, A.M., H. Zhang, S. Landis, R.J. Smeyne, I. Silos-Santiago, and M. Barbacid. 1996. TrkA, but not TrkC, receptors are essential for survival of sympathetic neurons *in vivo*. J Neurosci 16: 6208–6218.

Fahnestock, M., B. Michalski, B. Xu, and M.D. Coughlin. 2001. The precursor pro-nerve growth factor is the predominant form of nerve growth factor in brain and is increased in Alzheimer's disease. Mol Cell Neurosci 18: 210–220.

Fortress, A.M., M. Buhusi, K.L. Helke, and A.C. Granholm. 2011. Cholinergic Degeneration and Alterations in the TrkA and p75NTR Balance as a Result of Pro-NGF Injection into Aged Rats. J Aging Res 2011: 460543.

Friedman, W.J. 2010. Proneurotrophins, seizures, and neuronal apoptosis. Neuroscientist 16: 244–252.

Ge, Y.W., and D.K. Lahiri. 2002. Regulation of promoter activity of the APP gene by cytokines and growth factors: implications in Alzheimer's disease. An New York Acad Sci 973: 463–467.

Glerup, S., A. Nykjaer, and C.B. Vaegter. 2014. Sortilins in neurotrophic factor signaling. Handb Exp Pharmacol 220: 165–189.

Goate, A., and J. Hardy. 2012. Twenty years of Alzheimer's disease-causing mutations. J Neurochem 120 Suppl 1: 3–8.

Gong, B., O.V. Vitolo, F. Trinchese, S. Liu, M. Shelanski, and O. Arancio. 2004. Persistent improvement in synaptic and cognitive functions in an Alzheimer mouse model after rolipram treatment. J Clin Invest 114: 1624–1634.

Gong, C.X., T. Lidsky, J. Wegiel, L. Zuck, I. Grundke-Iqbal, and K. Iqbal. 2000. Phosphorylation of microtubule-associated protein tau is regulated by protein phosphatase 2A in mammalian brain. Implications for neurofibrillary degeneration in Alzheimer's disease. J Biol Chem 275: 5535–5544.

Gorski, J.A., S.A. Balogh, J.M. Wehner, and K.R. Jones. 2003. Learning deficits in forebrain-restricted brain-derived neurotrophic factor mutant mice. Neuroscience 121: 341–354.

Greenberg, M.E., B. Xu, B. Lu, and B.L. Hempstead. 2009. New insights in the biology of BDNF synthesis and release: implications in CNS function. J Neurosci 29: 12764–12767.

Gruss, H.J., and S.K. Dower. 1995. The TNF ligand superfamily and its relevance for human diseases. Cytokines Mol Ther 1: 75–105.

Hall, A.M., R.Y. Moore, O.L. Lopez, L. Kuller, and J.T. Becker. 2008. Basal forebrain atrophy is a presymptomatic marker for Alzheimer's disease. Alz Dementia 271–279.

Hardy, J., and D. Allsop. 1991. Amyloid deposition as the central event in the aetiology of Alzheimer's disease. Trends Pharmacol Sci 12: 383–388.

Harte-Hargrove, L.C., N.J. Maclusky, and H.E. Scharfman. 2013. Brain-derived neurotrophic factor-estrogen interactions in the hippocampal mossy fiber pathway: implications for normal brain function and disease. Neuroscience 239: 46–66.

Hashimoto, Y., Y. Kaneko, E. Tsukamoto, H. Frankowski, K. Kouyama, Y. Kita, T. Niikura, S. Aiso, D.E. Bredesen, M. Matsuoka, and I. Nishimoto. 2004. Molecular characterization of neurohybrid cell death induced by Alzheimer's amyloid-beta peptides via p75NTR/PLAIDD. J Neurochem 90: 549–558.

Hattiangady, B., M.S. Rao, G.A. Shetty, and A.K. Shetty. 2005. Brain-derived neurotrophic factor, phosphorylated cyclic AMP response element binding protein and neuropeptide Y decline as early as middle age in the dentate gyrus and CA1 and CA3 subfields of the hippocampus. Exp Neurol 195: 353–371.

Hefti, F., M.P. Armanini, K.D. Beck, I.W. Caras, K.S. Chen, P.J. Godowski, L.J. Goodman, R.G. Hammonds, M.R. Mark, P. Moran, M.C. Nishimura, H.S. Phillips, A. Shih, J. Valverde, and J.W. Winslow. 1996.

Development of neurotrophic factor therapy for Alzheimer's disease. Ciba Found Symp 196: 54–63; discussion 63–59.

Hempstead, B.L. 2006. Dissecting the diverse actions of pro- and mature neurotrophins. Curr Alzh Res 3: 19–24.

Hock, C., K. Heese, F. Muller-Spahn, P. Huber, W. Riesen, R.M. Nitsch, and U. Otten. 2000. Increased cerebrospinal fluid levels of neurotrophin 3 (NT-3) in elderly patients with major depression. Mol Psychiatry 5: 510–513.

Holsinger, R.M., J. Schnarr, P. Henry, V.T. Castelo, and M. Fahnestock. 2000. Quantitation of BDNF mRNA in human parietal cortex by competitive reverse transcription-polymerase chain reaction: decreased levels in Alzheimer's disease. Mol Brain Res 76: 347–354.

Howlett, D.R. 2011. APP transgenic mice and their application to drug discovery. Histol Histopathol 26: 1611–1632.

Hsiao, G.H., and H.C. Chiu. 1996. Livedoid vasculitis. Response to low-dose danazol. Archives of Dermatol 132: 749–751.

Huang, E.J., and L.F. Reichardt. 2001. Neurotrophins: roles in neuronal development and function. Annual Rev Neurosci 24: 677–736.

Iwasaki, Y., T. Negishi, M. Inoue, T. Tashiro, T. Tabira, and N. Kimura. 2012. Sendai virus vector-mediated brain-derived neurotrophic factor expression ameliorates memory deficits and synaptic degeneration in a transgenic mouse model of Alzheimer's disease. J Neurosci Res 90: 981–989.

Jang, S.W., X. Liu, M. Yepes, K.R. Shepherd, G.W. Miller, Y. Liu, W.D. Wilson, G. Xiao, B. Blanchi, Y.E. Sun, and K. Ye. 2010. A selective TrkB agonist with potent neurotrophic activities by 7,8-dihydroxyflavone. Proc Nat Aca Sci U S A 107: 2687–2692.

Jeanneteau, F., K. Deinhardt, G. Miyoshi, A.M. Bennett, and M.V. Chao. 2010. The MAP kinase phosphatase MKP-1 regulates BDNF-induced axon branching. Nat Neurosci 13: 1373–1379.

Kaplan, D.R., and F.D. Miller. 2000. Neurotrophin signal transduction in the nervous system. Curr Opinion Neurobiol 10: 381–391.

Kenchappa, R.S., C. Tep, Z. Korade, S. Urra, F.C. Bronfman, S.O. Yoon, and B.D. Carter. 2010. p75 neurotrophin receptor-mediated apoptosis in sympathetic neurons involves a biphasic activation of JNK and up-regulation of tumor necrosis factor-alpha-converting enzyme/ADAM17. J Biol Chem 285: 20358–20368.

Kimura, N., S. Okabayashi, and F. Ono. 2014. Dynein dysfunction disrupts beta-amyloid clearance in astrocytes through endocytic disturbances. Neuroreport 25: 514–520.

Laske, C., E. Stransky, T. Leyhe, G.W. Eschweiler, W. Maetzler, A. Wittorf, S. Soekadar, E. Richartz, N. Koehler, M. Bartels, G. Buchkremer, and K. Schott. 2007. BDNF serum and CSF concentrations in Alzheimer's disease, normal pressure hydrocephalus and healthy controls. J Psychiatric Res 41: 387–394.

Laske, C., K. Stellos, N. Hoffmann, E. Stransky, G. Straten, G.W. Eschweiler, and T. Leyhe. 2011. Higher BDNF serum levels predict slower cognitive decline in Alzheimer's disease patients. Int J Neuropsychopharmacol 14: 399–404.

Lee, B.F., C.K. Liu, C.T. Tai, N.T. Chiu, G.C. Liu, H.S. Yu, and M.C. Pai. 2001a. Alzheimer's disease: scintigraphic appearance of Tc-99m HMPAO brain spect. The Kaohsiung J Med Sci 17: 394–400.

Lee, M.H., N.D. Amin, A. Venkatesan, T. Wang, R. Tyagi, H.C. Pant, and A. Nath. 2013. Impaired neurogenesis and neurite outgrowth in an HIV-gp120 transgenic model is reversed by exercise via BDNF production and Cdk5 regulation. J Neurovirol 19: 418–431.

Lee, R., P. Kermani, K.K. Teng, and B.L. Hempstead. 2001b. Regulation of cell survival by secreted proneurotrophins. Science 294: 1945–1948.

Li, G., E.R. Peskind, S.P. Millard, P. Chi, I. Sokal, C.E. Yu, L.M. Bekris, M.A. Raskind, D.R. Galasko, and T.J. Montine. 2009. Cerebrospinal fluid concentration of brain-derived neurotrophic factor and cognitive function in non-demented subjects. PloS One 4: e5424.

Liu, M.Y., S. Wang, W.F. Yao, Z.J. Zhang, X. Zhong, L. Sha, M. He, Z.H. Zheng, and M.J. Wei. 2014. Memantine improves spatial learning and memory impairments by regulating NGF signaling in APP/ PS1 transgenic mice. Neuroscience 273: 141–151.

Longo, F.M., T. Yang, J.K. Knowles, Y. Xie, L.A. Moore, and S.M. Massa. 2007. Small molecule neurotrophin receptor ligands: novel strategies for targeting Alzheimer's disease mechanisms. Curr Alz Res 4: 503–506.

Lynch, G., E.A. Kramar, C.S. Rex, Y. Jia, D. Chappas, C.M. Gall, and D.A. Simmons. 2007. Brain-derived neurotrophic factor restores synaptic plasticity in a knock-in mouse model of Huntington's disease. J Neurosci 27: 4424–4434.

Ma, B., J.N. Savas, M.V. Chao, and N. Tanese. 2012. Quantitative analysis of BDNF/TrkB protein and mRNA in cortical and striatal neurons using alpha-tubulin as a normalization factor. Cytometry. Part A 81: 704–717.

Mandelkow, E.M., K. Stamer, R. Vogel, E. Thies, and E. Mandelkow. 2003. Clogging of axons by tau, inhibition of axonal traffic and starvation of synapses. Neurobiol Aging 24: 1079–1085.

Matrone, C., M.T. Ciotti, D. Mercanti, R. Marolda, and P. Calissano. 2008a. NGF and BDNF signaling control amyloidogenic route and Abeta production in hippocampal neurons. P Nat Acad Sci U S A 105: 13139–13144.

Matrone, C., A. Di Luzio, G. Meli, S. D'Aguanno, C. Severini, M.T. Ciotti, A. Cattaneo, and P. Calissano. 2008b. Activation of the amyloidogenic route by NGF deprivation induces apoptotic death in PC12 cells. J Alz Dis 13: 81–96.

Mendell, L.M. 1999. Neurotrophin action on sensory neurons in adults: an extension of the neurotrophic hypothesis. Pain Suppl 6: S127–132.

Minichiello, L. 2009. TrkB signalling pathways in LTP and learning. Nat Rev. Neurosci 10: 850–860.

Mufson, E.J., J.M. Conner, and J.H. Kordower. 1995. Nerve growth factor in Alzheimer's disease: defective retrograde transport to nucleus basalis. Neuroreport 6: 1063–1066.

Mufson, E.J., S.Y. Ma, E.J. Cochran, D.A. Bennett, L.A. Beckett, S. Jaffar, H.U. Saragovi, and J.H. Kordower. 2000. Loss of nucleus basalis neurons containing trkA immunoreactivity in individuals with mild cognitive impairment and early Alzheimer's disease. J Comp Neurol 427: 19–30.

Murray, P.S., and P.V. Holmes. 2011. An overview of brain-derived neurotrophic factor and implications for excitotoxic vulnerability in the hippocampus. Int J Peptides 2011: 654085.

Nagahara, A.H., T. Bernot, R. Moseanko, L. Brignolo, A. Blesch, J.M. Conner, A. Ramirez, M. Gasmi, and M.H. Tuszynski. 2009a. Long-term reversal of cholinergic neuronal decline in aged non-human primates by lentiviral NGF gene delivery. Exp Neurol 215: 153–159.

Nagahara, A.H., D.A. Merrill, G. Coppola, S. Tsukada, B.E. Schroeder, G.M. Shaked, L. Wang, A. Blesch, A. Kim, J.M. Conner, E. Rockenstein, M.V. Chao, E.H. Koo, D. Geschwind, E. Masliah, A.A. Chiba, and M.H. Tuszynski. 2009b. Neuroprotective effects of brain-derived neurotrophic factor in rodent and primate models of Alzheimer's disease. Nat Med 15: 331–337.

Niewiadomska, G., A. Mietelska-Porowska, and M. Mazurkiewicz. 2011. The cholinergic system, nerve growth factor and the cytoskeleton. Behav Brain Res 221: 515–526.

Nishitomi, K., G. Sakaguchi, Y. Horikoshi, A.J. Gray, M. Maeda, C. Hirata-Fukae, A.G. Becker, M. Hosono, I. Sakaguchi, S.S. Minami, Y. Nakajima, H.F. Li, C. Takeyama, T. Kihara, A. Ota, P.C. Wong, P.S. Aisen, A. Kato, N. Kinoshita, and Y. Matsuoka. 2006. BACE1 inhibition reduces endogenous Abeta and alters APP processing in wild-type mice. J of Neurochem 99: 1555–1563.

Numakawa, T. 2014. Possible protective action of neurotrophic factors and natural compounds against common neurodegenerative diseases. Neural Regen Res 9: 1506–1508.

Orefice, L.L., E.G. Waterhouse, J.G. Partridge, R.R. Lalchandani, S. Vicini, and B. Xu. 2013. Distinct roles for somatically and dendritically synthesized brain-derived neurotrophic factor in morphogenesis of dendritic spines. J Neurosci 33: 11618–11632.

Origlia, N., S. Capsoni, L. Domenici, and A. Cattaneo. 2006. Time window in cholinomimetic ability to rescue long-term potentiation in neurodegenerating anti-nerve growth factor mice. J Alz Dis : JAD 9: 59–68.

Peng, S., J. Wuu, E.J. Mufson, and M. Fahnestock. 2004. Increased proNGF levels in subjects with mild cognitive impairment and mild Alzheimer disease. J Neuropathol Exp Neurol 63: 641–649.

Peng, S., J. Wuu, E.J. Mufson, and M. Fahnestock. 2005. Precursor form of brain-derived neurotrophic factor and mature brain-derived neurotrophic factor are decreased in the pre-clinical stages of Alzheimer's disease. J Neurochem 93: 1412–1421.

Perry, G., A. Nunomura, A.D. Cash, M.A. Taddeo, K. Hirai, G. Aliev, J. Avila, T. Wataya, S. Shimohama, C.S. Atwood, and M.A. Smith. 2002. Reactive oxygen: its sources and significance in Alzheimer disease. J Neural Trans. Suppl: 69–75.

Picklo, M.J., T.J. Montine, V. Amarnath, and M.D. Neely. 2002. Carbonyl toxicology and Alzheimer's disease. Toxicol Appl Pharmacol 184: 187–197.

Poon, W.W., M. Blurton-Jones, C.H. Tu, L.M. Feinberg, M.A. Chabrier, J.W. Harris, N.L. Jeon, and C.W. Cotman. 2011. beta-Amyloid impairs axonal BDNF retrograde trafficking. Neurobiol Aging 32: 821–833.

Reichardt, L.F. 2006. Neurotrophin-regulated signalling pathways. Philosophical trans of the Royal Soc London. Series B, Biol Sci 361: 1545–1564.

Rex, C.S., J.C. Lauterborn, C.Y. Lin, E.A. Kramar, G.A. Rogers, C.M. Gall, and G. Lynch. 2006. Restoration of long-term potentiation in middle-aged hippocampus after induction of brain-derived neurotrophic factor. J Neurophysiology 96: 677–685.

Rohe, M., M. Synowitz, R. Glass, S.M. Paul, A. Nykjaer, and T.E. Willnow. 2009. Brain-derived neurotrophic factor reduces amyloidogenic processing through control of SORLA gene expression. J Neurosci 29: 15472–15478.

Roses, A.D. 1995. Apolipoprotein E genotyping in the differential diagnosis, not prediction, of Alzheimer's disease. Ann Neurol 38: 6–14.

Roses, A.D., A.M. Saunders, E.H. Corder, M.A. Pericak-Vance, S.H. Han, G. Einstein, C. Hulette, D.E. Schmechel, M. Holsti, and D. Huang. 1995. Influence of the susceptibility genes apolipoprotein E-epsilon 4 and apolipoprotein E-epsilon 2 on the rate of disease expressivity of late-onset Alzheimer's disease. Arzneimittelforschung 45: 413–417.

Sadot, E., D. Gurwitz, J. Barg, L. Behar, I. Ginzburg, and A. Fisher. 1996. Activation of m1 muscarinic acetylcholine receptor regulates tau phosphorylation in transfected PC12 cells. J Neurochem 66: 877–880.

Scarpi, D., D. Cirelli, C. Matrone, G. Castronovo, P. Rosini, E.G. Occhiato, F. Romano, L. Bartali, A.M. Clemente, G. Bottegoni, A. Cavalli, G. De Chiara, P. Bonini, P. Calissano, A.T. Palamara, E. Garaci, M.G. Torcia, A. Guarna, and F. Cozzolino. 2012. Low molecular weight, non-peptidic agonists of TrkA receptor with NGF-mimetic activity. Cell Death Dis 3: e389.

Schindowski, K., K. Belarbi, and L. Buee. 2008. Neurotrophic factors in Alzheimer's disease: role of axonal transport. Genes Brain Behav 7 Suppl 1: 43–56.

Scott Bitner, R. 2012. Cyclic AMP response element-binding protein (CREB) phosphorylation: a mechanistic marker in the development of memory enhancing Alzheimer's disease therapeutics. Biochem Pharmacol 83: 705–714.

Selkoe, D.J. 1997. Alzheimer's disease: genotypes, phenotypes, and treatments. Science 275: 630–631.

Silhol, M., V. Bonnichon, F. Rage, and L. Tapia-Arancibia. 2005. Age-related changes in brain-derived neurotrophic factor and tyrosine kinase receptor isoforms in the hippocampus and hypothalamus in male rats. Neuroscience 132: 613–624.

Sims, N.R., D.M. Bowen, S.J. Allen, C.C. Smith, D. Neary, D.J. Thomas, and A.N. Davison. 1983. Presynaptic cholinergic dysfunction in patients with dementia. J Neurochem 40: 503–509.

Skaper, S.D. 2012. The neurotrophin family of neurotrophic factors: an overview. Methods Mol Biol 846: 1–12.

Smeyne, R.J., R. Klein, A. Schnapp, L.K. Long, S. Bryant, A. Lewin, S.A. Lira, and M. Barbacid. 1994. Severe sensory and sympathetic neuropathies in mice carrying a disrupted Trk/NGF receptor gene. Nature 368: 246–249.

Smith, M.A., and G. Perry. 1997. The pathogenesis of Alzheimer disease: an alternative to the amyloid hypothesis. J Neuropathol Exp Neurol 56: 217.

Smith, P.A. 2014. BDNF: No gain without pain? Neuroscience 283C: 107–123.

Sobue, G., M. Yamamoto, M. Doyu, M. Li, T. Yasuda, and T. Mitsuma. 1998. Expression of mRNAs for neurotrophins (NGF, BDNF, and NT-3) and their receptors (p75NGFR, trk, trkB, and trkC) in human peripheral neuropathies. Neurochem Res 23: 821–829.

Song, W., M. Volosin, A.B. Cragnolini, B.L. Hempstead, and W.J. Friedman. 2010. ProNGF induces PTEN via p75NTR to suppress Trk-mediated survival signaling in brain neurons. J Neurosci 30: 15608–15615.

Steiner, J.P., and A. Nath. 2014. Neurotrophin strategies for neuroprotection: are they sufficient? J Neuroimmune Pharmacol 9: 182–194.

Su, S.C., and L.H. Tsai. 2011. Cyclin-dependent kinases in brain development and disease. Annl Rev Cell Dev Biol 27: 465–491.

Suhl, K.H., J.B. Park, E.Y. Park, and S.K. Rhee. 2012. Effect of nerve growth factor and its transforming tyrosine kinase protein and low-affinity nerve growth factor receptors on apoptosis of notochordal cells. Int Orthop 36: 1747–1753.

Svendsen, C.N., J.D. Cooper, and M.V. Sofroniew. 1991. Trophic factor effects on septal cholinergic neurons. Ann New York Acad Sci 640: 91–94.

Tapia-Arancibia, L., E. Aliaga, M. Silhol, and S. Arancibia. 2008. New insights into brain BDNF function in normal aging and Alzheimer disease. Brain Res Rev 59: 201–220.

Teng, K.K., S. Felice, T. Kim, and B.L. Hempstead. 2010. Understanding proneurotrophin actions: Recent advances and challenges. Dev Neurobiol 70: 350–359.

Thornton, E., R. Vink, P.C. Blumbergs, and C. Van Den Heuvel. 2006. Soluble amyloid precursor protein alpha reduces neuronal injury and improves functional outcome following diffuse traumatic brain injury in rats. Brain Research 1094: 38–46.

Trojanowski, J.Q., M.L. Schmidt, R.W. Shin, G.T. Bramblett, D. Rao, and V.M. Lee. 1993. Altered tau and neurofilament proteins in neuro-degenerative diseases: diagnostic implications for Alzheimer's disease and Lewy body dementias. Brain Pathol 3: 45–54.

Tuszynski, M.H., L. Thal, M. Pay, D.P. Salmon, H.S. U, R. Bakay, P. Patel, A. Blesch, H.L. Vahlsing, G. Ho, G. Tong, S.G. Potkin, J. Fallon, L. Hansen, E.J. Mufson, J.H. Kordower, C. Gall, and J. Conner. 2005. A phase 1 clinical trial of nerve growth factor gene therapy for Alzheimer disease. Nat Med 11: 551–555.

Tyler, W.J., S.P. Perrett, and L.D. Pozzo-Miller. 2002. The role of neurotrophins in neurotransmitter release. Neuroscientist 8: 524–531.

Venero, J.L., K.D. Beck, and F. Hefti. 1994. Intrastriatal infusion of nerve growth factor after quinolinic acid prevents reduction of cellular expression of choline acetyltransferase messenger RNA and trkA messenger RNA, but not glutamate decarboxylase messenger RNA. Neuroscience 61: 257–268.

Walsh, D.M., and D.J. Selkoe. 2007. A beta oligomers—a decade of discovery. J Neurochem 101: 1172–1184.

Weiner, H.L., and D.J. Selkoe. 2002. Inflammation and therapeutic vaccination in CNS diseases. Nature 420: 879–884.

Williams, B., A.C. Granholm, and K. Sambamurti. 2007. Age-dependent loss of NGF signaling in the rat basal forebrain is due to disrupted MAPK activation. Neurosci Lett 413: 110–114.

Williams, B.J., M. Eriksdotter-Jonhagen, and A.C. Granholm. 2006. Nerve growth factor in treatment and pathogenesis of Alzheimer's disease. Prog Neurobiol 80: 114–128.

Wong, J., M. Higgins, G. Halliday, and B. Garner. 2012. Amyloid beta selectively modulates neuronal TrkB alternative transcript expression with implications for Alzheimer's disease. Neuroscience 210: 363–374.

Woo, N.H., H.K. Teng, C.J. Siao, C. Chiaruttini, P.T. Pang, T.A. Milner, B.L. Hempstead, and B. Lu. 2005. Activation of p75NTR by proBDNF facilitates hippocampal long-term depression. Nat Neurosci 8: 1069–1077.

Yan, Q., M.J. Radeke, C.R. Matheson, J. Talvenheimo, A.A. Welcher, and S.C. Feinstein. 1997. Immunocytochemical localization of TrkB in the central nervous system of the adult rat. J Comp Neurol 378: 135–157.

Yang, J., C.J. Siao, G. Nagappan, T. Marinic, D. Jing, K. McGrath, Z.Y. Chen, W. Mark, L. Tessarollo, F.S. Lee, B. Lu, and B.L. Hempstead. 2009. Neuronal release of proBDNF. Nat Neurosci 12: 113–115.

Yang, J., L.C. Harte-Hargrove, C.J. Siao, T. Marinic, R. Clarke, Q. Ma, D. Jing, J.J. Lafrancois, K.G. Bath, W. Mark, D. Ballon, F.S. Lee, H.E. Scharfman, and B.L. Hempstead. 2014a. proBDNF negatively regulates neuronal remodeling, synaptic transmission, and synaptic plasticity in hippocampus. Cell Rep 7: 796–806.

Yang, J.L., Y.T. Lin, P.C. Chuang, V.A. Bohr, and M.P. Mattson. 2014b. BDNF and exercise enhance neuronal DNA repair by stimulating CREB-mediated production of apurinic/apyrimidinic endonuclease 1. Neuromol Med 16: 161–174.

Yang, T., J.K. Knowles, Q. Lu, H. Zhang, O. Arancio, L.A. Moore, T. Chang, Q. Wang, K. Andreasson, J. Rajadas, G.G. Fuller, Y. Xie, S.M. Massa, and F.M. Longo. 2008. Small molecule, non-peptide p75 ligands inhibit Abeta-induced neurodegeneration and synaptic impairment. PloS One 3: e3604.

Yoshii, A., and M. Constantine-Paton. 2010. Postsynaptic BDNF-TrkB signaling in synapse maturation, plasticity, and disease. Dev Neurobiol 70: 304–322.

Yuen, E.C., C.L. Howe, Y. Li, D.M. Holtzman, and W.C. Mobley. 1996. Nerve growth factor and the neurotrophic factor hypothesis. Brain Dev 18: 362–368.

Zhang, J., I. Sokal, E.R. Peskind, J.F. Quinn, J. Jankovic, C. Kenney, K.A. Chung, S.P. Millard, J.G. Nutt, and T.J. Montine. 2008. CSF multianalyte profile distinguishes Alzheimer and Parkinson diseases. Am J Clin Pathol 129: 526–529.

Zuccato, C., and E. Cattaneo. 2009. Brain-derived neurotrophic factor in neurodegenerative diseases. Nat Rev Neurol 5: 311–322.

Zweifel, L.S., R. Kuruvilla, and D.D. Ginty. 2005. Functions and mechanisms of retrograde neurotrophin signalling. Nat Rev Neurosci 6: 615–625.

14

Neurotransmitters and Receptors

Vesna Jevtovic-Todorovic

Introduction

Over the past century the field of neuroscience has enjoyed unimaginable growth and numerous breakthrough discoveries. In particular, the domain of neurotransmitters and their receptor systems has been evolving with the speed of light not only in terms of detecting novel ones but also in terms of deciphering their biological functions with a keen eye on developing novel therapeutic targets. Hence, the true comprehensive review of presently available knowledge regarding all neurotransmitters and the role of their receptor systems during the central nervous system (CNS) development could not be accomplished in earnest with a chapter but rather would require a full-length book. Nevertheless, what could be accomplished is up-to-date overview of major neurotransmitter systems that are known to control and modulate important biological, physiological and psychological functions during brain development.

Ever since scientists have discovered that neuronal communication is not only electrical in nature (*via* gap junctions) but could be carried between neurons with endogenous chemicals, i.e., neurotransmitters, the quest for understanding the mechanisms as to how neurotransmitters get synthesized, delivered to the synapse, released, engaged with postsynaptic receptors and get up-taken has been relentless. Although the precise number of endogenous neurotransmitters is not known, it is believed that there are potentially more than one hundred. The first clue that neurons

Professor and Chair Department of Anesthesiology, University of Colorado School of Medicine, 12801 E 17th Place, Research 1, South L18-4100, Aurora, CO 80045.
E-mail: vesna.jevtovic-todorovic@ucdenver.edu

communicate by releasing neurotransmitters was presented in the early 1900's when Dr. Cajal reported the existence of synaptic cleft, a tiny gap between neurons, thus suggesting that a chemical messenger must be released to assure the communication between neurons. This notion was indeed confirmed in the early 1920's when Dr. Loewi discovered the presence of the first known neurotransmitter, acetylcholine (Saladin, 2009).

Although the criteria important to identifying an endogenous chemical as a neurotransmitter are evolving over time some initial considerations remain an important guide: (1) they have to be present or synthesized in the neuron; (2) activation of a neuron must promote the release of it and cause a response in its target; (3) when the neurotransmitter is applied exogenously on the target it must trigger the same response as the endogenously released one; and, (4) some type of uptake mechanism or metabolic processing has to occur to remove the neurotransmitter ones a desired effect was achieved. These well-established criteria are being expanded to include chemicals that have an influence on the release or uptake of the neurotransmitters or have an effect on changing the structure of the synapse. Interestingly, there are neurons that can release more than one neurotransmitter (especially neuropeptides) from their terminals (Breedlove *et al.*, 2013). The requirement that the neurotransmitter should be present or synthesized in the neuron is now being challenged to include chemicals that can carry messages between neurons *via* the influence on the postsynaptic membrane regardless of their site of origin.

Neurotransmitter Release and Recycling

Neurotransmitters are normally synthesized in the cytosol of axon terminals and are stored in the organelles called synaptic vesicles typically about 40–50 nm in diameter (Bonanomi *et al.*, 2006). Although any given neuron receives inputs from numerous others, each neuron typically synthesizes one kind of neurotransmitters. The process by which the neurotransmitters are released into the synaptic cleft of a 'chemical' synapse is called exocytosis. The neurotransmitter exerts its desired effect by binding to a specific receptor located on the plasma membrane of a post-synaptic cell (most commonly another neuron). The release of neurotransmitters *via* exocytosis is a complex process that is referred to as 'vesicle-targeting and fusion' and has some unique properties. Namely, it depends on the action potential that travels to the axon terminus and, once the synaptic vesicles fuse with the plasma membrane and the neurotransmitter is released into the synaptic cleft, they are quickly recycled (Schweizer and Ryan, 2006). The action potential serves an important purpose—it activates voltage-gated Ca^{2+} channels localized in the plasma membrane adjacent to the synaptic vesicles. Their activation leads to an influx of Ca^{2+} ions from the extracellular space into the cytosol and this sudden and localized rise in the cytosolic Ca^{2+} from its resting level of < 0.1 nM to 1–100 nM promotes the fusion of the synaptic vesicles to the plasma membrane and exocytosis of the neurotransmitter. The key to this process being conducted in a synchronized and 'on command' fashion is for extra Ca^{2+} ions to be quickly pumped out of the cytosol by Ca^{2+} ATPases so that the next signal can be processed in timely fashion. It is important to note that two pools of neurotransmitter-containing synaptic vesicles have been recognized (Schweizer and Ryan, 2006). One is "docked" at the plasma membrane and is considered to be immediately releasable pool and the other is a reserve pool located in the active zone serving as replenishment when needed. It is believed that only about 10% of the docked vesicles get released with a spike in the cytosolic Ca^{2+}.

To assure the availability of a neurotransmitter after a wave of exocytosis occurs, the process of rapid recycling of synaptic-vesicle membrane has to occur *via* the process called endocytosis. Once released into the synaptic cleft, the neurotransmitter undergoes rapid metabolism (e.g., acetylcholine, neuropeptides), uptake (most common for the majority of the neurotransmitters) or diffusion away from the synaptic cleft in order to prevent prolonged stimulation of the postsynaptic cell.

The most well-known metabolic process for a rapid removal of the neurotransmitter from the synaptic cleft once its action on the postsynaptic receptor has been achieved is the one described for acetylcholine and it involves its hydrolysis to acetate and choline by the enzyme acethylcholinesterase. The enzyme located in the synaptic cleft is anchored to the plasma membrane by a glycosylphosphatidylinositol anchor thus being in close proximity to the action site of the acetylcholine (Amenta and Tayebati, 2008) assure that acetylcholine gets replenished, choline is transported from the synaptic cleft back into the nerve terminal by a Na^+/choline symporter, so that it could be used for *de novo* synthesis.

Unlike acetylcholine, the majority of other neurotransmitters are recycled intact, *via* an active uptake mechanism, from the synaptic cleft into the axon terminals that released them. Transporters are unique for each neurotransmitter; GABA, norepinephrine, dopamine, and serotonin were the first to be cloned and studied. They are transmembrane proteins and Na^+/neurotransmitter symporters. Like with all Na^+ symporters, the movement of Na^+ into the cell down its electrochemical gradient provides the energy for uptake of the neurotransmitters (Kaila *et al.*, 2014). The fact that the transporters play such an important role in removing the neurotransmitter from the synaptic cleft has been used for the development of novel therapeutic strategies. In particular, antidepressant medications such as fluoxetine and imipramine are based on blockage of serotonin uptake (Abbas *et al.*, 2012) and desipramine on blockage of norepinephrine uptake (Brunello *et al.*, 2002).

Classification of Neurotransmitters

Neurotransmitters can be classified based on their basic chemical composition and based on their mode of activity. Five main categories based on the chemical composition of a neurotransmitter are: amino acids, monoamines, peptides, gasotransmitters and 'others'. The examples of the major neurotransmitters that are amino acids are: glutamate, N-methyl-D-aspartate (NMDA), D-serine, γ-aminobutyric acid (GABA) and glycine. The examples of the major neurotransmitters that are monoamines are dopamine, epinephrine (adrenaline), norepinephrine (noradrenaline), histamine and serotonin. Also trace amines such as phenethylamine, *N*-methylphenethylamine, tyramine, 3-iodothyronamine, octopamine, tryptamine, etc., should be considered in this category. The examples of the major neurotransmitters that are peptides are: somatostatin, substance P, cocaine and amphetamine regulated transcript, opioid peptides (e.g., endorphins and enkephalins). In addition to the major ones, there are over 50 neuroactive peptides that are often released in tandem with another transmitter (usually small-molecule ones) to help fine tune neuronal regulation. The examples of the major neurotransmitters that are gasotransmitters are: nitric oxide, carbon monoxide, hydrogen sulfide. The gaseous neurotransmitter are difficult to study since once synthesized they are immediately dispersed into the extracellular space and into the cell where they induce the cascade of events mediated *via* second messenger system. They are quickly broken down which is why their half-lives are measured in seconds. Although not belonging to any of these categories and thus considered as 'others' the important neurotransmitters that should be included are acetylcholine, adenosine, anandamide, nucleotides (e.g., ATP) and the corresponding nucleosides. It is noteworthy that single ions could be considered neurotransmitters as well. For example, synaptically released zinc is sometimes considered to be a transmitter (Marchetti, 2014).

Two of the most abundant neurotransmitters in the mammalian brain are glutamate and GABA, one considered to be the most prevalent excitatory neurotransmitter and the other most prevalent inhibitory neurotransmitter, respectively. This brings us to another classification based on functional properties of a given neurotransmitter and its ability to enhance or suppress neuronal electrical activity—the category of the excitatory and the inhibitory neurotransmitters. The excitatory neurotransmitters act *via* the excitatory (type I) synapses that are typically located on the neuronal processes (shafts and dendritic spines) (Matus and Walters, 1976). They promote neuronal excitation and the generation of an action potential. Once a neurotransmitter binds to an excitatory receptor located on the post-synaptic membrane, either Na^+ or Na^+/K^+ channel transforms into an open state allowing the influx of ions and membrane depolarization resulting in the generation of an action potential and activation of the neuron. The inhibitory neurotransmitters act *via* the inhibitory (type II) synapses and are typically located on the neuronal soma (Matus and Walters, 1976). The activation of an inhibitory receptor on the postsynaptic cell leads to opening of K^+ or Cl^- channels and membrane hyperpolarization which in turn inhibits generation of an action potential. Neuronal communication achieved *via* neurotransmitter release is very complex and involves the reception of thousands of excitatory and inhibitory signals delivered to each neuron. Thus, it is known that a neurotransmitter can induce an excitatory response in some postsynaptic neurons and an inhibitory in others. The fact that any given neuron can receive inputs from numerous other neurons (especially interneurons) *via* inhibitory and excitatory receptors brings up a phenomenon called 'signal computation' (Bhalla, 2003). This is the mechanism by which the neuron continuously 'averages' incoming signals to decide whether action potential should be generated. By passively moving along the plasma membrane depolarization and hyperpolarization are summed together usually at the axon hillock. If the axon hillock becomes depolarized to the so-called threshold potential, the action potential is generated; otherwise the cell remains inactive. So even though the membrane potential undergoes constant changes with each quantum of neurotransmitter release and subsequent activation of its receptor, the generation of the action potential is an all-or-none phenomenon, i.e., it occurs only when the threshold potential is reached.

Physiological Functions and Main Receptors Systems of the Major Neurotransmitters

The basic definition of a receptor is that it is a membrane protein that is embedded in the phospholipid bilayer of a postsynaptic membrane and is activated by a neurotransmitter. To begin to understand the role of post-synaptic receptors we can divide them into two main categories: ligand-gated ion channels (ionotropic receptors) and G protein—coupled receptors (metabotropic receptors). Either type of receptors can be excitatory or inhibitory, but the main distinction is based on the speed of their response. Although the majority of receptors in the CNS require rapid response (in seconds and minutes), there are many roles that slow synapses and receptors coupled to G-protein play in modulating neuronal activity. Once the neurotransmitter binds to this type of synapse on a post-synaptic membrane, G protein is activated and in many cases directly binds to a separate ion-channel protein. This in turn modulates its ion conductance (Takahashi, 2000). If the G protein action is not mediated *via* ion channels, it could also be modulated *via* activating adenylatecyclase or phospholipase C, triggering a rise in cytosolic cAMP or Ca^{2+}, respectively (Takahashi, 2000). As second messengers, cytosolic cAMP or Ca^{2+} then modulate the ion conductance of an ion channel. An important feature of G-protein linked receptors is that their response is intrinsically slower and lasts longer compared to ligand-gated channels. Good examples of slow acting receptors that are coupled to G protein are the ones modulated by neuropeptide since they often need to exert their effect over a prolonged period of time (Zhang *et al.*, 2011).

It is noteworthy that neurotransmitter receptors could become desensitized or downregulated if exposed to their neurotransmitter in prolonged fashion. It is considered to be a consequence of the protein allosteric modulation since each receptor has several discrete conformations, more or less amenable to the binding of its neurotransmitter.

The Physiology of Neurotransmitters and their Receptors during Brain Development

It is important to note that all key elements of neuronal development depend on a fine balance between various neurotransmitters. In particular, it has been known that two major neurotransmitters, glutamate and GABA, control all aspects of neuronal migration, differentiation, maturation and synaptogenesis, the key components of mammalian brain development (Komuro and Rakic, 1993). Synaptogenesis involves massive dendritic branching and formation of trillion synaptic contacts between neurons thus enabling the formation of meaningful neuronal circuitries and orderly neuronal maps. The processes by which axonal and dendritic projections find the right 'target' and the most appropriate pathways for growth are very complex and not within the scope of this chapter. However, it is worth mentioning that the synaptogenesis and the development of neuronal processes are based on activity-dependent remodeling suggesting that neuronal firing and interneuronal communications are crucial for timely and proper synaptogenesis (Brown *et al.*, 1997; Dobbing and Sands, 1979).

All aspects of developmental synaptogenesis are tightly controlled by glia, which actively participate in neuron-glia signaling while providing an appropriate milieu for neuron-neuron interaction (Allen and Barres, 2005). The electrical activity and synaptic signaling are strategically important during synaptogenesis, so much so that the major inhibitory neurotransmitter, GABA serves as an excitatory neurotransmitter during early stages of synaptogenesis (Ben-Ari, 2002). Although synapses are very pliable and undergo constant remodeling where new synapses are being formed and others are being pruned away throughout ones life, the bulk of fundamental neuronal networks and synaptic contacts are formed during developmental synaptogenesis. In humans that time period occurs during the last trimester of *in utero* life and the first few years of post-natal life while being most intense during the first several months of post-natal life (Dobbing and Sands, 1979).

Unphysiologic Modulation of the Neurotransmitters and their Receptor Systems during Brain Development causes Neurotoxic Damage to the Immature Neurons

The fact that neuronal activity and communication are crucial for proper formation of synaptic contacts and for the establishment of stable receptor structures which form the foundation for cognitive and behavioral development brings into focus a class of drugs with main goal of unphysiologically 'switching off' or 'turning down' neuronal communication for the purpose of achieving amnesia, analgesia and hypnosis. This class of drugs is called general anesthetics. Both gaseous (e.g., nitrous oxide, isoflurane, sevoflurane, desflurane, etc.) and intravenous (e.g., benzodiazepines, barbiturates, propofol, etomidate, etc.) anesthetics are frequently used for the purpose of assuring the patients' comfort during painful intervention.

General anesthetics are very potent and effective in transiently inhibiting neuronal communication (Hudetz, 2012). However, despite their widespread use, the mechanisms of their anesthetic action are not fully understood. Based on the studies published over the last few decades, it is becoming increasingly evident that there are specific cellular targets through which general anesthetics act (Franks, 2008). In general, enhancement of inhibitory synaptic transmission and/or inhibition of excitatory synaptic transmission have been reported. In particular, it is becoming widely accepted that many intravenous anesthetics, among them barbiturates, benzodiazepines, propofol, and etomidate (Franks, 2008; Hirota *et al.*, 1998), as well as inhalational volatile anesthetics such as isoflurane, sevoflurane, desflurane, and halothane (Nishikawa and Harrison, 2003; Pearce, 2000), promote inhibitory neurotransmission by enhancing $GABA_A$-induced currents in neuronal tissue. For this reason, they are often referred to as GABAergic agents. On the other hand, a small number of intravenous anesthetics [e.g., phencyclidine (PCP) and its derivative, ketamine (Lodge and Anis, 1982)] and inhalational anesthetics, nitrous oxide and xenon (Franks *et al.*, 1998; Jevtovic-Todorovic *et al.*, 1998) inhibit excitatory neurotransmission by blocking N-methyl-D-aspartate (NMDA) receptors, a subtype of glutamate receptors.

Although general anesthetics are powerful modulators of GABA and glutamate and thus cause significant imbalance in their functioning, only recently has it been recognized that general anesthetics, given in clinically relevant concentrations and combinations, are potentially damaging to neuronal cells in adult and immature brains thus suggesting that these agents are potentially deleterious. This newly developing knowledge suggests the importance of maintaining a fine balance in neurotransmitters release and their receptor activation during critical stages of synaptogenesis. Since general anesthetics are becoming rapidly recognized as potentially powerful neurotoxins for the developing brain the remainder of this chapter will review presently available animal and human findings regarding the neurotoxic potential of commonly used intravenous and inhalational anesthetics.

The neurotoxic potential of general anesthetics in the developing mammalian brain

Due to skillful and sophisticated pediatric anesthesia management, the exposure of very young children, including premature babies as young as 24 weeks postconception, to general anesthesia is becoming common resulting in the annual administration of more than 3 million anesthetics (Calb, 2008; Sun *et al.*, 2008). The frequency of operating suite visits has grown exponentially, as has the length of stays in intensive care units. Heroic attempts to save premature and very ill infants has resulted in prolonged, deep sedation and repeated anesthesia during an extremely delicate period of human development.

As stated earlier the excitatory neurotransmitter glutamate promotes all key aspects of neuronal development (Komuro and Rakic, 1993) and the fine balance between GABA and glutamate neurotransmission is important for proper and timely formation of neuronal circuitries. Neurons that are not successful in making meaningful connections are considered redundant and are destined to die by programmed cell death, i.e., apoptosis, which occurs naturally during normal development of the mammalian CNS. Although this is a physiological process, apoptosis during normal development is tightly controlled, resulting in the removal of only a small percentage of neurons (Jevtovic-Todorovic *et al.*, 2003). A disturbance of the fine balance between glutamatergic and GABAergic neurotransmission by excessive depression of neuronal activity and unphysiological changes in the synaptic environment during a crucial stage of brain development may constitute a generic signal for developing neurons to "commit suicide". Since the goal of surgical anesthesia is to render the patient unconscious and insentient to pain, the ultimate question becomes whether general anesthetics at doses that ensure substantial receptor occupancy and profound depression of neuronal activity could be promoting excessive activation of neuroapoptosis and the death of large populations of developing neurons.

Based on recently published findings (Jevtovic-Todorovic *et al.*, 2003; Loepke *et al.*, 2009; Rizzi *et al.*, 2008; Young *et al.*, 2005), it is becoming widely accepted that common general anesthetics do indeed cause significant and widespread neuroapoptotic degeneration of developing neurons in various mammalian species, including rats, mice, guinea pigs, and nonhuman primates. The peak of vulnerability to anesthesia-induced neuroapoptosis in each species coincides with its peak of synaptogenesis, with much less vulnerability observed during late stages of synaptogenesis (Rizzi *et al.*, 2008; Yon *et al.*, 2005). Aside from prominent caspase-3 staining, a hallmark of apoptotic death which could be detected at the light microscopic level, detailed examination at the ultrastructural level suggests that the initial insult, visible mainly in the nucleus, is marked by the clamping of chromatin, followed by disruption of the nuclear membrane, intermixing of the cytoplasm and nucleoplasm, and the formation of apoptotic bodies (Jevtovic-Todorovic *et al.*, 2003) (Figure 14.1). In recent years, we have put forward a considerable effort into elucidating the mechanisms of anesthesia-induced developmental neuroapoptosis and found it to involve several cascades of cellular events that ultimately lead to neuronal deletion and impairment of proper synapse formation.

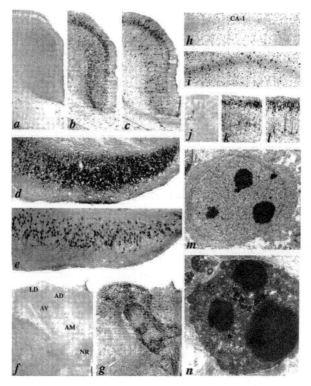

Figure 14.1. Triple anesthetic cocktail induces apoptotic neurodegeneration.

(*a–l*) are light micrographic views of various brain regions of either a control rat (*a, f, h, j*) or a rat exposed to the triple anesthetic cocktail (0.75-vol% isoflurane with midazolam at 9 mg/kg, s.c., and nitrous oxide at 75-vol% for 6 hrs) (*b-e, g, i, k, l*). Some sections were stained by the DeOlmos silver method (*a, b, d, f, g, k*); the others were immunocytochemically stained to reveal caspase-3 activation (*c, e, h–j, l*). The regions illustrated are the posterior cingulate/retrosplenial cortex (*a–c*), subiculum (*d, e*), anterior thalamus (*f, g*), rostral CA1 hippocampus (*h, i*), and parietal cortex (*j–l*). The individual nuclei shown in the anterior thalamus (*f, g*) are laterodorsal (*LD*), anterodorsal (*AD*), anteroventral (*AV*), anteromedial (*AM*), and nucleus reuniens (*NR*). *m* and *n* are electron micrographic scenes depicting the ultrastructural appearance of neurons undergoing apoptosis. The cell in *m* displays an early stage of apoptosis in which dense spherical chromatin balls are forming in the nucleus while the nuclear membrane remains intact; few changes are evident in the cytoplasm. The cell in *n* exhibits a much later stage of apoptosis in which the entire cell is condensed, the nuclear membrane is absent, and there is intermixing of nuclear and cytoplasmic constituents. These are hallmark characteristics of neuronal apoptosis as it occurs in the *in-vivo* mammalian brain [Reproduced with permission from Journal of Neuroscience, 2003; 23(3): 876–882].

The importance of cellular organelles in anesthesia-induced developmental modulation of neuronal receptors

Apoptosis can occur *via* different biochemical pathways, resulting in activation of effector caspases as the final step. Mitochondria and endoplasmic reticuli became primary targets of interest for initiating the cascade of events ultimately leading to neuronal death by apoptosis.

Apoptosis can occur *via* different biochemical pathways that result in activation of effector caspases as the final step. The mitochondria-dependent pathway, also called the intrinsic pathway, involves the down-regulation of anti-apoptotic proteins from the bcl-2 family (e.g., bcl-x_L), an increase in mitochondrial membrane permeability, followed by an increase in cytochrome c release into the cytoplasm. The cytochrome c, in turn, activates caspases-9 and -3, resulting in apoptosis. We found that the mitochondria-dependent cascade is activated within the first 2 h when general anesthesia exposure occurs at the peak of synaptogenesis. This activation is characterized by a significant decrease in levels of bcl-x_L protein, a significant rise in cytochrome c, and activation of caspase-9 (Yon *et al.*, 2006; Yon *et al.*, 2005).

In addition to activation of the intrinsic apoptotic cascade, general anesthetics cause a significant and long lasting-disturbance in mitochondrial morphogenesis. Two weeks after anesthesia exposure, mitochondrial enlargement and deranged, fragmented cristae and inner membranes were noted, suggesting significant impairment of mitochondrial membrane integrity (Sanchez *et al.*, 2011). Although anesthesia-induced mitochondrial enlargement could be due to swelling, mitochondrial regeneration in neurons depends on fine dynamics between mitochondrial fusion and fission (Chan, 2006); deranged fusion leads to mitochondrial fragmentation while deranged fission leads to mitochondrial enlargement. An imbalance between fission and fusion appears to have a causal role in initiating several adult neurodegenerative diseases (Bossy-Wetzel *et al.*, 2003; Wang *et al.*, 2009). For example, in Parkinson's and Alzheimer's diseases, impaired fission/fusion can lead to an increase in a large mitochondrial pool, with a medium-sized population remaining relatively intact (Trimmer *et al.*, 2000). General anesthetics also may disturb mitochondrial dynamics favoring excessive mitochondrial fusion and impaired fission since the anesthesia-induced changes in mitochondrial size distribution were found to be similar to those observed in these neurodegenerative diseases (Sanchez *et al.*, 2011). Disturbances in fission-fusion balance may not be tolerated well by immature and functionally busy mammalian neurons that are in need of adequate metabolic support. Indeed, impairment of mitochondrial morphogenesis may be, at least in part, the cause of the reported anesthesia-induced developmental neurodegeneration (Ikonomidou *et al.*, 2000; Ikonomidou *et al.*, 1999; Lu *et al.*, 2006; Slikker *et al.*, 2007; Yon *et al.*, 2005).

Although mitochondrial enlargement is noted in very immature neurons, the existing literature often refers to large (and "giant") mitochondria in aging neurons (Navarro and Boveris, 2010). Might there be a parallel between certain elements of mitochondrial neurodegeneration unique to aging and anesthesia-induced mitochondrial neurodegeneration unique to the developing brain? This possibility was examined further by determining how anesthesia affects developmental fusion and fission, especially by focusing on two important GTPase proteins—Drp1, and mitofusin 2, which are critical for proper fusion and fission pathway activation and that are known to be modulated in aging neurons (Smirnova *et al.*, 2001; Zorzano *et al.*, 2010). It was reported that indeed anesthesia in the developing neurons causes acute sequestration of the main fission protein, Drp 1, from the cytoplasm to mitochondria, and its oligomerization on the outer mitochondrial membrane which in turn leads to the formation of the ring-like structures that are required for mitochondrial fission (Chan, 2006). The fission was further promoted by anesthesia-induced mitofusin-2, a protein necessary for maintaining the opposing process, mitochondrial fusion (Boscolo *et al.*, 2013).

Since mitochondria are generated in the neuronal body, initial work was focused on the morphology of mitochondria located at this site; however, neurons have multiple compartments (e.g., dendrites, axons, and synapses) that are located far from the cell body and hence depend heavily on proper mitochondrial distribution (Morris and Hollenbeck, 1993). As the main regulators of ATP production, mitochondria are frequently found in the vicinity of active growth cones in developing neurons (Morris and Hollenbeck, 1993) and in terminals with active synapses (Rowland *et al.*, 2000; Shepherd and Harris, 1998). Thus, it is of equal importance to understand how anesthesia-induced morphological changes affect mitochondrial migration (Yaffe, 1999). Our recent work suggests that significantly fewer mitochondria are located in presynaptic neuronal profiles in anesthesia-treated brain than in controls (Sanchez *et al.*, 2011). Since these mitochondria also are significantly larger than those in controls, it was proposed that anesthesia-induced mitochondrial enlargement causes mitochondria to be sluggish and 'stuck' in more proximal cellular compartments, thus shifting their regional distribution away from very distant thin and highly arborized dendritic branches at a time when their presence is necessary for normal synapse formation and development. In fact, we and others have reported that anesthesia impairs plasticity of dendritic spines and the formation, stability and function of developing synapses (Briner *et al.*, 2010; Briner *et al.*, 2011; Head *et al.*, 2009; Lunardi *et al.*, 2010). Since mitochondrial ATP production at the vicinity of an active synapse regulates all the elements of neurotransmitter synthesis, release (*via* exocytosis) and uptake it is clear that mitochondrial dysfunction during critical stages of synaptogenesis may be devastating for the developing neurons.

Since neurons are highly dependent on ATP synthesis and produce reactive oxygen species (ROS) as byproducts of oxidative phosphorylation in mitochondria it is important to understand how excessive ROS production may impair normal neurotransmitter release and receptor function. Because of their high oxygen requirements and relative deficiency in oxidative defenses—in particular, low to moderate activity of catalase and Mn-superoxide dismutase (SOD)—neurons are highly sensitive to excessive ROS production. This vulnerability, combined with their high content of polyunsaturated fatty acids, makes them susceptible to excessive lipid peroxidation and cellular damage (Halliwell, 1992). Morphological distortion and impaired regional distribution of mitochondria are accompanied by production of excessive ROS and significant peroxidation of cellular and subcellular lipid membranes. These effects are important in the development and progression of several neurological diseases that are marked by severe cognitive decline (Bennett, 2005; Reddy, 2006; Reddy, 2007; Trushina *et al.*, 2004). Early exposure to anesthesia makes developing neurons susceptible to mitochondria-induced ROS up-regulation, lipid peroxidation, and protein oxidation. This in turn may contribute to the observed impairment in receptors function and could ultimately be responsible for the reported cognitive impairments. To address a potential functional link between disturbances in receptor function, ROS up-regulation and cognitive disturbances, we administered EUK-134, a synthetic ROS scavenger having both Mn-SOD and catalase activity (Baker *et al.*, 1998; Liu *et al.*, 2003), or R(+) pramipexole [R(+) PPX], a synthetic aminobenzothiazol derivative that blocks permeability transition pores, restores the integrity of mitochondrial membranes (Sayeed *et al.*, 2006), and limits ROS production (Cassarino *et al.*, 1998; Le *et al.*, 2000). We found that curtailing ROS up-regulation and lipid peroxidation with these inhibitors preserved mitochondrial morphogenesis and neuronal viability and also prevented the development of learning and memory impairment in adolescent rats that were exposed to general anesthesia at the peak of synaptogenesis (at post-natal day 7) (Boscolo *et al.*, 2012). This result suggests that anesthesia-induced developmental neurotoxicity is, in part, the result of mitochondria-induced ROS that ultimately leads to neuronal damage and behavioral impairment.

Significant inhibition of anesthesia-induced developmental neurodegeneration also has been provided by other strategies to protect mitochondria. For example, melatonin, a naturally occurring sleep hormone that up-regulates bcl-xL (Yon *et al.*, 2006) and prevents cytochrome c leak, and carnitine, a nutritional supplement that protects mitochondrial integrity (Zou *et al.*, 2008), both provide significant protection against neuronal apoptosis. Although it remains to be determined whether melatonin and carnitine also protect against anesthesia-induced cognitive impairment, it is clear that mitochondria play an integral role in proper development of the immature neurons and their synaptic connections. Collectively, this line of work suggests that preventing excessive lipid peroxidation and protecting mitochondria may permit safe use of general anesthesia during early stages of brain development.

The upstream trigger for impairment of mitochondrial function may be excessive increase in intracellular Ca^{2+}. Since the endoplasmic reticulum (ER) is the primary source of releasable Ca^{2+} in neurons, it plays an important role in neuronal function and survival. As discussed earlier, intracellular Ca^{2+} regulates many aspects of neuronal development, including synapse development and functioning, membrane excitability, protein synthesis, neuronal apoptosis and autophagy—in short, all important elements of neuronal survival (Berridge, 2009; Decuypere *et al.*, 2011; Hanson *et al.*, 2004). A moderate increase in Ca^{2+} release *via* activation of inositol 1,4,5-trisphosphate receptors may provide neuroprotection in some forms of brain injury (Wei *et al.*, 2007); however, excessive activation of these receptors may lead to elevation of intracellular Ca^{2+} into the toxic range. A disturbance in Ca^{2+} homeostasis has been implicated in some forms of learning and memory deficits (Power *et al.*, 2002; Rosenzweig and Barnes, 2003).

The elevated Ca^{2+} may cause cytochrome c leak (Hanson *et al.*, 2004), which could promote further mitochondrial dysfunction. Thus, the ER could be an important initial target of anesthesia-induced developmental neurotoxicity. Indeed, Zhao and colleagues have shown that the inhalational anesthetic isoflurane activates inositol 1,4,5-trisphosphate receptors to induce significant Ca^{2+} release from the ER, resulting in modulation of mitochondrial bcl-xL protein, which then promotes apoptotic neuronal death in the immature rat brain (Zhao *et al.*, 2010). Similar modulation of inositol 1,4,5-trisphosphate receptors was reported with the general anesthetics propofol, desflurane and sevoflurane, with resultant cytosolic Ca^{2+} overload and an increase in mitochondrial permeability transition pore activity resulting in mitochondrial swelling and uncontrolled release of pro-apoptotic factors (Inan and Wei, 2010).

The Importance of Anesthesia-induced Modulation of Neurotransmitters in the Formation of Neuronal Networks

The morphological changes described thus far represent substantial changes in neuronal structure that can be detected easily using histological assessments. An important issue that has been brought to light on several occasions is that seemingly subtle changes that cannot be detected morphologically remain in surviving 'normal' neurons after the grossly damaged neurons have been removed. Based on presently available evidence, these neurons may not be truly functional; i.e., their communications may be faulty. We first noted that an early exposure to general anesthesia causes long-term impairment in synaptic transmission in the hippocampus of adolescent rats (postnatal day 27–33) exposed to anesthesia at the peak of their synaptogenesis (postnatal day 7) (Jevtovic-Todorovic *et al.*, 2003). In particular, long-term potentiation was impaired significantly despite the presence of robust short-term potentiation. This observation suggested a long-lasting disturbance in neuronal circuitries in the young hippocampus, a brain region that is crucial for proper learning and memory development. A deficit in long-term potentiation was confirmed when synaptic transmission was examined using patch clamp recordings of evoked inhibitory post synaptic current (eIPSC) and evoked excitatory post synaptic current (eEPSC) by recording from the pyramidal layer of control and anesthesia-treated rat subiculum, an important component of the hippocampal complex. Again it was noted that anesthesia-treated animals suffered from impaired synaptic transmission with inhibitory transmission affected significantly (Sanchez *et al.*, 2011).

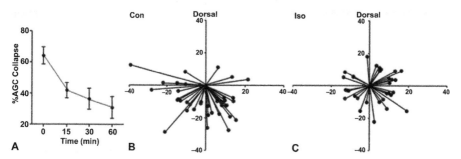

Figure 14.2. Isoflurane causes immediate effects on axonal growth cone sensitivity to Semaphorin3A but requires a longer exposure for effects on axon guidance.

A time-response curve is shown depicting the percent axonal growth cone collapse to Semaphorin3A at increasing time of exposure to isoflurane, which shows a loss of collapse in as little as 15 min (A). Trajectory diagrams from a slice overlay assay conducted at 5h, the minimum time for appropriate targeting of controls (B) shows that a disruption of axon guidance is seen with 1.8% isoflurane at this time point. Axes in (B) and (C) in μm. [Reproduced with permission from *Anesthesiology*, 2013; 118(4): 825–833].

Although the precise mechanisms responsible for the long-lasting changes in synaptic communication post-anesthesia remain to be deciphered, some recent findings suggest that anesthetics impair axon targeting and inhibit axonal growth cone collapse, resulting in lack of proper response to guidance cues, thus causing errors in axon targeting (Mintz *et al.*, 2013) (Figure 14.2).

Is there a Correlation between Excessive Modulation of Neurotransmitters during Critical Stages of Brain Development and Long-term Effects on Animal Behavior?

The above pathomorphological findings make it clear that anesthesia exposure results in neuronal deletion (Nikizad *et al.*, 2007; Rizzi *et al.*, 2008) and long-lasting impairment of synapses in vulnerable brain regions (Briner *et al.*, 2010; Briner *et al.*, 2011; Head *et al.*, 2009; Lunardi *et al.*, 2010). The ultimate question is whether and how these observations translate to lasting effects on behavior. The short answer appears to be that they do. Development of cognitive abilities of animals exposed to general anesthetics at the peak of synaptogenesis, lagged behind those of controls, with the gap widening into adulthood (Figure 14.3). Even intravenous general anesthetics like propofol or thiopental in combination with ketamine (but not singly), at post-natal day 10, alters mouse behavior later in young adulthood (Fredriksson *et al.*, 2007). Similar adult behavioral deficits were noted when mice were exposed at post-natal day 10 to a cocktail containing ketamine and diazepam (Fredriksson *et al.*, 2004).

Although anesthesia cocktails seem to be most detrimental, ketamine given alone during early stages of brain development in rats also caused later deficits in habituation, and in learning and memory (Fredriksson *et al.*, 2004). When anesthetic agents with GABAergic and N-methyl-D-aspartate (NMDA) antagonist properties are combined, which is done frequently in the clinical setting (for example, nitrous oxide and volatile anesthetics or propofol and ketamine), cognitive deficits are more profound (Fredriksson *et al.*, 2004; Fredriksson *et al.*, 2007; Jevtovic-Todorovic *et al.*, 2003). Although causality is difficult to establish, it is reasonable to propose that anesthesia-induced neuroapoptosis (discussed above), is at least in part, responsible for the observed cognitive deficits. It remains to be established whether multiple, longer-lasting exposures to anesthesia cocktails during vulnerable periods have greater effects on neurocognitive development than were initially predicted.

General anesthesia is rarely administered in the absence of surgery, and its associated pain and tissue injury. Thus, the reported neurotoxic potential of general anesthesia during brain development needs to be confirmed in the setting of surgical stimulation. Using clinically relevant concentrations of nitrous oxide and isoflurane Shu and his colleagues found that nociception enhanced neuroapoptosis and worsened long-term cognitive impairments when compared to anesthesia alone (Figure 14.4) (Shu *et al.*, 2012). These results would seem to undermine the hope that surgical stimulation may somehow be 'protective' against anesthesia-induced neurotoxicity.

Results from rodent studies are known to translate poorly to humans, but newly emerging behavioral studies with non-human primates are beginning to suggest that translation is likely. Dr. Paule and his colleagues have examined the effects of continuous neonatal (age 5 or 6 days) ketamine infusion (24 h) sufficient to maintain a light surgical plane of anesthesia on behavioral development in primates (Paule *et al.*, 2011). They observed that ketamine-treated primates exhibit long-term disturbances in all important aspects of cognitive development, such as learning, psychomotor speed, concept formation and motivation. These effects occurred despite an absence of physiological or metabolic trespass. Although 24 hours of anesthetic exposure would be considered unusual, it certainly occurs, especially in critically ill patients of all ages. These results suggest further translation to humans.

MORRIS WATER MAZE

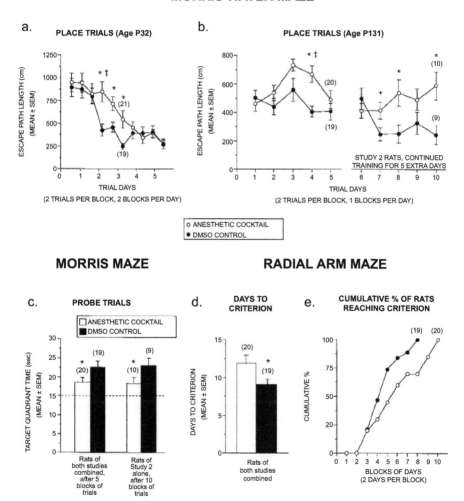

Figure 14.3. contd....

Is there any Relevance to Cognitive and Behavioral Development in Humans after Childhood Surgery?

The effects of perioperative events on the psychological and emotional development of children have been known for many decades. In 1945 Levy's retrospective study was the first to make an association between relatively short surgical procedures, such as tonsillectomy, adenoidectomy, or appendectomy, in otherwise healthy children and the development of new behavioral problems such as terrors, dependency, destructiveness, and disobedience within the first six postoperative months(Levy, 1945). Most of the sensitive children were in the youngest age group; 33%–58% of them were between 0 and 2 years of age. Several studies over the succeeding decades reported that the incidence of surgery-associated psychological disturbances ranged from 9% to 20%, and that children under 2 years of age were at most risk. Further, the risk was

---

Figure 14.3. contd.

Figure 14.3. Effects of neonatal triple anesthetic cocktail treatment on spatial learning.

(*a*) Rats were tested at P32 for their ability to learn the location of a submerged (not visible) platform. ANOVA analysis of data on the length of escape paths yielded a significant main effect of treatment ($p = 0.032$) and a significant treatment by blocks of trials interaction ($p = 0.024$). These results indicated that the place-training performance of rats given an anesthetic cocktail (0.75-vol% isoflurane with midazolam at 9 mg/kg, s.c., and nitrous oxide at 75-vol% for 6 hrs) was significantly inferior to that of control rats. Subsequent pairwise comparisons indicated that differences were greatest during blocks 4, 5, and 6 ($p = 0.003, 0.012$, and 0.019, respectively). However, rats given the anesthetic cocktail improved their performance to control-like levels during the last four blocks of trials. (*b*) Rats were retested as adults (P131) for their ability to learn a different location of the submerged platform. The graph on the left shows the path-length data from the first five place trials when all rats were tested. ANOVA analysis of these data yielded a significant main effect of treatment ($p = 0.013$), indicating that the control rats, in general, used significantly shorter paths in swimming to the platform than did rats treated with the anesthetic cocktail. Subsequent pairwise comparisons showed that differences were greatest during block 4 ($p = 0.001$).

The graph on the right shows the data from rats given 5 additional training days as adults. During these trials, rats in the control group improved their performance and appeared to reach asymptotic levels, whereas the rats given the anesthetic cocktail showed no improvement. ANOVA analysis of these data yielded a significant main effect of treatment ($p = 0.045$), as well as a significant treatment by blocks of trials interaction ($p = 0.001$). Additional pairwise comparisons showed that group differences were greatest during blocks 7, 8, and 10 ($p = 0.032, 0.013$, and 0.017, respectively). (*c*) Probe trial performance of rats given the anesthetic cocktail and control rats during adult testing. Search behavior of the rats was quantified when the submerged platform was removed from the pool after the last place trials in blocks 5 and 10. The histogram on the left presents data for rats in both studies 1 and 2 combined after five blocks of place trials were completed. The histogram on the right presents data for rats in study 2 alone, after 10 blocks of place trials were completed. The dotted line represents the amount of time that animals would be expected to spend in the target quadrant based on chance alone. Both histograms show that the control rats spent significantly more time in the target quadrant than did the anesthesia-exposed rats, regardless of whether the probe tests were done on both study groups after five blocks or only on the study 2 rats after 10 trials. (*d, e*) Data are shown from the radial arm maze test done on P53 to evaluate spatial working memory capabilities. (d) This histogram shows that rats given the anesthetic cocktail rats required significantly more days to reach a criterion demonstrating learning (8 correct responses out of the first 9 responses for 4 consecutive days) compared with controls. (*e*) This plot shows the days to criterion data as the cumulative percentage of rats reaching criterion in each group as a function of blocks of training days. The acquisition rate of rats given the anesthetic cocktail began to slow around the fourth block of trials and remained slower throughout the rest of the experiment. Numbers in parentheses in each graph indicate sample sizes. *$p < 0.05$; Bonferroni corrected level: †$p < 0.005$ in *a*; †$p < 0.01$ in *b*. [Reproduced with permission from *Journal of Neuroscience*, 2003; 23(3): 876–882].

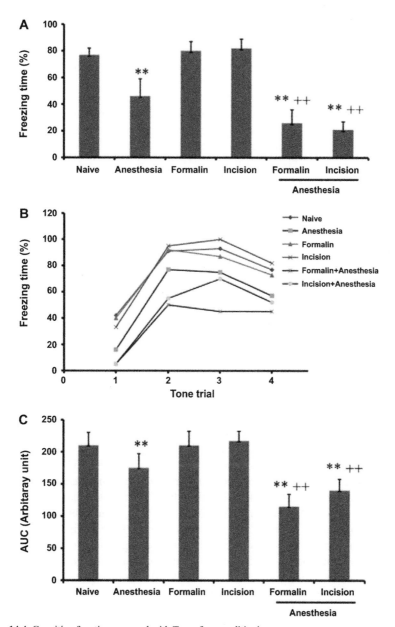

Figure 14.4. Cognitive function assessed with Trace fear conditioning.

Seven day old rat pups received 70% N_2O + 0.75% isoflurane anesthesia with or without the plantar surface of left paw injection with 5% formalin or skin incision for 6 h. Other cohorts received 6 h of 70% nitrogen balanced with oxygen (naïve control), or combined with left paw plantar surface injection of 5% formalin (formalin group) or skin incision (incision group). They were then allowed to live up to the age of 40 days for cognitive function test. (A) The mean data of percentage of freezing time after context testing. (B) The mean data of percentage of freezing time after tone testing. (C) The average data of the area under curve (AUC) derived from panel B. Mean ± SD, n = 6. **p < 0.05 vs. Naïve; ++p < 0.01 vs. Anesthesia. [Reproduced with permission from *Neurobiology of Disease*, 2012; 45: 743–750.]

independent of the anesthetic drug used (Backman and Kopf, 1986; Eckenhoff, 1953; Jackson, 1951; Standley *et al.*, 1978), and in some case, the behavioral disturbances persisted for months and even years. In a prospective multi-center survey of 551 children, Campbell *et al.* found a 47% overall incidence of new behavioral problems, including anxiety, nightmares, and attention seeking, and again noted a relatively higher risk in those having procedures in the first two years of life (Campbell, 1988).

Because the behavioral changes were independent of drug or technique, it was initially assumed that the emotional shocks of hospitalization, separation from family, and the physical traumas of surgical intervention, including pain, fluid imbalance, nutritional changes, and blood loss, caused the regressive behavioral changes. However, the relationship between anesthesia and acute personality changes was suggested in 1953, when Eckenhoff published a retrospective study of 612 patients under the age of 12 years, having either tonsillectomy or appendectomy under a variety of anesthetic drugs (cyclopropane, nitrous oxide, morphine, and pentobarbital). Behavioral changes and new-onset bed-wetting, occurred, on average, two months after surgery with highest incidence (57%) in those younger than 3 years, and lowest incidence (8%), in those older than 8 years.

Backman and Kopf were the first to suggest a relationship between anesthesia and long-term cognitive delay (Backman and Kopf, 1986). In this report children were anesthetized with ketamine and halothane for removal of congenital nevocytic nevi, a fairly minor procedure. Nevertheless, the authors reported an increased incidence of cognitive impairment, described as regressive behavioral changes, lasting up to 18 months after the procedure. Again, children younger than 3 years were the most sensitive. These authors clearly voiced concern that general anesthetics may cause long-term cognitive effects.

Clinical investigations are still at an early stage, but the evidence that has emerged over the last few years consistently points to detrimental effects of anesthetic exposure in young children on subsequent behavioral and cognitive development.

In a population-based retrospective birth-cohort study of 5,357 children, Wilder *et al.* found that children who received two or more general anesthetics before the age of 4 years were at increased risk of learning disability as adolescents (Figure 14.5) (Wilder *et al.*, 2009). Moreover, the risk increased with longer cumulative exposure duration (more than 2 hours). Of particular concern is their finding that the cohort exposed to anesthesia before the age of 4 years had cognitive scores that were two standard deviations below predicted. This implies that early exposure to general anesthesia may have prevented achievement of full cognitive potential. In an even larger population study, Sun *et al.* assessed learning disabilities in 228,961 individuals (Sun *et al.*, 2008). Children who had procedures requiring anesthesia before the age of 3 years required more Medicaid services for learning disability than did children who did not have these procedures. Procedures in premature infants have also been associated with an excess of behavioral disabilities later in life. For example, surgically treated premature infants with patent ductusarteriosus (Chorne *et al.*, 2007; Rees *et al.*, 2007) or necrotizing enterocolitis (Hintz *et al.*, 2005) had worse neurological outcomes than did premature infants who were treated medically.

A recent retrospective clinical study examined the association between procedures requiring general anesthesia prior to the age of 2 and the occurrence of attention deficit

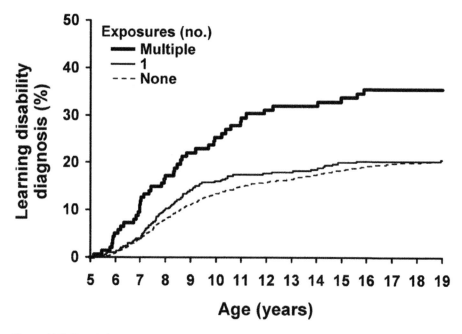

Figure 14.5. Cumulative percentage of learning disabilities diagnosis by the age at exposure shown separately for those that have zero, one, or multiple anesthetic exposures before age 4 yr. [Reproduced with permission from *Anesthesiology*, 2009; 110(4): 796–804.]

hyperactivity disorder (ADHD) up to the age of 19 years (Sprung *et al.*, 2012). As found in the prior Wilder *et al.* study, only the cohort having two or more exposures before the age of 2 years was associated with ADHD (Wilder *et al.*, 2009); the lack of association in single exposures was not explained by anesthetic type or comorbidities (Figure 14.6).

Despite the limitations inherent to retrospective clinical studies, such as lack of randomization, unmatched controls, and numerous uncontrolled (and unknown) variables, this study points to yet another behavioral impairment that could be associated with early exposure to anesthesia. But association does not imply causality. A measure of caution is advisable since the effects of surgery, or the illness requiring surgery, cannot be clearly separated from the effects of anesthesia in retrospective clinical studies. However, the complex issues associated with the design of randomized, double-blinded prospective clinical studies of very young patients cannot be underestimated. These issues include, but are not limited to: ethical considerations; the lack of biomarkers of apoptosis that can be used safely in a living organism; the complexity and significance of various clinical outcomes, especially neurocognitive ones; and the lack of appropriate controls.

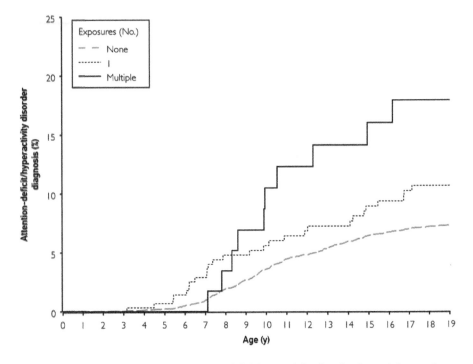

Figure 14.6. Cumulative percentage of attention-deficit/hyperactivity disorder diagnosis by age shown separately for those that had 0, 1, or multiple procedures performed with general anesthesia under the age of 2 years.

The number of individuals at risk at 5, 10, and 15 years of age is 4995, 4009, and 3576 for those with no exposure; 284, 230, and 214 for those with 1 exposure; and 64, 50, and 44 for those with 2 or more exposures. [Reproduced with permission from *Mayo ClinProc*, 2012; 87(2): 120–129.]

Conclusion Remarks

Modern neuroscience reminds us of the importance of proper functioning and fine modulation of neurotransmitters not only in fully developed brain but importantly during various stages of brain development. Any degree of 'tempering' with timely neurotransmitter synthesis, release, receptor activation or inhibition could have substantial consequences on synaptogenesis, neuronal network formation and on proper cognitive and behavioral development. Although it might be hard to accept that commonly used drugs such as general anesthetics are perhaps neurotoxins, it shows us how powerful modulators of major inhibitory and excitatory neurotransmitters which govern all the key elements of neuronal development could permanently alter the course of normal brain development.

References

Abbas, G., S. Naqvi, and A. Dar. 2012. Comparison of monoamine reuptake inhibitors for the immobility time and serotonin levels in the hippocampus and plasma of sub-chronically forced swim stressed rats. Pak J Pharm Sci 25: 441–445.

Allen, N.J., and B.A. Barres. 2005. Signaling between glia and neurons: focus on synaptic plasticity. Curr Opin Neurobiol 15: 542–548.

Amenta, F., and S.K. Tayebati. 2008. Pathways of acetylcholine synthesis, transport and release as targets for treatment of adult-onset cognitive dysfunction. Curr Med Chem 15: 488–498.

Backman, M.E., and A.W. Kopf. 1986. Iatrogenic effects of general anesthesia in children: considerations in treating large congenital nevocytic nevi. J Dermatol Surg Oncol 12: 363–367.

Baker, K., C.B. Marcus, K. Huffman, H. Kruk, B. Malfroy, and S.R. Doctrow. 1998. Synthetic combined superoxide dismutase/catalase mimetics are protective as a delayed treatment in a rat stroke model: a key role for reactive oxygen species in ischemic brain injury. J Pharmacol Exp Ther 284: 215–221.

Ben-Ari, Y. 2002. Excitatory actions of gaba during development: the nature of the nurture. Nat Rev Neurosci 3: 728–739.

Bennett, M.C. 2005. The role of alpha-synuclein in neurodegenerative diseases. Pharmacol Ther 105: 311–331.

Berridge, M.J. 2009. Inositol trisphosphate and calcium signalling mechanisms. Biochim Biophys Acta 1793: 933–940.

Bhalla, U.S. 2003. Temporal computation by synaptic signaling pathways. J Chem Neuroanat 26: 81–86.

Bonanomi, D., F. Benfenati, and F. Valtorta. 2006. Protein sorting in the synaptic vesicle life cycle. Prog Neurobiol 80: 177–217.

Boscolo, A., J.A. Starr, V. Sanchez, N. Lunardi, M.R. DiGruccio, C. Ori, A. Erisir, P. Trimmer, J. Bennett, and V. Jevtovic-Todorovic. 2012. The abolishment of anesthesia-induced cognitive impairment by timely protection of mitochondria in the developing rat brain: the importance of free oxygen radicals and mitochondrial integrity. Neurobiol Dis 45: 1031–1041.

Boscolo, A., D. Milanovic, J.A. Starr, V. Sanchez, A. Oklopcic, L. Moy, C.C. Ori, A. Erisir, and V. Jevtovic-Todorovic. 2013. Early exposure to general anesthesia disturbs mitochondrial fission and fusion in the developing rat brain. Anesthesiology 118: 1086–1097.

Bossy-Wetzel, E., M.J. Barsoum, A. Godzik, R. Schwarzenbacher, and S.A. Lipton. 2003. Mitochondrial fission in apoptosis, neurodegeneration and aging. Curr Opin Cell Biol 15: 706–716.

Breedlove, S.M., N.V. Watson, and S.F. 2013. Biological psychology: an introduction to behavioral, cognitive, and clinical neuroscience (Seventh ed.). Sunderland, MA: Sinauer Associates.

Briner, A., M. De Roo, A. Dayer, D. Muller, W. Habre, and L. Vutskits. 2010. Volatile anesthetics rapidly increase dendritic spine density in the rat medial prefrontal cortex during synaptogenesis. Anesthesiology 112: 546–556.

Briner, A., I. Nikonenko, M. De Roo, A. Dayer, D. Muller, and L. Vutskits. 2011. Developmental Stage-dependent persistent impact of propofol anesthesia on dendritic spines in the rat medial prefrontal cortex. Anesthesiology 115: 282–293.

Brown, J.K., T. Omar, and M. O'Regan. 1997. Brain development and the development of tone and muscle. pp. 1–41. In: Kevin J. Connolly, and Hans Forssberg (eds.). Neurophysiology and the Neuropsychology of Motor Development. Mackeith Press, London.

Brunello, N., J. Mendlewicz, S. Kasper, B. Leonard, S. Montgomery, J. Nelson, E. Paykel, M. Versiani, and G. Racagni. 2002. The role of noradrenaline and selective noradrenaline reuptake inhibition in depression. Eur Neuropsychopharmacol 12: 461–475.

Calb, C. 2008. A new study raises questions about the risks to young children. Newsweek (Health Section, October 21).

Campbell, I.R., J.M. Scaife, and J.M. Johnstone. 1988. Physiological effects of day-case surgery compared with inpatient surgery. Archs Dis Childhood 63: 415–417.

Cassarino, D.S., C.P. Fall, T.S. Smith, and J.P. Bennett, Jr. 1998. Pramipexole reduces reactive oxygen species production *in vivo* and *in vitro* and inhibits the mitochondrial permeability transition produced by the parkinsonian neurotoxin methylpyridinium ion. J Neurochem 71: 295–301.

Chan, D.C. 2006. Mitochondrial fusion and fission in mammals. Annu Rev Cell Dev Biol 22: 79–99.

Chorne, N., C. Leonard, R. Piecuch, and R.I. Clyman. 2007. Patent ductus arteriosus and its treatment as risk factors for neonatal and neurodevelopmental morbidity. Pediatrics 119: 1165–1174.

Decuypere, J.P., G. Monaco, G. Bultynck, L. Missiaen, H. De Smedt, and J.B. Parys. 2011. The IP(3) receptor-mitochondria connection in apoptosis and autophagy. Biochim Biophys Acta 1813: 1003–1013.

Dobbing, J., and J. Sands. 1979. Comparative aspects of the brain growth spurt. Early Hum Dev 3: 79–83.

Eckenhoff, J.E. 1953. Relationship of anesthesia to postoperative personality changes in children. AMA Am J Dis Child 86: 587–591.

Franks, N.P., R. Dickinson, S.L. de Sousa, A.C. Hall, and W.R. Lieb. 1998. How does xenon produce anaesthesia? Nature 396: 324.

Franks, N.P. 2008. General anaesthesia: from molecular targets to neuronal pathways of sleep and arousal. Nat Rev Neurosci 9: 370–386.

Fredriksson, A., T. Archer, H. Alm, T. Gordh, and P. Eriksson. 2004. Neurofunctional deficits and potentiated apoptosis by neonatal NMDA antagonist administration. Behav Brain Res 153: 367–376.

Fredriksson, A., E. Ponten, T. Gordh, and P. Eriksson. 2007. Neonatal exposure to a combination of N-methyl-D-aspartate and gamma-aminobutyric acid type A receptor anesthetic agents potentiates apoptotic neurodegeneration and persistent behavioral deficits. Anesthesiology 107: 427–436.

Halliwell, B. 1992. Reactive oxygen species and the central nervous system. J Neurochem 59: 1609–1623.

Hanson, C.J., M.D. Bootman, and H.L. Roderick. 2004. Cell signalling: IP3 receptors channel calcium into cell death. Curr Biol 14: R933–935.

Head, B.P., H.H. Patel, I.R. Niesman, J.C. Drummond, D.M. Roth, and P.M. Patel. 2009. Inhibition of p75 neurotrophin receptor attenuates isoflurane-mediated neuronal apoptosis in the neonatal central nervous system. Anesthesiology 110: 813–825.

Hintz, S.R., D.E. Kendrick, B.J. Stoll, B.R. Vohr, A.A. Fanaroff, E.F. Donovan, W.K. Poole, M.L. Blakely, L. Wright, R. Higgins, and N.N.R. Network. 2005. Neurodevelopmental and growth outcomes of extremely low birth weight infants after necrotizing enterocolitis. Pediatrics 115: 696–703.

Hirota, K., S.H. Roth, J. Fujimura, A. Masuda, and Y. Ito. 1998. GABAergic mechanisms in the action of general anesthetics. Toxicol Lett 100-101: 203–207.

Hudetz, A.G. 2012. General anesthesia and human brain connectivity. Brain Connect 2: 291–302.

Ikonomidou, C., F. Bosch, M. Miksa, P. Bittigau, J. Vockler, K. Dikranian, T.I. Tenkova, V. Stefovska, L. Turski, and J.W. Olney. 1999. Blockade of NMDA receptors and apoptotic neurodegeneration in the developing brain. Science 283: 70–74.

Ikonomidou, C., P. Bittigau, M.J. Ishimaru, D.F. Wozniak, C. Koch, K. Genz, M.T. Price, V. Stefovska, F. Horster, T. Tenkova, K. Dikranian, and J.W. Olney. 2000. Ethanol-induced apoptotic neurodegeneration and fetal alcohol syndrome. Science 287: 1056–1060.

Inan, S., and H. Wei. 2010. The cytoprotective effects of dantrolene: a ryanodine receptor antagonist. Anesth Analg 111: 1400–1410.

Jackson, K. 1951. Psychologic preparation as a method of reducing the emotional trauma of anesthesia in children. Anesthesiology 12: 293–300.

Jevtovic-Todorovic, V., S.M. Todorovic, S. Mennerick, S. Powell, K. Dikranian, N. Benshoff, C.F. Zorumski, and J.W. Olney. 1998. Nitrous oxide (laughing gas) is an NMDA antagonist, neuroprotectant and neurotoxin. Nat Med 4: 460–463.

Jevtovic-Todorovic, V., R.E. Hartman, Y. Izumi, N.D. Benshoff, K. Dikranian, C.F. Zorumski, J.W. Olney, and D.F. Wozniak. 2003. Early exposure to common anesthetic agents causes widespread neurodegeneration in the developing rat brain and persistent learning deficits. J Neurosci 23: 876–882.

Kaila, K., T.J. Price, J.A. Payne, M. Puskarjov, and J. Voipio. 2014. Cation-chloride cotransporters in neuronal development, plasticity and disease. Nat Rev Neurosci 15: 637–654.

Komuro, H., and P. Rakic. 1993. Modulation of neuronal migration by NMDA receptors. Science 260: 95–97.

Le, W.D., J. Jankovic, W. Xie, and S.H. Appel. 2000. Antioxidant property of pramipexole independent of dopamine receptor activation in neuroprotection. J Neural Transm (Vienna) 107: 1165–1173.

Levy, D.M. 1945. Psychic trauma of operations in children and a note on combat neurosis. American Journal of Disease of Children 69: 7–25.

Liu, R., I.Y. Liu, X. Bi, R.F. Thompson, S.R. Doctrow, B. Malfroy, and M. Baudry. 2003. Reversal of age-related learning deficits and brain oxidative stress in mice with superoxide dismutase/catalase mimetics. Proc Natl Acad Sci U S A 100: 8526–8531.

Lodge, D., and N.A. Anis. 1982. Effects of phencyclidine on excitatory amino acid activation of spinal interneurones in the cat. Eur J Pharmacol 77: 203–204.

Loepke, A.W., G.K. Istaphanous, J.J. McAuliffe, 3rd, L. Miles, E.A. Hughes, J.C. McCann, K.E. Harlow, C.D. Kurth, M.T. Williams, C.V. Vorhees, and S.C. Danzer. 2009. The effects of neonatal isoflurane exposure in mice on brain cell viability, adult behavior, learning, and memory. Anesth Analg 108: 90–104.

Lu, L.X., J.H. Yon, L.B. Carter, and V. Jevtovic-Todorovic. 2006. General anesthesia activates BDNF-dependent neuroapoptosis in the developing rat brain. Apoptosis 11: 1603–1615.

Lunardi, N., C. Ori, A. Erisir, and V. Jevtovic-Todorovic. 2010. General anesthesia causes long-lasting disturbances in the ultrastructural properties of developing synapses in young rats. Neurotox Res 17: 179–188.

Marchetti, C. 2014. Interaction of metal ions with neurotransmitter receptors and potential role in neurodiseases. Biometals 27: 1097–1113.

Matus, A.I., and B.B. Walters. 1976. Type 1 and 2 synaptic junctions: differences in distribution of concanavalin A binding sites and stability of the junctional adhesion. Brain Res 108: 249–256.

Mintz, C.D., K.M. Barrett, S.C. Smith, D.L. Benson, and N.L. Harrison. 2013. Anesthetics interfere with axon guidance in developing mouse neocortical neurons *in vitro* via a gamma-aminobutyric acid type A receptor mechanism. Anesthesiology 118: 825–833.

Morris, R.L., and P.J. Hollenbeck. 1993. The regulation of bidirectional mitochondrial transport is coordinated with axonal outgrowth. J Cell Sci 104 (Pt 3): 917–927.

Navarro, A., and A. Boveris. 2010. Brain mitochondrial dysfunction in aging, neurodegeneration, and Parkinson's disease. Front Aging Neurosci 2.

Nikizad, H., J.H. Yon, L.B. Carter, and V. Jevtovic-Todorovic. 2007. Early exposure to general anesthesia causes significant neuronal deletion in the developing rat brain. Ann N Y Acad Sci 1122: 69–82.

Nishikawa, K., and N.L. Harrison. 2003. The actions of sevoflurane and desflurane on the gamma-aminobutyric acid receptor type A: effects of TM2 mutations in the alpha and beta subunits. Anesthesiology 99: 678–684.

Paule, M.G., M. Li, R.R. Allen, F. Liu, X. Zou, C. Hotchkiss, J.P. Hanig, T.A. Patterson, W. Slikker, Jr., and C. Wang. 2011. Ketamine anesthesia during the first week of life can cause long-lasting cognitive deficits in rhesus monkeys. Neurotoxicol Teratol 33: 220–230.

Pearce, R.A. 2000. Effects of volatile anesthetics on GABAA receptors: electrophysiological studies. pp. 245–272. *In*: E.J. Moody, and P. Skolnick (eds.). Molecular Basis of Anesthesia. CRC Press, Boca Raton, FL.

Power, J.M., W.W. Wu, E. Sametsky, M.M. Oh, and J.F. Disterhoft. 2002. Age-related enhancement of the slow outward calcium-activated potassium current in hippocampal CA1 pyramidal neurons *in vitro*. J Neurosci 22: 7234–7243.

Reddy, P.H. 2006. Mitochondrial oxidative damage in aging and Alzheimer's disease: implications for mitochondrially targeted antioxidant therapeutics. J Biomed Biotechnol 2006: 31372.

Reddy, P.H. 2007. Mitochondrial dysfunction in aging and Alzheimer's disease: strategies to protect neurons. Antioxid Redox Signal 9: 1647–1658.

Rees, C.M., A. Pierro, and S. Eaton. 2007. Neurodevelopmental outcomes of neonates with medically and surgically treated necrotizing enterocolitis. Arch Dis Child Fetal Neonatal Ed 92: F193–198.

Rizzi, S., L.B. Carter, C. Ori, and V. Jevtovic-Todorovic. 2008. Clinical anesthesia causes permanent damage to the fetal guinea pig brain. Brain Pathol 18: 198–210.

Rosenzweig, E.S., and C.A. Barnes. 2003. Impact of aging on hippocampal function: plasticity, network dynamics, and cognition. Prog Neurobiol 69: 143–179.

Rowland, K.C., N.K. Irby, and G.A. Spirou. 2000. Specialized synapse-associated structures within the calyx of Held. J Neurosci 20: 9135–9144.

Saladin, K.S. 2009. Anatomy and Physiology: The Unity of Form and Function. McGraw Hill.

Sanchez, V., S.D. Feinstein, N. Lunardi, P.M. Joksovic, A. Boscolo, S.M. Todorovic, and V. Jevtovic-Todorovic. 2011. General anesthesia causes long-term impairment of mitochondrial morphogenesis and synaptic transmission in developing rat brain. Anesthesiology 115: 992–1002.

Sayeed, I., S. Parvez, K. Winkler-Stuck, G. Seitz, I. Trieu, C.W. Wallesch, P. Schonfeld, and D. Siemen. 2006. Patch clamp reveals powerful blockade of the mitochondrial permeability transition pore by the D2-receptor agonist pramipexole. FASEB J 20: 556–558.

Schweizer, F.E., and T.A. Ryan. 2006. The synaptic vesicle: cycle of exocytosis and endocytosis. Curr Opin Neurobiol 16: 298–304.

Shepherd, G.M., and K.M. Harris. 1998. Three-dimensional structure and composition of CA3-->CA1 axons in rat hippocampal slices: implications for presynaptic connectivity and compartmentalization. J Neurosci 18: 8300–8310.

Shu, Y., Z. Zhou, Y. Wan, R.D. Sanders, M. Li, C.K. Pac-Soo, M. Maze, and D. Ma. 2012. Nociceptive stimuli enhance anesthetic-induced neuroapoptosis in the rat developing brain. Neurobiol Dis 45: 743–750.

Slikker, W., Jr., X. Zou, C.E. Hotchkiss, R.L. Divine, N. Sadovova, N.C. Twaddle, D.R. Doerge, A.C. Scallet, T.A. Patterson, J.P. Hanig, M.G. Paule, and C. Wang. 2007. Ketamine-induced neuronal cell death in the perinatal rhesus monkey. Toxicol Sci 98: 145–158.

Smirnova, E., L. Griparic, D.L. Shurland, and A.M. van der Bliek. 2001. Dynamin-related protein Drp1 is required for mitochondrial division in mammalian cells. Mol Biol Cell 12: 2245–2256.

Sprung, J., R.P. Flick, S.K. Katusic, R.C. Colligan, W.J. Barbaresi, K. Bojanic, T.L. Welch, M.D. Olson, A.C. Hanson, D.R. Schroeder, R.T. Wilder, and D.O. Warner. 2012. Attention-deficit/hyperactivity disorder after early exposure to procedures requiring general anesthesia. Mayo Clin Proc 87: 120–129.

Standley, K., A.B. Soule, S.A. Copans, and R.P. Klein. 1978. Multidimensional sources of infant temperament. Genet Psychol Monogr 98: 203–231.

Sun, L.S., G. Li, C. Dimaggio, M. Byrne, V. Rauh, J. Brooks-Gunn, A. Kakavouli, A. Wood, and N. Coinvestigators of the Pediatric Anesthesia Neurodevelopment Assessment Research. 2008. Anesthesia and neurodevelopment in children: time for an answer? Anesthesiology 109: 757–761.

Takahashi, T. 2000. Synaptic modulation mediated by G-Protein-Coupled presynaptic receptors. pp. 147–153. *In*: Kenji Kuba, Haruhiro Higashida, David A. Brown, and Tohru Yoshioka (eds.). Slow Synaptic Responses and Modulation. Springer, Japan.

Trimmer, P.A., R.H. Swerdlow, J.K. Parks, P. Keeney, J.P. Bennett, Jr., S.W. Miller, R.E. Davis, and W.D. Parker, Jr. 2000. Abnormal mitochondrial morphology in sporadic Parkinson's and Alzheimer's disease cybrid cell lines. Exp Neurol 162: 37–50.

Trushina, E., R.B. Dyer, J.D. Badger, 2nd, D. Ure, L. Eide, D.D. Tran, B.T. Vrieze, V. Legendre-Guillemin, P.S. McPherson, B.S. Mandavilli, B. Van Houten, S. Zeitlin, M. McNiven, R. Aebersold, M. Hayden, J.E. Parisi, E. Seeberg, I. Dragatsis, K. Doyle, A. Bender, C. Chacko, and C.T. McMurray. 2004. Mutant huntingtin impairs axonal trafficking in mammalian neurons *in vivo* and *in vitro*. Mol Cell Biol 24: 8195–8209.

Wang, X., B. Su, H.G. Lee, X. Li, G. Perry, M.A. Smith, and X. Zhu. 2009. Impaired balance of mitochondrial fission and fusion in Alzheimer's disease. J Neurosci 29: 9090–9103.

Wei, H., G. Liang, and H. Yang. 2007. Isoflurane preconditioning inhibited isoflurane-induced neurotoxicity. Neurosci Lett 425: 59–62.

Wilder, R.T., R.P. Flick, J. Sprung, S.K. Katusic, W.J. Barbaresi, C. Mickelson, S.J. Gleich, D.R. Schroeder, A.L. Weaver, and D.O. Warner. 2009. Early exposure to anesthesia and learning disabilities in a population-based birth cohort. Anesthesiology 110: 796–804.

Yaffe, M.P. 1999. The machinery of mitochondrial inheritance and behavior. Science 283: 1493–1497.

Yon, J.H., J. Daniel-Johnson, L.B. Carter, and V. Jevtovic-Todorovic. 2005. Anesthesia induces neuronal cell death in the developing rat brain via the intrinsic and extrinsic apoptotic pathways. Neuroscience 135: 815–827.

Yon, J.H., L.B. Carter, R.J. Reiter, and V. Jevtovic-Todorovic. 2006. Melatonin reduces the severity of anesthesia-induced apoptotic neurodegeneration in the developing rat brain. Neurobiol Dis 21: 522–530.

Young, C., V. Jevtovic-Todorovic, Y.Q. Qin, T. Tenkova, H. Wang, J. Labruyere, and J.W. Olney. 2005. Potential of ketamine and midazolam, individually or in combination, to induce apoptotic neurodegeneration in the infant mouse brain. Br J Pharmacol 146: 189–197.

Zhang, Y., Z. Wang, G.S. Parks, and O. Civelli. 2011. Novel neuropeptides as ligands of orphan G protein-coupled receptors. Curr Pharm Des 17: 2626–2631.

Zhao, Y., G. Liang, Q. Chen, D.J. Joseph, Q. Meng, R.G. Eckenhoff, M.F. Eckenhoff, and H. Wei. 2010. Anesthetic-induced neurodegeneration mediated via inositol 1,4,5-trisphosphate receptors. J Pharmacol Exp Ther 333: 14–22.

Zorzano, A., M. Liesa, D. Sebastian, J. Segales, and M. Palacin. 2010. Mitochondrial fusion proteins: dual regulators of morphology and metabolism. Semin Cell Dev Biol 21: 566–574.

Zou, X., N. Sadovova, T.A. Patterson, R.L. Divine, C.E. Hotchkiss, S.F. Ali, J.P. Hanig, M.G. Paule, W. Slikker, Jr., and C. Wang. 2008. The effects of L-carnitine on the combination of, inhalation anesthetic-induced developmental, neuronal apoptosis in the rat frontal cortex. Neuroscience 151: 1053–1065.

15

Neuronal Ion Channels

Slobodan M. Todorovic

Introduction

Neuronal signaling can be conveyed over long distances within the central and peripheral nervous system because neurons generate and conduct action potentials that propagate away from the site of initiation. In essence, action potentials in the nervous system and other excitable cells (e.g., cardiac) are generated by the flow of different ions through voltage-gated and ligand-gated channels that are formed by specific proteins that form pores within lipid domains of neuronal membranes.

Voltage-gated ion channels

On the basis of the membrane potentials at which they activate, as well as on the basis of their specificity for certain ions, voltage-gated ion channels are subdivided into three large families such as voltage-gated sodium channels, voltage-gated potassium channels and voltage-gated calcium channels. In the resting state, potassium ions are mostly confined to the intracellular space while sodium and calcium ions are more prevalent in the extracellular space. Our knowledge of the function of various ion channels depends mainly on the use of the patch-clamp variant of voltage-clamp techniques for measuring ionic currents (Hamill *et al.*, 1981). When the patch-clamp technique is used it is possible to deliver strong depolarizing stimuli to the neurons of interest and evoke the flow of sodium, calcium and potassium ions through the neuronal membrane that will in turn induce specific waveforms that comprise action potentials

Professor of Anesthesiology, University of Colorado School of Medicine, Anschutz Medical Campus, 12801 E 17th Place, Research 1, South L18-4100, Aurora, CO 80045.
E-mail: slobodan.todorovic@ucdenver.edu

as described in classical experiments in giant squid axons (Hodgkin and Huxley, 1952). At resting state, the neuronal membrane potential is largely determined by the activity of background potassium channels. Typically, the initial event to depolarizing pulse is the activation of conductances carried by inward sodium and calcium ions whose flow tends to depolarize membrane potential by increasing the positive charge inside and reducing the positive charge outside the neuron (Katz and Miledi, 1970). This depolarization is terminated by the activation of outward voltage-gated potassium currents in the neuronal membrane which tend to hyperpolarize membrane potential by reducing the positive charge inside and increasing the positive charge outside the neuronal membrane. This will generate counterbalance to the movement of sodium and calcium ions and eventually reset the original resting membrane potential. Figure 15.1 depicts typical events during action potential in neurons and the sequence of activation of voltage-gated sodium channels which generate a fast upstroke of action potentials, voltage-gated calcium channels which generate a plateau of action potentials and underlie repetitive spike firing, and finally voltage-gated potassium channels that terminate membrane depolarization.

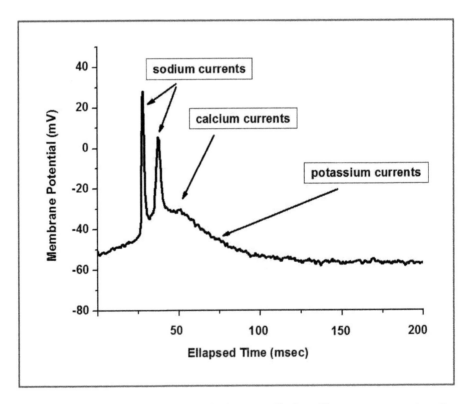

Figure 15.1. Action potential waveform in thalamic neurons. The figure illustrates a representative action potential waveform in patch-clamp recordings from nRT (reticular thalamic neurons) in acute thalamic slices. Values of membrane potentials are plotted over time as a response to brief depolarization via recording electrode. Note that different parts of action potential are carried by different ionic conductances.

Ligand-gated ion channels

In contrast to the superfamily of voltage-gated ion channels, these proteins are not sensitive to voltage but are activated upon the binding of specific ligands that are typically released during synaptic activity. In general these receptor-channel complexes are made of transmembrane heterooligomers of five subunits arranged around a central ion channel. Two major families of ligand-gated ion channels that are expressed in the central nervous system (CNS) can be activated by main inhibitory neurotransmitters such as γ-aminobutyric acid (GABA) or main excitatory neurotransmitters such as glutamate and N-Methyl-D-Aspartate (NMDA). Activation of $GABA_A$ subtypes of receptors induces the hyperpolarization of the neuronal membrane and the inhibition of neuronal activity by inducing movement of negatively charged chloride ions from outside to the inside of the membrane. In contrast, the activation of NMDA receptors induces neuronal excitation by depolarizing membrane potential due to increased influx of sodium and calcium ions through the non-selective cation channel. Thus, through the modulation of the function of $GABA_A$ and NMDA receptors in concert with voltage-gated ion channel, it is possible to regulate neuronal excitability and consequently fine tune neuronal activity.

In this chapter we will focus on the expression of voltage-gated and ligand-gated ion channels in the cortical and thalamic circuitry and their modulation with general anesthetics. We will particularly focus on the subclass of voltage-gated calcium channels that play a crucial role in neuronal oscillation. Furthermore, we will discuss long-term plasticity of the channels in the thalamus that can contribute to the neurotoxic effects of general anesthetics in the developing brain.

Neuronal ion Channels as Cellular Targets for the Acute Effects of General Anesthetics

General anesthetics have been clinically used for nearly two centuries, but the mechanisms whereby different classes of these agents achieve different clinical effects are not well understood. A complete anesthetic state involves the loss of consciousness (hypnosis) and movement (immobilization), as well as the loss of both pain sensation (analgesia) and recollection of the event (amnesia). An early theory proposed that nonspecific alteration of the lipid membrane in nerve cells accounts for the anesthetic state (Meyer, 1899; Overton, 1901). However, research advances in the last few decades suggest that general anesthetics act through specific sites on the neuronal membrane and that different ion channels mediate their effects. Potentiation of inhibitory $GABA_A$ currents occurs when most anesthetic agents (e.g., barbiturates, propofol, etomidate, halothane, and isoflurane) are administered in clinically relevant concentrations (Franks, 2008; Franks and Lieb, 1994; Rudolph and Antkowiak, 2004; Urban, 2002). However, a group of agents called dissociative anesthetics, like ketamine and nitrous oxide (N_2O, laughing gas) (Jevtovic-Todorovic *et al.*, 1998; Mennerick *et al.*, 1998), as well as xenon (Franks *et al.*, 1998), do not significantly affect $GABA_A$ currents but inhibit a major excitatory drive in the CNS via the blockade of N-methyl-D-aspartate receptors (NMDAr). Anesthetics can affect not only these ligand-gated ion channels, but also neuronal voltage-gated ion channels. An increase in K^+ leak conductance (I_{LEAK}) and the ensuing hyperpolarization of resting membrane potentials (RMP) in motor neurons by volatile anesthetics (Franks, 2008; Sirois *et al.*, 1998) could contribute to the loss of movement during general anesthesia. In addition, it has been shown that clinically relevant concentrations of ketamine can also inhibit cortical HCN1 pacemaker channels that underlie a neuronal hyperpolarization-activated cationic current (I(h)) (Chen *et al.*, 2009). It was recognized more than three decades ago that volatile anesthetics block cardiac calcium currents (Bosnjak *et al.*, 1991; Lynch *et al.*, 1981; Terrar and Victory, 1988). Furthermore, several studies (Camara *et al.*, 2001; Eckle *et al.*, 2012; Herrington and Lingle, 1992; Herrington *et al.*, 1991; McDowell *et al.*, 1996; Orestes *et al.*, 2009; Ries and Puil, 1999; Study, 1994; Takenoshita and Steinbach, 1991; Todorovic and Lingle, 1998; Todorovic *et al.*, 2000) indicate that voltage-gated calcium channels in both central and peripheral neurons are inhibited by some general anesthetics at concentrations that occur under clinical conditions.

Calcium Current Inhibition may Contribute to the Spectrum of Clinical Effects of Some General Anesthetics

Voltage-gated calcium channels, which are heteromeric complexes found in the plasma membrane of virtually all cell types, show a high level of electrophysiological and pharmacological diversity. These channels consist of a pore-forming alpha1 subunit and ancillary subunits β, γ, and α2-δ (Catterall, 2000; Khosravani and Zamponi, 2006; Miller, 1998). Figure 15.2 shows the complexity and diversity of these channels with multiple subtypes of α1 and ancillary subunits. On the basis of the membrane potential at which they activate, these channels are subdivided into high voltage-activated (HVA) and low voltage-activated (LVA) or transient T-type calcium channels (T-channels). These channels in nerve tissue have a central function in sensory, cognitive, and motor pathways, and in controlling cell excitability and neurotransmitter release. Recent studies have shown that, in general, T-channels are particularly sensitive to inhibition by clinically relevant concentrations of general anesthetics (Eckle *et al.*, 2012; Orestes *et al.*, 2009). An important issue is that even a small blockade of a particular calcium channel may produce profound physiological effects. Where this has been examined, even small changes in calcium influx into presynaptic terminals can result in profound changes in transmitter release and synaptic efficacy (Wu and Saggau, 1997; Zucker and Regehr, 2002). Indeed, presynaptic transmitter release is proportional up to the 4th power of calcium entry (Catterall and Few, 2008). Therefore, even if calcium channels are only partially inhibited by anesthetics in the clinically relevant range, this can profoundly alter neuronal signaling.

There are Multiple Subtypes of HVA Calcium Currents

Pharmacological and physiological experiments on native calcium currents (Catterall, 2000; Khosravani and Zamponi, 2006) support the existence of at least five types of HVA calcium currents (L-, N-, P-, Q-, R- currents). These channels, which are products

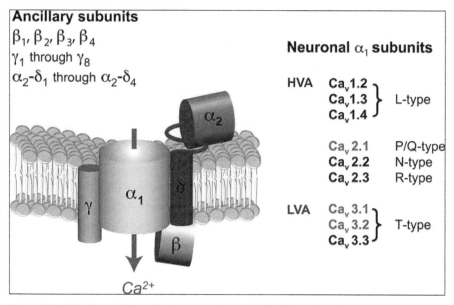

Figure 15.2. Subunit assembly and subtypes of voltage-gated calcium channels. Graphic representation of the HVA calcium channel complex consisting of the main pore-forming α_1-subunit plus ancillary, β-, γ-, and α_2-δ-subunits. T-type channels may consist of only the α_1-subunit (not separately shown). Different neuronal α_1-subunits correspond to different calcium channel isoforms identified in native neurons. (Used with permission, Khosravani and Zamponi, 2006.)

of different genes, give rise to $\alpha 1$ subunits that form the pores of the neuronal calcium channels (Figure 15.2). They are members of different families: Ca$_v$1 (former $\alpha 1$C) encoding L-type, Ca$_v$2.1 ($\alpha 1$A) encoding P/Q-type, Ca$_v$2.2 ($\alpha 1$B) encoding N-type, and Ca$_v$2.3 ($\alpha 1$E) encoding R-type HVA current (Ertel *et al.*, 2000). Most HVA current components activate over a similar voltage range and are completely blocked with divalent metals such as cadmium. Thus, subtype identification relies on the selectivity of particular pharmacological agents. L-current is blocked reversibly by 5 μM nifedipine; N-current is irreversibly blocked by ω-conotoxin GVIA; P-current is blocked essentially irreversibly (recovery time constant ~ 50 min) by ω-agatoxin (Aga) GVIA (< 100 nM) (Mintz *et al.*, 1992); Q-current is blocked by ω-conotoxin MVIIC (McDonough *et al.*, 1996). With all the other currents blocked, there remains a cadmium-sensitive inward current, the R-(resistant) current. SNX-482 is a peptide isolated from tarantula toxin that selectively and irreversibly blocks Ca$_v$2.3 currents in mammalian cell lines (Newcomb *et al.*, 1998). The usefulness of this toxin in studies of native R-type currents was suggested by some studies and it was considered for a while to be the only selective blocker of neuronal Ca$_v$2.3 currents (Breustedt *et al.*, 2003; Tottene *et al.*, 2000; Williams *et al.*, 1994). However, a recent study reported that this drug inhibits A-type potassium currents with a greater potency than Ca$_v$2.3 currents (Kimm and Bean, 2014).

Depending on the synaptic system being examined, P-, L-, N- and Q-subtypes of HVA currents may contribute to the calcium influx and support transmitter release (Catterall, 2000; Miller, 1998). Because of the paucity of selective blockers, the role of R-current remains poorly understood, but it was shown that this subtype can support excitatory transmitter release in brainstem neurons (Wu *et al.*, 1998) and has a role in synaptic plasticity (Breustedt *et al.*, 2003). Mice lacking the $Ca_V2.3$ gene show abnormal pain responses (Saegusa *et al.*, 2000) and altered sensitivities to general anesthetics (Takei *et al.*, 2003a). Furthermore, mice lacking $Ca_V2.2$ gene also show altered sensitivity to general anesthetics (Takei *et al.*, 2003b).

The diversity of HVA calcium channels in neurons is complicated by the presence of ancillary subunits that can modulate the channels' functions. Associated with the pore-forming α1 subunit is the membrane-anchored, largely extracellular α2-δ, the cytoplasmic β, and sometimes a transmembrane γ subunit; these subunits dramatically influence the properties and surface expression of these channels (Catterall, 2000; Khosravani and Zamponi, 2006).

T-channels are a Family of Distinct Voltage-gated Calcium Channels

Whereas it has been known that HVA calcium currents arise from multiple forms of calcium channels having distinct pharmacological properties, the extent to which T-current arises from multiple calcium channel subtypes became clear more recently. The cloning of α1 subunits of T-channels has revealed the existence of at least three subtypes, G ($Ca_V3.1$) (Perez-Reyes *et al.*, 1998), H ($Ca_V3.2$) (Cribbs *et al.*, 1998) and I ($Ca_V3.3$) (Lee *et al.*, 1999), which are likely to contribute to the heterogeneity of T-currents observed in native cells (Herrington and Lingle, 1992; Todorovic and Lingle, 1998). T-currents are thought to have a unique function in neuronal excitability (Huguenard, 1996; Llinas, 1988; Perez-Reyes, 2003). The major functions of T-channels in neurons include promoting calcium-dependent burst firing, regulating low-amplitude intrinsic neuronal oscillations, promoting calcium entry, and boosting synaptic signals. In cardiac pacemaker cells, T-channels contribute towards maintaining control of the firing rate. In vascular smooth muscle cells, T-currents have a role in calcium-dependent contractions and cell proliferation under pathological conditions (Ertel *et al.*, 1997). T-currents in the thalamus appear to function in seizure susceptibility and initiation (Kim *et al.*, 2001; Tsakiridou *et al.*, 1995). T-currents typically share several properties. First, while barium ions permeate HVA channels better than calcium ions, permeation by calcium and barium is similar in T-channels. Second, T-currents are activated at relatively negative potentials (i.e., –60 mV), deactivate slowly, and inactivate completely and relatively rapidly, although there are differences in kinetic behavior among T-currents (Perez-Reyes, 2003). Although T-currents are relatively easy to study in isolation from other calcium current components, by virtue of their unique activation, deactivation, and inactivation properties, there are limited pharmacological tools for the identification and investigation of T-currents. Unlike the family of HVA calcium currents, no natural toxins or venom components have been identified that selectively block T-currents. A scorpion toxin, kurtoxin, blocks potently recombinant T-type but not HVA currents; however, the usefulness of kurtoxin is limited because it also blocks sodium currents (Chuang *et al.*, 1998) and multiple subtypes of native HVA calcium currents (Sidach and Mintz, 2002). Of the compounds with known effects on T-currents, including ethosuximide (Coulter *et al.*, 1989), amiloride (Tang *et al.*, 1988), and valproicacid (Kelly *et al.*, 1990), most are reported to block with IC_{50}s (concentrations that produce 50% inhibition of current amplitude) in excess of 100 μM. At these concentrations, effects on other ion channels also occur, reducing the usefulness of such agents as specific probes of T-current function. Among the available pharmacological agents, nickel (Perez-Reyes, 2003) and N_2O (Todorovic *et al.*, 2001a) are probably the most useful for the definition of T-currents over HVA currents since both block T-currents in several-fold higher concentrations. Furthermore, both nickel and N_2O are preferential blockers of the $Ca_V3.2$ isoform of T-channels, making these agents valuable tools for use in studies of native calcium channels. One exception to this rule is that nickel also blocks potently native and recombinant R-type calcium channels (Zamponi *et al.*, 1996).

The different sensitivity of distinct T-current variants may underlie the potential clinical effects of T-current blockers. In this regard, the action of some anticonvulsants is thought to result from partial inhibition of T-current in specific CNS neurons in the thalamus (Coulter *et al.*, 1989), but this finding has been challenged by a study reporting the insensitivity of thalamic relay neurons to ethosuximide (Leresche *et al.*, 1998). Some newer agents promise to provide more selective and potent blockade of T-channels. A neuroactive steroid with a 5α configuration at the steroid A,B ring fusion [(+)-ECN] [(3β,5α,17β)-17-hydroxyestrane-3-carbonitrile], is a potent voltage-dependent partial blocker of T-channels in rat sensory neurons (IC_{50} of 300 nM). This channel blocker does not affect voltage-gated Na^+, K^+, or HVA calcium channels or glutamate and GABA-gated channels, even in concentrations several-folds higher than those that block isolated T-currents in rat sensory neurons (Todorovic *et al.*, 1998). Also, another novel steroid was described as a voltage-dependent blocker of T-channels in rat sensory neurons, which has a 5β configuration at the steroid A,B ring fusion (3β-OH). This steroid completely blocks T-currents with an IC_{50} of 0.8 μM, but, at 10–20-fold higher concentrations, has little effect on HVA calcium channels and voltage-gated Na^+ and K^+ channels (Todorovic *et al.*, 2004b). Furthermore, it was reported that 3β-OH also blocks native T-currents and underlying burst firing in thalamic neurons without affecting background potassium conductance (Joksovic *et al.*, 2007). Therefore, neuroactive steroids that selectively block neuronal T-channels may be used as new tools in studies of the functions of these channels in native systems. The recent identification of novel T-channel antagonists such as 4-aminomethyl-4-fluoropiperdine is also very promising. This agent completely blocks recombinant T-currents with high potency (IC_{50} of about 100 nM) and has a minimal effect on recombinant HVA calcium currents at 10 μM (Shipe *et al.*, 2008). Furthermore, the selectivity and potency of this blocker in native thalamic and peripheral sensory neurons has been validated (Choe *et al.*, 2011; Dreyfus *et al.*, 2010).

T-channels are new considerations in anesthesia research, but the part they play in pain transmission and the neuronal sleep pathway make them important targets for further exploration. T-channels were initially discovered in peripheral sensory neurons (Carbone and Lux, 1984) but their role in pain transmission was not discovered until recently. Pharmacological (Kim *et al.*, 2003; Nelson *et al.*, 2005; Nelson *et al.*, 2007a; Nelson *et al.*, 2007b; Pathirathna *et al.*, 2005a; Pathirathna *et al.*, 2005b; Todorovic *et al.*, 2001b; Todorovic *et al.*, 2004a; Todorovic *et al.*, 2004b), gene knockout (Choi *et al.*, 2007), and *in vivo* knock-down studies (Bourinet *et al.*, 2005) have firmly established a prominent role for $Ca_V3.2$ T-channels in amplifying nociceptive signals in the periphery and identified them as possible contributors to the development of central sensitization in the dorsal horn of the spinal cord (Ikeda *et al.*, 2003; Ikeda *et al.*, 2006). It is therefore not surprising that anesthetic agents that inhibit T-channels, such as isoflurane, also often have notable analgesic qualities. In agreement with this, a decrease in MAC (minimal alveolar concentration) requirements for isoflurane in $Ca_V3.2$ knock-out mice likely results from the prominent function of these channels in pain signaling (Orestes *et al.*, 2009). In contrast, studies of $Ca_V3.1$ knock-out mice found no changes in MAC for isoflurane or other volatile anesthetics (Petrenko *et al.*, 2007). Although $Ca_V3.1$ and $Ca_V3.2$ isoforms are similarly inhibited by isoflurane (Orestes *et al.*, 2009), and their mRNA transcripts are expressed in peripheral sensory neurons and the dorsal horn of the spinal cord (Talley *et al.*, 1999), it appears that $Ca_V3.2$ is a more important contributor to anesthetic-induced analgesia. These results strongly suggest that any future pharmaceutical development of drugs that selectively inhibit $Ca_V3.2$ T-channels would be a useful adjuvant for general anesthesia since they would reduce MAC.

Ion Channels in the Thalamus as Major Targets of Hypnotic effects of General Anesthetics

There is growing recognition that thalamic nuclei are important for awareness and cognitive functions (Alkire *et al.*, 2000; Kinney *et al.*, 1994; Llinas *et al.*, 2005; McCormick and Bal, 1997). The rhythmicity of this complex circuitry depends to a great extent on the ability of thalamic cells to burst in oscillatory patterns. The bursting firing mode is largely supported by abundant T-channels in the various thalamic nuclei (Destexhe *et al.*, 1996; Destexhe *et al.*, 1998). A key element in rhythm generation within the thalamus is the GABAergic nucleus reticularis thalami (nRT), which is not only reciprocally connected to the thalamocortical (TC) relay neurons of dorsal thalamic nuclei, but also receives collateral excitatory connections from corticothalamic and thalamocortical fibers (Cox *et al.*, 1996; Jones, 1985). In addition, nRT neurons create numerous axonal and dendro-dendritic synapses that can inhibit one another (Sanchez-Vives *et al.*, 1997; Shu and McCormick, 2002) and may receive GABAergic inputs from other forebrain structures as well (Asanuma, 1994; Jourdain *et al.*, 1989; Pare *et al.*, 1990). Earlier *in vivo* extracellular recordings have shown that volatile general anesthetics modulate synaptic transmission and depress the excitability of thalamic neurons, which in turn causes blockade of thalamo-cortical information transfer (Angel, 1991; Detsch *et al.*, 2002; Vahle-Hinz *et al.*, 2007a; Vahle-Hinz *et al.*, 2007b). At least part of this inhibitory effect of anesthetics on the thalamic sensory processing can be attributed to the direct inhibition of thalamic T-channels (Eckle *et al.*, 2012; Joksovic *et al.*, 2005a; Joksovic *et al.*, 2005b). Furthermore, functional imaging studies in humans and other animals have led to the theory that the direct and indirect depression of thalamo-cortical neurons provides a convergent point for neural pathways of anesthetic action leading to a sleep-like state (Alkire *et al.*, 2000).

Mutually interconnected cortical, nRT, and TC relay neurons exhibit phasic behaviors such as tonic and burst firing that represent different functional modes (Contreras and

Steriade, 1996; McCormick and Bal, 1997). During tonic firing, which predominates during conscious states, there is a faithful transfer of sensory information to cortical neurons with characteristic low-amplitude, high-frequency electroencephalographic (EEG) patterns. In contrast, during slow oscillations that occur with a T-channel-dependent burst firing pattern of these neurons, there is impairment of sensory transfer and gradual transition to sleep or unconscious states. This is manifested on EEG recordings by slow-frequency, high-amplitude wave patterns such as sleep spindles and δ-waves. Figure 15.3 illustrates the current theory that during natural sleep and early stages of anesthesia, burst firing in the thalamo-cortical loop indicates impairment of sensory transfer. In agreement with this, $Ca_V3.1$ T-channel knockout mice not only show slower anesthetic induction (Petrenko *et al.*, 2007), but also have an abnormal natural sleep phenotype that likely results from the lack of T-channels and underlying burst firing in TC neurons (Anderson *et al.*, 2005; Lee *et al.*, 2004). Furthermore, it was shown that, similarly, $Ca_V3.2$ knockout mice display delayed anesthetic induction with isoflurane (Orestes *et al.*, 2009). This is likely due to the fact that at sub-anesthetic concentrations, the activation of $GABA_A$ receptors and resting potassium I_{LEAK} sufficiently hyperpolarizes TC neurons, which in turn deinactivate T-channels, thus

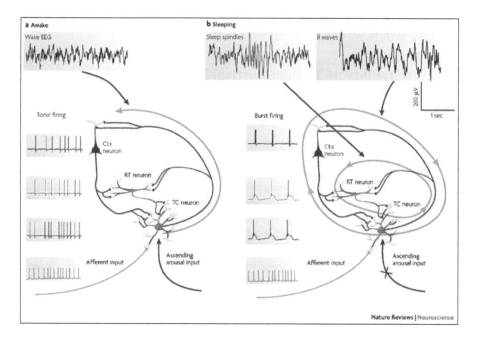

Figure 15.3. Thalamic oscillation. (a) During wakefulness ascending excitatory input from brainstem nuclei to thalamo-cortical (TC) neurons (red) provides a depolarizing drive that causes TC and reticular (nRT) neurons (blue) to exhibit single-spike tonic firing and allows faithful transfer of sensory information from the periphery up to cortical (Ctx) neurons (black). (b) During deep or rapid-eye-movement (NREM) sleep and/or early stages of anesthesia, TC and RT neurons switch to a burst-firing mode that they adopt by default in the absence of external depolarizing input. Because of extensive connectivity of neurons in this loop, synchrony of burst-firing neurons induces characteristic slow δ waves (7–14 Hz) and sleep spindles (1–4 Hz) on EEG. This pattern is associated with an impairment of sensory information transfer. (Used with permission, Franks, 2008.)

promoting a burst firing pattern. However, with the progressive hyperpolarization of the neuronal membrane and the consequent shunting of action potentials associated with deeper anesthetic states, this could contribute to a burst-suppression pattern and eventually lead to a completely flat line on EEG, which generally correlates with a complete block of sensory information at the level of the thalamic sensory gate (Akrawi *et al.*, 1996; Franks, 2008; Steriade *et al.*, 1994; Steriade *et al.*, 1990). This effect would be consistent with a strong direct inhibition of T-channels and more profound hyperpolarization of the neuronal membrane, both of which would lead to eventual cessation of thalamo-cortical rhythmicity.

Furthermore, a prominent role of isoflurane in inhibition of presynaptic $Ca_V2.3$ channels which support GABAergic transmission in nRT cells was reported (Joksovic *et al.*, 2009). It was hypothesized that, given that nRT neurons are uniformly inhibitory, it is likely that decreasing feedback inhibition to nRT cells with isoflurane with a concomitant enhancement of inhibitory transmission in TC cells would initially favor a slow oscillatory mode of thalamic neurons, which could subsequently contribute to a suppression pattern on EEG. In support of this view, decrease in the duration of burst suppression episodes in EEG under *in vivo* isoflurane anesthesia was reported in $Ca_V2.3$ knockout mice (Joksovic *et al.*, 2009). Thus, it is important to understand how general anesthetics affect different classes of voltage-gated calcium channels that contribute to different functional modes of thalamic neurons.

Hyperexcitability of Thalamocortical Networks after Exposure to General Anesthesia during Brain Development

As mentioned above, most currently used general anesthetic agents either have NMDA current-blocking or/and $GABA_A$ current-enhancing properties, which are thought to be essential for their sedative/hypnotic properties (Franks, 2008; Rudolph and Antkowiak, 2004). Unfortunately, it has been well-documented that increased activation of $GABA_A$ receptors and/or blockade of NMDA receptors can trigger widespread neurodegeneration in developing rodent and non-human primate brains including the thalamus (Brambrink *et al.*, 2012; Brambrink *et al.*, 2010; Jevtovic-Todorovic *et al.*, 2003; Rizzi *et al.*, 2008; Zou *et al.*, 2011; Zou *et al.*, 2009). Although human studies addressing the issue of safety of clinical anesthesia in the developing brain still are not conclusive, at least some concerns have been raised based on the findings that children exposed to general anesthesia at an early age show increased incidence in learning disability later in life (Wilder *et al.*, 2009).

In addition to well-documented evidence that general anesthetics cause neurodegenerative changes in the developing brain, recent evidence supports the idea that early exposure to general anesthetics may alter the function of ion channels that control neuronal excitability as well. For example, it has been shown that exposure to general anesthetics at postnatal day 7 (P7) produces persisting alterations in brain structure and function, including diminished long-term potentiation (Jevtovic-Todorovic *et al.*, 2003) and decreased hippocampal inhibitory transmission (Sanchez *et al.*, 2011). In another more recent study the authors provided the first description of synaptic and intrinsic ion channel plasticity in the nRT after single exposure to clinically relevant anesthetics during brain development (DiGruccio *et al.*, 2015). The authors exposed for 6 hours P7 rat pups to anesthesia with a clinically relevant cocktail consisting of 0.75% isoflurane, 75% N_2O and 9 mg/kg midazolam since these agents often are used in combination in clinical anesthesia in order to provide hypnosis, analgesia and amnesia, respectively. The single 6-hour long treatment with anesthetic cocktail in P7 rat pups induced lasting alterations (2–3 weeks) in several aspects of neuronal signaling such as: (1) decrease in synaptic inhibition mediated by $GABA_A$ receptors; (2) increase in glutamate-mediated excitatory synaptic transmission as evidenced by increased strength in both evoked and action potential-independent synaptic currents; (3) up-regulation of T-type calcium currents with associated increase in intrinsic excitability manifested by an increased ability to fire tonic action potentials and bursts of action potentials; and (4) increased intensity of pharmacologically induced spike-and-wave discharges (SWDs) in intact thalamocortical circuits as measured by EEG recordings *in vivo*. The SWD discharges are important since they are a hallmark of absence seizures. Surprisingly, similar alterations in inhibitory synaptic currents in the thalamus are described in a rat model of absence seizures (Bessaih *et al.*, 2006). One possible explanation for the changes is that they represent adaptive responses to neuronal loss induced by general anesthetics. It is reasonable to hypothesize that nRT neurons increase their activity to compensate for neuronal loss induced by general anesthetics in order to maintain balance in the circuitry. Unfortunately, this homeostatic plasticity may become maladaptive and could contribute to chronic hyperexcitability of the affected networks. Further comprehensive preclinical and clinical studies are needed to establish a possible link between anesthetic-induced plasticity of synaptic and intrinsic ion channels in the nRT neurons and abnormal oscillations in thalamocortical networks that underlay many neurological disorders such as tinnitus, absence seizures, sleep disturbances and cognitive impairment collectively termed "thalamocortical dysrhythmias" (Llinas *et al.*, 2005). These and future studies may provide a rationale for new strategies to prevent abnormalities and/or to normalize function of the nRT, the main inhibitory structure in the thalamus, and thus to provide safer anesthesia.

Summary/Conclusions

Recent work indicates that studies of ion channels expressed in the thalamus, as well as the neuronal circuits that regulate its activity, can advance our knowledge of how anesthetics cause loss of consciousness. Specifically, voltage-gated calcium channels have the important function of controlling cellular excitability and transmitter release in CNS neurons. General anesthetics are particularly volatile agents that depress currents arising from these channels in the thalamus at clinically relevant concentrations and this effect may contribute to the useful hypnotic, amnesic, anticonvulsant and analgesic effects of general anesthetics. Furthermore, voltage-gated calcium channels expressed in nRT and TC neurons in concert with inhibitory $GABA_A$ channels and excitatory ion channels activated by glutamate have a central role in the generation of rhythmic thalamocortical oscillations, which in turn control consciousness and cognition (Contreras and Steriade, 1996; Kinney *et al.*, 1994; Krnjevic and Puil, 1997; McAlonan and Brown, 2002; Sherman, 2005; Steriade *et al.*, 1990), as well as abnormal excitability that can contribute to seizures, neurogenic pain, and other neuropsychiatric disorders termed "thalamo-cortical dysrhythmias" (Llinas *et al.*, 2005). Hence, knowledge of the functions of different classes of ligand-gated and voltage-gated ion channels in thalamocortical circuitry and their lasting alterations with general anesthetics during brain development can be used to reduce the potentially serious side effects of clinical anesthesia.

Acknowledgments

This chapter was supported in part by NIH grant GM-102525 (SMT) and funds from the Department of Anesthesiology at The University of Virginia, Charlottesville, VA. We thank Dr. Michael DiGruccio for providing data for Figure 15.1.

References

Akrawi, W.P., J.C. Drummond, C.J. Kalkman, and P.M. Patel. 1996. A comparison of the electrophysiologic characteristics of EEG burst-suppression as produced by isoflurane, thiopental, etomidate, and propofol. J Neurosurg Anesthesiol 8: 40–46.

Alkire, M.T., R.J. Haier, and J.H. Fallon. 2000. Toward a unified theory of narcosis: brain imaging evidence for a thalamocortical switch as the neurophysiologic basis of anesthetic-induced unconsciousness. Conscious Cogn 9: 370–386.

Anderson, M.P., T. Mochizuki, J. Xie, W. Fischler, J.P. Manger, E.M. Talley, T.E. Scammell, and S. Tonegawa. 2005. Thalamic Cav3.1 T-type Ca2+ channel plays a crucial role in stabilizing sleep. Proc Natl Acad Sci U S A 102: 1743–1748.

Angel, A. 1991. The G. L. Brown lecture. Adventures in anaesthesia. Exp Physiol 76: 1–38.

Asanuma, C. 1994. GABAergic and pallidal terminals in the thalamic reticular nucleus of squirrel monkeys. Exp Brain Res 101: 439–451.

Bessaih, T., L. Bourgeais, C.I. Badiu, D.A. Carter, T.I. Toth, D. Ruano, B. Lambolez, V. Crunelli, and N. Leresche. 2006. Nucleus-specific abnormalities of GABAergic synaptic transmission in a genetic model of absence seizures. J Neurophysiol 96: 3074–3081.

Bosnjak, Z.J., F.D. Supan, and N.J. Rusch. 1991. The effects of halothane, enflurane, and isoflurane on calcium current in isolated canine ventricular cells. Anesthesiology 74: 340–345.

Bourinet, E., A. Alloui, A. Monteil, C. Barrere, B. Couette, O. Poirot, A. Pages, J. McRory, T.P. Snutch, A. Eschalier, and J. Nargeot. 2005. Silencing of the Cav3.2 T-type calcium channel gene in sensory neurons demonstrates its major role in nociception. EMBO J 24: 315–324.

Brambrink, A.M., A.S. Evers, M.S. Avidan, N.B. Farber, D.J. Smith, X. Zhang, G.A. Dissen, C.E. Creeley, and J.W. Olney. 2010. Isoflurane-induced neuroapoptosis in the neonatal rhesus macaque brain. Anesthesiology 112: 834–841.

Brambrink, A.M., A.S. Evers, M.S. Avidan, N.B. Farber, D.J. Smith, L.D. Martin, G.A. Dissen, C.E. Creeley, and J.W. Olney. 2012. Ketamine-induced neuroapoptosis in the fetal and neonatal rhesus macaque brain. Anesthesiology 116: 372–384.

Breustedt, J., K.E. Vogt, R.J. Miller, R.A. Nicoll, and D. Schmitz. 2003. Alpha1E-containing Ca2+ channels are involved in synaptic plasticity. Proc Natl Acad Sci U S A 100: 12450–12455.

Camara, A.K., Z. Begic, W.M. Kwok, and Z.J. Bosnjak. 2001. Differential modulation of the cardiac L- and T-type calcium channel currents by isoflurane. Anesthesiology 95: 515–524.

Carbone, E., and H.D. Lux. 1984. A low voltage-activated, fully inactivating Ca channel in vertebrate sensory neurones. Nature 310: 501–502.

Catterall, W.A. 2000. Structure and regulation of voltage-gated Ca2+ channels. Annu Rev Cell Dev Biol 16: 521–555.

Catterall, W.A., and A.P. Few. 2008. Calcium channel regulation and presynaptic plasticity. Neuron 59: 882–901.

Chen, X., S. Shu, and D.A. Bayliss. 2009. HCN1 channel subunits are a molecular substrate for hypnotic actions of ketamine. J Neurosci 29: 600–609.

Choe, W., R.B. Messinger, E. Leach, V.S. Eckle, A. Obradovic, R. Salajegheh, V. Jevtovic-Todorovic, and S.M. Todorovic. 2011. TTA-P2 is a potent and selective blocker of T-type calcium channels in rat sensory neurons and a novel antinociceptive agent. Mol Pharmacol 80: 900–910.

Choi, S., H.S. Na, J. Kim, J. Lee, S. Lee, D. Kim, J. Park, C.C. Chen, K.P. Campbell, and H.S. Shin. 2007. Attenuated pain responses in mice lacking Ca(V)3.2 T-type channels. Genes Brain Behav 6: 425–431.

Chuang, R.S., H. Jaffe, L. Cribbs, E. Perez-Reyes, and K.J. Swartz. 1998. Inhibition of T-type voltage-gated calcium channels by a new scorpion toxin. Nat Neurosci 1: 668–674.

Contreras, D., and M. Steriade. 1996. Spindle oscillation in cats: the role of corticothalamic feedback in a thalamically generated rhythm. J Physiol 490 (Pt 1): 159–179.

Coulter, D.A., J.R. Huguenard, and D.A. Prince. 1989. Characterization of ethosuximide reduction of low-threshold calcium current in thalamic neurons. Ann Neurol 25: 582–593.

Cox, C.L., J.R. Huguenard, and D.A. Prince. 1996. Heterogeneous axonal arborizations of rat thalamic reticular neurons in the ventrobasal nucleus. J Comp Neurol 366: 416–430.

Cribbs, L.L., J.H. Lee, J. Yang, J. Satin, Y. Zhang, A. Daud, J. Barclay, M.P. Williamson, M. Fox, M. Rees, and E. Perez-Reyes. 1998. Cloning and characterization of alpha1H from human heart, a member of the T-type Ca2+ channel gene family. Circ Res 83: 103–109.

Destexhe, A., D. Contreras, M. Steriade, T.J. Sejnowski, and J.R. Huguenard. 1996. *In vivo, in vitro,* and computational analysis of dendritic calcium currents in thalamic reticular neurons. J Neurosci 16: 169–185.

Destexhe, A., M. Neubig, D. Ulrich, and J. Huguenard. 1998. Dendritic low-threshold calcium currents in thalamic relay cells. J Neurosci 18: 3574–3588.

Detsch, O., E. Kochs, M. Siemers, B. Bromm, and C. Vahle-Hinz. 2002. Differential effects of isoflurane on excitatory and inhibitory synaptic inputs to thalamic neurones *in vivo*. Br J Anaesth 89: 294–300.

DiGruccio, M.R., S. Joksimovic, P.M. Joksovic, N. Lunardi, R. Salajegheh, V. Jevtovic-Todorovic, M.P. Beenhakker, H.P. Goodkin, and S.M. Todorovic. 2015. Hyperexcitability of rat thalamocortical networks after exposure to general anesthesia during brain development. J Neurosci 35: 1481–1492.

Dreyfus, F.M., A. Tscherter, A.C. Errington, J.J. Renger, H.S. Shin, V.N. Uebele, V. Crunelli, R.C. Lambert, and N. Leresche. 2010. Selective T-type calcium channel block in thalamic neurons reveals channel redundancy and physiological impact of I(T)window. J Neurosci 30: 99–109.

Eckle, V.S., M.R. Digruccio, V.N. Uebele, J.J. Renger, and S.M. Todorovic. 2012. Inhibition of T-type calcium current in rat thalamocortical neurons by isoflurane. Neuropharmacology 63: 266–273.

Ertel, E.A., K.P. Campbell, M.M. Harpold, F. Hofmann, Y. Mori, E. Perez-Reyes, A. Schwartz, T.P. Snutch, T. Tanabe, L. Birnbaumer, R.W. Tsien, and W.A. Catterall. 2000. Nomenclature of voltage-gated calcium channels. Neuron 25: 533–535.

Ertel, S.I., E.A. Ertel, and J.P. Clozel. 1997. T-type Ca2+ channels and pharmacological blockade: potential pathophysiological relevance. Cardiovasc Drugs Ther 11: 723–739.

Franks, N.P., and W.R. Lieb. 1994. Molecular and cellular mechanisms of general anaesthesia. Nature 367: 607–614.

Franks, N.P., R. Dickinson, S.L. de Sousa, A.C. Hall, and W.R. Lieb. 1998. How does xenon produce anaesthesia? Nature 396: 324.

Franks, N.P. 2008. General anaesthesia: from molecular targets to neuronal pathways of sleep and arousal. Nat Rev Neurosci 9: 370–386.

Hamill, O.P., A. Marty, E. Neher, B. Sakmann, and F.J. Sigworth. 1981. Improved patch-clamp techniques for high-resolution current recording from cells and cell-free membrane patches. Pflugers Arch 391: 85–100.

Herrington, J., R.C. Stern, A.S. Evers, and C.J. Lingle. 1991. Halothane inhibits two components of calcium current in clonal (GH3) pituitary cells. J Neurosci 11: 2226–2240.

Herrington, J., and C.J. Lingle. 1992. Kinetic and pharmacological properties of low voltage-activated Ca2+ current in rat clonal (GH3) pituitary cells. J Neurophysiol 68: 213–232.

Hodgkin, A.L., and A.F. Huxley. 1952. A quantitative description of membrane current and its application to conduction and excitation in nerve. J Physiol 117: 500–544.

Huguenard, J.R. 1996. Low-threshold calcium currents in central nervous system neurons. Annu Rev Physiol 58: 329–348.

Ikeda, H., B. Heinke, R. Ruscheweyh, and J. Sandkuhler. 2003. Synaptic plasticity in spinal lamina I projection neurons that mediate hyperalgesia. Science 299: 1237–1240.

Ikeda, H., J. Stark, H. Fischer, M. Wagner, R. Drdla, T. Jager, and J. Sandkuhler. 2006. Synaptic amplifier of inflammatory pain in the spinal dorsal horn. Science 312: 1659–1662.

Jevtovic-Todorovic, V., S.M. Todorovic, S. Mennerick, S. Powell, K. Dikranian, N. Benshoff, C.F. Zorumski, and J.W. Olney. 1998. Nitrous oxide (laughing gas) is an NMDA antagonist, neuroprotectant and neurotoxin. Nat Med 4: 460–463.

Jevtovic-Todorovic, V., R.E. Hartman, Y. Izumi, N.D. Benshoff, K. Dikranian, C.F. Zorumski, J.W. Olney, and D.F. Wozniak. 2003. Early exposure to common anesthetic agents causes widespread neurodegeneration in the developing rat brain and persistent learning deficits. J Neurosci 23: 876–882.

Joksovic, P.M., D.A. Bayliss, and S.M. Todorovic. 2005a. Different kinetic properties of two T-type Ca2+ currents of rat reticular thalamic neurones and their modulation by enflurane. J Physiol 566: 125–142.

Joksovic, P.M., B.C. Brimelow, J. Murbartian, E. Perez-Reyes, and S.M. Todorovic. 2005b. Contrasting anesthetic sensitivities of T-type Ca2+ channels of reticular thalamic neurons and recombinant Ca(v)3.3 channels. Br J Pharmacol 144: 59–70.

Joksovic, P.M., D.F. Covey, and S.M. Todorovic. 2007. Inhibition of T-type calcium current in the reticular thalamic nucleus by a novel neuroactive steroid. Ann N Y Acad Sci 1122: 83–94.

Joksovic, P.M., M. Weiergraber, W. Lee, H. Struck, T. Schneider, and S.M. Todorovic. 2009. Isoflurane-sensitive presynaptic R-type calcium channels contribute to inhibitory synaptic transmission in the rat thalamus. J Neurosci 29: 1434–1445.

Jones, E.G. 1985. The Thalamus. Plenum, New York.

Jourdain, A., K. Semba, and H.C. Fibiger. 1989. Basal forebrain and mesopontine tegmental projections to the reticular thalamic nucleus: an axonal collateralization and immunohistochemical study in the rat. Brain Res 505: 55–65.

Katz, B., and R. Miledi. 1970. Further study of the role of calcium in synaptic transmission. J Physiol 207: 789–801.

Kelly, K.M., R.A. Gross, and R.L. Macdonald. 1990. Valproic acid selectively reduces the low-threshold (T) calcium current in rat nodose neurons. Neurosci Lett 116: 233–238.

Khosravani, H., and G.W. Zamponi. 2006. Voltage-gated calcium channels and idiopathic generalized epilepsies. Physiol Rev 86: 941–966.

Kim, D., I. Song, S. Keum, T. Lee, M.J. Jeong, S.S. Kim, M.W. McEnery, and H.S. Shin. 2001. Lack of the burst firing of thalamocortical relay neurons and resistance to absence seizures in mice lacking alpha(1G) T-type Ca(2+) channels. Neuron 31: 35–45.

Kim, D., D. Park, S. Choi, S. Lee, M. Sun, C. Kim, and H.S. Shin. 2003. Thalamic control of visceral nociception mediated by T-type Ca2+ channels. Science 302: 117–119.

Kimm, T., and B.P. Bean. 2014. Inhibition of A-type potassium current by the peptide toxin SNX-482. J Neurosci 34: 9182–9189.

Kinney, H.C., J. Korein, A. Panigrahy, P. Dikkes, and R. Goode. 1994. Neuropathological findings in the brain of Karen Ann Quinlan. The role of the thalamus in the persistent vegetative state. N Engl J Med 330: 1469–1475.

Krnjevic, K., and E. Puil. 1997. Cellular mechanisms of general anesthesia. Princ Med Biol: 811–828.

Lee, J., D. Kim, and H.S. Shin. 2004. Lack of delta waves and sleep disturbances during non-rapid eye movement sleep in mice lacking alpha1G-subunit of T-type calcium channels. Proc Natl Acad Sci U S A 101: 18195–18199.

Lee, J.H., A.N. Daud, L.L. Cribbs, A.E. Lacerda, A. Pereverzev, U. Klockner, T. Schneider, and E. Perez-Reyes. 1999. Cloning and expression of a novel member of the low voltage-activated T-type calcium channel family. J Neurosci 19: 1912–1921.

Leresche, N., H.R. Parri, G. Erdemli, A. Guyon, J.P. Turner, S.R. Williams, E. Asprodini, and V. Crunelli. 1998. On the action of the anti-absence drug ethosuximide in the rat and cat thalamus. J Neurosci 18: 4842–4853.

Llinas, R., F.J. Urbano, E. Leznik, R.R. Ramirez, and H.J. van Marle. 2005. Rhythmic and dysrhythmic thalamocortical dynamics: GABA systems and the edge effect. Trends Neurosci 28: 325–333.

Llinas, R.R. 1988. The intrinsic electrophysiological properties of mammalian neurons: insights into central nervous system function. Science 242: 1654–1664.

Lynch, C., 3rd, S. Vogel, and N. Sperelakis. 1981. Halothane depression of myocardial slow action potentials. Anesthesiology 55: 360–368.

McAlonan, K., and V.J. Brown. 2002. The thalamic reticular nucleus: more than a sensory nucleus? Neuroscientist 8: 302–305.

McCormick, D.A., and T. Bal. 1997. Sleep and arousal: thalamocortical mechanisms. Annu Rev Neurosci 20: 185–215.

McDonough, S.I., K.J. Swartz, I.M. Mintz, L.M. Boland, and B.P. Bean. 1996. Inhibition of calcium channels in rat central and peripheral neurons by omega-conotoxin MVIIC. J Neurosci 16: 2612–2623.

McDowell, T.S., J.J. Pancrazio, and C. Lynch, 3rd. 1996. Volatile anesthetics reduce low-voltage-activated calcium currents in a thyroid C-cell line. Anesthesiology 85: 1167–1175.

Mennerick, S., V. Jevtovic-Todorovic, S.M. Todorovic, W. Shen, J.W. Olney, and C.F. Zorumski. 1998. Effect of nitrous oxide on excitatory and inhibitory synaptic transmission in hippocampal cultures. J Neurosci 18: 9716–9726.

Meyer, H. 1899. Zur theorie der alcoholnarkose. Arch Exp Pathol Pharmacol 42: 109–118.

Miller, R.J. 1998. Presynaptic receptors. Annu Rev Pharmacol Toxicol 38: 201–227.

Mintz, I.M., M.E. Adams, and B.P. Bean. 1992. P-type calcium channels in rat central and peripheral neurons. Neuron 9: 85–95.

Nelson, M.T., P.M. Joksovic, E. Perez-Reyes, and S.M. Todorovic. 2005. The endogenous redox agent L-cysteine induces T-type Ca2+ channel-dependent sensitization of a novel subpopulation of rat peripheral nociceptors. J Neurosci 25: 8766–8775.

Nelson, M.T., P.M. Joksovic, P. Su, H.W. Kang, A. Van Deusen, J.P. Baumgart, L.S. David, T.P. Snutch, P.Q. Barrett, J.H. Lee, C.F. Zorumski, E. Perez-Reyes, and S.M. Todorovic. 2007a. Molecular mechanisms of subtype-specific inhibition of neuronal T-type calcium channels by ascorbate. J Neurosci 27: 12577–12583.

Nelson, M.T., J. Woo, H.W. Kang, I. Vitko, P.Q. Barrett, E. Perez-Reyes, J.H. Lee, H.S. Shin, and S.M. Todorovic. 2007b. Reducing agents sensitize C-type nociceptors by relieving high-affinity zinc inhibition of T-type calcium channels. J Neurosci 27: 8250–8260.

Newcomb, R., B. Szoke, A. Palma, G. Wang, X. Chen, W. Hopkins, R. Cong, J. Miller, L. Urge, K. Tarczy-Hornoch, J.A. Loo, D.J. Dooley, L. Nadasdi, R.W. Tsien, J. Lemos, and G. Miljanich. 1998. Selective peptide antagonist of the class E calcium channel from the venom of the tarantula Hysterocrates gigas. Biochemistry 37: 15353–15362.

Orestes, P., D. Bojadzic, R.M. Chow, and S.M. Todorovic. 2009. Mechanisms and functional significance of inhibition of neuronal T-type calcium channels by isoflurane. Mol Pharmacol 75: 542–554.

Overton, S. 1901. Studien uber die narkose zugleich ein beitrag zur allgemeinen pharmacologie. Verlag von Gustav Fischer, Jena.

Pare, D., L.N. Hazrati, A. Parent, and M. Steriade. 1990. Substantia nigra pars reticulata projects to the reticular thalamic nucleus of the cat: a morphological and electrophysiological study. Brain Res 535: 139–146.

Pathirathna, S., B.C. Brimelow, M.M. Jagodic, K. Krishnan, X. Jiang, C.F. Zorumski, S. Mennerick, D.F. Covey, S.M. Todorovic, and V. Jevtovic-Todorovic. 2005a. New evidence that both T-type calcium channels and GABAA channels are responsible for the potent peripheral analgesic effects of 5alpha-reduced neuroactive steroids. Pain 114: 429–443.

Pathirathna, S., S.M. Todorovic, D.F. Covey, and V. Jevtovic-Todorovic. 2005b. 5alpha-reduced neuroactive steroids alleviate thermal and mechanical hyperalgesia in rats with neuropathic pain. Pain 117: 326–339.

Perez-Reyes, E., L.L. Cribbs, A. Daud, A.E. Lacerda, J. Barclay, M.P. Williamson, M. Fox, M. Rees, and J.H. Lee. 1998. Molecular characterization of a neuronal low-voltage-activated T-type calcium channel. Nature 391: 896–900.

Perez-Reyes, E. 2003. Molecular physiology of low-voltage-activated t-type calcium channels. Physiol Rev 83: 117–161.

Petrenko, A.B., M. Tsujita, T. Kohno, K. Sakimura, and H. Baba. 2007. Mutation of alpha1G T-type calcium channels in mice does not change anesthetic requirements for loss of the righting reflex and minimum alveolar concentration but delays the onset of anesthetic induction. Anesthesiology 106: 1177–1185.

Ries, C.R., and E. Puil. 1999. Mechanism of anesthesia revealed by shunting actions of isoflurane on thalamocortical neurons. J Neurophysiol 81: 1795–1801.

Rizzi, S., L.B. Carter, C. Ori, and V. Jevtovic-Todorovic. 2008. Clinical anesthesia causes permanent damage to the fetal guinea pig brain. Brain Pathol 18: 198–210.

Rudolph, U., and B. Antkowiak. 2004. Molecular and neuronal substrates for general anaesthetics. Nat Rev Neurosci 5: 709–720.

Saegusa, H., T. Kurihara, S. Zong, O. Minowa, A. Kazuno, W. Han, Y. Matsuda, H. Yamanaka, M. Osanai, T. Noda, and T. Tanabe. 2000. Altered pain responses in mice lacking alpha 1E subunit of the voltage-dependent Ca2+ channel. Proc Natl Acad Sci U S A 97: 6132–6137.

Sanchez, V., S.D. Feinstein, N. Lunardi, P.M. Joksovic, A. Boscolo, S.M. Todorovic, and V. Jevtovic-Todorovic. 2011. General anesthesia causes long-term impairment of mitochondrial morphogenesis and synaptic transmission in developing rat brain. Anesthesiology 115: 992–1002.

Sanchez-Vives, M.V., T. Bal, and D.A. McCormick. 1997. Inhibitory interactions between perigeniculate GABAergic neurons. J Neurosci 17: 8894–8908.

Sherman, S.M. 2005. The role of the thalamus in cortical function: not just a simple relay. Thalamus & Related Systems 3: 205–216.

Shipe, W.D., J.C. Barrow, Z.Q. Yang, C.W. Lindsley, F.V. Yang, K.A. Schlegel, Y. Shu, K.E. Rittle, M.G. Bock, G.D. Hartman, C. Tang, J.E. Ballard, Y. Kuo, E.D. Adarayan, T. Prueksaritanont, M.M. Zrada, V.N. Uebele, C.E. Nuss, T.M. Connolly, S.M. Doran, S.V. Fox, R.L. Kraus, M.J. Marino, V.K. Graufelds, H.M. Vargas, P.B. Bunting, M. Hasbun-Manning, R.M. Evans, K.S. Koblan, and J.J. Renger. 2008. Design, synthesis, and evaluation of a novel 4-aminomethyl-4-fluoropiperidine as a T-type Ca2+ channel antagonist. J Med Chem 51: 3692–3695.

Shu, Y., and D.A. McCormick. 2002. Inhibitory interactions between ferret thalamic reticular neurons. J Neurophysiol 87: 2571–2576.

Sidach, S.S., and I.M. Mintz. 2002. Kurtoxin, a gating modifier of neuronal high- and low-threshold ca channels. J Neurosci 22: 2023–2034.

Sirois, J.E., J.J. Pancrazio, C. Lynch, 3rd, and D.A. Bayliss. 1998. Multiple ionic mechanisms mediate inhibition of rat motoneurones by inhalation anaesthetics. J Physiol 512 (Pt 3): 851–862.

Steriade, M., E.G. Jones, and R.R. Llinas. 1990. Thalamic Oscillations and Signaling. Wiley, New York.

Steriade, M., F. Amzica, and D. Contreras. 1994. Cortical and thalamic cellular correlates of electroencephalographic burst-suppression. Electroencephalogr Clin Neurophysiol 90: 1–16.

Study, R.E. 1994. Isoflurane inhibits multiple voltage-gated calcium currents in hippocampal pyramidal neurons. Anesthesiology 81: 104–116.

Takei, T., H. Saegusa, S. Zong, T. Murakoshi, K. Makita, and T. Tanabe. 2003a. Anesthetic sensitivities to propofol and halothane in mice lacking the R-type (Cav2.3) Ca2+ channel. Anesth Analg 97: 96–103, table of contents.

Takei, T., H. Saegusa, S. Zong, T. Murakoshi, K. Makita, and T. Tanabe. 2003b. Increased sensitivity to halothane but decreased sensitivity to propofol in mice lacking the N-type Ca2+ channel. Neurosci Lett 350: 41–45.

Takenoshita, M., and J.H. Steinbach. 1991. Halothane blocks low-voltage-activated calcium current in rat sensory neurons. J Neurosci 11: 1404–1412.

Talley, E.M., L.L. Cribbs, J.H. Lee, A. Daud, E. Perez-Reyes, and D.A. Bayliss. 1999. Differential distribution of three members of a gene family encoding low voltage-activated (T-type) calcium channels. J Neurosci 19: 1895–1911.

Tang, C.M., F. Presser, and M. Morad. 1988. Amiloride selectively blocks the low threshold (T) calcium channel. Science 240: 213–215.

Terrar, D.A., and J.G. Victory. 1988. Effects of halothane on membrane currents associated with contraction in single myocytes isolated from guinea-pig ventricle. Br J Pharmacol 94: 500–508.

Todorovic, S.M., and C.J. Lingle. 1998. Pharmacological properties of T-type Ca2+ current in adult rat sensory neurons: effects of anticonvulsant and anesthetic agents. J Neurophysiol 79: 240–252.

Todorovic, S.M., M. Prakriya, Y.M. Nakashima, K.R. Nilsson, M. Han, C.F. Zorumski, D.F. Covey, and C.J. Lingle. 1998. Enantioselective blockade of T-type Ca2+ current in adult rat sensory neurons by a steroid that lacks gamma-aminobutyric acid-modulatory activity. Mol Pharmacol 54: 918–927.

Todorovic, S.M., E. Perez-Reyes, and C.J. Lingle. 2000. Anticonvulsants but not general anesthetics have differential blocking effects on different T-type current variants. Mol Pharmacol 58: 98–108.

Todorovic, S.M., V. Jevtovic-Todorovic, S. Mennerick, E. Perez-Reyes, and C.F. Zorumski. 2001a. Ca(v)3.2 channel is a molecular substrate for inhibition of T-type calcium currents in rat sensory neurons by nitrous oxide. Mol Pharmacol 60: 603–610.

Todorovic, S.M., V. Jevtovic-Todorovic, A. Meyenburg, S. Mennerick, E. Perez-Reyes, C. Romano, J.W. Olney, and C.F. Zorumski. 2001b. Redox modulation of T-type calcium channels in rat peripheral nociceptors. Neuron 31: 75–85.

Todorovic, S.M., A. Meyenburg, and V. Jevtovic-Todorovic. 2004a. Redox modulation of peripheral T-type Ca2+ channels *in vivo*: alteration of nerve injury-induced thermal hyperalgesia. Pain 109: 328–339.

Todorovic, S.M., S. Pathirathna, B.C. Brimelow, M.M. Jagodic, S.H. Ko, X. Jiang, K.R. Nilsson, C.F. Zorumski, D.F. Covey, and V. Jevtovic-Todorovic. 2004b. 5beta-reduced neuroactive steroids are novel voltage-dependent blockers of T-type Ca2+ channels in rat sensory neurons *in vitro* and potent peripheral analgesics *in vivo*. Mol Pharmacol 66: 1223–1235.

Tottene, A., S. Volsen, and D. Pietrobon. 2000. alpha(1E) subunits form the pore of three cerebellar R-type calcium channels with different pharmacological and permeation properties. J Neurosci 20: 171–178.

Tsakiridou, E., L. Bertollini, M. de Curtis, G. Avanzini, and H.C. Pape. 1995. Selective increase in T-type calcium conductance of reticular thalamic neurons in a rat model of absence epilepsy. J Neurosci 15: 3110–3117.

Urban, B.W. 2002. Current assessment of targets and theories of anaesthesia. Br J Anaesth 89: 167–183.

Vahle-Hinz, C., O. Detsch, C. Hackner, and E. Kochs. 2007a. Corresponding minimum alveolar concentrations of isoflurane and isoflurane/nitrous oxide have divergent effects on thalamic nociceptive signalling. Br J Anaesth 98: 228–235.

Vahle-Hinz, C., O. Detsch, M. Siemers, and E. Kochs. 2007b. Contributions of GABAergic and glutamatergic mechanisms to isoflurane-induced suppression of thalamic somatosensory information transfer. Exp Brain Res 176: 159–172.

Wilder, R.T., R.P. Flick, J. Sprung, S.K. Katusic, W.J. Barbaresi, C. Mickelson, S.J. Gleich, D.R. Schroeder, A.L. Weaver, and D.O. Warner. 2009. Early exposure to anesthesia and learning disabilities in a population-based birth cohort. Anesthesiology 110: 796–804.

Williams, M.E., L.M. Marubio, C.R. Deal, M. Hans, P.F. Brust, L.H. Philipson, R.J. Miller, E.C. Johnson, M.M. Harpold, and S.B. Ellis. 1994. Structure and functional characterization of neuronal alpha 1E calcium channel subtypes. J Biol Chem 269: 22347–22357.

Wu, L.G., and P. Saggau. 1997. Presynaptic inhibition of elicited neurotransmitter release. Trends Neurosci 20: 204–212.

Wu, L.G., J.G. Borst, and B. Sakmann. 1998. R-type Ca2+ currents evoke transmitter release at a rat central synapse. Proc Natl Acad Sci U S A 95: 4720–4725.

Zamponi, G.W., E. Bourinet, and T.P. Snutch. 1996. Nickel block of a family of neuronal calcium channels: subtype- and subunit-dependent action at multiple sites. J Membr Biol 151: 77–90.

Zou, X., T.A. Patterson, R.L. Divine, N. Sadovova, X. Zhang, J.P. Hanig, M.G. Paule, W. Slikker, Jr., and C. Wang. 2009. Prolonged exposure to ketamine increases neurodegeneration in the developing monkey brain. Int J Dev Neurosci 27: 727–731.

Zou, X., F. Liu, X. Zhang, T.A. Patterson, R. Callicott, S. Liu, J.P. Hanig, M.G. Paule, W. Slikker, Jr., and C. Wang. 2011. Inhalation anesthetic-induced neuronal damage in the developing rhesus monkey. Neurotoxicol Teratol 33: 592–597.

Zucker, R.S., and W.G. Regehr. 2002. Short-term synaptic plasticity. Annu Rev Physiol 64: 355–405.

16

Brain Injury-Neuroprotection

Koichi Yuki,[1] *Mary Ellen McCann*[2] and *Sulpicio G. Soriano*[3],*

Introduction

Brain protection during the perioperative period has historically been the focus of research in anesthesiology. Maintaining a balance between the metabolic demands of cellular homeostasis with the supply of substrates is essential in preventing neurological injury (Fukuda and Warner, 2007). A clear distinction between neuroprotection and neuroresuscitation should be established in the context of a patient undergoing surgery. This review will focus on neuroprotection, which implies optimizing physiological conditions to prevent neurological injury.

On the other hand, neuroresuscitation refers to limiting the progression of and enhancing recovery from a primary neurological injury. Traumatic brain injury, cardiac arrest, perinatal asphyxia and acute stroke are common conditions that have published evidence-based guidelines on neuroresuscitation that are beyond the scope of this review (Brain Trauma *et al.*, 2007; Jauch *et al.*, 2013; Kochanek *et al.*, 2012). Given its clinical relevance, the issue of the potential neurotoxic effects of anesthetic and sedative drugs will also be addressed.

[1] Assistant Professor of Anaesthesia, Department of Anesthesiology, Perioperative and Pain Medicine, Boston Children's Hospital, 300 Longwood Avenue, Boston, Massachusetts, USA 02115.
E-mail: koichi.yuki@childrens.harvard.edu
[2] Associate Professor of Anaesthesia, Department of Anesthesiology, Perioperative and Pain Medicine, Boston Children's Hospital, 300 Longwood Avenue, Boston, Massachusetts, USA 02115.
E-mail: mary.mccann@childrens.harvard.edu
[3] Professor of Anaesthesia, BCH Endowed Chair in Pediatric Neuroanesthesia, Department of Anesthesiology, Perioperative and Pain Medicine, Boston Children's Hospital, 300 Longwood Avenue, Boston, Massachusetts, USA 02115.
E-mail: sulpicio.soriano@childrens.harvard.edu
* Corresponding author

Patients undergoing surgery and interventional procedures typically receive general anesthetic drugs to render them insensate and amnestic. However, these procedures place patients at risk for central nervous system (CNS) injury (Mashour *et al.*, 2015). Cerebral ischemia, hypoxia and procedure-induced trauma can occur during neurological, cardiac and vascular surgery. Despite these known catastrophic possibilities, evolving neurological injuries are extremely difficult to be detected under general anesthesia. Therefore, it is essential that cellular homeostasis is maintained despite the lack of feedback from an anesthetized patient. This chapter will review the etiology of brain injury in surgical patients receiving general anesthesia. Compelling preclinical evidence demonstrates the neuroprotective properties of anesthetic drugs and induced hypothermia. The neurodepressant effects of anesthetic drugs and hypothermia make them reasonable candidates for reducing the catabolic processes leading to and during an evolving CNS injury. However it should be noted that to date, clinical trials assessing the neuroprotective impact of anesthetic drugs and hypothermia on surgical patients have yet to demonstrate clinical efficacy (Warner *et al.*, 2014).

Etiology

Neuronal metabolism

The CNS has the highest requirement for oxygen and glucose of all the organ systems (Zauner *et al.*, 2002). Neuronal function is fueled by adenosine triphosphate (ATP) produced by oxidative metabolism of glucose. The brain has high metabolic demand (7 mg glucose/100 g brain tissue/min in infants and 5 mg glucose/100 g brain tissue/min in adults) and requires a continual supply of glucose, since the brain has extremely limited glycogen storage. Likewise, cerebral metabolic rate of oxygen ($CMRO_2$) of the pediatric and adult brain is 5.2 and 3.2 ml/100 g brain tissue/min, respectively (Udomphorn *et al.*, 2008). It is this high demand and limited storage for the substrates that make the cellular components of the CNS extremely vulnerable to hypoxic and ischemic events. Of note, the majority of glucose is converted into ATP, but approximately 10% of glucose goes through aerobic glycolysis to be converted to lactate instead, despite the presence of oxygen (Vaishnavi *et al.*, 2010). How the brain differentially uses these two pathways for glucose metabolism is an interesting issue to understand from the standpoint of brain protection, but this may need further investigation. The basal metabolic rate of a neuron is approximately 40% of total cellular energy expenditure and fuels the electrochemical gradient across the cell membrane, intracellular activity and maintenance of cellular structure. The transmembrane gradient of the major physiologic cations sodium, potassium and calcium, controls the generation of neuronal action potentials. Functional neuronal activity that drives neurotransmitter and cellular protein synthesis consumes the remaining 60%. Failure to maintain the electrochemical gradient leads to disruption of the cell membranes and cell death. Given these energy requirements, senescent neurons can tolerate a theoretical reduction of 60% of substrate delivery. Therefore any attempts to reduce cerebral metabolic rate for glucose (CMRglu) or $CMRO_2$ is limited to 60%. The potential to reduce activation and synaptic transmission in neurons serves as the basis of neuroprotective strategies.

Etiology of neuronal injury

Neuronal injury primarily is the result of mismatching delivery and utilization of substrate. Hypoxia and ischemia lead to a reduction in cellular membrane ion pump function due to impaired oxidative phosphorylation. This leads to neuronal depolarization and ion gradient collapse (i.e., an increase in intracellular calcium concentration via voltage-gated calcium channel) and release of glutamate. Unregulated release of glutamate further depolarizes nutrient-deprived neurons. This results in oxidative mitochondrial stress with increased levels of reactive oxygen species (ROS) and calcium into the cytosol via N-methyl-D-aspartate receptor (NMDA receptor). This excitotoxic state ultimately impairs protein and lipid production and activates inflammatory cascade, which ultimately triggers necrotic and apoptotic cell death pathways. Suppression of these processes should also increase the survivability of neurons.

Inadequate delivery of substrate and pathological increases in cellular metabolism are precipitating conditions for neuronal injury. Oxygen delivery is a product of the cardiac output and arterial blood oxygen content. In adults, the brain takes approximately 15% of the cardiac output. In the pediatric population, the number is higher. Tissue delivery of glucose follows in parallel fashion. Pathological conditions that impair cardiac output, glucose levels or oxygenation will ultimately cause neuronal cell death. Likewise, pathological increases in neuronal activation, with seizures as the prototype, will increase metabolic demand and overwhelm substrate reserve. Therefore, neuroprotective strategies should focus on mitigating pathologic conditions in order to maintain cellular homeostasis.

Despite the restoration of tissue perfusion and sufficient delivery of substrates, activation of parallel signaling pathways worsens the primary insult during this reperfusion phase. Aptly named reperfusion injury, activation of inflammation, liberation of ROS and apoptosis all lead to worsening of the primary injury. Neuroresuscitative management strategies are designed to mitigate these secondary processes.

Ischemia

Cerebral ischemia is one of the leading causes of brain injury. Cessation or interruption of blood flow to the brain initiates a cascade of processes that trigger CNS damage. There are two distinct patterns of cerebral ischemia. Global cerebral ischemia occurs during a cardiac arrest or transient periods of cerebral hypoperfusion. Systemic hypotension can occur during surgery and general anesthesia and may account for most neurological morbidity in the intraoperative period. Several factors can mediate intraoperative hypotension. Massive blood loss due to trauma or inadequate hemostasis is a clinically significant cause for intraoperative hypotension. Despite the reestablishment of blood flow to the brain, district regions such as the hippocampus are more susceptible to neuronal damage.

Focal ischemia follows disruption of blood flow to specific vascular beds. This may occur after an embolic occlusion or vasospasm of the proximal segment of the artery. Restoration of regional blood flow does not completely reverse the primary ischemic injury because the anoxia may trigger apoptotic cell death (Du *et al.*, 1996). Mild hypotension can also cause focal ischemia in brain regions supplied by fixed vascular lesions, such is the case in patients with carotid artery disease.

Prevention

Prevention of acute brain injury in the clinical setting is primarily based on two tenets, delivery of substrate and reduction of $CMRO_2$ and CMRglu. Maintaining this delicate balance between supply and demand is difficult to gauge under general anesthesia (Figure 16.1). Brain function can be assessed with electroencephalography (EEG), near infrared spectroscopy (NIRS), and transcranial Doppler (TCD) ultrasonography. However the positive impact of these experimental monitors on neurological outcomes is yet to be determined (Hirsch *et al.*, 2012). Given this lack of physiological feedback, meticulous maintenance of cerebral perfusion pressure, glucose, carbon dioxide and temperature is essential in the intraoperative management of patients at risk for neurological injury. Nonphysiologic levels of glucose and oxygen can independently worsen cerebral ischemic changes. Hyperthermia can also increase the $CMRO_2$, which can injure the brain in a low perfusion state.

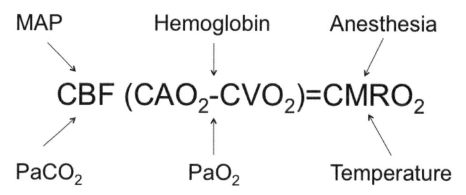

Figure 16.1. The Flick equation stipulates that the supply of oxygen, and other substrates, should match the metabolic demand of the end organ (Brain Trauma *et al.*, 2007).

Cerebral perfusion

The risk of intraoperative cerebral hypoperfusion leading to CNS injury during routine general anesthesia is unknown. Cerebral autoregulation maintains that adequate cerebral blood flow (CBF) is preserved over a range of blood pressures and arterial concentrations of oxygen and carbon dioxide (Figure 16.2). This cerebral autoregulatory range is regulated by intrinsic vascular factors that preserve flow-metabolism coupling. Endothelium and smooth muscle work in tandem to dilate when blood pressure is low and conversely constrict when it is elevated. Below the lower

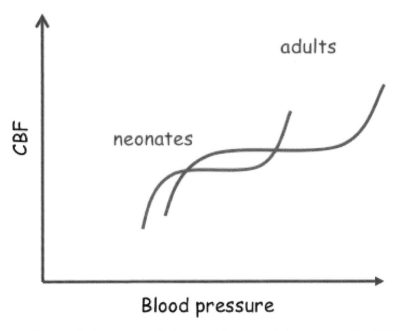

Figure 16.2. The theoretical concept of cerebral autoregulation dictates that cerebral blood flow (CBF) stay constant over a range of blood pressure. Since neonatal blood pressures are lower than adult, the neonatal cerebral autoregulatory range is shifted to the left.

limit of the autoregulatory range, cerebral perfusion is pressure passive, meaning that the cerebral blood flow will decrease in concert with systemic hypotension. Oxygen and carbon dioxide can also regulate CBF (Figure 16.2). Cerebral perfusion can be decreased by intraoperative hypotension and hypocapnia, which can occur during a routine general anesthetic. Cerebral perfusion pressure (CPP) is the pressure gradient that drives CBF. It is calculated by subtracting jugular venous pressure or intracranial pressure from the mean arterial blood pressure. Low CPP leads to cerebral ischemia. Intraoperative hemodynamic monitors do not directly measure CPP. Therefore blood pressure serves as a surrogate measure of CPP.

A review of the impact of intraoperative blood pressure in adult patients undergoing non-cardiac, non-neurosurgical surgery revealed that 30% reduction in mean blood pressure from baseline, was associated with postoperative stroke (Bijker *et al.*, 2012). Therefore maintenance of appropriate blood pressure is paramount.

Appropriate management of intraoperative arterial pressure is hampered by assumptions on a range of blood pressures that maintain adequate CBF and cerebral autoregulation. This hypothetical range of blood pressure is simplified and not necessarily supported by the original reports (Drummond, 1997). Cerebral autoregulatory range in adults is also affected by coexisting disease processes that may impair cerebral perfusion due to vasculopathies, hypertension and congestive heart failure. Furthermore, investigations in adults on cardiopulmonary bypass reveal startling heterogeneity of the lower limit of autoregulation (Joshi *et al.*, 2012). Recognition of the heterogeneity and potential absence of cerebral autoregulation is critical in the physiologic management of patients under general anesthesia.

This variability also exists in pediatric patients. Although the resting blood pressures are lower in infants and young children, TCD studies demonstrate that the lower limit of cerebral autoregulation is equivalent in these age groups (Vavilala *et al.*, 2003). Children younger than 2 years may have lower autoregulatory reserve because of their relatively low baseline mean arterial pressures and may be at greater risk of cerebral ischemia. Multimodal analysis of cerebral perfusion in infants and children undergoing cardiopulmonary bypass surgery reveals a wide range of lower limits of autoregulation (Brady *et al.*, 2010). This observation highlights the limitations of current monitors to optimize cerebral perfusion (Lee, 2014).

Oxygen

Impaired oxygen delivery to tissue beds initiates a cascade of energy failure and subsequent activation of cell death pathways. Therefore, it is imperative to maintain adequate oxygenation as a neuroprotective measure. However, cerebral reperfusion after a period of ischemia and hypoxia may also worsen the primary insult. Hyperoxia activate inflammation, reduces nitric oxide release and increases ROS production, which increases secondary ischemic injury. Clinical trials of hyperoxia in brain injured patients have not demonstrated any clinical benefit (Longhi and Stocchetti, 2004).

Glucose

Glucose is a major fuel for brain metabolism and hypoglycemia clearly poses a risk for neuronal injury. However, abundance of this energy source is also considered to enhance brain injury (Lipshutz and Gropper, 2009). Hyperglycemia is a common stress response to surgery and critical illness and may be associated with worse neurological outcome. More severe ischemic lesions were seen in hyperglycemic than in normoglycemic patients following stroke (Parsons *et al.*, 2002). Furthermore, normalization of blood glucose during the first 48 hours of hospitalization following acute stroke demonstrated a better survival, suggesting that there is a role in glycemic control (Gentile *et al.*, 2006). The underlying mechanism to link hyperglycemia with worsening neurological outcome is unclear. One potential explanation is that relative abundance of glucose over oxygen leads to anaerobic glycolysis and tissue acidosis. Regardless of the mechanism, intensive insulin therapy to maintain tight glucose control (80–110 mg/dL) favors improved survival in surgical intensive care unit patients (van den Berghe *et al.*, 2001). However, tight glucose control is not innocuous because of the additional risk of hypoglycemia. Symptoms from hypoglycemia are not appreciated under general anesthesia. Tight glucose control has been tried in cardiac surgery patients without any apparent benefit (Gandhi *et al.*, 2007). The data from non-cardiac surgeries are quite limited. The guidelines from the American Diabetes Association and the American Association of Clinical Endocrinologists suggest maintaining a glucose range of 140–180 mg/dL in critically ill patients (Moghissi *et al.*, 2009).

Carbon dioxide

Cerebral blood flow is sensitive to the partial arterial pressure of carbon dioxide ($PaCO_2$) (Meng and Gelb, 2015). While the impact of PaO_2 on CBF is likely insignificant because very little changes in CBF occur at $PaO_2 > 50$ mmHg (Udomphorn *et al.*, 2008), there is a linear relationship between CBF and $PaCO_2$. Within physiologically relevant range, CBF increases by 2–4% per mmHg $PaCO_2$. Hyperventilation can induce reduction of cerebral blood volume due to cerebral vasoconstriction. This may provide a temporary benefit in the presence of intracranial hypertension, but prophylactic hyperventilation is not advocated. In the setting of traumatic brain injury, hyperventilation can also increase the ischemic burden on the injured brain (Coles *et al.*, 2007).

Temperature

Hypothermia reduces cellular metabolic rates and oxygen consumption linearly and is an effective way to protect organ function (Gravlee *et al.*, 2007). Metabolic rates are determined by temperature sensitive enzyme-controlled chemical reactions. The Q_{10} temperature coefficient is the ratio of metabolic rates at two temperatures 10°C apart, and is 3.6 for infants and 2.6 for adults, suggesting that hypothermia more effectively reduces metabolic rates in children versus adults. Intracellular pH and enzyme activity are preserved and metabolic demand is reduced at low temperature. While cerebral blood flow decreases linearly with the decrease in temperature, cerebral metabolic demand ($CMRO_2$) drops exponentially with decreasing temperature so the metabolic demands of the brain are met. At normothermia, the mean ratio of CBF to $CMRO_2$ is 20:1 and at deep hypothermia, the ratio increases to 75:1.

Additional effects of hypothermia include preservation of high-energy phosphate stores, reduction of excitatory neurotransmitter release, suppression of microglial activation and attenuation of neutrophil migration into ischemic tissue (Busto *et al.*, 1989). Attenuation of the ischemia-mediated release of excitatory neurotransmitters such as glutamate reduces the downstream activation of lytic enzymes that lead to further tissue injury (Rothman and Olney, 1986). Hypothermia reduces apoptosis and tissue necrosis by decreasing cytochrome-c release and caspase-3 activation.

Hypothermia has shown clear benefit against brain injury in the laboratory (Yenari and Hemmen, 2010). In patients, deep hypothermic circulatory arrest (DHCA) (15–22°C) clearly protects the brain, allowing cessation of cerebral circulation for about 40 minutes. In addition, systemic hypothermia reduces metabolic demand in other organ systems

Unfortunately, hypothermia is not without disadvantages. These include the induced dysfunction of the blood-brain barrier, which increases cerebral hydrostatic pressure leading to cerebral edema, impaired coagulation system, and increase in systemic vascular resistance and reduction in blood flow to organs. Blood flow is reduced to the greatest extent in skeletal muscle and the extremities, followed by the kidney, and the splanchnic bed. Given these disadvantages, mild hypothermia (32–35°C) is often implemented in non-cardiac procedures. Unfortunately, the evidence of mild hypothermia in the intraoperative period is rather elusive. The neuroprotective effect of prophylactic hypothermia was tested in a randomized control trial (RCT) comparing mild hypothermia (33°C) to normothermia (36.5°C) in patients undergoing aneurysm surgery (Todd *et al.*, 2005). Mild hypothermia did not improve neurological outcome in this patient population. A recent Cochrane report confirmed the lack of improved neurological outcome in a meta analysis of 4 similarly conducted RCTs (Galvin *et al.*, 2015). In contrast, phase III clinical trials demonstrated that there was a clear benefit of mild hypothermia in post-cardiac arrest survivors and neonatal hypoxic encephalopathy (Ramani, 2006). Although there is no convincing evidence to suggest that mild hypothermia should be used to prevent or attenuate potential brain injury under general anesthesia, it is clear that hyperthermia is detrimental in the post-ischemic brain. In fact, spontaneous hyperthermia is associated with poor outcomes following acute stroke (Kammersgaard *et al.*, 2002). Therefore, hyperthermia should be quickly recognized and aggressively treated.

Brain protection strategies during cardiac and neurological surgery

Cardiac surgery has a relatively association with brain injury (Hogue *et al.*, 2006). Several attempts have been made to mitigate this devastating complication. A landmark intervention in cardiac surgery was the use of DHCA, but was unfortunately linked to poor neurologic outcomes. Since then, the technique of DHCA has improved by maintaining the hematocrit greater than 30%, uniform brain cooling by actively cooling for longer than 20 minutes, using pH-stat acid-base management and placing ice on the head. In pH-stat management, the addition of carbon dioxide to the cardiopulmonary bypass (CPB) circuit preserves a constant partial pressure of carbon dioxide and neutral pH at decreasing temperatures. This approach allows cerebral vasodilation and increased cerebral blood flow, more homogenous cooling, and a greater reduction in cerebral oxygen consumption compared with alpha-stat management. Patients undergoing surgical correction of d-transposition of the great arteries with pH-stat management under DHCA were shown to have shorter endotracheal intubation times, shorter length of intensive care unit (ICU) stay and lower mortality (du Plessis *et al.*, 1997). However, children who underwent reparative surgeries using either alpha-stat or pH-stat during DHCA had no detectable differences in early neurodevelopmental outcomes (Bellinger *et al.*, 2003).

Alternatives to DHCA include intermittent reperfusion and regional cerebral perfusion (RCP). RCP refers to the technique of directing arterial pump blood flow to the innominate artery, thus selectively perfusing the right subclavian and right common carotid arteries. The flow rates necessary to provide adequate cerebral and somatic perfusion during RLFP have yet to be completely elucidated although flow rates greater than 30 ml/kg/min are generally used. In pediatric literature, there was no difference in neurodevelopmental outcomes in single ventricle patients undergoing the Norwood procedure who received RCP as compared to DHCA (Goldberg *et al.*, 2007).

In neurosurgery, the neuroprotective effect of prophylactic hypothermia was tested in a randomized control trial (RCT) comparing mild hypothermia (33°C) to normothermia (36.5°C) in patients undergoing aneurysm surgery without clear benefit (Todd *et al.*, 2005). A recent Cochrane report also supported this observation (Galvin *et al.*, 2015).

Pharmacology

Reduction of neuronal activation metabolism is the goal of neuroprotective strategies. General anesthetics suppress neuronal activity by inhibiting brain activation (Brown *et al.*, 2011; Rudolph and Antkowiak, 2004). Deep levels of general anesthesia produces a characteristic pattern of a burst of sharp waves followed by isoelectric periods on the electroencephalogram (EEG) (Swank and Watson, 1949). The phenomenon is called burst suppression and is associated with a diminution in neural activity during suppressed brain function and neuronal metabolism (Ching *et al.*, 2012). The EEG is utilized in the operating room and intensive care units to monitor burst suppression in patients with or who are at risk for brain injury. The most recognized molecular targets of general anesthetics are the gamma-amino butyric acid (GABA) and N-methyl D-aspartate (NMDA) receptors. Several classes of general anesthetics and sedatives activate or antagonize these receptors respectively. GABA agonists include the volatile anesthetics (isoflurane, sevoflurane and desflurane) and intravenous drugs (phenobarbital, thiopental and propofol). Ketamine is the clinically available NMDA antagonist. Suppression of activity-dependent $CRMO_2$ by these anesthetic drugs are the primary mode of neuroprotection for vulnerable neural cells (Nellgard *et al.*, 2000). However, if neuronal activity is completely suppressed, as in the case of global ischemia and anoxia, pharmacological and hypothermic interventions are not neuroprotective. These drugs also activate and suppress other biological processes that exacerbate the primary neurological insult.

Barbiturates

Barbiturates produce an isoelectric EEG and reduce the $CMRO_2$ in laboratory animals and humans. The administration of thiopental resulted in a reduction in ATP and lactate concentrations in dogs subjected to a progressive arterial hypoxemia, which is model of global cerebral ischemia (Michenfelder and Theye, 1973). These findings demonstrate that barbiturates reduce requirement for energy substrates (ATP) and accumulation of lactate, which are consistent with a suppressed $CMRO_2$ (Michenfelder and Theye, 1973). Barbiturates also have a protective effect on rodent models focal cerebral ischemia (Warner *et al.*, 1996; Warner *et al.*, 1991). However, barbiturates have no efficacy in protecting against global or anoxic brain injury (Steen *et al.*, 1979; Todd *et al.*, 1982).

Multiple clinical investigations have examined the clinical efficacy of prophylactic barbiturate administration on the neurocognitive outcomes in patients undergoing surgical procedures that are associated with intraoperative neurological morbidity. Patients undergoing open-heart surgery with cardiopulmonary bypass (CPB) have long been associated with postoperative neurocognitive deficits. The incidence of new neurological deficits following CPB was common and was attributed to microemboli and low CBF. An initial report demonstrated an improvement in neurocognitive scores in patients receiving thiopental (Nussmeier *et al.*, 1986). However, subsequent reports showed no substantial impact of thiopental when compared to hypothermia or volatile anesthetics (Newman *et al.*, 1998; Zaidan *et al.*, 1991). A notable difference in these reports is the use of membrane oxygenation on the cardiopulmonary bypass circuit to reduce microemboli and the standard use of induced hypothermia in the later studies.

Propofol

Propofol is a hypnotic sedative drug and is primarily a GABA receptor agonist. It reduces $CMRO_2$ by inhibiting synaptic transmission. However, it can inhibit NMDA receptor at high concentrations (Grasshoff and Gillessen, 2005). Propofol also optimizes $CBF-CRMO_2$ coupling by maintaining CBF reactivity to changes in $PaCO_2$ (Newman *et al.*, 1995). Several reports attribute the neuroprotective properties of propofol to inhibition of, mitochondrial swelling, apoptosis and inflammation (Adembri *et al.*, 2006; Engelhard *et al.*, 2004; Yuki *et al.*, 2011).

Experimental models of transient focal cerebral ischemia in rodents have consistently demonstrated reductions in infarct volumes and neurobehavioral deficits (Engelhard *et al.*, 2004; Gelb *et al.*, 2002).

Volatile anesthetics

Volatile anesthetics, isoflurane, sevoflurane and desflurane, activate inhibitory and inhibit of excitatory neurotransmission by direct interaction with GABA and NMDA receptors respectively. These pleotropic agents also regulate ligand-gated ionotropic receptors and co-transporter systems on the cell membrane and mitochondrial function, which ultimately leads to suppression on neuronal activity and reduction in $CMRO_2$. Volatile anesthetics have been shown to protect against focal cerebral ischemia and improve neurological outcome in rodents (Sakai *et al.*, 2007; Zhao *et al.*, 2007). However, there is a lack of neuroprotection for anoxic injury and when the volatile agent is administered after the inciting injury.

Volatile anesthetics may be neuroprotective by inducing cellular preconditioning. Anesthetic-induced preconditioning activates sarcolemmal and mitochondrial K_{ATP} channels, adenosine receptors and cell survival pathways (PKC, Akt, ERK1/2 and p38) (Kitano *et al.*, 2007). Volatile anesthetics have inherent anti-inflammatory properties, which hold the potential of attenuating the inflammatory injury that ensues after an ischemic event (Soriano *et al.*, 1999; Soriano *et al.*, 1996; Yuki *et al.*, 2010; Yuki *et al.*, 2008; Yuki *et al.*, 2011).

Prior to these preclinical investigations, the use of a general anesthetic regiment of cyclopropane, thiopental, hypercarbia and high inspired concentrations of oxygen in patients undergoing temporary occlusion of the carotid artery resulted in a reduction on ischemic patterns on the EEG and postoperative neurological morbidity (Wells *et al.*, 1963). This observation suggests that anesthetic increases tolerance to cerebral ischemia during temporary carotid artery occlusion by reducing $CMRO_2$, while hypercarbia and hyperoxia maximized oxygen delivery. Subsequent clinical reports on the neuroprotective impact of the modern volatile anesthetics have been inconclusive (Kitano *et al.*, 2007).

Taken together, preclinical data support the notion that specific anesthetic drugs are neuroprotective in the setting of transient focal ischemia where neuronal activity is still present at low levels. However, in conditions where anoxia and global cerebral ischemia occur as in the case of cardiac arrest, or profound hypotension, neuronal activity is completely nonexistent, and anesthetic drugs do not offer any protection.

Neurotoxicity

General anesthetics *in vivo* and *in vitro* experiments have long been shown to have neurotoxic effects on developing central nervous system (Loepke and Soriano, 2008; Steen and Michenfelder, 1979). These effects include decreased neurogenesis, abnormal dendrite formation, decreased glial cell formation and increase in neuroapoptosis in both the brain and spinal cord (Brambrink *et al.*, 2010; De Roo *et al.*, 2009; Loepke and Soriano, 2008). Neuroapoptosis or programmed cell death occurs normally during fetal development as part of cerebral and neuronal maturation. It differs histologically from ischemic cell death in that there is no inflammation and it tends to occur in isolated cells rather than regionally. However, anesthetic exposure during vulnerable periods in laboratory animal has been shown to lead to a marked increase in apoptotic cell death and subsequent learning deficits especially in the domain of executive function in animals allowed to mature. Parallel investigations in older rodents link anesthetic exposure to Alzheimer Disease (Baranov *et al.*, 2009; Xie *et al.*, 2007). Extrapolations of these preclinical studies to humans are fraught with uncertainty because of physiological differences between species and difficulties in physiologic monitoring for glucose, blood pressure and respiration in very young and small mammals.

Retrospective epidemiologic evidence clearly supports an association between surgery at a young age and later poor neurodevelopmental outcomes in humans (DiMaggio *et al.*, 2009; Ing *et al.*, 2012; Wilder *et al.*, 2009). However, the patient's underlying pathology, the type, duration and scope of the surgery and the conduct of the general anesthetics, inevitably confound these studies (Blakely *et al.*, 2006; Hintz *et al.*, 2005; Rees *et al.*, 2007; Simon *et al.*, 1993; Tobiansky *et al.*, 1995; Walsh *et al.*, 1989). Given the inconclusive nature of these reports, the Food and Drug Administration published an advisory highlighting the need to for additional investigations into the impact of anesthetics on neurocognition (Rappaport *et al.*, 2015).

Conclusion

The public burden of brain injury will continue to fuel both preclinical and clinical investigations on neuroprotection. The initial success of anesthetic drugs and hypothermia in laboratory models of neural injury is compelling and points to the multiple molecular pathways and cellular processes involved in this phenomenon. However, extrapolation of these findings to the bedside is not supported by subsequent clinical trials. Furthermore the adverse effects of anesthetic drugs on neurodevelopment also provide insight to the yet unknown mechanisms of anesthetic drug action. Only rigorous clinical trials will determine the clinical efficacy of the various applications of anesthetic drugs and management strategies in the vulnerable surgical patients.

References

Adembri, C., L. Venturi, A. Tani, A. Chiarugi, E. Gramigni, A. Cozzi, T. Pancani, R.A. De Gaudio, and D.E. Pellegrini-Giampietro. 2006. Neuroprotective effects of propofol in models of cerebral ischemia: inhibition of mitochondrial swelling as a possible mechanism. Anesthesiology 104: 80–89.

Baranov, D., P.E. Bickler, G.J. Crosby, D.J. Culley, M.F. Eckenhoff, R.G. Eckenhoff, K.J. Hogan, V. Jevtovic-Todorovic, A. Palotas, M. Perouansky, E. Planel, J.H. Silverstein, H. Wei, R.A. Whittington, Z. Xie, Z. Zuo, A. First International Workshop on, and D. Alzheimer's. 2009. Consensus statement: First International Workshop on Anesthetics and Alzheimer's disease. Anesth Analg 108: 1627–1630.

Bellinger, D.C., D. Wypij, A.J. duPlessis, L.A. Rappaport, R.A. Jonas, G. Wernovsky, and J.W. Newburger. 2003. Neurodevelopmental status at eight years in children with dextro-transposition of the great arteries: the Boston Circulatory Arrest Trial. J Thorac Cardiovasc Surg 126: 1385–1396.

Bijker, J.B., S. Persoon, L.M. Peelen, K.G. Moons, C.J. Kalkman, L.J. Kappelle, and W.A. van Klei. 2012. Intraoperative hypotension and perioperative ischemic stroke after general surgery: a nested case-control study. Anesthesiology 116: 658–664.

Blakely, M.L., J.E. Tyson, K.P. Lally, S. McDonald, B.J. Stoll, D.K. Stevenson, W.K. Poole, A.H. Jobe, L.L. Wright, R.D. Higgins, and N.N.R. Network. 2006. Laparotomy versus peritoneal drainage for necrotizing enterocolitis or isolated intestinal perforation in extremely low birth weight infants: outcomes through 18 months adjusted age. Pediatrics 117: e680–687.

Brady, K.M., J.O. Mytar, J.K. Lee, D.E. Cameron, L.A. Vricella, W.R. Thompson, C.W. Hogue, and R.B. Easley. 2010. Monitoring cerebral blood flow pressure autoregulation in pediatric patients during cardiac surgery. Stroke 41: 1957–1962.

Brain Trauma, F., S. American Association of Neurological, and S. Congress of Neurological. 2007. Guidelines for the management of severe traumatic brain injury. J Neurotrauma 24 Suppl 1: S1–106.

Brambrink, A.M., A.S. Evers, M.S. Avidan, N.B. Farber, D.J. Smith, X. Zhang, G.A. Dissen, C.E. Creeley, and J.W. Olney. 2010. Isoflurane-induced neuroapoptosis in the neonatal rhesus macaque brain. Anesthesiology 112: 834–841.

Brown, E.N., P.L. Purdon, and C.J. Van Dort. 2011. General anesthesia and altered states of arousal: a systems neuroscience analysis. Annu Rev Neurosci 34: 601–628.

Busto, R., M.Y. Globus, W.D. Dietrich, E. Martinez, I. Valdes, and M.D. Ginsberg. 1989. Effect of mild hypothermia on ischemia-induced release of neurotransmitters and free fatty acids in rat brain. Stroke 20: 904–910.

Ching, S., P.L. Purdon, S. Vijayan, N.J. Kopell, and E.N. Brown. 2012. A neurophysiological-metabolic model for burst suppression. Proc Natl Acad Sci U S A 109: 3095–3100.

Coles, J.P., T.D. Fryer, M.R. Coleman, P. Smielewski, A.K. Gupta, P.S. Minhas, F. Aigbirhio, D.A. Chatfield, G.B. Williams, S. Boniface, T.A. Carpenter, J.C. Clark, J.D. Pickard, and D.K. Menon. 2007. Hyperventilation following head injury: effect on ischemic burden and cerebral oxidative metabolism. Crit Care Med 35: 568–578.

De Roo, M., P. Klauser, A. Briner, I. Nikonenko, P. Mendez, A. Dayer, J.Z. Kiss, D. Muller, and L. Vutskits. 2009. Anesthetics rapidly promote synaptogenesis during a critical period of brain development. PLoS One 4: e7043.

DiMaggio, C., L.S. Sun, A. Kakavouli, M.W. Byrne, and G. Li. 2009. A retrospective cohort study of the association of anesthesia and hernia repair surgery with behavioral and developmental disorders in young children. J Neurosurg Anesthesiol 21: 286–291.

Drummond, J.C. 1997. The lower limit of autoregulation: time to revise our thinking? Anesthesiology 86: 1431–1433.

Du, C., R. Hu, C.A. Csernansky, C.Y. Hsu, and D.W. Choi. 1996. Very delayed infarction after mild focal cerebral ischemia: a role for apoptosis? J Cereb Blood Flow Metab 16: 195–201.

du Plessis, A.J., R.A. Jonas, D. Wypij, P.R. Hickey, J. Riviello, D.L. Wessel, S.J. Roth, F.A. Burrows, G. Walter, D.M. Farrell, A.Z. Walsh, C.A. Plumb, P. del Nido, R.P. Burke, A.R. Castaneda, J.E. Mayer, Jr., and J.W. Newburger. 1997. Perioperative effects of alpha-stat versus pH-stat strategies for deep hypothermic cardiopulmonary bypass in infants. J Thorac Cardiovasc Surg 114: 991–1000; discussion 1000–1001.

Engelhard, K., C. Werner, E. Eberspacher, M. Pape, U. Stegemann, K. Kellermann, R. Hollweck, P. Hutzler, and E. Kochs. 2004. Influence of propofol on neuronal damage and apoptotic factors after incomplete cerebral ischemia and reperfusion in rats: a long-term observation. Anesthesiology 101: 912–917.

Fukuda, S., and D.S. Warner. 2007. Cerebral protection. Br J Anaesth 99: 10–17.

Galvin, I.M., R. Levy, J.G. Boyd, A.G. Day, and M.C. Wallace. 2015. Cooling for cerebral protection during brain surgery. Cochrane Database Syst Rev 1: CD006638.

Gandhi, G.Y., G.A. Nuttall, M.D. Abel, C.J. Mullany, H.V. Schaff, P.C. O'Brien, M.G. Johnson, A.R. Williams, S.M. Cutshall, L.M. Mundy, R.A. Rizza, and M.M. McMahon. 2007. Intensive intraoperative insulin therapy versus conventional glucose management during cardiac surgery: a randomized trial. Ann Intern Med 146: 233–243.

Gelb, A.W., N.A. Bayona, J.X. Wilson, and D.F. Cechetto. 2002. Propofol anesthesia compared to awake reduces infarct size in rats. Anesthesiology 96: 1183–1190.

Gentile, N.T., M.W. Seftchick, T. Huynh, L.K. Kruus, and J. Gaughan. 2006. Decreased mortality by normalizing blood glucose after acute ischemic stroke. Acad Emerg Med 13: 174–180.

Goldberg, C.S., E.L. Bove, E.J. Devaney, E. Mollen, E. Schwartz, S. Tindall, C. Nowak, J. Charpie, M.B. Brown, T.J. Kulik, and R.G. Ohye. 2007. A randomized clinical trial of regional cerebral perfusion versus deep hypothermic circulatory arrest: outcomes for infants with functional single ventricle. J Thorac Cardiovasc Surg 133: 880–887.

Grasshoff, C., and T. Gillessen. 2005. Effects of propofol on N-methyl-D-aspartate receptor-mediated calcium increase in cultured rat cerebrocortical neurons. Eur J Anaesthesiol 22: 467–470.

Gravlee, G., R. Davis, A. Stammers, and R. Ungerleider. 2007. Cardiopumonary Bypass: Principles and Practice. Lippincott Williams & Wilkins, Philadelphia, PA.

Hintz, S.R., D.E. Kendrick, B.J. Stoll, B.R. Vohr, A.A. Fanaroff, E.F. Donovan, W.K. Poole, M.L. Blakely, L. Wright, R. Higgins, and N.N.R. Network. 2005. Neurodevelopmental and growth outcomes of extremely low birth weight infants after necrotizing enterocolitis. Pediatrics 115: 696–703.

Hirsch, J.C., M.L. Jacobs, D. Andropoulos, E.H. Austin, J.P. Jacobs, D.J. Licht, F. Pigula, J.S. Tweddell, and J.W. Gaynor. 2012. Protecting the infant brain during cardiac surgery: a systematic review. Ann Thorac Surg 94: 1365–1373; discussion 1373.

Hogue, C.W., Jr., C.A. Palin, and J.E. Arrowsmith. 2006. Cardiopulmonary bypass management and neurologic outcomes: an evidence-based appraisal of current practices. Anesth Analg 103: 21–37.

Ing, C., C. DiMaggio, A. Whitehouse, M.K. Hegarty, J. Brady, B.S. von Ungern-Sternberg, A. Davidson, A.J. Wood, G. Li, and L.S. Sun. 2012. Long-term differences in language and cognitive function after childhood exposure to anesthesia. Pediatrics 130: e476–485.

Jauch, E.C., J.L. Saver, H.P. Adams, Jr., A. Bruno, J.J. Connors, B.M. Demaerschalk, P. Khatri, P.W. McMullan, Jr., A.I. Qureshi, K. Rosenfield, P.A. Scott, D.R. Summers, D.Z. Wang, M. Wintermark, H. Yonas, C. American Heart Association Stroke, N. Council on Cardiovascular, D. Council on Peripheral Vascular, and C. Council on Clinical. 2013. Guidelines for the early management of patients with acute ischemic stroke: a guideline for healthcare professionals from the American Heart Association/American Stroke Association. Stroke 44: 870–947.

Joshi, B., M. Ono, C. Brown, K. Brady, R.B. Easley, G. Yenokyan, R.F. Gottesman, and C.W. Hogue. 2012. Predicting the limits of cerebral autoregulation during cardiopulmonary bypass. Anesth Analg 114: 503–510.

Kammersgaard, L.P., H.S. Jorgensen, J.A. Rungby, J. Reith, H. Nakayama, U.J. Weber, J. Houth, and T.S. Olsen. 2002. Admission body temperature predicts long-term mortality after acute stroke: the Copenhagen Stroke Study. Stroke 33: 1759–1762.

Kitano, H., J.R. Kirsch, P.D. Hurn, and S.J. Murphy. 2007. Inhalational anesthetics as neuroprotectants or chemical preconditioning agents in ischemic brain. J Cereb Blood Flow Metab 27: 1108–1128.

Kochanek, P.M., N. Carney, P.D. Adelson, S. Ashwal, M.J. Bell, S. Bratton, S. Carson, R.M. Chesnut, J. Ghajar, B. Goldstein, G.A. Grant, N. Kissoon, K. Peterson, N.R. Selden, R.C. Tasker, K.A. Tong, M.S. Vavilala, M.S. Wainwright, C.R. Warden, S. American Academy of Pediatrics-Section on Neurological, S. American Association of Neurological Surgeons/Congress of Neurological, S. Child Neurology, P. European Society of, C. Neonatal Intensive, S. Neurocritical Care, G. Pediatric Neurocritical Care Research, M. Society of Critical Care, U.K. Paediatric Intensive Care Society, A. Society for Neuroscience in, C. Critical, I. World Federation of Pediatric, and S. Critical Care. 2012. Guidelines for the acute medical management of severe traumatic brain injury in infants, children, and adolescents—second edition. Pediatr Crit Care Med 13 Suppl 1: S1–82.

Lee, J.K. 2014. Cerebral perfusion pressure: how low can we go? Paediatr Anaesth 24: 647–648.

Lipshutz, A.K., and M.A. Gropper. 2009. Perioperative glycemic control: an evidence-based review. Anesthesiology 110: 408–421.

Loepke, A.W., and S.G. Soriano. 2008. An assessment of the effects of general anesthetics on developing brain structure and neurocognitive function. Anesth Analg 106: 1681–1707.

Longhi, L., and N. Stocchetti. 2004. Hyperoxia in head injury: therapeutic tool? Curr Opin Crit Care 10: 105–109.

Mashour, G.A., D.T. Woodrum, and M.S. Avidan. 2015. Neurological complications of surgery and anaesthesia. Br J Anaesth 114: 194–203.

Meng, L., and A.W. Gelb. 2015. Regulation of cerebral autoregulation by carbon dioxide. Anesthesiology 122: 196–205.

Michenfelder, J.D., and R.A. Theye. 1973. Cerebral protection by thiopental during hypoxia. Anesthesiology 39: 510–517.

Moghissi, E.S., M.T. Korytkowski, M. DiNardo, D. Einhorn, R. Hellman, I.B. Hirsch, S.E. Inzucchi, F. Ismail-Beigi, M.S. Kirkman, G.E. Umpierrez, E. American Association of Clinical, and A. American Diabetes. 2009. American Association of Clinical Endocrinologists and American Diabetes Association consensus statement on inpatient glycemic control. Diabetes Care 32: 1119–1131.

Nellgard, B., G.B. Mackensen, J. Pineda, J.C. Wellons, 3rd, R.D. Pearlstein, and D.S. Warner. 2000. Anesthetic effects on cerebral metabolic rate predict histologic outcome from near-complete forebrain ischemia in the rat. Anesthesiology 93: 431–436.

Newman, M.F., J.M. Murkin, G. Roach, N.D. Croughwell, W.D. White, F.M. Clements, and J.G. Reves. 1995. Cerebral physiologic effects of burst suppression doses of propofol during nonpulsatile cardiopulmonary bypass. CNS Subgroup of McSPI. Anesth Analg 81: 452–457.

Newman, M.F., N.D. Croughwell, W.D. White, I. Sanderson, W. Spillane, and J.G. Reves. 1998. Pharmacologic electroencephalographic suppression during cardiopulmonary bypass: a comparison of thiopental and isoflurane. Anesth Analg 86: 246–251.

Nussmeier, N.A., C. Arlund, and S. Slogoff. 1986. Neuropsychiatric complications after cardiopulmonary bypass: cerebral protection by a barbiturate. Anesthesiology 64: 165–170.

Parsons, M.W., P.A. Barber, P.M. Desmond, T.A. Baird, D.G. Darby, G. Byrnes, B.M. Tress, and S.M. Davis. 2002. Acute hyperglycemia adversely affects stroke outcome: a magnetic resonance imaging and spectroscopy study. Ann Neurol 52: 20–28.

Ramani, R. 2006. Hypothermia for brain protection and resuscitation. Curr Opin Anaesthesiol 19: 487–491.

Rappaport, B.A., S. Suresh, S. Hertz, A.S. Evers, and B.A. Orser. 2015. Anesthetic neurotoxicity—clinical implications of animal models. N Engl J Med 372: 796–797.

Rees, C.M., A. Pierro, and S. Eaton. 2007. Neurodevelopmental outcomes of neonates with medically and surgically treated necrotizing enterocolitis. Arch Dis Child Fetal Neonatal Ed 92: F193–198.

Rothman, S.M., and J.W. Olney. 1986. Glutamate and the pathophysiology of hypoxic—ischemic brain damage. Ann Neurol 19: 105–111.

Rudolph, U., and B. Antkowiak. 2004. Molecular and neuronal substrates for general anaesthetics. Nat Rev Neurosci 5: 709–720.

Sakai, H., H. Sheng, R.B. Yates, K. Ishida, R.D. Pearlstein, and D.S. Warner. 2007. Isoflurane provides long-term protection against focal cerebral ischemia in the rat. Anesthesiology 106: 92–99; discussion 98–10.

Simon, N.P., N.R. Brady, R.L. Stafford, and R.W. Powell. 1993. The effect of abdominal incisions on early motor development of infants with necrotizing enterocolitis. Dev Med Child Neurol 35: 49–53.

Soriano, S.G., S.A. Lipton, Y.F. Wang, M. Xiao, T.A. Springer, J.C. Gutierrez-Ramos, and P.R. Hickey. 1996. Intercellular adhesion molecule-1-deficient mice are less susceptible to cerebral ischemia-reperfusion injury. Ann Neurol 39: 618–624.

Soriano, S.G., A. Coxon, Y.F. Wang, M.P. Frosch, S.A. Lipton, P.R. Hickey, and T.N. Mayadas. 1999. Mice deficient in Mac-1 (CD11b/CD18) are less susceptible to cerebral ischemia/reperfusion injury. Stroke 30: 134–139.

Steen, P.A., and J.D. Michenfelder. 1979. Neurotoxicity of anesthetics. Anesthesiology 50: 437–453.

Steen, P.A., J.H. Milde, and J.D. Michenfelder. 1979. No barbiturate protection in a dog model of complete cerebral ischemia. Ann Neurol 5: 343–349.

Swank, R.L., and C.W. Watson. 1949. Effects of barbiturates and ether on spontaneous electrical activity of dog brain. J Neurophysiol 12: 137–160.

Tobiansky, R., K. Lui, S. Roberts, and M. Veddovi. 1995. Neurodevelopmental outcome in very low birthweight infants with necrotizing enterocolitis requiring surgery. J Paediatr Child Health 31: 233–236.

Todd, M.M., H.S. Chadwick, H.M. Shapiro, B.J. Dunlop, L.F. Marshall, and R. Dueck. 1982. The neurologic effects of thiopental therapy following experimental cardiac arrest in cats. Anesthesiology 57: 76–86.

Todd, M.M., B.J. Hindman, W.R. Clarke, J.C. Torner, and I. Intraoperative Hypothermia for Aneurysm Surgery Trial. 2005. Mild intraoperative hypothermia during surgery for intracranial aneurysm. N Engl J Med 352: 135–145.

Udomphorn, Y., W.M. Armstead, and M.S. Vavilala. 2008. Cerebral blood flow and autoregulation after pediatric traumatic brain injury. Pediatr Neurol 38: 225–234.

Vaishnavi, S.N., A.G. Vlassenko, M.M. Rundle, A.Z. Snyder, M.A. Mintun, and M.E. Raichle. 2010. Regional aerobic glycolysis in the human brain. Proc Natl Acad Sci U S A 107: 17757–17762.

van den Berghe, G., P. Wouters, F. Weekers, C. Verwaest, F. Bruyninckx, M. Schetz, D. Vlasselaers, P. Ferdinande, P. Lauwers, and R. Bouillon. 2001. Intensive insulin therapy in critically ill patients. N Engl J Med 345: 1359–1367.

Vavilala, M.S., L.A. Lee, and A.M. Lam. 2003. The lower limit of cerebral autoregulation in children during sevoflurane anesthesia. J Neurosurg Anesthesiol 15: 307–312.

Walsh, M.C., R.M. Kliegman, and M. Hack. 1989. Severity of necrotizing enterocolitis: influence on outcome at 2 years of age. Pediatrics 84: 808–814.

Warner, D.S., J.G. Zhou, R. Ramani, and M.M. Todd. 1991. Reversible focal ischemia in the rat: effects of halothane, isoflurane, and methohexital anesthesia. J Cereb Blood Flow Metab 11: 794–802.

Warner, D.S., S. Takaoka, B. Wu, P.S. Ludwig, R.D. Pearlstein, A.D. Brinkhous, and F. Dexter. 1996. Electroencephalographic burst suppression is not required to elicit maximal neuroprotection from pentobarbital in a rat model of focal cerebral ischemia. Anesthesiology 84: 1475–1484.

Warner, D.S., M.L. James, D.T. Laskowitz, and E.F. Wijdicks. 2014. Translational research in acute central nervous system injury: lessons learned and the future. JAMA Neurol 71: 1311–1318.

Wells, B.A., A.S. Keats, and D.A. Cooley. 1963. Increased tolerance to cerebral ischemia produced by general anesthesia during temporary carotid occlusion. Surgery 54: 216–223.

Wilder, R.T., R.P. Flick, J. Sprung, S.K. Katusic, W.J. Barbaresi, C. Mickelson, S.J. Gleich, D.R. Schroeder, A.L. Weaver, and D.O. Warner. 2009. Early exposure to anesthesia and learning disabilities in a population-based birth cohort. Anesthesiology 110: 796–804.

Xie, Z., Y. Dong, U. Maeda, R.D. Moir, W. Xia, D.J. Culley, G. Crosby, and R.E. Tanzi. 2007. The inhalation anesthetic isoflurane induces a vicious cycle of apoptosis and amyloid beta-protein accumulation. J Neurosci 27: 1247–1254.

Yenari, M.A., and T.M. Hemmen. 2010. Therapeutic hypothermia for brain ischemia: where have we come and where do we go? Stroke 41: S72–74.

Yuki, K., N.S. Astrof, C. Bracken, R. Yoo, W. Silkworth, S.G. Soriano, and M. Shimaoka. 2008. The volatile anesthetic isoflurane perturbs conformational activation of integrin LFA-1 by binding to the allosteric regulatory cavity. FASEB J 22: 4109–4116.

Yuki, K., N.S. Astrof, C. Bracken, S.G. Soriano, and M. Shimaoka. 2010. Sevoflurane binds and allosterically blocks integrin lymphocyte function-associated antigen-1. Anesthesiology 113: 600–609.

Yuki, K., S.G. Soriano, and M. Shimaoka. 2011. Sedative drug modulates T-cell and lymphocyte function-associated antigen-1 function. Anesth Analg 112: 830–838.

Zaidan, J.R., A. Klochany, W.M. Martin, J.S. Ziegler, D.M. Harless, and R.B. Andrews. 1991. Effect of thiopental on neurologic outcome following coronary artery bypass grafting. Anesthesiology 74: 406–411.

Zauner, A., W.P. Daugherty, M.R. Bullock, and D.S. Warner. 2002. Brain oxygenation and energy metabolism: part I-biological function and pathophysiology. Neurosurgery 51: 289–301; discussion 302.

Zhao, P., L. Peng, L. Li, X. Xu, and Z. Zuo. 2007. Isoflurane preconditioning improves long-term neurologic outcome after hypoxic-ischemic brain injury in neonatal rats. Anesthesiology 107: 963–970.

Index

Printed and bound by CPI Group (UK) Ltd, Croydon, CR0 4YY

01/11/2024

01782624-0009